Electricity and Magnetism

# Electricity and Magnetism

## Volume 2

B. I. BLEANEY

*Formerly Fellow of St. Hugh's College, Oxford*

and

B. BLEANEY

*Dr. Lee's Professor Emeritus of Experimental Philosophy,
University of Oxford*

**THIRD EDITION**

OXFORD UNIVERSITY PRESS

Oxford University Press, Walton Street, OX2 6DP

Oxford New York Toronto
Delhi Bombay Calcutta Madras Karachi
Petaling Jaya Singapore Hong Kong Tokyo
Nairobi Dar es Salaam Cape Town
Melbourne Auckland

and associated companies in
Berlin Ibadan

Oxford is a trade mark of Oxford University Press

First edition 1957
Second edition 1965
Third edition 1976
Reprinted (with corrections) 1978
Reprinted (with corrections) 1983
Reprinted 1985, 1987
First published as a two volume edition 1989
Reprinted 1990

British Library Cataloguing in Publication Data
Bleaney, B. I. (Betty Isabelle)
Electricity and magnetism.—3rd ed.
Vol. 2
2. Electricity 2. Magnetism
I. Title II. Bleaney, B. (Brebis)
537
ISBN 0–19–851173-6

Library of Congress Cataloging in Publication Data

(Data available)

Phototypeset by Cotswold Typesetting Limited, Gloucester

Printed in Great Britain by
J. W. Arrowsmith Ltd, Bristol

# Preface to the third edition

*De manera que acordé, aunque contra mi voluntad, meter segunda vez la pluma en tan estraña lavor é tan agena de mi facultad, hurtando algunos ratos á mi principal estudio, con otras horas destinadas para recreación, puesto que no han de faltar nuevos detractores á la nueva edicion.*
*1499*

*Fernando de Rojas*

So I agreed, albeit unwillingly (since there cannot fail to be fresh critics of a new edition), again to exercise my pen in so strange a labour, and one so foreign to my ability, stealing some moments from my principal study, together with other hours destined for recreation.

FOR the third edition of this textbook the material has been completely revised and in many parts substantially rewritten. S.I. units are used throughout; references to c.g.s. units have been almost wholly eliminated, but a short conversion table is given in Appendix D. The dominance of solid-state devices in the practical world of electronics is reflected in a major change in the subject order.

Chapters 1–9 set out the macroscopic theory of electricity and magnetism, with only minor references to the atomic background, which is discussed in Chapters 10–17. A simple treatment of lattice vibrations is introduced in Chapter 10 in considering the dielectric properties of ionic solids. The discussion of conduction electrons and metals has been expanded into two chapters, and superconductivity, a topic previously excluded, is the subject of Chapter 13. Minor changes have been made in the three chapters (14–16) on magnetism. The discussion of semiconductor theory precedes new chapters on solid-state devices, but we have endeavoured to present such devices in a manner which does not presuppose a knowledge in depth of the theory. The remaining chapters, on amplifiers and oscillators, vacuum tubes, a.c. measurements, noise, and magnetic resonance, bring together the discussion of electronics and its applications.

The authors are grateful to many colleagues in Oxford and readers elsewhere for helpful comments on previous editions which have been incorporated in the present volume. In particular we are indebted to Dr. G. A. Brooker for numerous and detailed comments and suggestions; to Drs F. V. Price and J. W. Hodby, whose reading of new material on electronics in draft form resulted in substantial improvement of the presentation; to Drs F. N. H. Robinson and R. A. Stradling for several helpful suggestions; and to Messrs

C. A. Carpenter and J. Ward for the considerable trouble taken in producing Fig. 23.3. We are indebted to Professors M. Tinkham and O. V. Lounasmaa for generously sending us material in advance of publication; and to Professor L. F. Bates, F.R.S., Drs R. Dupree, and R. A. Stradling for their kindness in providing the basic diagrams for Figs 15.6, 6.15, and 17.9. We wish to thank Miss C. H. Bleaney for suggesting the quotation which appears above.

*Clarendon Laboratory,*                                        B, I. B.
*Oxford*                                                            B. B.
*February 1975*

# Acknowledgements

THE authors are indebted to the following for permission to use published diagrams as a basis for figures in the text: the late Sir K. S. Krishnan; G. Benedek; R. Berman; A. H. Cooke; B. R. Cooper; G. Duyckaerts, D. K. Finnemore; M. P. Garfunkel; R. V. Jones; A. F. Kip; C. Kittel; D. N. Langenberg; D. E. Mapother; K. A. G. Mendelssohn; R. W. Morse; D. E. Nagle; N. E. Phillips; H. M. Rosenberg; J. W. Stout; R. A. Stradling; W. Sucksmith; W. P. Wolf; American Institute of Physics; American Physical Society; Institution of Electrical Engineers; Institute of Physics and the Physical Society (London); Royal Society (London); Bell Telephone Laboratories; G. E. C. Hirst Research Centre.

# Note added in 1989

The opportunity has been taken of dividing this textbook into two volumes.

Volume 1: Chapters 1 to 9 inclusive, covering the basic theory of electricity and magnetism.

Volume 2: Chapters 10 to 24 inclusive, covering electrical and magnetic properties of matter, including semiconductors and their applications in electronics, alternating current measurements, fluctuations and noise, magnetic resonance.

A number of minor errors have been corrected, and a section (20.8) has been added on Operational Amplifiers. We wish to thank Dr. F. N. H. Robinson for suggesting this, and Dr. J. F. Gregg, I. D. Morris, and J. C. Ward for help in its preparation. We are indebted to Dr. L. V. Morrison of the Royal Greenwich Observatory, Cambridge (Stellar Reference Frame Group), for the up-to-date plot of the variations in the length of the day, measured by the caesium clock, that now appears as Fig. 24.12. It is based on data published by the Bureau de l'Heure, Paris.

B. I. B.
B. B.

# Contents

# CONTENTS OF VOLUME 1

# Volume 2

# 10. Dielectrics

This chapter is the first of several which approach electric and magnetic phenomena from an atomic viewpoint. For dielectrics the discussion is confined to linear phenomena, where the polarization is linearly proportional to the applied field. High-field effects, and effects involving spontaneous polarization (ferro- and antiferroelectricity) are excluded.

## 10.1. Macroscopic quantities in an atomic medium

A dielectric medium consists of an assembly of negatively charged particles (electrons) and positively charged particles (nuclei). We assume it to be electrically neutral, the total negative charge being equal to the total positive charge, and that all the charges are 'bound'. By this is meant that in zero electric field each charge occupies its equilibrium position, from which it may be slightly displaced when an electric field is applied, but that it is unable to move continuously from place to place: there is no 'free charge' and no conduction current.

In electromagnetic theory, a dielectric is a continuous medium which becomes electrically polarized under the action of an electric field. The concepts involved are the macroscopic quantities, charge density, and electric polarization, defined as continuous functions. The charge density $\rho_e$ is the ratio of an infinitesimal charge to an infinitesimal volume, taken to the limit as the infinitesimals are reduced to zero. The electric polarization is similarly defined, with net dipole moment instead of net charge. These concepts are clearly not immediately applicable to a set of discontinuous charges, and a method of relating the electrical properties of such an assembly to the corresponding properties of a continuous medium is required.

In the electron theory of Lorentz macroscopic quantities such as charge density and polarization are regarded as averages over loosely defined volumes which are small on a macroscopic scale but large enough to contain very many elementary charges. In considering the propagation of an electromagnetic wave through the medium, such averages must be taken over regions whose linear dimensions are small compared with the wavelength. For optical waves ($\lambda \sim 5 \times 10^{-7}$ m) such dimensions can be over 100 times the atomic spacing in a solid ($\sim 2 \times 10^{-10}$ m), so that the number of charges involved in the average may be $10^6$–$10^9$. Nevertheless there is a conceptual difficulty. The total charge in any such volume must be an integral multiple of the fundamental unit of charge, but if the boundaries are moved to include or exclude just one electronic charge, an apparent change in the charge density is obtained which is clearly not meaningful. (In fact one electronic charge of $1\cdot6\times10^{-19}$ C in a cube of dimensions $10^{-7}$ m gives a charge density of $160\,\mathrm{C\,m^{-3}}$, which is enormous compared with any charge density realizable in practice.) This difficulty is overcome by taking averages over a large number of such volumes of random size, so that fluctuations in volumes of size small compared with $\lambda^3$ are smoothed out.

The averages which we require may be defined more precisely by the use of a three-dimensional Fourier analysis to represent the distribution of charge in space (Robinson 1973). The method may be illustrated by considering a simple one-dimensional example: an infinite line of charges $q$ placed at regular intervals $a_0$, for example at the points 0, $\pm a_0$, and all integral multiples of $\pm a_0$. The charge density, expressed as a Fourier series, is easily shown to be

$$\rho_e = \frac{q}{a_0} + \frac{q}{2a_0} \sum_{n=1}^{\infty} \cos(2\pi nx/a_0). \tag{10.1}$$

Clearly the average charge density is $q/a_0$, and the spatial repetition rates of the oscillating terms are $a_0^{-1}$ and multiples thereof. In any length $l$ reaching from $x_0 - \frac{1}{2}l$ to $x_0 + \frac{1}{2}l$ the total charge is found by integration over this interval, and the corresponding charge density is

$$\frac{1}{l}\int \rho_e\,dx = \frac{q}{a_0} + \frac{q}{2a_0 l}\sum_{n=1}^{\infty}\int_{x_0-\frac{1}{2}l}^{x_0+\frac{1}{2}l} \cos(2\pi nx/a_0)\,dx$$

$$= \frac{q}{a_0} + \frac{q}{2\pi nl}\sum_{n=1}^{\infty} \cos(2\pi nx_0/a_0)\sin(\pi nl/a_0). \tag{10.2}$$

For arbitrary values of $x_0$ and $l$ the oscillatory terms are as often positive as negative, and the macroscopic charge density is $q/a_0$ (which in general is not an exact multiple of $q$). The Lorentz approach is in fact equivalent to taking an average over ranges of values of $x_0$ and of $l$ which are large compared with $a_0$.

This one-dimensional example illustrates the result that with a regular array of charges, as in the lattice of a crystal, Fourier analysis produces a constant component (which is the macroscopic average we require), together with components which repeat in space at the same rate as the lattice or higher. Such oscillating components cause strong scattering, the lattice acting as a three-dimensional diffraction grating for wavelengths simply related to the lattice spacing, that is, X-rays of wavelengths $\sim 2\times10^{-10}$ m. In contrast, much longer wavelengths are propagated freely through the lattice and can be treated using the macroscopic theory of

Chapter 8. Imperfections on an atomic scale, such as displacements of lattice ions under the action of thermal vibrations or the irregularities of molecular position in a liquid, do not much affect this result. This is true even in a gas, since at N.T.P. the average intermolecular distance ($\sim 5 \times 10^{-9}$ m) is still small compared with optical wavelengths. The randomness in the molecular spacing will cause some scattering, but this is small for optical waves and quite negligible for radio waves.

## 10.2. Macroscopic polarization and the local field

Experimentally no net charge on an atom or molecule has ever been detected, and Dylla and King (1973) have shown that the fractional difference in the numerical values of the charge on electron and proton is less than $1 \cdot 3 \times 10^{-21}$. However, application of an electric field displaces the positive charges in the direction of the field and the negative charges in the opposite direction, so that every substance acquires an 'induced' electric dipole moment. It is usual to distinguish between two effects: (1) an 'electronic contribution', arising from the displacement of electrons relative to positively charged nuclei; (2) an 'ionic contribution', arising from the displacement of heavy positive ions relative to heavy negative ions, which occurs in molecules and in ionic solids containing ions of different charge.

In addition to induced moments there may be permanent electric dipole moments. In an atom or monatomic ion the distribution of charge is symmetric about the centre and, as discussed in § 2.3, no permanent electric dipole moment can be present. Some molecules also fall into this symmetry classification; simple examples are the homonuclear diatomic molecules such as $H_2$, $N_2$, $O_2$ and symmetric linear polyatomic molecules such as $CO_2$, which has the structure O—C—O. Such molecules are known as 'non-polar molecules'. Molecules which do not satisfy this symmetry requirement normally possess permanent electric dipole moments, and are called 'polar molecules'.

For a finite body, the macroscopic electric dipole moment is the vector sum of the individual dipole moments. We may distinguish between three types of contributions.

(1) Each atom, ion, or non-polar molecule has zero dipole moment in zero field, but acquires a dipole moment when a field is applied; such moments are known as 'induced dipoles'.

(2) When polar molecules are present, their permanent dipole moments are completely randomly oriented in zero field, so that the vector sum is zero, at any rate on a time average; in the presence of a field, orientations in which the dipole moment is parallel to the

field have a lower energy and so are statistically more favourable, giving a resultant dipole moment in the direction of the field.

(3) In an ionic lattice, the positive ions are displaced in the direction of an applied field while the negative ions are displaced in the opposite direction, giving a resultant dipole moment to the whole body.

The techniques of Fourier analysis that were illustrated above for the charge density can also be applied to the components of the dipole moments, and under similar conditions yield a macroscopic dipole moment per unit volume. This is the polarization $\mathbf{P}$, which appears in the macroscopic equations (cf. § 1.5)

$$\mathbf{D} = \epsilon_0 \mathbf{E} + \mathbf{P} = \epsilon_r \epsilon_0 \mathbf{E} \tag{10.3}$$

and

$$\mathbf{P} = (\epsilon_r - 1) \epsilon_0 \mathbf{E}. \tag{10.4}$$

In performing the Fourier analysis, we use the dipole components given by eqns (2.25). For neutral atoms and neutral molecules the result is independent of the origin of $\mathbf{r}$ in the equation $\mathbf{p} = \int \rho_e \mathbf{r} \, d\tau$, as pointed out in § 2.3. For charged ions this is not true; for the induced dipoles arising from the displacement of the electrons relative to the nucleus we must measure $\mathbf{r}$ from the nucleus, while if the ion is bodily displaced in the lattice $\mathbf{r}$ must be measured from the point occupied by the ion when no electric field is applied.

To evaluate the dipole moment, we must calculate the effect of an electric field on an electron bound in a molecule. This is a quantum-mechanical problem, whose solution is only approximate except in a simple case such as the hydrogen atom, and which we shall not pursue further. Instead we regard the medium as composed of atoms, ions, or molecules which acquire an induced dipole moment $\mathbf{p}_i$ under the action of the electric field, and which sometimes possess also a permanent dipole moment. The size of the induced dipole $\mathbf{p}_i$ is determined by the magnitude of the field $\mathbf{E}_{local}$ acting on the particle; we can write

$$\mathbf{p}_i = \alpha \mathbf{E}_{local}, \tag{10.5}$$

where $\alpha$ is a quantity known as the 'polarizability' of the atom, ion, or molecule. At ordinary field strengths $\mathbf{p}_i$ is linearly proportional to $\mathbf{E}_{local}$, and $\alpha$ is therefore a constant, independent of field strength. The polarization arising from the induced dipoles, if all are identical and the number per unit volume is $n_0$, is

$$\mathbf{P} = n_0 \alpha \mathbf{E}_{local}, \tag{10.6}$$

where we have assumed that each induced dipole $\mathbf{p}_i$ is parallel to $\mathbf{E}_{local}$ as would be the case in the absence of any anisotropy.

The field strength $\mathbf{E}_{local}$ is the actual field acting on a particle (atom, ion, or molecule) in the dielectric medium. It is not necessarily the same as the macroscopic field strength $\mathbf{E}$, which is the average field in the medium. The reason for this is that in calculating $\mathbf{E}$ we must take the average of the vector sum of the electric fields set up by all the dipoles in the medium, including that of the particle for which we require the value of $\mathbf{E}_{local}$. Obviously, the field of this dipole does not act upon itself, and it must therefore be excluded from the vector summation which gives the value of $\mathbf{E}_{local}$. The difference between the two can be written as $\mathbf{E}_{corr}$, where

$$\mathbf{E}_{local} = \mathbf{E} + \mathbf{E}_{corr}. \qquad (10.7)$$

In forming these vector sums it is found that, although the electric field of a dipole falls off as the inverse cube of the distance, the sums do not converge rapidly. There are larger numbers of more distant dipoles, so that their total contribution is as important as those of nearby dipoles, and the 'dipole field' is said to be 'long-range'. We shall consider first the case of a dielectric of infinite extent in which there exists a uniform macroscopic field of strength $\mathbf{E}$. Such a field exists inside a parallel-plate capacitor completely filled with dielectric, and then $\mathbf{E} = V/t$, where $V$ is the voltage between the plates and $t$ their separation. In capacitors of different geometry $\mathbf{E}$ is not uniform, but this does not affect the results provided that $\mathbf{D}$, $\mathbf{P}$ in eqns (10.3)–(10.4) are linearly proportional to $\mathbf{E}$.

## 10.3. The Lorentz correction and the Clausius–Mossotti relation

An approximate solution to the problem of computing $\mathbf{E}_{corr}$ was obtained by Lorentz; it applies only to induced dipoles which are all parallel to the applied field. In an infinite dielectric, the more distant dipoles can be treated as a continuous medium with a polarization $\mathbf{P}$, producing a field strength $\mathbf{E}'$, while the net field strength $\mathbf{E}''$ of the nearby dipoles must be calculated exactly (excluding the field of the dipole at which we need the value of $\mathbf{E}_{local}$). The sum of the two contributions is

$$\mathbf{E}' + \mathbf{E}'' = \mathbf{E}_{corr}. \qquad (10.8)$$

An imaginary spherical surface is drawn centred on the point at which we require $\mathbf{E}_{local}$. The radius of this sphere is sufficiently large compared with the intermolecular distance that the effects of dipoles outside it may be evaluated by macroscopic theory, while it also contains so many molecules that its centre is a representative point and summation of the fields of the dipoles inside it gives a correct average.

To evaluate $\mathbf{E}'$ we use the results embodied in eqn (1.16). The potential, and hence also the field strength $\mathbf{E}'$, of the dipoles outside the spherical surface are the same as those of a volume distribution $-\mathrm{div}\,\mathbf{P}$

outside the sphere, and an apparent surface charge distribution on the sphere. Since $\mathbf{P}$ is uniform, div $\mathbf{P} = 0$; the surface charge density is $-P \cos \theta$, where the minus sign appears because we are considering the effect of the dielectric *outside* the sphere. From the result in the first part of Problem 2.1, we have

$$\mathbf{E}' = +\mathbf{P}/3\epsilon_0, \tag{10.9}$$

where $\mathbf{P}$ is the actual polarization of the medium. We must not include any allowance for the distortion of the field by a spherical cavity, (as in the second part of Problem 2.1) because we have not excavated a real cavity in the medium.

The value of $\mathbf{E}''$ is clearly more difficult to calculate, and it depends on how the dipoles are arranged. Lorentz showed that for a cubical array of parallel dipoles, $\mathbf{E}'' = 0$. This result is valid for induced dipoles, which are all parallel to the field, but not for permanent dipoles which are nearly randomly oriented. With $\mathbf{E}'' = 0$, the Lorentz correction is simply

$$\mathbf{E}_{\text{corr}} = \mathbf{E}' = \mathbf{P}/3\epsilon_0, \tag{10.10}$$

and the local field is given by

$$\mathbf{E}_{\text{local}} = \mathbf{E} + \mathbf{E}_{\text{corr}} = \mathbf{E} + \mathbf{P}/3\epsilon_0. \tag{10.11}$$

From eqn (10.6) the polarization is

$$\mathbf{P} = n_0 \alpha (\mathbf{E} + \mathbf{P}/3\epsilon_0), \tag{10.12}$$

and on combining this with eqn (10.4) we have the equation

$$\frac{\epsilon_r - 1}{\epsilon_r + 2} = \frac{n_0 \alpha}{3\epsilon_0}. \tag{10.13}$$

This result was first derived by Clausius and Mossotti, and is generally known by their names.

If the dielectric is not infinite in extent, the vector sum of the dipole fields depends on the shape of the body. This result has already been encountered in macroscopic theory (§ 2.4), which shows that the field inside the body differs from the external applied field. In Fig. 10.1 a dielectric body B is placed in a uniform field of strength $\mathbf{E}_0$. The field acting on a dipole at the centre of an imaginary Lorentz sphere L is $\mathbf{E}_{\text{local}}$, which is the sum of $\mathbf{E}_0$ and the field strengths due to all the surrounding dipoles. Those inside the sphere contribute a field strength $\mathbf{E}''$ which is taken (as above) to be zero. Those outside L are equivalent to a volume distribution of charge $-\text{div } \mathbf{P}$, together with distributions of charge on the surfaces L and B, by eqn (1.16). If the susceptibility $\chi_e$ is uniform, isotropic, and independent of field strength, then div $\mathbf{P} = \chi_e \epsilon_0 \text{ div } \mathbf{E} = 0$. The inner surface charge on the sphere L contributes the field strength $\mathbf{E}'$,

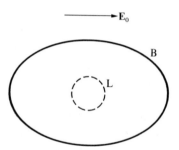

FIG. 10.1. A dielectric body B in a uniform field of strength $\mathbf{E}_0$. Inside B the field $\mathbf{E}$ differs from $\mathbf{E}_0$ because of the depolarizing field. L is the imaginary surface of a Lorentz sphere.

as above. The outer surface charge on B produces the 'depolarization field' which causes the field strength $\mathbf{E}$ in the dielectric to be different from $\mathbf{E}_0$. For a dielectric sphere in a vacuum, from § 2.4 we have $\mathbf{E} = \mathbf{E}_0 - \mathbf{P}/3\epsilon_0$, so that $\mathbf{E}_{\text{local}} = \mathbf{E} + \mathbf{E}' = \mathbf{E}_0$. This is obviously a special case; it is just the result we should obtain by increasing the size of the Lorentz sphere until it contains the whole body. There is then no dielectric outside this sphere so that the Lorentz field must vanish, that is, $\mathbf{E}_{\text{local}} = \mathbf{E}_0$ since $\mathbf{E}'' = 0$. For a dielectric of infinite extent the question of a depolarizing field does not arise; eqns (10.3) and (10.4) relate to the field $\mathbf{E}$ in the medium, which is assumed to be known.

## 10.4. Static permittivity of non-polar gases

In a gas the molecules are randomly placed and moving with molecular velocities, but it can be shown that the average value of $\mathbf{E}''$ is again zero for induced dipole moments which are all parallel to the applied field. Hence for non-polar gases, in which the molecules have no permanent dipole moments, we have $\mathbf{E}' = \mathbf{P}/3\epsilon_0$; the value of $\mathbf{E}_{\text{local}}$ is given by eqn (10.11), and we should expect the Clausius–Mossotti relation (10.13) to hold. Since the number of molecules $n_0$ per unit volume varies with the density, it is convenient to multiply both sides of (10.13) by $M/\rho$, where $M$ is the molecular weight and $\rho$ the density. This gives

$$\frac{\epsilon_r - 1}{\epsilon_r + 2} \cdot \frac{M}{\rho} = \frac{N_A \alpha}{3\epsilon_0}, \tag{10.14}$$

where $N_A$ is Avogadro's number, a universal constant. The quantity $N_A \alpha / 3\epsilon_0$ is sometimes called the 'molar polarizability'.

Values of the relative permittivity of a number of common gases at N.T.P. are given in Table 10.1. The difference between $\epsilon_r$ and unity is of order $10^{-3}$, and for non-polar gases it increases with the complexity and hence with the size of the molecule. The permittivity has been used as a method of estimating molecular size. This makes the approximation that

TABLE 10.1

*Relative permittivity of some common gases at atmospheric pressure* $(10^5 \text{ Pa})$ *and 273 K*

| Gas | $(\epsilon_r - 1)10^3$ | Dipole moment (debyes) |
|---|---|---|
| He | 0·071 | 0 |
| $H_2$ | 0·270 | 0 |
| $O_2$ | 0·531 | 0 |
| $N_2$ | 0·588 | 0 |
| $CO_2$ | 0·988 | 0 |
| $CH_4$ | 0·948 | 0 |
| $C_2H_4$ | 1·38 | 0 |
| CO | 0·692 | 0·10 |
| $N_2O$ | 1·08 | 0·17 |
| $NH_3$ | 8·34 | 1·45 |
| $SO_2$ | 9·93 | 1·59 |

TABLE 10.2

*Relative permittivity of $CO_2$ and the Clausius–Mossotti relation*

| Experimenters | Pressure (Pa) | Relative permittivity (at 373 K) | $\dfrac{\epsilon_r - 1}{\epsilon_r + 2} \cdot \dfrac{M}{\rho}$ $(m^3)$ |
|---|---|---|---|
| Keyes and Kirkwood (1930)[†] | $1\cdot0 \times 10^6$ | 1·00753 | $7\cdot49 \times 10^{-6}$ |
|  | $3\cdot0 \times 10^6$ | 1·0240 | $7\cdot53 \times 10^{-6}$ |
|  | $5\cdot0 \times 10^6$ | 1·0431 | $7\cdot57 \times 10^{-6}$ |
|  | $7\cdot0 \times 10^6$ | 1·0645 | $7\cdot60 \times 10^{-6}$ |
|  | $10\cdot0 \times 10^6$ | 1·1041 | $7\cdot69 \times 10^{-6}$ |
|  | $15\cdot1 \times 10^6$ | 1·1912 | $7\cdot73 \times 10^{-6}$ |
| Michels and Michels (1932)[‡] | $10\cdot32 \times 10^6$ | 1·1086 | $7\cdot71 \times 10^{-6}$ |
|  | $19\cdot45 \times 10^6$ | 1·2695 | $7\cdot75 \times 10^{-6}$ |
|  | $29\cdot54 \times 10^6$ | 1·3895 | $7\cdot70 \times 10^{-6}$ |
|  | $36\cdot50 \times 10^6$ | 1·4375 | $7\cdot68 \times 10^{-6}$ |
|  | $47\cdot66 \times 10^6$ | 1·4900 | $7\cdot67 \times 10^{-6}$ |
|  | $58\cdot83 \times 10^6$ | 1·5274 | $7\cdot66 \times 10^{-6}$ |
|  | $70\cdot02 \times 10^6$ | 1·5570 | $7\cdot66 \times 10^{-6}$ |
|  | $81\cdot23 \times 10^6$ | 1·5812 | $7\cdot65 \times 10^{-6}$ |
|  | $97\cdot06 \times 10^6$ | 1·6097 | $7\cdot62 \times 10^{-6}$ |

The value of $\dfrac{\epsilon_r - 1}{\epsilon_r + 2} \cdot \dfrac{M}{\rho}$ at N.T.P. is $7\cdot33 \times 10^{-6} \text{ m}^3$ for 1 mol.

The data indicate a slight rise in the molar polarizability with pressure, followed by a small decrease at the highest pressures ($\sim$1000 atmospheres).

† Keyes, F. G. and Kirkwood, J. G. (1930). *Phys. Rev.* **36**, 754.
‡ Michels, A. and Michels, C. (1932). *Phil. Trans. R. Soc.* A**231**, 409.

each molecule can be regarded as a perfectly conducting sphere of radius $a$; in a field of strength $E$ such a sphere acquires a dipole moment $4\pi\epsilon_0 a^3 E$, giving a value for $\alpha = 4\pi\epsilon_0 a^3$. Values of the molecular radius $a$ obtained in this way are generally of the same order but somewhat smaller than those obtained from kinetic theory (cf. Problem 10.2).

The validity of the Clausius–Mossotti relation has been tested by various experimenters up to pressures of $10^8$ Pa. For the non-polar gas $CO_2$ the constancy of the quantity $N_A\alpha/3\epsilon_0$, calculated using eqn (10.14), is shown in Table 10.2. At low pressures the value of $\epsilon_r - 1$ is so small that the approximation $\epsilon_r + 2 = 3$ is quite good; this is equivalent to neglecting the Lorentz correction and ignoring the difference between $\mathbf{E}_{local} = \mathbf{E}(\epsilon_r + 2)/3$ and $\mathbf{E}$. At higher pressures this approximation is obviously not valid, and the constancy of the values in the last column of Table 10.2 justifies the assumptions made in the Lorentz method. The Clausius–Mossotti relation also depends on the assumption that the polarizability $\alpha$ is independent of $n_0$; this implies that it does not depend on the distance to other molecules, and that short-range forces between the molecules can be neglected. This is a reasonable assumption in a gas, but we must not be surprised if it fails in a liquid or a solid.

## 10.5. Static permittivity of polar gases

The last four molecules listed in Table 10.1 have permanent electric dipole moments, and the values of $\epsilon_r - 1$ are noticeably greater for $NH_3$ and $SO_2$, whose molecules have dipole moments of order 1 debye. This unit is defined as

$$1 \text{ debye} = 3 \cdot 336 \times 10^{-30} \text{ C m.}$$

It is equivalent to an electronic charge of $1 \cdot 6 \times 10^{-19}$ C multiplied by a distance of about $2 \times 10^{-11}$ m, which is about one-tenth of the internuclear distance in a typical molecule.

The simplest type of polar molecule is a diatomic molecule formed from two dissimilar atoms. For a molecule such as KCl, an oversimplified picture is that of two ions, $K^+$ and $Cl^-$, for which the electric dipole moment should equal the electronic charge multiplied by the internuclear distance. The observed dipole moments listed in Table 10.3 show that this crude model is not far wrong for the alkali halides; the model can be improved by allowing for the fact that the field of each charged ion polarizes the other ion (see Fig. 10.2), producing induced moments $\mathbf{p}_i$ which have the opposite direction to the main dipole moment $\mathbf{p}$. The ionic picture is clearly much less good for the hydrogen halide molecules HCl, HBr, HI, where in progression the dipole moments actually decrease while the internuclear distance increases. This shows that the picture of such a molecule as a pair of charged ions is a gross oversimplification; in

TABLE 10.3

*Internuclear distances and electric dipole moments
of some diatomic molecules*

| Molecule | Internuclear distance $r$ (nm) | Electronic charge$\times r$ (debye) | Observed dipole moment (debye) |
|---|---|---|---|
| CsF | 0·2345 | 11·2 | 7·88 |
| CsCl | 0·2906 | 14·0 | 10·46 |
| CsI | 0·3315 | 15·9 | 12·1 |
| KF | 0·255 | 12·2 | 7·33 |
| KCl | 0·2667 | 12·8 | 10·48 |
| KBr | 0·2821 | 13·5 | 10·41 |
| KI | 0·3048 | 14·6 | 11·05 |
| HCl | 0·127 | 6·1 | 1·03 |
| HBr | 0·142 | 6·8 | 0·78 |
| HI | 0·162 | 7·8 | 0·38 |

fact the electronic charge tends to be shared more equally between the two nuclei. This tendency increases as we go from HCl to HI, corresponding to a progressive change towards covalent binding in which the valence electrons are wholly shared between the two atoms.

For polyatomic molecules the problem is more complicated, and we illustrate it by reference to some triatomic molecules. Both $SO_2$ (1·59 debye) and $H_2O$ (1·84 debye) have large dipole moments, and this shows they cannot be symmetrical linear molecules like $CO_2$, which is O—C—O and has no dipole moment. It does not exclude the possibility that they are asymmetrical linear molecules like $N_2O$, whose structure is N—N—O, but the latter has only a small moment of 0·17 debye. In fact both $H_2O$ and $SO_2$ have the shape of isosceles triangles, as in Fig. 10.3, the angles at the apex (the O or S nucleus) being 105° and 120° respectively. The direction of the dipole moment is clearly normal to the

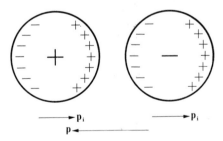

FIG. 10.2. The induced dipoles $p_i$ on each ion are in the opposite direction to the main dipole $p$ formed by the charges on the two ions, so that the total dipole moment is less than $p$.

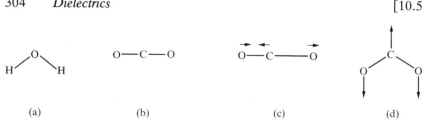

(a)              (b)                    (c)                    (d)

FIG. 10.3. (a) The $H_2O$ molecule has the shape of an isosceles triangle, with a large electric dipole moment. (b) The $CO_2$ molecule is linear and symmetrical, with no electric dipole moment. The $CO_2$ molecule acquires an instantaneous moment in a bending (d) or asymmetrical vibration (c). In a symmetrical vibration (not shown) where the oxygen atoms move in and out together and the carbon atom is stationary, the shape remains as in (b) and there is no oscillating dipole moment.

base of the triangle, and in the plane of the triangle, since reflection symmetry excludes other possibilities.

When an electric field is applied to a polar gas, in addition to the induced moments considered in the last section there will be a contribution to the polarization from the permanent dipole moments. The potential energy of a dipole $\mathbf{p}$ in a field of strength $\mathbf{E}$ is $-\mathbf{p}.\mathbf{E} = -pE \cos \theta$, where $\theta$ is the angle between $\mathbf{p}$ and $\mathbf{E}$. In a classical treatment using the Boltzmann distribution, since the parallel direction has the lower energy, there will be more dipoles pointing with the field than pointing against it, giving a net polarization in the direction of the field. The problem has already been treated in detail for the corresponding magnetic case, and the result can be adapted to the electrical case by writing $\mathbf{p}$ and $\mathbf{E}_{\text{local}}$ instead of $\mathbf{m}$ and $\mathbf{B}$. For small field strengths, the contribution $\mathbf{P}_d$ from the permanent dipoles is (cf. eqn (6.13))

$$\mathbf{P}_d = \left(\frac{n_0 p^2}{3kT}\right)\mathbf{E}_{\text{local}},\qquad(10.15)$$

where $k$ is Boltzmann's constant and $T$ the absolute temperature.

The relation between the local field $\mathbf{E}_{\text{local}}$ and the macroscopic field $\mathbf{E}$ in the medium is more complicated than for induced dipoles because the permanent dipoles are not all oriented parallel to the field (see § 10.7). For gases at such low densities that the difference between $\mathbf{E}_{\text{local}}$ and $\mathbf{E}$ can be neglected, the static relative permittivity $\epsilon_s$ is given by the relation

$$\epsilon_s - \epsilon_i = \frac{\mathbf{P}_d}{\epsilon_0 \mathbf{E}} = \frac{n_0 p^2}{3\epsilon_0 kT}.\qquad(10.16)$$

Here $\epsilon_i$ is that part of the relative permittivity due to the induced dipoles alone; and, in the low-density limit, $\epsilon_i - 1 = n_0 \alpha / \epsilon_0$.

In this derivation we have used the formula appropriate to the limit of $pE/kT \ll 1$, a condition which is generally readily fulfilled. At room temperature $kT = 4 \times 10^{-21}$ J, so that even for a dipole with a moment of 4 debye a field of $3 \times 10^7$ V m$^{-1}$ is required to make $pE/kT = 0.1$. We cannot increase this ratio for a gas by going to low temperatures, since polar gases tend to have high liquefaction- and freezing-points owing to the large intermolecular forces between the permanent dipoles. These same intermolecular forces make eqn (10.16) quite inapplicable to a liquid or a solid. This is very different from the corresponding magnetic case, because in many paramagnetic salts the magnetic dipoles remain relatively free to orient themselves in a magnetic field, and Curie's law may hold accurately over a wide range of temperature.

The derivation of the contribution to the polarization from the permanent dipoles which we have given is a purely classical one, and the reader may wonder to what extent it is confirmed by wave mechanics. The answer to this is that exactly the same result is obtained, but in a surprisingly different way. This may be illustrated by reference to a diatomic molecule. The rotational states of such a molecule are distinguished by having quantized values of the angular momentum equal to $J(h/2\pi)$, where $J$ is zero or a positive integer and $h$ is Planck's constant. The calculation shows that in small fields the states for which $J \neq 0$ contribute nothing to the polarization in respect of the permanent dipole moment of the molecule. This is reasonable because when the molecule is turning end over end, the average projection of the dipole moment on any direction in space is zero. In the state $J = 0$, however, the molecule is not rotating, and the whole of the contribution comes from this state. At high temperatures a large number of rotational states are occupied, and the fraction of molecules which are in the state $J = 0$ is proportional to $1/T$. This gives the same temperature variation as the classical theory, and detailed calculation shows that the numerical constant is also the same (see Pauling and Wilson 1935).

Eqn (10.16) may be used to determine the electric dipole moment from measurements of the permittivity of a gas (for experimental methods see § 22.6). To separate the contributions from the induced polarization and the permanent dipoles, measurements are made over as wide a temperature range as possible. A plot of $\epsilon_r$ against $T^{-1}$ should give a straight line, from the slope of which the dipole moment can be found using eqn (10.16). The intercept at $T^{-1} = 0$ gives the value of $\epsilon_i$.

## 10.6. Dispersion in gases

The theory of electromagnetic waves (Chapter 8) shows that the refractive index $n$ of a substance should be equal to the square-root of $\epsilon_r$, if the magnetic permeability can be taken as unity, as is usually the case. A comparison of the values of $\epsilon_r$ measured at low frequencies with the

TABLE 10.4

| Gas | $(\epsilon_r - 1)\,10^6$ at N.T.P. | | | | |
|---|---|---|---|---|---|
| | 0·1 MHz[†] | 1 MHz[‡] | 9000 MHz[§] | 24 000 MHz[‖] | Optical[††] |
| Air | 570±0·7 | 567·0±1·0 | 575·4±1·4 | 576·0±0·2 | 575·7±0·2 |
| Nitrogen | 578±0·7 | 579·6±1·0 | 586·9±2·9 | 588·3±0·2 | 581·3 |
| Oxygen | 528±1 | 523·3±1 | 530·0±1·9 | 531·0±0·4 | 532·7 |
| Argon | 545±1 | 545·1±0·5 | — | 555·7±0·4 | 554·7 |
| Carbon dioxide | 987±1 | 987·5±2 | 985·5±3 | 988 ±2 | — |
| Hydrogen | 270±1 | 272 | — | — | 272 |

[†] Lovering W. F. and Wiltshire L. (1951), *Proc Inst. Elect. Engrs.* 98, Part II, 557.
[‡] Hector L. G. and Woernley D. L. (1946), *Phys. Rev.* **69**, 101.
[§] Birnbaum G., Kryder S. J., and Lyons L. (1951), *J. Appl. Phys.* **22**, 95.
[‖] Essen L., and Froome K. D. (1951), *Proc. phys. Soc.* B**64**, 862.
[††] $(n^2 - 1)10^6$ (various authors), extrapolated to infinite wavelength.

square of the refractive indices measured in the optical region (that is, at frequencies of the order of $10^{14}$ Hz) gives very poor agreement with this relation except in the case of simple non-polar gases. Values of $\epsilon_r$ of a few such gases measured over a wide range of frequencies are given in Table 10.4 together with the square of the optical refractive index. The latter is extrapolated to 'infinite wavelengths' to correct for dispersion in the optical region. The agreement is seen to be excellent in the cases quoted.

In the optical region, variation of the refractive index with wavelength has been known for a very long time, and is called dispersion. In general the refractive index increases as the wavelength decreases, and this is known as 'normal dispersion'. The reverse case, where the refractive index decreases with decreasing wavelength, occurs only in the vicinity of an absorption line and is difficult to observe because of the absorption. This is known as 'anomalous dispersion', but both types have a simple explanation in terms of classical theory, based on the assumption that an atom contains electrons vibrating at certain natural frequencies characteristic of the type of atom, and that the application of an alternating electric field sets such electrons into forced vibration.

Let us take the simplest possible case of a gas subjected to an oscillating electric field $\mathbf{E}_{\text{local}} = \mathbf{E}^{(0)} \exp j\omega t$. We shall assume that the wavelength of the incident radiation is very large compared with atomic dimensions (which is true up to the region of hard X-rays), so that the field acting on an electron in a given atom is independent of its position with respect to the nucleus, which is assumed to be stationary. Each electron of charge $-e$ is displaced a distance $\mathbf{s}$ by the field, and the restoring force is written as $-m\omega_p^2\mathbf{s}$, where $\omega_p/2\pi$ is the natural frequency of oscillation of the electron and $m$ its mass. In addition there will be damping due to collisions, radiation of energy, etc., which may be represented by a term $-m\gamma(ds/dt)$. Hence we have

$$m\left(\frac{d^2s}{dt^2} + \gamma\frac{ds}{dt} + \omega_p^2 s\right) = -e\,E^{(0)}\exp j\omega t. \qquad (10.17)$$

The solution of this is

$$s = -\frac{e\,E_{\text{local}}}{m\{(\omega_p^2-\omega^2)+j\gamma\omega\}} + \exp(-\tfrac{1}{2}\gamma t)[\mathbf{A}\cos\{(\omega_p^2-\tfrac{1}{4}\gamma^2)^{\frac{1}{2}}t\}+\mathbf{B}\sin\{(\omega_p^2-\tfrac{1}{4}\gamma^2)^{\frac{1}{2}}t\}].$$

The terms in **A** and **B** average to zero over many atoms since **A** and **B** depend on the initial conditions and are as often positive as negative. The instantaneous electric dipole moment due to the displacement of the electron is $\mathbf{p} = -e\mathbf{s}$, and, if there are $n_0$ molecules per unit volume, the polarization **P** is

$$\mathbf{P} = n_0\mathbf{p} = \frac{n_0 e^2 \mathbf{E}_{\text{local}}}{m}\frac{1}{(\omega_p^2-\omega^2)+j\gamma\omega}. \qquad (10.18)$$

For gases at higher density a correction for the difference between the local field and the external field may be applied in the same way as in § 10.3, leading to the formula

$$\frac{\epsilon_r-1}{\epsilon_r+2} = \frac{n^2-1}{n^2+2} = \frac{n_0 e^2}{3m\epsilon_0}\frac{1}{(\omega_p^2-\omega^2)+j\gamma\omega}. \qquad (10.19)$$

This formula shows that both $\epsilon_r$ and **n** must be regarded as complex. Writing $\epsilon_r = \epsilon'-j\epsilon'' = (n-jk)^2$, where $n$ is the real part of the refractive index and $k$ is the absorption coefficient, we may separate the real and imaginary parts of eqn (10.19). The formula is clumsy to handle, however, and we shall assume that we are dealing only with gases at such low pressures that we can neglect the Lorentz correction. When the value of $k$ is small, and negligible except near an absorption line, we may also make the approximation, if the line is narrow, of writing

$$(\omega_p^2-\omega^2) = (\omega_p+\omega)(\omega_p-\omega) \approx 2\omega(\omega_p-\omega).$$

Then we obtain the formulae

$$\left.\begin{array}{l} \epsilon' = n^2-k^2 \approx n^2 = 1 + \dfrac{n_0 e^2}{2m\omega\epsilon_0}\left\{\dfrac{\omega_p-\omega}{(\omega_p-\omega)^2+\Delta\omega^2}\right\}, \\[3mm] \epsilon'' = 2nk \approx 2k = \dfrac{n_0 e^2}{2m\omega\epsilon_0}\left\{\dfrac{\Delta\omega}{(\omega_p-\omega)^2+\Delta\omega^2}\right\}, \end{array}\right\} \qquad (10.20)$$

where the symbol $\Delta\omega$ has been used for $\gamma/2$, and we have assumed $n \approx 1$, $k \ll n$.

The variation of $n$ and $k$ in the neighbourhood of a weak absorption line is shown in Fig. 10.4. The absorption coefficient reaches a maximum at the resonant frequency where $\omega = \omega_p$, and falls to half its maximum

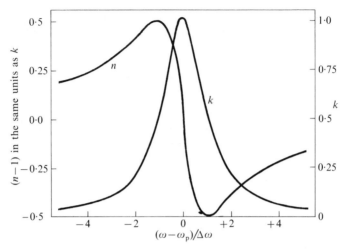

FIG. 10.4. Variation of $n$ and $k$ near a narrow absorption line (from eqn (10.20)). $(n-1)$ and $k$ are in units of $n_0 e^2/4m\omega\epsilon_0 \Delta\omega$.

value at $\omega_p - \omega = \pm\Delta\omega$. In optical usage, the quantity $2\,\Delta\nu = \Delta\omega/\pi$ is called the 'half-width' of the line, meaning the frequency difference between the points at which the absorption has dropped to half the maximum value. Microwave spectroscopists, however, prefer to call $\Delta\nu$ the half-width.

In general, each atom or molecule possesses a number of characteristic resonant frequencies, and the expressions given above for the refractive index and absorption coefficient should be replaced by others with summations over the various values of $\omega_p$. If the number of electrons per molecule which have a resonant frequency $\omega_p$ is denoted by $f_p$, we may write eqn (10.19) in the form

$$\frac{\epsilon_r - 1}{\epsilon_r + 2} = \frac{n^2 - 1}{n^2 + 2} = \frac{n_0 e^2}{3m\epsilon_0} \sum_p \frac{f_p}{(\omega_p^2 - \omega^2) + j\gamma\omega}. \tag{10.21}$$

The value of $f_p$ is known as the 'oscillator strength' of an absorption line, and on classical theory we should expect it to be unity. In practice it generally has values less than unity, and the quantum-mechanical explanation shows that this corresponds to the fact that each electron possesses a number of possible frequencies of oscillation, and its total oscillator strength is divided between them. Then $\sum_p f_p = 1$ for each electron.

At frequencies far from resonance the absorption coefficient is negligible. At very low frequencies, where $\omega \ll$ any value of $\omega_p$, we have

$$\frac{\epsilon_r - 1}{\epsilon_r + 2} = \frac{n_0 e^2}{3m\epsilon_0} \sum_p \frac{f_p}{\omega_p^2} = \frac{n_0 \alpha}{3\epsilon_0} \tag{10.22}$$

by comparison with (10.13). This shows that the molecular polarizability $\alpha$ is intimately connected with the oscillator strengths and absorption lines. This important result follows from a simple picture in which the displacement **s** of a particle is assumed to set up a restoring force which is proportional to **s**. The dynamical equation of motion is $m(d^2s/dt^2) + m\omega_p^2 s = 0$, corresponding to harmonic vibrations of frequency $\omega_p/2\pi$. In a static field of strength **E**, if the charge on the particle is $q$, we have $q\mathbf{E} = m\omega_p^2 \mathbf{E}$, and the induced dipole moment is

$$\mathbf{p}_i = q^2 \mathbf{E}/m\omega_p^2 = \alpha_p \mathbf{E},$$

where $\alpha_p$ is the polarizability associated with the vibration at frequency $\omega_p/2\pi$. On writing $q^2 = e^2 f_p$, and summing over all resonant frequencies, we obtain

$$\alpha = \frac{e^2}{m} \sum_p \frac{f_p}{\omega_p^2}. \tag{10.23}$$

As the frequency is raised to pass through an absorption line at $\omega_p/2\pi$, the refractive index goes through the anomalous variation shown in Fig. 10.4 and approaches a smaller limiting value on the high-frequency side than it had on the low-frequency side. When there are a number of absorption lines, the behaviour is as shown in Fig. 10.5, and finally, when $\omega$ is greater than all values of $\omega_p$, $n$ approaches unity, but the value of $(n-1)$ is slightly negative. This is the well-known anomaly in the refractive index in the X-ray region, and the value of $n$ is then generally calculated by assuming that the electrons are free, so that eqn (10.17)

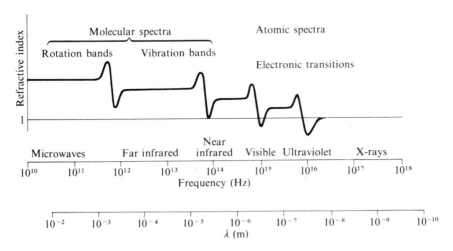

FIG. 10.5. Schematic diagram showing variation of refractive index with frequency.

reduces to

$$m\frac{d^2\mathbf{s}}{dt^2} = -e\mathbf{E}' \exp j\omega t.$$

This is equivalent to the assumption that $\omega \gg \omega_p, \gamma$.

Electronic resonant frequencies lie in the visible or ultraviolet regions of the spectrum; for colourless compounds they lie entirely in the ultraviolet, and the refractive index increases with frequency in the visible region, corresponding to 'normal dispersion'. The electronic contribution to the polarizability may therefore be deduced from the optical refractive index, with a correction to 'infinite wavelength' to allow for dispersion. The electronic molar polarizability is defined, by analogy with eqn (10.14), as the quantity

$$\frac{n^2-1}{n^2+2} \cdot \frac{M}{\rho} = \frac{N_A \alpha_{el}}{3\epsilon_0}. \tag{10.24}$$

For a compound this quantity is approximately equal to the sum of the contributions from the individual atoms or ions, and, from measurements on a large range of compounds, it is possible to deduce a set of atomic or ionic polarizabilities. The additive relation is not exact because the polarizability is sensitive to the nature of the chemical binding. However, a general rule is that the polarizability is larger for negative ions than for positive ions with the same number of electrons (for example, $Cl^-$ has a polarizability several times greater than $K^+$), and in any column of the periodic table it increases with atomic number (for example, the polarizability of xenon is some 20 times greater than that of helium).

In addition to the electronic contributions to the polarizability, there are 'ionic' contributions arising from the motion of heavy ions, with resonant frequencies in the infrared. We consider first internal vibrations in molecules which involve changes in the internuclear distances, with resonances in the region $\lambda \sim 3\text{--}100\ \mu\text{m}$. Only vibrations in which an oscillating dipole moment is set up are important, as can be illustrated for the molecule $CO_2$. In equilibrium this has a linear symmetric structure with no dipole moment, but if it is bent or asymmetrically distorted as in Fig. 10.3 it acquires an electric dipole moment. Such distortions may be induced by a static electric field, and contribute to the static polarizability; when vibrating, the oscillating dipole moment causes absorption bands in the near infrared. On the other hand, the symmetric vibration in $CO_2$ gives no dipole moment and no absorption band; it is said to be 'inactive'.

If a molecule has a permanent electric dipole moment, there is an additional term (eqn (10.16)) in the static permittivity. Classically this arises from a net change in the orientation of the dipoles; in quantum mechanics it is connected with the rotational energy levels. For such a molecule, an electromagnetic field exerts a couple which, in classical mechanics, changes the rate of rotation. In quantum mechanics, transitions occur between the rotational levels which give rise to a 'pure rotational' spectrum in the far infrared (typically, $\lambda \sim 100$–$10^4\ \mu$m). Contributions to the permittivity occur only if the dipole moment is electric, but a few molecules such as $O_2$, NO, and $NO_2$ have permanent magnetic dipole moments of the order of a Bohr magneton ($1\ \mu_B$). The contribution to the static permeability from such magnetic dipoles is very small compared with the permittivity due to electric dipoles, involving $\mu_0 m^2$ (eqn (6.13)) instead of $p^2/\epsilon_0$ (eqn (10.16)).

## 10.7. Static permittivity of liquids

The static values of $\epsilon_r$ for liquids are related to absorption lines in a similar way to that outlined for gases. In simple atomic substances, such as the condensed phases of the rare gases of the atmosphere, there are no absorption bands in the infrared and $\epsilon_r$ does not differ greatly from that deduced from the Clausius–Mossotti formula, using the value of $\alpha$ measured in the gas phase. This is particularly true for liquid helium, which has an exceptionally low density and permittivity; at the boiling-point, the density is 125 kg m$^{-3}$ (one-eighth that of water), and $\epsilon_r = 1\cdot048$. This corresponds to a molar polarizability ($N_A\alpha/3\epsilon_0$) of $0\cdot12 \times 10^{-6}$, very close to the value of $0\cdot123 \times 10^{-6}$ for the gas.

Organic liquids such as benzene ($C_6H_6$) or carbon tetrachloride ($CCl_4$) are assemblies of non-polar molecules, and similar relations hold for $\epsilon_r$, though there may be small contributions from vibrations internal to the molecules.

The situation is quite different for a 'polar liquid' in which the molecules have permanent electrical dipole moments. Because of the high density, a small departure from random orientation of the dipoles produces a large contribution $\mathbf{P}_d$ to the polarization, and if this is included in the calculation of the local field by the Lorentz method, it yields non-sensical results. If we write

$$\mathbf{E}_{local} = \mathbf{E} + \mathbf{P}_d/3\epsilon_0$$

in eqn (10.11), on combining this with (10.15) we obtain

$$\frac{\mathbf{P}_d}{\mathbf{E}} = \frac{\mathbf{P}_d}{\mathbf{E}_{local} - \mathbf{P}_d/3\epsilon_0} = \frac{n_0 p^2}{3k(T - T_c)}, \tag{10.25}$$

with $kT_c = n_0 p^2/3\epsilon_0$. For water this gives $T_c \sim 1000$ K, which would mean that below this temperature water should be spontaneously polarized, as in a ferromagnet (Chapter 15). This unrealistic result arises as follows. The interaction of a dipole with its surroundings produces a 'reaction field' which is parallel to the dipole (cf. Problem 2.2); this field may induce an added moment parallel to the permanent moment but it cannot change the orientation of the moment. In calculating the net polarization arising from reorientation of the permanent moments in a field, it is therefore wrong to use a value of the local field which includes the contribution from the permanent dipoles, as we did above.

This problem has been treated by Onsager (1936) using an approach basically different from that of Lorentz. Each dipole is regarded as being at the centre of a *real* spherical cavity, whose size is assumed equal to the average volume occupied by each molecule. Standard electrostatic theory is then used to evaluate separately the effect of the surrounding dielectric on the randomly oriented permanent dipole, and on the induced dipole (which is parallel to the applied field). Some rather complicated algebra yields the formula

$$\frac{(\epsilon_s - \epsilon_i)(2\epsilon_s + \epsilon_i)}{\epsilon_s(\epsilon_i + 2)^2} = \frac{n_0 p^2}{9\epsilon_0 kT},\tag{10.26}$$

where $\epsilon_s$ is the total static relative permittivity, while $\epsilon_i$ is that part arising solely from the induced dipoles. For water, using $\epsilon_i = 4.9$ (see § 10.8) and $p = 1.94$ debye, this formula gives $\epsilon_s = 105$ at $T = 273$ K, falling to $\epsilon_s = 78$ at 373 K. The measured values are 88 and 55 respectively; although the formula gives values on the high side, they correspond much better to reality than does the Lorentz prediction. Better agreement is hardly to be expected; $H_2O$ is a triangular molecule, while a spherical cavity is assumed, and short-range forces are neglected. Only long-range forces (forces between dipoles) are included in methods using formulae based on electrostatic theory, and, in spite of its apparent difference of approach, Onsager's theory gives the same result as that of Lorentz for induced dipoles.

In dilute solutions of polar molecules in non-polar solvents, the permanent dipoles are far apart on average and their mutual interactions can be neglected. Onsager's theory can be extended to allow for interaction with the solvent molecules, giving an effective value of the dipole moment which must be allowed for when using eqn (10.16). The permittivity for a range of concentrations is measured by a standard method (see § 22.6), and extrapolated to infinite dilution. The solvents normally employed are carefully purified benzene and carbon tetrachloride. Electric dipole moments measured in this way agree fairly well with those found using the gaseous method (§ 10.5); the latter is the more satisfactory method when it can be used, the solvent method being employed for substances of very low vapour pressure. For a number of simple molecules (such as those in Table 10.3) accurate values of the dipole moment have been obtained from microwave spectroscopy (see Townes and Schawlow 1955) or electric resonance in molecular beams (see Ramsey 1956), by measurement of the splitting of the rotational lines in an electric field.

## 10.8. Dispersion in polar liquids

For polar gases $\epsilon_r$ is greater than the square of the optical refractive index; the difference is due to dispersion in the infrared, mainly associated with the rotational spectrum of the molecules. In a polar liquid such as water, the high value of $\epsilon_r$ is due to reorientation of the electric dipoles in the applied field, and in an alternating field this reorientation must occur sufficiently quickly for the dipoles to follow the field reversals if they are to make their full contribution to the polarization. If reorientation takes a finite time $\sim \tau$, the dipoles cannot follow a field whose angular frequency is such that $\omega\tau \gg 1$. In the region where $\omega\tau \sim 1$, the permittivity falls, and energy is absorbed from the alternating field into the dielectric, finally appearing as heat.

The mechanism which inhibits reorientation is collisions between the molecules. This results from the molecular motion, which is visible in the form of Brownian motion, and results in the phenomenon of viscosity. For a spherical particle of radius $a$ suspended in a liquid of viscosity $\eta$, the mean-square value $\overline{\theta^2}$ of the rotational displacement angle $\theta$ in a time $t$ is

$$\overline{\theta^2} = \frac{kT}{4\pi\eta a^3}\, t = \frac{t}{\tau}, \tag{10.27}$$

where $\tau = 4\pi\eta a^3/kT$ is a characteristic time for the Brownian motion. If we apply this to the molecules of a liquid such as water, taking $a = 2\cdot3\times10^{-10}$ m, the value found from the viscosity of the vapour, and $\eta = 0\cdot001$ S.I. units at 290 K, we find $\tau = 3\cdot7\times10^{-11}$ s. Since this time is longer than any of the characteristic periodic times of rotation of the free water molecule, it follows that the molecule cannot rotate at any of its natural frequencies in the liquid state. Instead, the dispersion associated with the permanent dipoles will take place at frequencies such that $\omega \approx 1/\tau$, that is at a wavelength of the order of 10 mm.

In order to introduce $\tau$ into our treatment of the permittivity, we consider the effect of maintaining a steady field on a polar liquid, and then suddenly removing it. Under the influence of the Brownian motion, the preferred orientations of the dipoles will gradually disappear. It is reasonable to suppose that the rate of decay of the polarization is proportional to the instantaneous value of the polarization, and we write

$$d\mathbf{P}/dt = -\mathbf{P}/\tau,$$

giving

$$\mathbf{P} = \mathbf{P}_0 e^{-t/\tau},$$

where $\tau$ is a characteristic 'relaxation time' which we would expect to be of the same order as that found above for the Brownian motion. Here $\mathbf{P}$ is, of course, only that part of the polarization associated with the permanent dipoles. If the field is not switched off but changed suddenly to a value for which the equilibrium polarization is $\mathbf{P}_0$ then the rate of change of $\mathbf{P}$ is given by the equation

$$\frac{d\mathbf{P}}{dt} = \frac{(\mathbf{P}_0 - \mathbf{P})}{\tau}, \quad \text{or} \quad \mathbf{P} + \tau\frac{d\mathbf{P}}{dt} = \mathbf{P}_0.$$

When an alternating field $\mathbf{E} \exp j\omega t$ is applied, we may write our equation for the polarization in the form (cf. eqn (10.15))

$$\mathbf{P} + \tau\frac{d\mathbf{P}}{dt} = \mathbf{P}_0 = \frac{n_0 \mathbf{p}^2}{3kT}\mathbf{E} \exp j\omega t$$

giving

$$\mathbf{P} = \frac{n_0 \mathbf{p}^2}{3kT}\frac{\mathbf{E} \exp j\omega t}{1 + j\omega\tau}. \tag{10.28}$$

Here $\mathbf{E}$ is the amplitude of the local alternating field, and $\mathbf{P}$ is that part of the polarization due only to the permanent dipoles. Allowing for these effects, we find for the relative permittivity $\epsilon_r$ the expression

$$\frac{\epsilon_r - \epsilon_i}{\epsilon_s - \epsilon_i} = \frac{1}{1 + j\omega\tau}, \tag{10.29}$$

where $\epsilon_i$ is that part of the relative permittivity due to induced polarization, and $\epsilon_s$ is the static relative permittivity. This result holds for the Onsager treatment for the local field, but it can be shown that the formula is similar if the Lorentz correction is used, except that we must use a modified relaxation time $\tau' = \tau(\epsilon_s + 2)/(\epsilon_i + 2)$. This difference is significant only in a comparison of the relaxation time determined from the dispersion of the relative permittivity with that from the Brownian motion.

Eqn (10.28) above shows that $\mathbf{P}$, and hence $\epsilon_r$ is complex, and we may either write $\epsilon_r = \epsilon_r' - j\epsilon_r''$, or $\epsilon_r = (n - jk)^2$, where $n$ is the refractive index and $k$ the absorption coefficient. Then

$$n^2 - k^2 = \epsilon_r' = \frac{\epsilon_s - \epsilon_i}{1 + \omega^2\tau^2} + \epsilon_i, \tag{10.30}$$

$$2nk = \epsilon_r'' = \frac{(\epsilon_s - \epsilon_i)\omega\tau}{1 + \omega^2\tau^2}. \tag{10.31}$$

The variation of these quantities with frequency is easily seen from Fig. 10.6. $\epsilon_r'$ falls from $\epsilon_s$ at low frequencies to $\epsilon_i$ at high frequencies, the transition taking place near $\omega = 1/\tau$. $\epsilon_r''$ has a maximum in this region at $\omega = 1/\tau$, and falls to zero at both low and high frequencies.

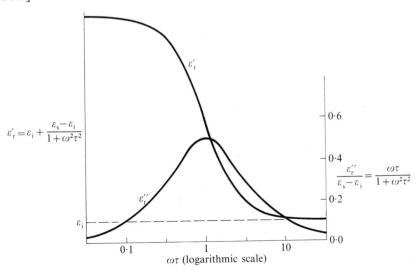

$$\varepsilon_r' = \varepsilon_i + \frac{\varepsilon_s - \varepsilon_i}{1 + \omega^2 \tau^2}$$

$$\frac{\varepsilon_r''}{\varepsilon_s - \varepsilon_i} = \frac{\omega\tau}{1 + \omega^2\tau^2}$$

$\omega\tau$ (logarithmic scale)

FIG. 10.6. Variation of $\epsilon_r'$ and $\epsilon_r''$ for a polar liquid.

The theory of the dispersion in polar liquids was given by Debye (1929), and that above is a simplified version of his treatment. It was first verified for glycerine and a number of alcohols, for which $\tau$ is much greater than for water owing to their higher viscosity and larger molecular radius. For example, at 295 K the values of $\epsilon_r'$ and $\epsilon_r''$ for glycerine at a wavelength of 9·5 m are 42 and 8·6 respectively, showing that we are already well into the region of anomalous dispersion at this wavelength. Vacuum-tube oscillators and detectors were available for such wavelengths, but the dispersion in water could not be measured accurately until centimetre-wave technique was established, owing to the small value of $\tau$. We shall here describe the measurements of Collie, Hasted, and Ritson (1948).

An accurate method of determining the permittivity of low-loss liquids, using resonant cavities, is described in § 22.6. This method cannot be used with water, since the absorption is so great if the cavity is filled with water that no resonance can be observed. This difficulty can be surmounted by using a cavity partly filled with water, the degree of filling being adjusted to give a measurable change in the resonant frequency and $Q_F$ of the cavity.

An alternative method, used by the same workers, is to determine the propagation constant in a waveguide filled with water. This constant depends on both the real and imaginary parts of the relative permittivity and on the linear dimensions of the guide. By using two guides of different sizes, the values of $\epsilon_r'$ and $\epsilon_r''$ can be found separately, since their

relative contributions to the propagation constant depend on the size of
the guide. From § 9.6 the field in the guide varies as

$$\exp(-\gamma x) = \exp\{-(\alpha + j\beta)x\},$$

where (writing $\lambda_0$ for the wavelength in free space)

$$\gamma = 2\pi\left(\frac{1}{\lambda_c^2} - \frac{f^2}{v^2}\right)^{\frac{1}{2}} = 2\pi\left(\frac{1}{\lambda_c^2} - \frac{\epsilon_r}{\lambda_0^2}\right)^{\frac{1}{2}} = 2\pi\left(\frac{1}{\lambda_c^2} - \frac{\epsilon_r' - j\epsilon_r''}{\lambda_0^2}\right)^{\frac{1}{2}}.$$

Hence

$$\alpha^2 - \beta^2 = 4\pi^2\left(\frac{1}{\lambda_c^2} - \frac{\epsilon_r'}{\lambda_0^2}\right),$$

$$\alpha\beta = 2\pi^2\frac{\epsilon_r''}{\lambda_0^2}. \tag{10.32}$$

Thus two separate measurements of $\alpha$, with different values of $\lambda_c$, suffice
to determine $\epsilon_r'$ and $\epsilon_r''$. The value of $\alpha$ can be found by moving a detector
through the liquid, and the measurement of phase (that is, $\beta$), which is
very difficult when large attenuation is present, is avoided. Essentially the
method adopted was to use two waveguides beyond cut-off, which act as
attenuators in series, one being filled with water and the other not. The
two attenuators are adjusted in opposite directions so as to keep the
power reaching a receiver constant; thus no calibration of the receiver is
required. The attenuation in the water is calculated from the known law
of the air-filled attenuator.

The results obtained may be fitted accurately to the theory using a
value of $\tau = 1{\cdot}01 \times 10^{-11}$ s, as shown in Fig. 10.7, where both the calculated curves and the experimental points are given. The great intensity of

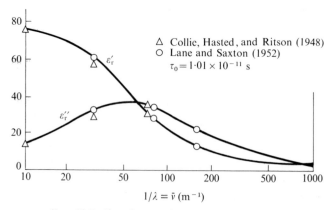

FIG. 10.7. Complex permittivity of water at 293 K.

the absorption is illustrated by the fact that at a wavelength of 12·4 mm, the power in an incident wave would be diminished by a factor of $e^{-2(2\pi k)} = e^{-36}$ or about $10^{-15 \cdot 5}$ in passing through a thickness of 12·4 mm. Thus water is quite 'black' at such wavelengths. It should be noted that in fitting these results, the best value of $\epsilon_i$, that part of the permittivity due to induced polarization, is found to be 4·9. This is appreciably higher than the square of the optical refractive index ($n = 1 \cdot 33$), showing that there must be other strong absorption bands in liquid water in the infrared; these are associated with internal vibrations of the $H_2O$ molecule. Rather similar results have been obtained by Lane and Saxton (1952) for methyl and ethyl alcohols.

The values of $\epsilon_r'$ and $\epsilon_r''$ in the region of dispersion vary quite rapidly with temperature, corresponding to a variation in the relaxation time $\tau$. Saxton (1952) has shown that $\tau$ varies from about $27 \times 10^{-12}$ s at 263 K (in supercooled water) down to $4 \cdot 7 \times 10^{-12}$ s at 323 K. This variation is very closely parallel to that of the viscosity, and indeed the value of $\tau$ is surprisingly close to that which would be obtained using the simple formula $\tau = 4\pi\eta a^3/kT$.

## 10.9. Ionic solids

For a simple atomic solid (such as diamond) or a molecular solid (most organic compounds) the permittivity is similar to that outlined at the beginning of § 10.7. When polar molecules are present, rotation of the permanent dipoles is completely inhibited except near the melting-point. For example, the value of $\epsilon_r$ for ice, measured at $10^{10}$ Hz and low temperatures, is about 3. Near the melting-point there is some dispersion in the region of $10^6$ Hz, $\epsilon_r$ rising from the value just quoted to about 80 at zero frequency. This results from some residual dipole rotation similar to that in liquid water, but with much larger values of $\tau$, associated with a high effective viscosity which increases rapidly at lower temperatures.

The position is different for another important class of solids: the ionic solids. These consist of regular lattices of positive and negative ions; the strong binding forces between such ions result in high melting-points. In an electric field the forces on the ions cause a small displacement of the lattice of positive ions relative to the negative ions. This gives a rather large electric polarization and high relative permittivity (see Table 10.5). The values of $\epsilon_r$ are the same as the static relative permittivity at radio-frequencies, but are appreciably greater than $n^2$ at a wavelength of 1 $\mu$m. Resonant frequencies must therefore exist in the infrared; these are connected with vibrations of the ions about their equilibrium positions. For a three-dimensional lattice this is a rather complex problem, but the main feature can be understood from a one-dimensional example.

*Values of the static relative permittivity $\epsilon_s$, the square of the optical refractive index $n^2$, and the frequency of the transverse optical mode $\nu_T$ for various simple ionic solids (all cubic)*

|      | $\epsilon_s$ | $n^2$ | $\nu_T$ (Hz)          |
| ---- | ------------ | ----- | -------------------- |
| LiF  | 9·0          | 1·9   | $9·1 \times 10^{12}$ |
| NaF  | 5·1          | 1·7   | $7·4 \times 10^{12}$ |
| NaCl | 6·0          | 2·3   | $4·9 \times 10^{12}$ |
| KCl  | 5·0          | 2·2   | $4·2 \times 10^{12}$ |
| KBr  | 4·8          | 2·4   | $3·5 \times 10^{12}$ |
| KI   | 4·9          | 2·7   | $3·0 \times 10^{12}$ |
| TlCl | 32           | 5·1   | $2·0 \times 10^{12}$ |

*The linear diatomic chain*

A linear diatomic chain (Fig. 10.8) consists of alternate masses $m_1$, $m_2$; in equilibrium the separation between successive masses is $a$. When a mass is displaced longitudinally by a small amount, a restoring force is set up which is assumed to be proportional to its instantaneous displacement relative to its immediate neighbours. For typical ions of mass $m_1$ at position $2n$ and $m_2$ at $2n+1$, the force equations are

$$m_1 \ddot{x}_{2n} = F(x_{2n+1}+x_{2n-1}-2x_{2n}),$$
$$m_2 \ddot{x}_{2n+1} = F(x_{2n+2}+x_{2n}-2x_{2n+1}). \tag{10.33}$$

These equations are similar to those for the electric filter (§ 9.1), and we assume that there exist solutions of a wave-like nature with the form

$$x_{2n} = A \exp j\{\omega t + 2nqa\},$$
$$x_{2n+1} = B \exp j\{\omega t + (2n+1)qa\}. \tag{10.34}$$

Substitution into (10.33) gives

$$-\omega^2 m_1 A - FB(e^{jqa} + e^{-jqa}) + 2FA = 0,$$
$$-\omega^2 m_2 B - FA(e^{jqa} + e^{-jqa}) + 2FB = 0. \tag{10.35}$$

Apart from the trivial case of $A = B = 0$, these two simultaneous equations have a solution only if the determinant formed from the coefficients of $A$, $B$ is zero:

$$\begin{vmatrix} 2F - m_1\omega^2 & -2F\cos qa \\ -2F\cos qa & 2F - m_2\omega^2 \end{vmatrix} = 0. \tag{10.36}$$

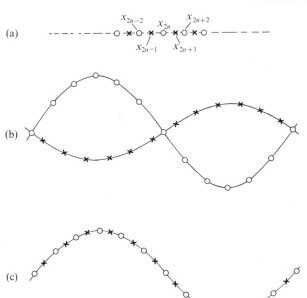

FIG. 10.8. A linear diatomic chain (a) of masses $m_1$ (at points $2n$, $2n\pm2$, etc.) and $m_2$ (at points $2n\pm1$, etc.), and the displacements of the masses in (b) the optical mode and (c) the acoustic mode.

The solution of this determinantal equation is

$$\frac{\omega^2}{F} = \left(\frac{1}{m_1}+\frac{1}{m_2}\right) \pm \left\{\left(\frac{1}{m_1}+\frac{1}{m_2}\right)^2 - \frac{4\sin^2 qa}{m_1 m_2}\right\}^{\frac{1}{2}}, \qquad (10.37)$$

and the ratio $A/B$ can be written as

$$\frac{A}{B} = \frac{\cos qa}{1-(\omega^2 m_1/2F)} = \frac{1-(\omega^2 m_2/2F)}{\cos qa}. \qquad (10.38)$$

Limiting values occur at $qa = 0$ and $qa = \pm\pi/2$, and are summarized in Table 10.6, where it is assumed that $m_1 > m_2$.

TABLE 10.6

|  | $qa = 0$ | $qa = \pm\pi/2$ |
|---|---|---|
| Lower or 'acoustic' branch | $\omega = 0$ <br> $A = B$ | $\omega = (2F/m_1)^{\frac{1}{2}}$ <br> $B = 0$ |
| Upper or 'optical' branch | $\omega = \left\{2F\left(\frac{1}{m_1}+\frac{1}{m_2}\right)\right\}^{\frac{1}{2}}$ <br> $A/B = -m_1/m_2$ | $\omega = (2F/m_2)^{\frac{1}{2}}$ <br> $A = 0$ |

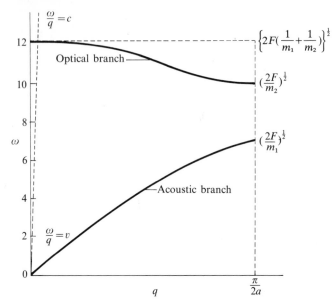

FIG. 10.9. The relation between $\omega$ and $q$ for a diatomic linear chain, given by eqn (10.37). In three dimensions the slope of the acoustic branch near $q = 0$ corresponds to the velocity of sound; the slope of the broken line $\omega/q = c$ is that of electromagnetic waves, and is larger by a factor of about $10^5$.

The two solutions corresponding to eqn (10.37) are plotted in Fig. 10.9 for the case $m_1 = 2m_2$. The lower branch starts at the origin. For values of $qa$ so small that we can replace $\sin qa$ by $qa$, the value of $\omega$ is proportional to $q$ and the phase velocity $\omega/q$ is constant; since $A = B$, adjacent masses move together (Fig. 10.8c) with the same amplitude of vibration. The wave motion is similar to that in a low-frequency elastic wave (a sound wave), and this branch of the spectrum is known as the 'acoustic' branch.

As $qa$ increases, $\omega$ no longer rises as fast as $q$, and the phase velocity $\omega/q$ becomes frequency-dependent. This corresponds to 'dispersion', and the way in which $\omega$ varies with $q$ is known as the 'dispersion relation'. At $\sin qa = \pm 1$, $B = 0$ while $A$ is finite. The masses $m_2$ are all stationary, and the frequency depends only on the masses $m_1$; successive masses $m_1$ vibrate with equal amplitude but opposite phase, since $2qa = \pi$. This is a standing wave in which no energy is propagated; the group velocity $d\omega/dq \rightarrow 0$ as $\sin qa \rightarrow \pm 1$.

For the upper branch at these points we have a similar result, except that the roles of the masses $m_1$ and $m_2$ are interchanged. The behaviour of the upper branch at $q = 0$ is of especial interest, as we shall see later. Masses $m_1$, $m_2$ vibrate in opposite directions with relative amplitudes

$A/B = -m_1/m_2$, keeping the centre of mass constant (Fig. 10.8(b)). The relative displacement of adjacent particles is large, so that the restoring forces are large, and so is $\omega$. In contrast, at $q = 0$ for the lower branch adjacent particles move together; the relative displacement is small and $\omega \to 0$ as $q \to 0$. The wavemotion resembles that on a continuous string.

In this analysis longitudinal wave motion has been assumed, but it is equally valid for transverse waves, where the vibrations can take place in two mutually perpendicular planes. The displacements of the particles shown in Fig. 10.8, are for a transverse wave. The restoring forces differ for longitudinal and transverse waves, so that different dispersion relations must be expected. Each has an upper and a lower branch; the branches are known as 'longitudinal acoustic' (LA), 'longitudinal optic' (LO), 'transverse acoustic' (TA), and 'transverse optic' (TO). In general there are two transverse dispersion relations corresponding to vibrations in orthogonal planes, each normal to the direction of propagation, but these are clearly identical for the linear chain.

So far no restriction has appeared on the number of values of $q$ which can be present. Such a restriction appears only when we consider a chain of finite length, and impose boundary conditions at the ends. A simple approach to this problem is to divide the infinite chain into sections each containing $N$ masses $m_1$ and $N$ masses $m_2$, each section being therefore of length $2Na$. The method of 'periodic boundary conditions' requires that the solution be identical at each equivalent point in each section, that is,

$$x_{2n} = x_{2n+2N}.$$

Then, for any value of $q$, we must have

$$\exp \mathrm{j}(2nqa) = \exp \mathrm{j}(2nqa+2Nqa)$$

which is satisfied provided that $2Nqa$ is an integral multiple of $2\pi$. The allowed values of $q$ are therefore

$$q = l(\pi/Na), \tag{10.39}$$

where $l$ is an integer with values lying between $\pm 1$ and $\pm\frac{1}{2}N$, the latter values corresponding to the zone boundaries at $q = \pm\pi/2a$. This gives altogether $N$ acoustic modes and $N$ optical modes, that is, a total of $2N$ modes (for displacements in one single direction), as we should expect for a total of $2N$ masses.

The range of values of $q$ between $-\pi/2a$ and $+\pi/2a$ is known as the 'first Brillouin zone'. The 'zone boundaries' at $\pm\pi/2a$ occur because we have a chain of discrete masses repeating at distances $2a$, and not a continuous string. It is possible to represent the vibrational motion by means of waves of shorter wavelength, but it is unnecessary; only the displacement of the masses can be observed, and this can always be represented by wavelengths in the first zone. Note also that the points $q = \pm \pi/2a$ represent identical states of motion of the masses.

*Interaction with an electromagnetic wave*

In the theory of forced oscillations the amplitude is largest at resonance; the frequency of the driving force must coincide with a natural resonant frequency of the driven system. For a system which extends over distances large compared with the wavelength, this is part of a more general requirement of a constant-phase relation between the driving force and the whole of the driven system. This implies that both $\omega$ and $q$ must be the same for an electromagnetic wave and for the vibration in the chain to produce an appreciable interaction. For an electromagnetic wave $\omega/q = c$, represented by a straight line in Fig. 10.9; its slope is the velocity of light, which is about $10^5$ times that of sound, given by $\omega/q$ for the acoustic branch at $q = 0$. A resonant interaction between a light wave and the vibrational motion can occur only at the points of intersection of $\omega/q = c$ with a dispersion curve. One such point is at $\omega = q = 0$, and corresponds to a static electric field; the other occurs at the intersection with the upper branch. This may also be taken as $q = 0$ without serious error, so that its angular frequency is

$$\omega_{\mathrm{T}} = \left\{ 2F\left(\frac{1}{m_1} + \frac{1}{m_2}\right) \right\}^{\frac{1}{2}}. \qquad (10.40)$$

Interaction with an electromagnetic wave requires that the masses carry electric charges, and for an electromagnetic wave travelling along the chain, the forces exerted are perpendicular to the chain. Thus interaction occurs only with 'transverse modes'; this is denoted by the subscript T in eqn (10.40).

If $m_1 = m_2$, the displacements of the two sets of masses are equal and opposite in the transverse optical mode at $q = 0$. If the masses have the same charge, no net dipole moment is created by the displacements, and the vibration is not excited by an electromagnetic wave. Correspondingly, no absorption occurs in a monatomic solid through interaction with the ionic vibrations, provided that the lattice is perfectly regular.

For a diatomic chain with unequal masses, we can deduce the interaction with an electromagnetic wave by assigning charges of $+Q$ and $-Q$ to alternate masses, and adding terms $+Q\mathbf{E}_{\mathrm{local}}$ and $-Q\mathbf{E}_{\mathrm{local}}$ to eqn (10.33). Only waves for which $q$ is virtually zero are excited, so that in eqns (10.34) we can put $q = 0$ and drop the subscript $n$. For an angular frequency $\omega$ we can write $\ddot{x} = -\omega^2 x$, and eqns (10.35) become

$$-\omega^2 A = (2F/m_1)(B-A) + (Q/m_1)\mathbf{E}_{\mathrm{local}},$$
$$-\omega^2 B = (2F/m_2)(A-B) - (Q/m_2)\mathbf{E}_{\mathrm{local}}. \qquad (10.41)$$

The quantity $(A-B)$ can be found by subtracting these two equations. For unit length of chain, containing $n_0$ masses of each type, the polarization $P = n_0 Q(A-B)$. With the help of eqn (10.40) we obtain

$$\mathbf{P} = \frac{n_0 Q^2 \mathbf{E}_{\text{local}}}{(\omega_T^2 - \omega^2)} \left( \frac{1}{m_1} + \frac{1}{m_2} \right). \tag{10.42}$$

This has a resonance at $\omega = \omega_T$, and is similar to eqn (10.18) without the damping term.

### Three dimensions

A real solid consists of an assembly of atoms or ions, vibrating in random directions about their equilibrium positions which, in a crystal, are points on a regular three-dimensional lattice. This is clearly an extremely complex problem for which no exact solution can be obtained. A simple analysis assumes that the vibrations can be represented by means of plane waves of the form

$$\exp \mathrm{j}(\omega t - q_x x - q_y y - q_z z) = \exp \mathrm{j}(\omega t - \mathbf{q} \cdot \mathbf{r}). \tag{10.43}$$

Here $\mathbf{q}$ is known as the 'wave vector'; its magnitude is $2\pi/\lambda$, and its direction is the direction of propagation. When boundary conditions are introduced, $q$ is restricted to certain allowed values, as in eqn (10.39). These values may be plotted as a set of points equally spaced along a line in 'reciprocal space'; the spacing is $\pi/Na$, which has the dimensions of (length)$^{-1}$. In three dimensions this gives a reciprocal space with an inverse relationship to the unit cell of the crystal lattice; just as the atoms lie on a lattice of discrete points in real space, the allowed values of $\mathbf{q}$ end on a discrete set of points in reciprocal space. In each case the spacing reflects the symmetry of the crystal lattice. For example, in a cubic lattice, directions parallel to the cube edges (axes of fourfold symmetry) must all be equivalent; a similar condition exists for the body diagonals (axes of threefold symmetry), and face diagonals (axes of twofold symmetry).

For the linear chain the zone boundaries correspond to the points at which travelling waves cannot be propagated, and the vibrations are represented by standing waves. The same is true in three dimensions; the boundaries now consist of sets of intersecting planes in reciprocal space, which form a surface enclosing a 'Brillouin zone'. A wave vector drawn from the origin which terminates on this surface corresponds to a wavelength which cannot propagate in the direction of the wave vector, but suffers an exponential attenuation similar to that of a wave whose frequency lies in the stop band of a filter or in the 'cut-off region' of a waveguide. All possible lattice vibrations can be represented by points in the 'first zone'; at the boundary the values of the wave vector are close to

$10^{10}$ m$^{-1}$, since the distance between identical atoms in a simple crystal is a few tenths of a nanometre ($10^{-10}$ m). For electromagnetic waves, such wave vectors lie in the X-ray region; in fact the condition that a wave vector terminates at the zone boundary is identical with the Bragg 'reflection condition' for diffraction of X-rays by the crystal lattice.

The simple equations for the diatomic linear chain are similar to those for a diatomic cubic crystal in which planes can be drawn containing only the one type of ion, alternating with planes containing the other type. The force constants are those between successive planes, but the dispersion relations are similar to Fig. 10.9, with acoustic and optical branches for both longitudinal and transverse modes. We may make a rough estimate of the frequency at the zone boundary by ignoring dispersion, and writing $\nu = \omega/2\pi = qv/2\pi$, where $v$ is the velocity of low-frequency sound waves. For most ionic solids this lies in the range $2-5\times10^3$ m s$^{-1}$, and for $q = 10^{10}$ m$^{-1}$ we obtain frequencies of $10^{12}-10^{13}$ Hz, in good agreement with the values of $\nu_T = \omega_T/2\pi$ listed in Table 10.5.

### Permittivity and dispersion

From dispersion theory we expect the static permittivity of an ionic solid to be related to the value of $\omega_T$. The actual relationship is complicated because of (1) local field corrections of the Lorentz type, arising from dipolar forces; and (2) short-range forces set up by the overlap of charge clouds on adjacent ions. Szigeti (1949) has derived the formula

$$\epsilon_s - \epsilon_{opt} = \left(\frac{\epsilon_{opt}+2}{3}\right)^2 \frac{n_0 Q^2}{m_r \epsilon_0 \omega_T^2}, \tag{10.44}$$

where $\epsilon_s$ is the static relative permittivity. $\epsilon_{opt}$ is the contribution from electrons within the ions, given by $n^2$ where $n$ is the extrapolated value of the optical refractive index. $m_r$ is a reduced mass, where for a simple diatomic solid with ionic masses $m_1$, $m_2$

$$\frac{1}{m_r} = \frac{1}{m_1} + \frac{1}{m_2}, \tag{10.45}$$

and $\pm Q$ is the effective charge on each ion, an empirical quantity but close to $ze$, where $z$ is the valency of the ion and $e$ the electronic charge.

Eqn (10.44) is clearly of the same form as eqn (10.42) when allowance is made for the difference between $\mathbf{E}$ and $\mathbf{E}_{local}$. If $\mathbf{P}_s$ is the static polarization at $\omega = 0$, (10.42) can be written in the form

$$\mathbf{P}/\mathbf{P}_s = \omega_T^2/(\omega_T^2 - \omega^2), \tag{10.46}$$

where $\mathbf{P}$ is the polarization at frequency $\omega/2\pi$ due to the ionic displacements. If $\epsilon_r$ is the relative permittivity at this frequency we have

$$\frac{\epsilon_r - \epsilon_{opt}}{\epsilon_s - \epsilon_{opt}} = \frac{\omega_T^2}{\omega_T^2 - \omega^2}. \tag{10.47}$$

This result clearly gives $\epsilon_r = \epsilon_s$ at zero frequency, while $\epsilon_r$ becomes infinite at $\omega = \omega_T$ because no damping is included. When $\omega$ just exceeds $\omega_T$, $\epsilon_r$ is negative; it diminishes in magnitude as $\omega$ increases, and passes through zero at an angular frequency $\omega_L$ such that

$$\epsilon_s/\epsilon_{opt} = (\omega_L/\omega_T)^2. \tag{10.48}$$

This result is known as the Lyddane–Sachs–Teller relationship, and it can be shown that $\omega_L$ corresponds to the limiting value of a longitudinal optical mode at $q = 0$. In a number of cases values of $\omega_L$ and $\omega_T$ have been determined by neutron diffraction, and the ratio is in good agreement with that obtained from dielectric measurements.

Since $\epsilon_r = (n - jk)^2$, a value of $\epsilon_r$ which is real but negative means $n = 0$ but $k$ is finite. The wave decays as $\exp(-\omega k z/c)$, and is not propagated in the frequency range between $\omega_T$ and $\omega_L$. This corresponds to the stop band in a filter, and the intrinsic impedance of the medium $Z_0$ is imaginary. It follows from eqn (8.53) that a wave incident on the dielectric within this frequency range is totally reflected. In early work this property was used to isolate various infrared wavelengths by successive reflections from the surfaces of ionic crystals; such wavelengths were known as 'reststrahlen' or 'residual rays'.

## 10.10. Ionized media and plasma oscillations

In a medium containing free particles with electric charges the theory of § 10.6 can be used, putting $\omega_p = 0$ since no restoring forces act on the particles. At low densities local field corrections can be neglected, and eqn (10.19) leads to a complex permittivity

$$\epsilon_r = 1 - \frac{n_0 q^2}{m\epsilon_0 \omega}\left(\frac{\omega + j\gamma}{\omega^2 + \gamma^2}\right) \tag{10.49}$$

for a medium containing $n_0$ particles (per unit volume) of charge $q$ and mass $m$.

For free particles an alternative approach is to work in terms of current. For an oscillating field $\mathbf{E}^{(0)} \exp j\omega t$, eqn (3.13a) becomes

$$\frac{d\mathbf{J}}{dt} + \frac{\mathbf{J}}{\tau} = \mathbf{J}\left(j\omega + \frac{1}{\tau}\right) = \left(\frac{n_0 q^2}{m}\right)\mathbf{E}^{(0)} \exp j\omega t$$

for that part of the current oscillating with angular frequency $\omega$. Since $(n_0 q^2/m)\tau = \sigma_0$, the conductivity at $\omega = 0$, we can write the equation for $\mathbf{J}$ as

$$\mathbf{J}(1 + j\omega\tau) = \sigma_0 \mathbf{E}^{(0)} \exp j\omega t. \tag{10.50}$$

This gives a complex conductivity $\mathbf{J}/\mathbf{E} = \boldsymbol{\sigma} = \sigma' - j\sigma''$, at frequency $\omega$, of

$$\boldsymbol{\sigma} = \sigma' - j\sigma'' = \sigma_0 \frac{1 - j\omega\tau}{1 + \omega^2\tau^2}. \tag{10.51}$$

These two different approaches emphasize the equivalence between a complex permittivity and a complex conductivity. For a given frequency eqn (8.7) becomes

$$\text{curl } \mathbf{H} = (\sigma' + j\omega\epsilon'_r\epsilon_0)\mathbf{E},$$

where $\sigma'$ and $\epsilon'_r$ are the real parts of the conductivity and relative permittivity. The coefficient of $\mathbf{E}$ in this equation can be expressed either as $\sigma' - j\sigma''$ with $\sigma'' = -\omega\epsilon'_r\epsilon_0$, or as $j\omega\epsilon_0(\epsilon'_r - j\epsilon''_r)$ with $\epsilon''_r = \sigma'/\omega\epsilon_0$.

In addition to oscillations driven by an applied field, natural oscillations may be present. In an assembly of charged and neutral particles, known as a plasma, the local charge density fluctuates because of the random motion of the charged particles. However, a momentary excess of charge in one region is counteracted by the mutual repulsion between the particles, which pushes them apart so that the excess quickly disappears. The momentum gained by the particles in this process causes them to overshoot; the excess is replaced by a defect in the charge density, and the particles are attracted back. Repetition of this process sets up a periodic disturbance known as a 'plasma oscillation'; the striations visible in gas discharges and electron or ion beams are due to standing waves set up by such oscillations.

A simple derivation of the frequency of such oscillations starts from the conservation of charge and Gauss's theorem. The charge motion creates a current density $\mathbf{J} = \rho_e\mathbf{v}$, and from eqn (3.3) we have

$$d\rho_e/dt = -\text{div } \mathbf{J} = -\text{div } \rho_e\mathbf{v}.$$

From Gauss's theorem $\epsilon_0 \text{ div } \mathbf{E} = \rho_e$, and hence

$$\epsilon_0 \text{ div}(d\mathbf{E}/dt) = d\rho_e/dt = -\text{div } \rho_e\mathbf{v} = -\text{div } \mathbf{J}.$$

Hence

$$\epsilon_0(d\mathbf{E}/dt) = -\rho_e\mathbf{v} = -\mathbf{J}, \tag{10.52}$$

where the constant of integration is zero since $\dot{\mathbf{E}} = 0$ where there is no motion of the particles.

In a plasma where both positive and negative ions are present, the negative ions are mostly electrons which move very much more rapidly than the more massive ionized atoms, and we can assume that all the current is carried by electrons, that is, by one type of particle, of mass $m$ and charge $q$. The equation of motion of these particles is $m(d\mathbf{v}/dt) = q\mathbf{E}$, so that (using eqn (10.52))

$$d\mathbf{J}/dt = \rho_e(d\mathbf{v}/dt) = (\rho_e q/m)\mathbf{E},$$
$$d^2\mathbf{J}/dt^2 = (\rho_e q/m)(d\mathbf{E}/dt) = -(\rho_e q/m\epsilon_0)\mathbf{J}.$$

If we assume that oscillations are of vanishingly small amplitude, the departure of $\rho_e$ from its mean value $\bar{n}q$, where $\bar{n}$ is the average number per unit volume, is negligible, and we can write

$$d^2\mathbf{J}/dt^2 = -(\bar{n}q^2/m\epsilon_0)\mathbf{J}. \qquad (10.53)$$

This is the equation of simple harmonic motion, showing that the current will oscillate at a frequency

$$\omega_p = 2\pi f_p = (\bar{n}q^2/m\epsilon_0)^{\frac{1}{2}}, \qquad (10.54)$$

known as the plasma frequency. We can estimate its magnitude by taking as an example a plasma of fully ionized hydrogen at a pressure of 1 Pa, for which $\bar{n} = 2 \cdot 7 \times 10^{20}$ m$^{-3}$. Then, with $q = -e$,

$$f_p = (\bar{n}e^2/4\pi^2\epsilon_0 m)^{\frac{1}{2}} = 9 \cdot 0\bar{n}^{\frac{1}{2}} = 1 \cdot 5 \times 10^{11}. \qquad (10.55)$$

This frequency corresponds to a wavelength of 2 mm for electromagnetic waves, and its measurement is an important tool in plasma physics since it gives the density of electrons. In ordinary gas discharges the degree of ionization is relatively low, and the plasma frequency may lie in the region $10^3$–$10^8$ Hz. In metals, on the other hand, the electron density is very much higher, and the electron plasma frequency is about $10^{15}$ Hz.

### References

COLLIE, C. H., HASTED, J. B., and RITSON, D. M. (1948). *Proc Phys. Soc.* **60,** 145.

DEBYE, P. (1929). *Polar molecules.* Dover Publications, New York.

DYLLA. H. R. and KING. J. G. (1973). *Phys. Rev.* A**7,** 1224.

LANE, J. A. and SAXTON, J. A. (1952). *Proc. R. Soc.* A**213,** 400.

ONSAGER, L. (1936). *J. Am. Chem. Soc.* **58,** 1486.

PAULING, L. and WILSON, E. B. (1935). *Introduction to quantum mechanics.* McGraw-Hill, New York.

RAMSEY, N. F. (1956). *Molecular beams.* Clarendon Press, Oxford.

ROBINSON. F. N. H. (1973). *Macroscopic electromagnetism.* Pergamon Press, Oxford.

SAXTON, J. A. (1952). *Proc. R. Soc.* A**213,** 473.

SZIGETI, C. (1949). *Trans. Faraday Soc.* **45,** 155.

TOWNES, C. H. and SCHAWLOW, A. L. (1955). *Microwave spectroscopy.* McGraw-Hill, New York.

### Problems

10.1.  The static relative permittivity of $CO_2$ and $NH_3$ is measured at 273 K and 373 K at a pressure of $10^5$ Pa (1 atm) and the values of $10^3(\epsilon_r - 1)$ are found to be:

|        | $CO_2$ | $NH_3$ |
|--------|--------|--------|
| 273 K  | 0·988  | 8·34   |
| 373 K  | 0·723  | 4·87   |

Calculate the permanent electric dipole moment for each gas, and also the radius of the molecule, assuming the polarizability to be the same as that of a conducting sphere.

(*Answer:* $p = 0$ debye and 1·45 debye; radius = 0·14 nm and 0·18 nm; from viscosity data the radii are 0·23 nm and 0·22 nm respectively.)

10.2. The relative permittivity of liquid helium at its boiling-point is 1·048, and its density is 125 kg m$^{-3}$. Calculate the refractive index of the gas at N.T.P., and estimate the radius of the helium atom, assuming it to behave like a conducting sphere. Compare the radius with (*a*) that given by the Bohr theory for an atom with a nuclear charge of two units in its ground state, (*b*) with that calculated from the diamagnetic susceptibility (see Problem 6.1).

(*Answer:* $n = 1.000034$; radius = 0·059 nm; Bohr-theory radius = 0·026 nm.)

10.3. In the upper regions of the atmosphere (the ionosphere) the gas molecules are ionized, mostly through the effects of ultraviolet radiation from the sun. Show that in a region where the number of free electrons per cubic metre is $n_0$, the refractive index for waves of frequency $f$ (Hz) is

$$\left(1 - \frac{n_0 e^2}{m\omega^2 \epsilon_0}\right)^{\frac{1}{2}} = \left(1 - \frac{\omega_p^2}{\omega^2}\right)^{\frac{1}{2}} \approx \left(1 - \frac{81 n_0}{f^2}\right)^{\frac{1}{2}},$$

where $\omega_p/2\pi$ is the plasma frequency.

At this frequency, the refractive index falls to zero, and the ionosphere is totally reflecting even at normal incidence. The frequency at which this occurs is called the critical frequency and at midday is about 4 MHz for the E-layer at latitude 40° N. Estimate the maximum value of $n_0$ in the E-layer from this figure.

(*Answer:* $n_0 = 2 \times 10^{11}$ m$^{-3}$. Neglect the Lorentz field.)

10.4. Show that in an ionized region such as that in the previous question the product of the group velocity and the phase velocity is equal to the square of the velocity in a vacuum.

10.5. A particle of charge $-e$ and mass $m$ performs a simple harmonic motion $s = s_0 \cos \omega t$ under the action of a restoring force. Show that through the radiation of energy (given by eqn (8.72)) the total energy of the particle falls as $W = W_0 e^{-\gamma t}$, where

$$\gamma = Z_0 e^2 \omega^2 / 6\pi mc^2 = 2\pi Z_0 e^2 f^2 / 3mc^2.$$

Here $Z_0 =$ intrinsic impedance of free space, and $f = \omega/2\pi$. On the quantum theory, the chance that an atom spends a time $t$ in an excited state has the probability of the order $\exp(-\gamma t)$; hence show that the mean lifetime $\tau = 1/\gamma$ of a sodium atom in an excited state before emitting one of the sodium D lines ($\lambda = 590$ nm) is roughly $1.6 \times 10^{-8}$ s.

10.6. By the uncertainty principle, the width $\Delta E$ of the upper energy level of the previous question is given by the relation $\tau \Delta E = h/2\pi$, where $h$ is Planck's constant. Use the relation $\Delta E = h(\Delta f)$ to show that this gives a linewidth of $\Delta f = \gamma/2\pi$. This is called the natural or radiation breadth of the line, and the same result is obtained on classical theory by a Fourier analysis of the spectrum of an oscillator whose energy is decaying exponentially as $W = W_0 \exp(-\gamma t)$.

Estimate the linewidths due to the Doppler effect and to collisions in a gas at 1000 K and 100 Pa pressure, and show that (*a*) at $f = 10^{10}$ Hz ($\lambda = 0.03$ m) collision-broadening is dominant, (*b*) at $f = 10^{15}$ Hz ($\lambda = 300$ nm) Doppler effect is dominant, (*c*) at $f = 10^{18}$ Hz ($\lambda = 0.3$ nm) natural line-breadth is dominant.

10.7. For a vibrating molecule, the polarizability $\alpha$ varies as $\alpha = \alpha_0(1+bx)$, where $x = a \cos pt$ is the change in the normal dimensions of the molecule due to the vibration. Show that if incident light of frequency $\omega/2\pi$ falls on the molecule, the scattered radiation will contain light of frequencies $(\omega \pm p)/2\pi$. (This is the classical explanation of the Raman effect.)

10.8. Show that for a narrow absorption line the maximum and minimum of the refractive index in the region of anomalous dispersion occur at the frequencies where the absorption coefficient has fallen to half its maximum value.

10.9. The conductivity of sea-water at 293 K is about $2\ \Omega^{-1}\,m^{-1}$. Show that the absorption at 0·01 m wavelength due to this conductivity is small compared with the Debye absorption, but the two are roughly equal at a wavelength of about 0·1 m. (Use the data given for pure water in Fig. 10.7; in fact the Debye relaxation time is somewhat altered by the salts dissolved.)

10.10. Discuss the propagation in and reflection from the surface of the sea of radio waves in the light of the data given in the previous question.

10.11. In a region of space containing $\bar{n}$ particles per unit volume of charge $q$ and mass $m$, where the pressure is so low that collisions may be neglected, eqns (8.3) and (8.7) may be written

$$\text{curl } \mathbf{E} = -\mu_0(\partial \mathbf{H}/\partial t); \qquad \text{curl } \mathbf{H} = \bar{n}q\mathbf{v} + \epsilon_0(\partial \mathbf{E}/\partial t).$$

Show that by using the equation of motion $q\mathbf{E} = m(\partial \mathbf{v}/\partial t)$, and eliminating $\mathbf{v}$ between these equations (taking $\text{grad div } \epsilon_0\mathbf{E} = \text{grad } \bar{n}q = 0$), we can obtain the equation

$$\frac{\partial^2 \mathbf{E}}{\partial t^2} + \frac{\bar{n}q^2}{\epsilon_0 m}\mathbf{E} = c^2\,\nabla^2\mathbf{E}.$$

This is the equation of Tonks and Langmuir for propagation of waves in an ionized medium. Verify that this wave equation is satisfied by

$$\mathbf{E} = \mathbf{E}_0 \exp \mathrm{j}(\omega t - kz)$$

provided that $\omega^2 = \omega_p^2 + c^2 k^2$, where $\omega_p/2\pi$ is the plasma frequency given by eqn (10.54).

Verify that this leads to the expression for the refractive index given in Problem 10.3.

# 11. Free electrons in metals

## 11.1. Quantum theory of free electrons in metals

ON the Drude theory of free electrons in a metal, outlined in § 3.11, the conduction electrons are confined to the volume of the metal but move freely inside like gas molecules in a container. It leads to formulae for the electrical conductivity (see eqn (3.12))

$$\sigma = n(q^2/m)\tau, \tag{11.1}$$

and for the thermal conductivity (see eqn (3.62))

$$K = \tfrac{1}{3}ncL(\mathrm{d}\bar{W}/\mathrm{d}T), \tag{11.2}$$

but makes no prediction as to the magnitude of the relaxation time $\tau$ or the mean free path $L = c\tau$, where $c$ is the random thermal velocity. This problem is avoided in considering the ratio $(K/\sigma)$, which is predicted to be of order $T(k/e)^2$ in accordance with the measurements of Wiedemann and Franz. There is one serious difficulty: the specific heat. On classical theory the kinetic energy of a free particle can have any value, and there is a continuous distribution of values, though some are more probable than others, corresponding to a Boltzmann distribution. As the temperature falls, the average energy decreases to zero at $T = 0$ K; at an absolute temperature $T$ the total translational energy of $N$ particles is $\tfrac{3}{2}NkT$. The differential of this gives a contribution of $\tfrac{3}{2}R$ to the molar heat, so that the molar heat of a metal should exceed that of an insulator by this amount. No such contribution is observed.

In quantum mechanics not every value of the energy is allowed, and the continuous distribution of energies is replaced by a discrete set of allowed energy levels. The spacing is extremely small, however, and the difference between this and the assumed classical continuous distribution produces no observable effect for real gases. It would become appreciable only at temperatures so low that ordinary substances have negligible vapour pressure, and at higher temperatures it is always masked by deviations from the perfect gas laws owing to the van der Waals forces. In the case of electrons in a metal, the spacing of the translational energy levels is larger because of the smaller mass of the electron, and the number per unit volume is much larger than in any real gas. The result of this is to emphasize the role played by the Pauli exclusion principle, which states that no two electrons in a given system can have the same set of quantum numbers. When allowance is made for the intrinsic spin angular momentum of the electron, this means that only two electrons can occupy any given translational energy level. Hence the kinetic energy of the electrons cannot be zero at 0 K, since this would mean that all the electrons were in one particular energy level. In fact the electrons occupy the lowest possible set of energy levels consistent with the Pauli exclusion principle, and their mean energy is very far from zero at 0 K. At a finite temperature the energy distribution can only be found using the Fermi–Dirac statistics, that is, the quantum statistics which take account of the Pauli exclusion principle. At ordinary temperatures it turns out that the energy distribution differs very little from that at 0 K; the latter can be found when the values of the allowed energy levels are known, and a simple method of computing the levels will now be given.

The fact that electrons and other particles have a wave-like aspect is well established from experiments on electron and neutron diffraction; the wavelength associated with a particle of linear momentum $\mathbf{p}$ is given by the de Broglie relation

$$2\pi/k = \lambda = h/p \quad \text{or} \quad \mathbf{k} = \mathbf{p}(2\pi/h) = \mathbf{p}/\hbar, \tag{11.3}$$

where $\mathbf{k}$ is the 'wave vector' which is parallel to the direction in which the wave is travelling, and $h$ is Planck's constant and $\hbar = h/2\pi$. For example, the wavelength is $1.23 \times 10^{-9}$ m for an electron of kinetic energy equal to 1 electronvolt (eV). (1 eV is the energy acquired by an electron in falling through a potential difference of 1 V). Free electrons confined within a metal rebound from the surface of the metal without losing any energy; on the wave aspect this means that the waves are totally reflected at the boundaries, and standing waves are set up. These standing waves are the allowed solutions of the wave equation for a particle in a box, in the same way that standing waves of certain wavelengths are allowed in a waveguide resonator (see § 9.7). By analogy with the theory of heat radiation, the number of allowed wavelengths in the range $\lambda$ to $\lambda + d\lambda$ is $V4\pi \, d\lambda/\lambda^4$, where $V$ is the volume of the box. For particles the wavelength is determined by the momentum, and from eqn (11.3) the number of possible values of the momentum in the range $p$ to $p + dp$ is found to be

$$(V/h^3)4\pi p^2 \, dp. \tag{11.4}$$

This relation may be derived in another way by means of the uncertainty relation, which states that the momentum component $p_x$ cannot be determined more precisely than to an amount $\Delta p_x$ such that $\Delta p_x \, \Delta x = h$, where $\Delta x$ is the uncertainty in the position coordinate. Similar relations hold for the other two axes, and hence

$$\Delta p_x \, \Delta p_y \, \Delta p_z = h^3/\Delta x \, \Delta y \, \Delta z = h^3/V$$

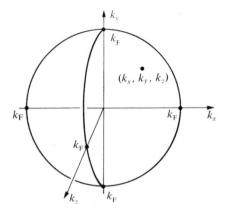

FIG. 11.1. The wave vector **k** of a particle is specified by its components $(k_x, k_y, k_z)$, corresponding to a point in 'k-space'. At 0 K all electrons are specified by points within the sphere of radius $k_F$, since this makes the energy a minimum.

if the particle is constrained to be within the volume $V$. If we write $p_x = \hbar k_x$, etc., by using eqn (11.3), we have

$$\Delta k_x \, \Delta k_y \, \Delta k_z = 8\pi^3/V.$$

The wave vector of a particle is specified in magnitude and direction by its components $k_x$, $k_y$, $k_z$ along three Cartesian axes, and the square of the wave vector is $k^2 = k_x^2 + k_y^2 + k_z^2$. A 'k-space' may be constructed, as in Fig. 11.1, where the coordinates are the components $k_x$, $k_y$, $k_z$ and the magnitude and direction of **k** are given by a vector drawn from the origin to the point $(k_x, k_y, k_z)$. All values of the wave vector which lie between **k** and **k**+**dk** are represented by points which lie within the spherical shell bounded by the radii $k$ and $k+dk$, and the volume of this shell is $4\pi k^2 \, dk$. Since $\Delta k_x \, \Delta k_y \, \Delta k_z$ is an element of volume in **k**-space, the Pauli exclusion principle may be stated in the form that only one point in **k**-space is allowed in an element of volume of size $8\pi^3/V$, so that in the volume $4\pi k^2 \, dk$ the allowed number of points is

$$di = (V/8\pi^3)4\pi k^2 \, dk. \tag{11.5}$$

In addition, we have to consider the angular momentum of the electron $\frac{1}{2}\hbar$ due to its intrinsic spin, which can have one of two components $\pm\frac{1}{2}\hbar$ along any axis (see § 14.2). The Pauli principle then allows two electrons, with opposite values of these angular-momentum components, to have the same translational energy, that is, we can assign a maximum of two electrons to each point in the (linear) momentum space.

The translational energy $W$ of an electron of momentum $\hbar k$ and mass $m$ is $\hbar^2 k^2/2m$, and the total kinetic energy will be smallest when $\sum (\hbar^2 k^2/2m)$ has its minimum value. It is readily seen that this occurs when the points in momentum space just fill a sphere of the least volume which will accommodate all the electrons, since if we replace any point within the sphere by one outside, its value of **k** and hence the value of the energy will be larger. If the radius of this sphere is $k_F$, its volume is $\frac{4}{3}\pi k_F^3$ and the number of possible points (that is, allowed values of the momentum) is $(V/8\pi^3)(\frac{4}{3}\pi k_F^3)$. Since only two electrons are allowed per point, if the total number of electrons in a volume $V$ of the metal is $N$, we must have

$$\tfrac{1}{2}N = (V/8\pi^3)\tfrac{4}{3}\pi k_F^3 \tag{11.6}$$

and hence

$$(W_F)_0 = \frac{\hbar^2 k_F^2}{2m} = \frac{\hbar^2}{2m}(3\pi^2 n)^{\frac{2}{3}}. \tag{11.7}$$

Here $n = N/V$ is the number of electrons per unit volume of the metal. $W_F$ is known as the 'Fermi energy' and at 0 K the energy of the highest occupied level is $(W_F)_0$.

The distribution may be expressed in terms of energy rather than momentum by using the relation $W = \hbar^2 k^2/2m$. The number $(2/V)\,\mathrm{d}i$ of states per unit volume which have energy in the range between $W$ and $W+\mathrm{d}W$ is

$$g(W)\,\mathrm{d}W = (2/V)\,\mathrm{d}i = \pi^{-2}k^2\,\mathrm{d}k = CW^{\frac{1}{2}}\,\mathrm{d}W, \tag{11.8}$$

where

$$C = \frac{1}{2\pi^2}\left(\frac{2m}{\hbar^2}\right)^{\frac{3}{2}} \tag{11.9}$$

has the value $1{\cdot}06\times10^{55}$ if $W$ is in joules, and $6{\cdot}79\times10^{27}$ if $W$ is expressed in electronvolts. Eqn (11.7) can be written as

$$n = \tfrac{2}{3}C\{(W_F)_0\}^{\frac{3}{2}} \tag{11.10}$$

and (11.8) as

$$g(W)\,\mathrm{d}W = \frac{3n}{2}\frac{W^{\frac{1}{2}}}{(W_F)_0^{\frac{3}{2}}}\,\mathrm{d}W. \tag{11.11}$$

A plot of $g(W)$ against $W$ gives a parabola, as in Fig. 11.2; at $T=0$ each state is occupied by an electron up to the sharp cut-off at the Fermi energy $(W_F)_0$, and the momentum $\mathbf{p}_F = \hbar \mathbf{k}_F$ corresponding to this point is known as the 'Fermi momentum'. The mean energy $\bar{W}$ per electron is

$$\bar{W} = \frac{1}{n}\int_0^{(W_F)_0} Wg(W)\,\mathrm{d}W = (W_F)_0^{-\frac{3}{2}}\int_0^{(W_F)_0} W^{\frac{3}{2}}\,\mathrm{d}W = \tfrac{3}{5}(W_F)_0, \tag{11.12}$$

and the total internal energy $U_0$ of $n$ electrons at 0 K is

$$U_0 = \tfrac{3}{5}n(W_F)_0. \tag{11.13}$$

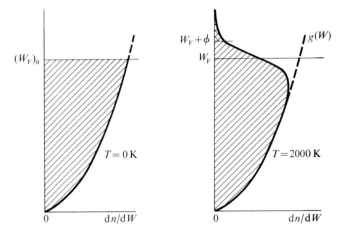

FIG. 11.2. The number of free electrons $dn$ with kinetic energy between $W$ and $W+dW$ in a metal, as given by the Fermi–Dirac distribution. At 0 K the electrons have energies only up to the Fermi level $(W_F)_0$; at a finite temperature $W_F$ is smaller than $(W)_{F0}$ (see eqn (11.19)). At room temperature the distribution is much closer to that on the left, since $dn/dW$ is only altered for electrons whose energy lies within $\approx kT$ of $W_F$.

The great difference between this type of energy distribution and a Maxwell–Boltzmann distribution is best demonstrated by considering numerical values. Eqn (11.7) shows that $(W_F)_0$ depends on the metal only through the quantity $n$, the number of electrons per unit volume. If we take sodium as an example, and assume that one valence electron per atom is released in the metal as a free electron, we have

$$n = 2 \cdot 5 \times 10^{28} \, \text{m}^{-3},$$

and $(W_F)_0$ is found to be $3 \cdot 1$ eV, the mean energy $\bar{W}$ being $1 \cdot 9$ eV. This is very large indeed. On the classical Maxwell-Boltzmann statistics, where the mean energy is $\frac{3}{2}kT$ at temperature $T$ in degree Kelvin, it is equivalent to $T = 20\,000$ K. The values for other metals calculated in the same way are given in Table 11.1.

For comparison, we calculate the value of $W_F$ for a monatomic gas of the rare isotope of mass 3 of helium, atoms of which should also obey the Fermi–Dirac statistics. The boiling-point of this gas is $3 \cdot 2$ K, and the number of atoms per cubic metre at atmospheric pressure in the gas at this temperature is $2 \cdot 3 \times 10^{27}$. This gives $W_F = 1 \cdot 15 \times 10^{-4}$ eV, which is equivalent only to a temperature of $1 \cdot 3$ K. This is small compared with the actual temperature, indicating that deviations from the perfect gas laws due to quantum effects will not be large. It is obvious that the low value of $W_F$ for this, or any other real gas, as compared with the electron gas in a metal, is due primarily to the difference in the mass of the

TABLE 11.1

*Values of* $W_F$ *(the 'Fermi' energy), the work function* $\phi$ *(as deduced from measurements of the photoelectric effect and thermionic emission), and the thermionic emission constant A*

| Substance | $W_F$ (eV) | Work function $\phi$ | | $A$ (A m$^{-2}$ K$^{-2}$) |
| | | Photoelectric effect (eV) | Thermionic emission (eV) | |
| --- | --- | --- | --- | --- |
| Lithium | 4·7 | 2·2 | | |
| Sodium | 3·1 | 1·9 | | |
| Potassium | 2·1 | 1·8 | | |
| Copper | 7·0 | 4·1 | 4·5 | 1·10×10⁶ |
| Silver | 5·5 | 4·7 | 4·3 | 1·07×10⁶ |
| Gold | 5·5 | 4·8 | 4·25 | 1·00×10⁶ |
| Molybdenum | 5·9 | | 4·2 | (0·50–1·15)×10⁶ |
| Tungsten | 5·8 | 4·49 | 4·5 | (0·20–0·60)×10⁶ |
| Platinum | 6·0 | >6·2 | 5·3 | 0·30×10⁶ |
| Nickel | 7·4 | 4·9 | 4·5 | 1·20×10⁶ |
| Thoriated tungsten | | | 2·6 | 0·60×10⁶ |
| (BaO, SrO) mixture | | | 1·8 | 0·03×10⁶ |

particles, which comes in the denominator of eqn (11.7), and also partly to the smaller number per unit volume. When $W_F \ll kT$, the particles have all energies up to those of order $kT$, and the density of occupied points in momentum space is low, that is, the chance of finding an occupied point in the fundamental volume $h^3/V$ is small. Under these circumstances the classical statistics are a valid approximation. For an electron gas, on the other hand, $W_F \gg kT$ at all ordinary temperatures, and every fundamental volume $8\pi^3/V$ of **k**-space contains an occupied point, only one such point (or two electrons, allowing for the spin) in each such volume being allowed by the exclusion principle. If we tried to use the classical picture, with the average energy of the order $kT$, this would correspond to putting a large number of electrons in each volume $8\pi^3/V$ of **k**-space, and the exclusion principle would be violated.

## 11.2. The Fermi–Dirac distribution function

The energy distribution at a finite temperature is given by Fermi–Dirac statistical mechanics, and we quote the results. Each 'point' in **k**-space corresponds to a quantized state of translational motion, and inclusion of the electron spin gives two quantum states to each point. The number of such states $g(W)\,dW$ per unit volume in the energy range $W$ to $W+dW$ is known as the 'density of states', and is given by eqns (11.8) and (11.9). The chance that a state of energy $W$ is occupied by an electron is given by the function

$$f = \frac{1}{\exp\{(W-W_F)/kT\}+1}.$$    (11.14)

At $T=0$ the denominator is infinite for $W>W_F$ and unity for $W<W_F$, so that $f=0$ in the former case and unity in the latter. This corresponds to the sharp cut-off in the occupation of states already discussed. At a finite temperature $f$ is still very close to zero for $W>W_F$ and to unity when $W<W_F$, except for the comparatively small range of energies which lie within a few $kT$ of $W_F$. The quantity $W_F$ is defined as the value of the energy at which the probability of a state being occupied is $f=\frac{1}{2}$, since then the denominator in (11.14) is $e^0+1=2$.

The number of electrons with energy between $W$ and $W+dW$ is

$$dn = fg(W)\,dW = \frac{CW^{\frac{1}{2}}}{\exp\{(W-W_F)/kT\}+1}\,dW,$$    (11.15)

and the total number of electrons per unit volume is

$$n = C\int_0^\infty \frac{W^{\frac{1}{2}}}{\exp\{(W-W_F)/kT\}+1}\,dW.$$    (11.16)

The distribution function $dn/dW$ appropriate to a temperature of about 2000 K is also shown in Fig. 11.2.

The value of $W_F$ is found from the integration in eqn (11.16), which can be done by numerical methods. In general we are interested only in the region where $kT \ll W_F$, and approximate solutions in the form of a power series in $kT/W_F$ can be obtained for this and a number of other integrals, using a method due to Sommerfeld. This gives

$$\int_0^\infty W^r f\,dW = \frac{1}{r+1}W_F^{r+1}\left\{1+\frac{\pi^2}{6}r(r+1)\left(\frac{kT}{W_F}\right)^2+\ldots\right\},$$    (11.17)

where successive terms are each smaller by factors of order $(kT/W_F)^2$. Use of this result with $r=\frac{1}{2}$ leads to

$$n = \frac{2}{3}CW_F^{\frac{3}{2}}\left\{1+\frac{\pi^2}{8}\left(\frac{kT}{W_F}\right)^2+\ldots\right\}.$$    (11.18)

Since $n$ is independent of temperature, it follows that $W_F$ must vary slightly with temperature, and by equating (11.10) and (11.18) we obtain

$$W_F = (W_F)_0 - \frac{\pi^2}{12}\frac{(kT)^2}{W_F}+\ldots$$    (11.19)

in the same approximation. At room temperature $kT$ is equivalent to

0·025 eV while $W_F$ is several volts, so that the approximation is a good one. In the following sections the difference between $W_F$ and $(W_F)_0$ can be neglected for most purposes. Obviously this does not apply to quantities involving differentials with respect to the temperature, such as the specific heat (§ 11.6).

## 11.3. Work function and contact potential

Hitherto we have considered only the kinetic energy of the electrons, but they may also possess potential energy. The Fermi–Dirac distribution function $f$ must then be written in a more general form, where $W$ is the total energy of a particle (kinetic plus potential energy). The probability that an electron occupies a quantum state of total energy $W$ is

$$f = \frac{1}{\exp\{(W-G)/kT\}+1},$$   (11.20)

where the quantity $G$ is the thermodynamic or chemical potential (sometimes known as the Gibbs function). It must be chosen in such a way that the total number of particles per unit volume is $n$, and the chance that a state with total energy $W = G$ is occupied by an electron is exactly $\frac{1}{2}$. If the potential energy is zero, then $W$ is just equal to the kinetic energy (which we now distinguish by writing it as $W_k$), and comparison with eqn (11.14) shows that $G$ is then equal to the Fermi energy $W_F$ and is determined by eqn (11.16). If a particle of charge $q$ has potential energy $qV$, its total energy is $W = W_k+qV$, and to reconcile eqn (11.20) with (11.14) we must have

$$W-G = W_k+qV-G = W_k-W_F,$$

which shows that

$$G = W_F+qV.$$   (11.21)

The value of $G$, like that of $W_F$, must be a function of temperature, but for a degenerate electron gas it is only a slowly varying function (cf. eqn (11.19)). The important property of $G$ is that for all systems in thermal equilibrium it must have the same value. We shall not present a detailed proof of this result, but it follows from the statistical result that systems are in dynamic equilibrium when all states of equal energy have the same probability of being occupied, combined with the thermodynamic result that systems are in thermal equilibrium when they are at the same temperature. Thus for two systems A and B for which the distribution functions are

$$f_A = \frac{1}{\exp\{(W-W_{FA})/kT\}+1},$$

$$f_B = \frac{1}{\exp\{(W-W_{FB})/kT\}+1},$$

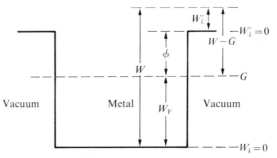

FIG. 11.3. The jump in potential energy at the surface of a metal corresponds to the difference in energy between the level for zero kinetic energy outside the metal ($W'_k = 0$) and the level for zero kinetic energy inside ($W_k = 0$). The jump is equal to $W_F + \phi$, the sum of the Fermi energy $W_F$ and the work function $\phi$.

it is clear that the condition which makes $f_A = f_B$ for all values of the energy $W$ and the temperature $T$ is simply $W_{FA} = W_{FB}$.

When there is no current flowing in a metal, $\mathbf{E} = 0$ from eqn (3.5), and the potential $V$ is constant inside the metal; so also are the values of $W_F$ and $G$. At the surface of the metal there must exist forces which restrain the charge carriers from leaving the metal. With no detailed knowledge of the nature of these forces, the simplest model we can adopt is to suppose that the electric potential rises sharply at the edge of the metal to a value well above the Fermi energy, so that electrons inside the metal have insufficient kinetic energy to surmount this barrier. In Fig. 11.3, the potential energy is assumed to rise vertically at the surface of the metal to a value which, outside the metal, is greater than the Fermi energy by an amount $\phi$. The quantity $\phi$ is known as the 'work function' and, at $T = 0$, it represents the least amount of energy which will suffice to raise an electron from the top of the Fermi distribution and remove it to a point outside the metal with zero kinetic energy.

When two metals are placed in contact, a potential difference is found to exist between them. This was discovered by Volta in 1797, and is sometimes known as the 'Volta effect'. It was not at first generally accepted as a fundamental property of metals, but it has a natural explanation in terms of the free-electron theory. In Fig. 11.4 it is seen that if metal A has a smaller work function $\phi_A$ than that $\phi_B$ of metal B, it is energetically possible for electrons from the top of the energy band in metal A to flow into metal B when contact is made, since the top of the band in B has a lower energy. The flow of charge sets up a potential difference between B and A which increases until the tops of the two energy distributions reach the same level. Equilibrium is then attained, and no more charge is transferred. The actual number of electrons transferred is an insignificant fraction of the total, so that the shaded areas in Figs 11.4(a) and (b) are equal. Thus the contact potential is just equal to the difference of the work functions.

At ordinary temperatures the chance that an electron has sufficient energy to surmount the 'potential barrier' between the two metals in Fig. 11.4(b) is extremely small. At $T = 0$ K it is zero, but no temperature-dependence is found experimentally in the flow of electrons between two metals in contact. This is attributed to the 'tunnel effect'. In quantum mechanics (but not in classical mechanics) there is a finite chance of an electron penetrating a barrier whose height is a few electronvolts provided its width is less than about 1 nm (a few atomic spacings). At any given energy the chance of crossing the barrier is the same in either direction. The net flow is therefore zero when the number of electrons approaching the barrier per unit time is the same on either side. This can be calculated from the Fermi–Dirac distribution function, and it leads again to the condition that the Fermi levels must be the same on either side.

The work function of a metal, and hence also the contact potential between two metals, is very sensitive to the state of the surface. For this reason, measurements of $\phi$ show rather a wide scatter; the best determinations are made with metallic films newly deposited by evaporation in a very high vacuum (for metals with low boiling-points), or (for a high-melting-point metal such as tungsten) with a surface cleaned by 'flashing' the metal at a temperature close to the melting-point in a high vacuum. The contact potential between two surfaces is measured indirectly as follows. A narrow beam of electrons is directed onto the metallic surface in a high vacuum, and the current reaching the surface is plotted as a function of the retarding potential applied to the surface. When a second

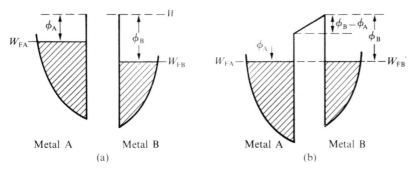

FIG. 11.4. Energy distribution of electrons in two metals A, B. (a) Before contact: $\phi_A$, $\phi_B$ are the energies required to remove an electron to rest at infinity. (b) After contact: electrons flow to metal B, changing its potential relative to A until the tops of the two energy distributions are level. The contact potential difference thus created $= \phi_B - \phi_A$ (in energy), and an electron released from B (for example by the photoelectric effect) would gain this amount of energy in moving from a point near B to a point near A.

metallic surface is substituted, the current to it gives a curve of similar shape but displaced by an amount equal to the contact potential difference between the two surfaces. The work of Mitchell and Mitchell (1951) gave the following mean values for the contact potential relative to a clean tungsten surface: copper $(-0\cdot05\pm0\cdot02)$V; silver $(+0\cdot23\pm0\cdot03)$V; aluminium $(+0\cdot31\pm0\cdot03)$ V.

## 11.4. The photoelectric effect and secondary emission

If an electron acquires an excess energy at least equal to the work function, it can escape from the metal and travel to an electrode held at a positive potential with respect to the emitting surface. A continuous flow of such electrons constitutes a current, and this is the basis of a number of applications. In this section we discuss first the photoelectric effect, in which the electrons acquire the excess energy from photons, and then 'secondary emission', in which the surface is bombarded with 'primary' electrons, some of whose energy is transferred by collision to electrons within the metal. Electrons are also emitted if an intense electric field is applied at the metal surface; this is closely related to thermionic emission, and is discussed in § 11.5.

*Photoelectric emission*

The energy associated with a quantum of radiation of frequency $v$ (a 'photon') is $hv$, where $h$ is Planck's constant, and in many respects the photons behave as particles with this energy. In an inelastic collision the whole of this energy may be transferred to an electron. If the electron in the metal originally had total energy $W = W_k + W_p$, where $W_k$ is its kinetic energy and $W_p$ its potential energy, then after collision it acquires an energy $W + hv$. On emission from the metal this energy remains the same but is divided differently between kinetic energy $W'_k$ and potential energy $W'_p$. Thus

$$W + hv = W_k + W_p + hv = W'_k + W'_p. \tag{11.22}$$

From Fig. 11.3 the jump in potential energy at the surface is $W_F + \phi$, and this must be equal to $W'_p - W_p$. Hence the kinetic energy of the emitted electron is

$$W'_k = hv + W_k - (W'_p - W_p)$$
$$= hv + W_k - (W_F + \phi) = hv - \phi - (W_F - W_k). \tag{11.23}$$

At any given frequency $v$, the maximum kinetic energy of the emitted electrons is $(hv - \phi)$, corresponding to electrons which come from the Fermi surface where $W_k = W_F$. No electrons are emitted unless $W'_k \geqslant 0$, so that the lowest frequency which can just cause photoelectric emission occurs at $hv = \phi$. Measurement of this frequency provides a method of

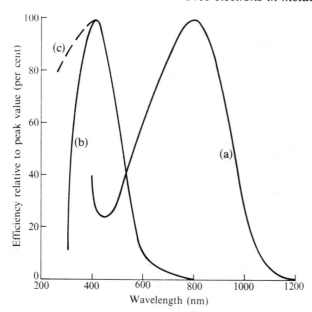

FIG. 11.5. Relative sensitivity of photoemissive tubes with cathodes of (a) caesium on oxidized silver; (b) caesium–antimony (glass window); (c) caesium–antimony (quartz window).

determining the work function $\phi$, knowing the value of $h$ (or $h/e$ if $\phi$ is expressed in volts); some values obtained by this method are given in Table 11.1. The cut-off frequency lies in the visible region only for barium, strontium, and the alkali metals, which have low work functions; for other metals it lies in the ultraviolet.

An important application of photoemission is in devices for the detection and quantitative measurement of the intensity of visible radiation. A photoemissive tube usually has a photosensitive area of a few square centimetres; a composite caesium–antimony surface reaches its maximum response at about $\lambda = 400$ nm, and with a quartz window it can be used into the ultraviolet (see Fig. 11.5). A surface of caesium on oxidized silver is 'red-sensitive', with peak sensitivity around $\lambda = 800$ nm. The photocurrent is attracted to an anode and reaches its saturation value of a few microamperes when the anode is at a positive potential of 50–100 V. The 'dark current' (the background current when the tube is in total darkness) is less than $0 \cdot 1\mu$A. For a vacuum tube the sensitivity is 20–40 $\mu$A per lumen; the response of the photoemissive surface is virtually instantaneous, the time lag for electron emission after light is switched on being less than $10^{-9}$ s. Gas-filled tubes have a sensitivity of over 100 $\mu$A per lumen, the additional current arising from ionization of the molecules of an inert gas by collision with the photoelectrons. Such tubes have the disadvantages that the current sensitivity depends on the anode voltage, and the response falls at frequencies above 10 kHz.

*Secondary emission*

When a solid surface is struck by electrons or ions of appreciable energy, secondary electrons are emitted from the surface. If the bombarding particles (primaries) are electrons, they must have an energy of a few electronvolts to eject an appreciable number of secondary electrons, but at higher energies the number of secondaries may be greater than the number of incident primaries. The secondary-emission ratio

$$\delta = \text{(number of secondary electrons/number of incident primaries)}$$

has maximum values of $0 \cdot 5$–$1 \cdot 6$ for pure metals, and is very sensitive to impurities and contamination of the surface. For composite surfaces, such as a layer of $Cs_2O$ on a base of silver, covered with a surface film of absorbed caesium, $\delta$ may be as high as 10.

At low energies there is an initial rise in $\delta$ because the number of secondaries released within the metal increases with the energy of the incident primary. The production of secondaries is most copious near the end of the primary path, so that the chance of escape declines rapidly as the primaries penetrate deeper into the metal at higher energies, causing $\delta$ to fall again after passing through a maximum at about 200–400 eV in the primary energy.

The chief use of secondary emission is for current amplification. In a 'photomultiplier', electrons from a photosensitive cathode are accelerated to strike a secondary emitting electrode, so arranged that these secondary electrons are accelerated to another secondary-emitting electrode; the process is repeated with a succession of electrodes to give an avalanche effect. An over-all current amplification of $10^5$ is readily attained with 10 electrodes, each with $\delta$ of 3–3·5. The potential of each successive electrode is maintained at 100–200 V higher than the previous one by means of a potentiometer system.

Another type of electron multiplier consists of a curved glass tube (Fig. 11.6), about 2 mm in outside diameter and 1 mm inside, some 50 mm long. The inside wall of the tube has a continuous, highly resistive coating; the output end is maintained at 3–4 kV positive with respect to the input, and the resistance between the ends is over $10^3$ M$\Omega$. When the tube is evacuated, an electron entering at the end where the potential is negative generates secondary electrons on collision with the wall; these are accelerated to produce further electrons on successive collisions, the over-all current gain being about $10^8$. A small horn, 5–10 mm in diameter, may be attached to the input to guide primary electrons into the tube.

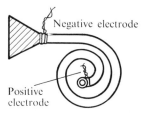

Fig. 11.6. The channel electron multiplier. Primary electrons, guided in by the horn, generate secondary electrons on collision with the inside wall of the tube; this process is repeated many times to give an over-all current gain of about $10^8$. The positive electrode is maintained some 3–4 kV above the negative electrode.

If the output current is passed through a resistance, so that each primary electron generates an output voltage pulse, this may be amplified and fed to a counting circuit.

## 11.5. Thermionic emission and field emission

When a metal is heated to a fairly high temperature, a small fraction of the conduction electrons will have kinetic energy (see Fig. 11.2) greater than $W_F + \phi$, and can therefore escape from the metal. This is known as 'thermionic emission'. If the metal (the 'cathode') is in a vacuum, and an electrode at a positive potential (the 'anode') is placed nearby to collect the emitted electrons, a continuous current ranging up to milliamperes or even amperes may be obtained. The magnitude of the current depends on the work function of the emitter, and varies very rapidly with temperature. For pure metals, copious emission can be obtained only at temperatures of order 2000 K, and metals with very high melting-points, such as tungsten or tantalum, must be used. Pure tungsten, with a work function of 4·5 eV (see Table 11.1), must be run at 2500 K, but thoriated tungsten ($\phi = 2.6$ eV) can operate at 1900 K. Most vacuum tubes, however, use an 'oxide-coated cathode' consisting of a barium–strontium oxide mixture, whose lower work function of 1·8 eV makes it possible to obtain current densities of order $10^4$ A m$^{-2}$ at 1100 K. Such oxide coatings, deposited on a nickel tube with an internal heater of tungsten wire (coated with a refractory insulator such as alumina), have the advantage of being equipotential surfaces. In contrast, along a tungsten filament directly heated by a current there will be a finite potential drop, so that the potential between cathode and anode varies along the cathode. This has the important disadvantage that the filament cannot be heated by alternating current, since this sets up an alternating voltage between cathode and anode which modulates the emitted current.

The close parallel between thermionic emission and evaporation was recognized by Richardson in 1901, but his formula for the emitted current density was based on classical statistics. We shall here use Fermi–Dirac statistics to consider the equilibrium between the electron gas in the metal and the 'electron gas' in the vacuum. Outside the metal, we assume there is no electrical field (since otherwise we do not have equilibrium) and the electrons are moving in random directions so that there is no net current flow. The electron distribution outside is described by the Fermi–Dirac distribution function of eqn (11.20), where in equilibrium the value of $G$ outside is just equal to the value of $W_F$ inside. From Fig. 11.3 it is clear that $W - G = W_k' + \phi$, where $W_k'$ is the kinetic energy outside the metal, and

$$f = \frac{1}{\exp\{(W_k' + \phi)/kT\} + 1} \tag{11.24}$$

Since the exponential term here is much larger than unity, we can write to a good approximation

$$f = \exp(-\phi/kT)\exp(-W_k'/kT). \tag{11.25}$$

The kinetic energy of an electron outside the metal is $W_k' = p^2/2m$; the density of states is thus given, as in eqn (11.8), by

$$g(W_k')\,dW_k' = C(W_k')^{\frac{1}{2}}\,dW_k' \tag{11.26}$$

and the number $dn$ of electrons with energies between $W_k'$ and $W_k' + dW_k'$ is found by multiplying this by eqn (11.25). We can now write simply $W$ for $W_k'$ (this is equivalent to taking $W_k' = 0$ as the zero of total energy), and

$$dn' = C' \exp(-W/kT) W^{\frac{1}{2}}\,dW, \tag{11.27}$$

where

$$C' = C \exp(-\phi/kT). \tag{11.28}$$

The total number of electrons is

$$n' = C' \int_0^\infty \exp(-W/kT) W^{\frac{1}{2}}\,dW$$

$$= C'(kT)^{\frac{3}{2}} \int_0^\infty \exp(-z) z^{\frac{1}{2}}\,dz = \tfrac{1}{2}\pi^{\frac{1}{2}} C'(kT)^{\frac{3}{2}}, \tag{11.29}$$

and their total kinetic energy is

$$\int_0^\infty W\,dn' = C' \int_0^\infty \exp(-W/kT) W^{\frac{3}{2}}\,dW = C'(kT)^{\frac{5}{2}} \int_0^\infty \exp(-z) z^{\frac{3}{2}}\,dz. \tag{11.30}$$

Since

$$\int_0^\infty \exp(-z) z^{\frac{3}{2}} \, dz = -[z^{\frac{3}{2}} \exp(-z)]_0^\infty + \tfrac{3}{2} \int_0^\infty \exp(-z) z^{\frac{1}{2}} \, dz,$$

where the term in square brackets is zero, the mean kinetic energy of the electrons which emerge from the metal is

$$\bar{W} = \frac{1}{n'} \int_0^\infty W \, dn' = \tfrac{3}{2} kT. \qquad (11.31)$$

This result is the same as in classical theory, where the chance of an electron having kinetic energy $W$ has the Boltzmann value $\exp(-W/kT)$; this corresponds to the Fermi–Dirac distribution when the chance $f$ of a quantum state with energy $W$ being occupied is extremely small. The electron gas outside the metal is said to be 'non-degenerate', in contrast with the 'degenerate' state inside the metal where $f$ is virtually equal to unity for all values of $W$ up to $W_F$.

The number of electrons with velocity $v = (2W/m)^{\frac{1}{2}}$ leaving the surface at an angle $\theta$ to the normal is $v \, dn'(\tfrac{1}{2} \sin \theta \, d\theta)$. This corresponds to a current density normal to the surface equal to $q(v \cos \theta) \, dn'(\tfrac{1}{2} \sin \theta \, d\theta)$, and on integrating over all angles we obtain

$$q \, dn' v \int_0^{\pi/2} \tfrac{1}{2} \sin \theta \cos \theta \, d\theta = \tfrac{1}{4} qv \, dn'. \qquad (11.32)$$

On substituting for $dn'$ and $v$ we obtain a total current density of

$$J = \tfrac{1}{4} qC' \int_0^\infty (2W/m)^{\frac{1}{2}} W^{\frac{1}{2}} \exp(-W/kT) \, dW$$

$$= \tfrac{1}{4} qC'(2/m)^{\frac{1}{2}} (kT)^2 \int_0^\infty z \exp(-z) \, dz. \qquad (11.33)$$

This last integral is unity, so that we can write

$$J = AT^2 \exp(-\phi/kT), \qquad (11.34)$$

where the numerical value of the constant $A$ is $1 \cdot 20 \times 10^6 \text{ A m}^{-2} \text{ K}^{-2}$.

This calculation assumes equilibrium conditions between the 'electron gas' outside the metal and that inside, the flow of electrons out of the metal being balanced by a flow back into the metal. If an electric field is applied to accelerate the electrons away from the surface, the return flow is removed, and $J$ represents the current density due to thermionic

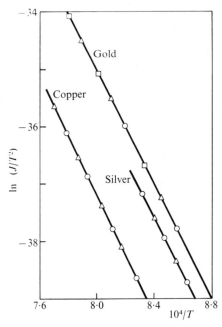

FIG. 11.7. Plot of $\ln(J/T^2)$ against $10^4/T$ for thermionic emission from copper, silver, and gold. (After Jain and Krishnan 1953.)

emission. Some experimental results are shown in Fig. 11.7, where $\ln(J/T^2)$ is plotted against $1/T$. The slope of the straight lines gives the value of the work function $\phi$; a number of values of $\phi$ and of $A$ are given in Table 11.1. In practice a considerable variation is found in the values of $A$, and this is attributed to various surface effects, though there is also a slight variation in $\phi$ with temperature, reflecting the variation in $W_F$ (eqn (11.19)); this affects the apparent value of $A$ obtained in fitting eqn (11.34). There is also evidence that the rate of emission is different from different surfaces of a single crystal, and that the values of $A$ obtained from a polycrystalline surface are too low.

*Emission of positive ions*

The emission of positive ions from a hot metal is much less probable than that of electrons since the ions are bound in the metal with no translational kinetic energy. An exception occurs in the case of some substances which have high vapour pressures and which would therefore be expected to 'boil off' from a hot surface. It was discovered by Langmuir and Kingdom (1925) that every atom of caesium (boiling-point of metal $\approx 950$ K) which strikes a hot tungsten wire comes off as a positive ion. The reason for this is that a caesium atom has an ionization potential

($W_I = 3\cdot87$ eV) lower than the work function of tungsten ($\phi = 4.5$ eV); thus it requires more energy to remove an electron from the tungsten metal to make a neutral caesium atom than to emit a positive caesium ion. The ratio of positive ions to neutral atoms is

$$s^+/s^0 = \exp\{(\phi - W_I)/kT\} \tag{11.35}$$

and for $(\phi - W_I) \approx 0\cdot6$ eV, virtually all the atoms emerge as ions from a tungsten wire at $T = 1200$ K ($\equiv 0\cdot1$ eV). This method of 'surface ioniza- tion' can be extended to atoms with rather higher ionization potentials by the use of oxidized tungsten, for which $\phi \approx 6$ eV. Neutral atoms and molecules are rather difficult to detect directly, and surface ionization provides a valuable method of detection for atoms (and for molecules which dissociate on the hot wire), allowing a positive monatomic ion to emerge. After acceleration, this may be passed through a mass spec- trometer to identify the ion (for example, when several isotopes are present) and eliminate contaminants. The positive ion current may be measured directly, or accelerated onto a surface to produce secondary electrons which are then passed through an electron multiplier. Further details can be found in Ramsey (1956).

*Field emission*

The emission of electrons from a metal under the influence of an applied electric field is closely related to thermionic emission. It occurs only when the direction of the field is such that negative charges are attracted away from the metal, and we must consider the effect of such a field on the potential barrier at the surface. Hitherto we have regarded the potential jump at the surface as infinitely sharp, and an applied field of the correct sign would cause the potential energy to drop linearly outside the surface, as indicated by the line OC in Fig. 11.8. However, the 'image force' (see § 2.5) outside a conducting surface gives a potential energy $W_p = -e^2/16\pi\epsilon_0 x$ for an electron at a distance $x$ from the surface, provided that $x$ is large compared with atomic separations. This potential energy corresponds to the broken line AA', and on including a field strength $E$ the potential energy outside the metal becomes

$$W_p = (-e^2/16\pi\epsilon_0 x) - Eex. \tag{11.36}$$

This function (the full line AB in Fig. 11.8) has a maximum, and electrons inside the metal whose energy exceeds the value corresponding to this maximum can easily escape. Effectively the work function is reduced, and the thermionic emission at a given temperature is increased. Emission will also occur by means of 'tunnelling' through the potential barrier, but for metals at ordinary temperatures this requires fields of order $10^8$ V m$^{-1}$ to produce appreciable emission. Field emission and thermionic emission

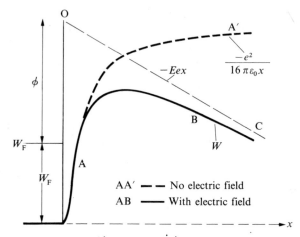

FIG. 11.8. Variation of potential energy $W$ outside a metal, allowing for the image force and an applied electric field.

may occur from a metal electrode into a dielectric, and electrons may tunnel through a thin dielectric film from one electrode to another (see, for example, O'Dwyer 1973).

## 11.6. Heat capacity of the conduction electrons

The heat capacity of the free electrons can be found from the Fermi–Dirac distribution function. The total energy per unit volume is

$$U = \int_0^\infty Wfg(W)\,dW = C\int_0^\infty W^{\frac{3}{2}}f\,dW$$

$$= \tfrac{2}{5}CW_F^{\frac{5}{2}}\left(1 + \frac{5\pi^2}{8}\left(\frac{kT}{W_F}\right)^2 + \dots\right) \qquad (11.37)$$

from eqn (11.17). Hence, using eqns (11.18) and (11.19), we find for the mean energy per electron $\bar{W} = U/n$,

$$\bar{W} = \tfrac{3}{5}(W_F)_0 + \frac{\pi^2}{4}\frac{(kT)^2}{W_F} + \dots . \qquad (11.38)$$

The heat capacity (at constant volume) is obtained by differentiation with respect to $T$, and for a mole of electrons we have

$$C_V/R = \tfrac{1}{2}\pi^2 kT/W_F, \qquad (11.39)$$

where $R$ is the gas constant and higher terms in the expansion are omitted.

Comparison with the classical value of $C_V/R = \frac{3}{2}$ shows that the quantum value is smaller by a factor of order $kT/W_F$. The heat capacity is proportional to the absolute temperature, and approaches the classical value only when $kT$ approaches $W_F$, a region where higher terms in the expansion leading to (11.38) can no longer be neglected. The reason for the small heat capacity given by the quantum theory is that only a small fraction $\approx kT/W_F$ of the electrons at the top of the energy distribution curve are able to increase their kinetic energy, as illustrated by the distribution curve of Fig. 11.2. The increase in energy of each electron is $\approx kT$, and so the total internal energy increases by an amount $\approx n(kT)^2/W_F$. Electrons in the middle of the band cannot be raised to higher energies unless they can reach energy levels above $W_F$, since all the available energy levels in the middle of the band are already occupied by electrons.

Since $kT \ll W_F$ at ordinary temperatures, the electronic heat capacity will be only a small fraction of that predicted by classical theory, and the difficulty of the large excess heat capacity predicted by that theory for metals is removed. An experimental test of eqn (11.39) is possible only at low temperatures, where the heat capacity associated with the lattice vibrations of a solid falls very rapidly. According to the theory of Debye, the lattice heat capacity is proportional to $T^3$ at sufficiently low temperatures, and eventually this becomes small compared with the electronic heat capacity, which falls only with $T$. The heat capacities of a number of metals have been measured, and below about 20 K they are found to follow a law of the form

$$C_V = \alpha T^3 + \gamma T. \tag{11.40}$$

Fig. 11.9 shows the relative magnitudes of the two contributions to the heat capacity of cobalt at temperatures below 20 K. This metal is ferromagnetic, and like a number of other transition-group metals, shows an abnormally high electronic heat capacity. In many metals the electronic heat capacity is predominant only below about 5 K, and rather precise measurements are required to determine it accurately. Eqn (11.40) may be written in the form

$$C_V/T = \alpha T^2 + \gamma, \tag{11.41}$$

and by plotting the quantity $C_V/T$ against $T^2$ a straight-line graph should be obtained whose intercept gives the value of $\gamma$. Fig. 11.10 shows a typical graph for copper.

The measured values of the electronic heat capacity of a number of metals are listed in Table 11.2. To compare them with theory we need to know the number $n$ of free electrons per unit volume, and $W_F$; for this purpose it is sufficiently accurate to use $(W_F)_0$, which, by eqn (11.10), also

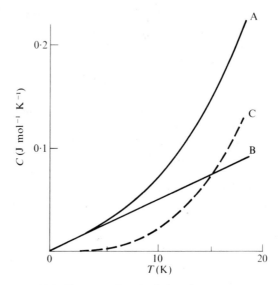

Fig. 11.9. Heat capacity of cobalt at low temperatures.
A   experimental curve.
B   electronic contribution; $C_V = 50·2 \times 10^{-4}T$.
C   lattice contribution; $C_V = 1944(T/443)^3$, where 443 is the
    Debye $\theta$ for cobalt.
                        (After Duyckaerts 1939.)

depends on $n$. For the alkali metals, and copper, silver, and gold, it is
reasonable to assume there is one conduction electron per atom, and
Table 11.2 shows that theory and experiment are in reasonable but by no
means exact agreement. The electronic heat capacities are appreciably
greater for the transition metals, of which iron, cobalt, and nickel belong
to the 3d group, palladium to the 4d, and platinum to the 5d group;
here it is difficult to predict the value of $n$, a point to which we shall
return in § 12.6.

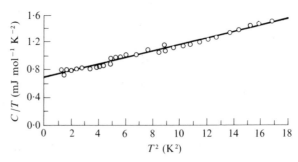

Fig. 11.10. Plot of $C/T$ against $T^2$ for copper. The intercept at $T^2 = 0$ gives the coefficient $\gamma$
of the electronic heat capacity. (Corak, Garfunkel, Satterthwaite, and Wexler 1955.)

TABLE 11.2

*Values of the coefficient γ of the electronic heat capacity γT, in*
$mJ\ K^{-2}\ mol^{-1}$, *as given by Kittel (1971).*

| Metal | γ (experiment) | γ (theory) | Metal | γ (experiment) | γ (theory) |
|-------|---------------|-----------|-------|---------------|-----------|
| Li | 1·63 | 0·75 | Fe | 4·98 | |
| Na | 1·38 | 1·09 | Co | 4·73 | |
| K | 2·08 | 1·67 | Ni | 7·02 | |
| Rb | 2·41 | 1·91 | Pd | 9·42 | |
| Cs | 3·20 | 2·24 | Pt | 6·8 | |
| Cu | 0·70 | 0·51 | Pb | 2·98 | 1·51 |
| Ag | 0·65 | 0·65 | | | |
| Au | 0·73 | 0·64 | | | |

Theoretical values are omitted for the transition metals, where the number of conduction electrons per atom is uncertain. One electron per atom is assumed for the alkali and noble metals, and two electrons per atom for lead.

## 11.7. Electrical and thermal conductivity

If all electrons are contained within the 'Fermi sphere' of Fig. 11.1, as should be the case in equilibrium at 0 K, the resultant current in any direction is zero; for every electron in a quantum state corresponding to momentum components $(p_x, p_y, p_z)$ there is another with components $(-p_x, -p_y, -p_z)$, so that the net momentum is zero. In the presence of an electric field of strength **E** a charge $q$ experiences an acceleration

$$d\mathbf{p}/dt = \hbar(d\mathbf{k}/dt) = q\mathbf{E} \tag{11.42}$$

whose effect is to displace the Fermi sphere at a constant rate $q\mathbf{E}/\hbar$ in the direction of the field. This acceleration is counteracted by the effects of collisions. The mean displacement in momentum is equal to the drift momentum $\bar{\mathbf{p}} = q\mathbf{E}\tau$ in the field **E** (cf. § 3.2), and in **k**-space this means that the Fermi sphere is displaced by an amount $q\mathbf{E}\tau/\hbar$, as shown in Fig. 11.11. It is readily seen from the figure that the essential change due to **E** is a transfer of a small number of electrons from one side of the Fermi sphere to the other, and it follows that the conductivity depends only on the electrons at the Fermi level. Thus we have the important result that in a metal the transport properties involve only electrons whose velocities correspond to the Fermi energy $W_F$. Provided that the Fermi surface is a sphere, this means that the electron velocity $c$ in a formula such as eqn (11.2) is fixed in magnitude, and random only in direction.

At a finite temperature the Fermi distribution does not give a sharp boundary at the surface of the Fermi sphere. At ordinary temperatures, however, the spread in energy (of order $kT$) is so small compared with

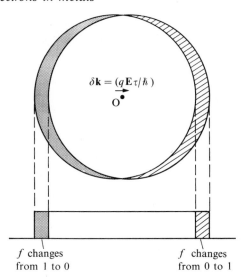

$$\delta \mathbf{k} = (q\mathbf{E}\tau/\hbar)$$

$f$ changes       $f$ changes
from 1 to 0      from 0 to 1

FIG. 11.11. Displacement of the Fermi sphere by an amount corresponding to the drift momentum $\delta\mathbf{k} = \mathbf{p}/\hbar = q\mathbf{E}\tau/\hbar$ in a field strength $\mathbf{E}$. Below is shown the change in the distribution function $f$ at $T = 0$. The current density $\mathbf{J} = n(q/m)\hbar\,\delta\mathbf{k}$.

$W_F$ that the velocity spread can be neglected, and the velocity $v_F$ at the Fermi surface assumed to be the same as at $T = 0$.

At first sight we might expect this to produce some rather dramatic changes in the formulae for the transport properties. On replacing $c$ by $v_F$ and writing $L = v_F\tau$, eqn (11.2) for the thermal conductivity becomes

$$K = \tfrac{1}{3}nv_F^2\tau(\mathrm{d}\bar{W}/\mathrm{d}T) = \tfrac{2}{3}n(W_F/m)\tau(\mathrm{d}\bar{W}/\mathrm{d}T). \tag{11.43}$$

From eqn (11.38) the value of $\mathrm{d}\bar{W}/\mathrm{d}T$ is $\tfrac{1}{2}\pi^2k^2T/W_F$, whence

$$K = \frac{\pi^2}{3} \cdot \frac{nk^2T}{m}\,\tau. \tag{11.44}$$

This equation shows that any temperature-dependence in the ratio $K/T$ must come from variation in the relaxation time $\tau$ with temperature, and it follows from eqn (11.1) that this is true also for the electrical conductivity $\sigma$. The ratio of these two conductivities is

$$\frac{K}{\sigma} = \frac{\pi^2}{3}\left(\frac{k}{e}\right)^2T = L_0T, \tag{11.45}$$

whose numerical value is in fact very close to the value given by classical theory (§ 3.11). However, in eqn (11.45) the $T$ arises from the fact that $\mathrm{d}\bar{W}/\mathrm{d}T$ is proportional to $T$, while $v_F^2 = 2W_F/m$ is independent of $T$; in the classical formulae it is just the other way round.

The numerical value of the Lorenz number $L_0$ given by eqn (11.45) is $2.45 \times 10^{-8} \, \text{W} \, \Omega \, \text{K}^{-2}$, and at room temperature the experimental results for most metals lie close to this value. This is not the case at low temperatures, as can be deduced from the variation of the electrical resistivity $\sigma^{-1}$ and thermal resistivity $K^{-1}$ shown for copper in Fig. 11.12. Near room temperature $K^{-1}$ is nearly constant, while $\sigma^{-1}$ increases almost linearly with $T$, but below 100 K each drops much faster than $T$. The quantity $K/\sigma T$, which should be equal to $L_0$, falls steadily in value as the temperature decreases. This behaviour is typical of many pure metals. It shows that

(1) $\tau$ must be a function of temperature;

(2) it is not correct to use the same value of $\tau$ in the equations for the electrical and the thermal conductivity.

In most metals the electron velocity at the Fermi surface is of the order of $10^6 \, \text{m s}^{-1}$, and since $\tau \approx 10^{-14} \, \text{s}$ at room temperature, the mean-free-path length must be of the order of $10^{-8} \, \text{m}$, or about 100 times the atomic spacing in a solid. At lower temperatures, where the conductivity is much higher, the mean-free-path length must be much longer. Classical physics offered no method of estimating the path length, and on the wave theory it was first pointed out by Houston that the mean-free-path length of an electron in a perfectly regular lattice should be infinite. If an electron is in an allowed energy state, then that is a stationary state; in the absence of a perturbation the electron should continue in its state of fixed energy and fixed wave vector, (that is, fixed momentum), indefinitely. Resistance to the flow of electrical or heat current must therefore arise from the fact

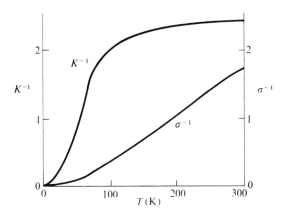

FIG. 11.12. Thermal resistivity $K^{-1}$ and electrical resistivity $\sigma^{-1}$ for pure copper as a function of temperature. The units of $K^{-1}$ are $10^{-3} \, \text{K} \, \text{m} \, \text{W}^{-1}$ and of $\sigma^{-1}$ are $10^{-8} \, \Omega \, \text{m}$.

that real metals do not have perfect lattices, two reasons for which are:

(1) the lattice contains foreign atoms (impurities) or atoms displaced from their normal positions (point defects and dislocations);

(2) the atoms deviate from their mean positions because of thermal vibrations in the lattice.

Each of these imperfections causes scattering of the electron waves in the same way that a defect in an insulating crystal scatters a light wave, whereas a perfect crystal does not.

An alloy is an outstanding example of a disordered lattice on which the different atoms are randomly located, and we would therefore expect its resistance to be higher than that of a pure metal. Furthermore, the scattering in this case should be independent of temperature, giving rise to the constant resistance which is characteristic of alloys. On the other hand, the quantized lattice vibrations cease to be excited at low temperatures, so that both the electrical and the thermal resistance should fall towards zero, being ultimately limited by defect scattering. This provides a qualitative explanation of the behaviour illustrated in Fig. 11.12; the subject will be discussed in more detail in §§ 12.7–12.8.

## 11.8. Thermoelectricity

In 1821 Seebeck discovered that in an electrical circuit made up of two different metals, a small current flows when the junctions between the two metals are not at the same temperature. On open-circuit, an e.m.f. of the order of microvolts per degree difference in temperature of the junctions is observed. The converse effect was discovered by Peltier in 1834. If the Seebeck e.m.f. is from metal A to metal B at the hot junction (Fig. 11.13), a current driven by an external e.m.f. of the same sign produces a cooling at this junction and a heating at the other junction. Both the Seebeck and the Peltier effect are completely reversible. The Seebeck e.m.f. for many metals can be represented by an equation of the form

$$V = at + \tfrac{1}{2}bt^2, \tag{11.46}$$

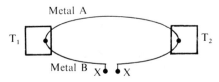

Fig. 11.13. The thermoelectric effect; the thermoelectric voltage is measured by insertion of a voltmeter at XX.

where $t$ is the temperature difference between the junctions. The Peltier coefficient $\Pi$ is defined as follows: the heat absorbed when a charge of $Q$ coulombs passes from metal A to metal B is $\Pi Q$ joules.

In 1854 Thomson (Lord Kelvin) realized that there should be a thermodynamic relation connecting the Seebeck and Peltier effects, but the relation he derived did not agree with experiment. He concluded that a further thermoelectric effect must exist, now known as the 'Thomson heat'. If an electric current of density $J_z$ flows along a conductor in which a temperature gradient $dT/dz$ is maintained, the heat produced in unit volume is

$$\rho J_z^2 - \mu_T J_z (dT/dz) + K(d^2T/dz^2).  \tag{11.47}$$

The first term is the Joule heating, determined by the resistivity $\rho$, and the last term is due to heat conduction. Both are irreversible effects. The second term is a reversible effect, which can be either a cooling or a heating, depending on the relative directions of the temperature gradient and the current flow. The sign of the effect also depends on the Thomson coefficient $\mu_T$, whose dimensions are watt ampere$^{-1}$ ($=$volt) per unit temperature gradient ($K\,m^{-1}$). $\mu_T$ is positive if heat is absorbed from the lattice when current flows up the temperature gradient.

The origins of the thermoelectric effects can be understood in the following way. The energy of the free carriers depends on the temperature, and the carriers at the hot end have more energy that those at the cold end of a metal. Because of their higher velocities, carriers from the hot end diffuse down the metal more rapidly than those from the cold end; they give up their excess energy on making collisions with the lattice, and this is the Thomson heat. When two metals are joined together, the rates of heat flow in the two metals are unequal; heat therefore appears at one junction and is absorbed at the other. This is the Peltier effect.

When the metal is on 'open-circuit', the flow of carriers continues only until a potential difference is set up which just counterbalances the flow. This gives an e.m.f. which is part of the Seebeck e.m.f. A second contribution arises in the following way. The contact potential between two metals with junctions at the same temperature is equal and acts in the opposite sense at the two junctions. However, the contact potential depends slightly on the temperature, leaving a residual e.m.f. when the two junctions are at different temperatures. The contact potential is the difference between the work functions of the two metals, and by considering the temperature variation of the latter we can again reduce the problem to a question of the bulk properties of each metal. In fact theory shows that the various effects are all so interconnected that we need consider only one, such as the Thomson heat, which may be measured as follows.

*Measurement of the Thomson heat*

A thin metal wire is stretched in a good vacuum between two heavy leads (Fig. 11.14) maintained at constant temperature. When a direct current is passed through the wire, the Joule heating raises the temperature at the centre and heat flows towards each end, irrespective of the direction of the current. The Thomson effect causes heat to be released in one half and absorbed in the" other; this process is reversed when the direction of current flow is reversed. The temperature of the wire is measured at two symmetrical points $P_1$, $P_2$ as in Fig. 11.14 by a pair of thermocouples, electrically insulated from the wire by means of thin paper. Two measurements are made. First the thermocouples are connected in series, so that the mean increase of temperature at the two points is found; this gives the temperature increase due to the Joule heating alone. Second, the couples are connected in opposition to give the difference in temperature between the two points due to the Thomson effect. Temperature differences arising from asymmetry in the positions of $P_1$, $P_2$ are eliminated by repeating the measurement with the direction of current flow reversed.

On the assumption that heat is lost only by thermal conduction along the wire, the differential equation including the Joule heating, the Thomson effect, and thermal conduction is

$$I^2R - I\mu_T \frac{dT}{dz} + KA \frac{d^2T}{dz^2} = 0, \tag{11.48}$$

where $I$ is the current flowing, $R$ the resistance per unit length, $A$ the cross-section of the wire, and $K$ its thermal conductivity. The solution of

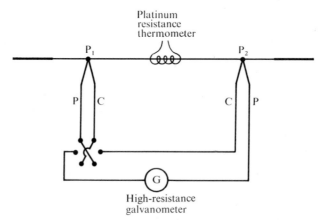

FIG. 11.14. Diagram of the apparatus used by Borelius, Keesom, and Johansson (1928) to measure the absolute value of the Thomson coefficient of a metal. $P_1$, $P_2$ thermocouple junctions of platinum (P) and constantan (C).

this equation is

$$cT = \exp fz - \cosh fa - (z/a)\sinh fa, \qquad (11.49)$$

where $f = I\mu_T/KA$ and $c = (\mu_T/IaR)\sinh fa$; the temperature at the ends of the wire $z = \pm a$ is taken to be zero. If the thermocouple readings are $T_1$ when connected in series and $T_2$ when in opposition then these are proportional to the sum and difference of the temperatures at $+b$ and $-b$ respectively. Hence

$$(fb)^{-1}\frac{T_2}{T_1} = \frac{(fb)^{-1}\sinh fb - (fa)^{-1}\sinh fa}{\cosh fb - \cosh fa}$$

which equals $\frac{1}{3}$ when $f$ is small. This gives

$$\mu_T = \frac{3T_2KA}{T_1Ib}. \qquad (11.50)$$

Here we have assumed the resistivity of the wire to be independent of temperature; in practice a small correction must be made for the slight change in temperature distribution caused by the varying resistance.

### Thermoelectric theory

When a charge passes round a circuit such as that in Fig. 11.13, there is an irreversible heating because of the resistance in the circuit, proportional to the square of the current. Thermal changes due to the Peltier and Thomson effects are proportional to the current and are reversible. By adopting the somewhat artificial device of assuming that we can neglect the irreversible heating when the current is sufficiently small, we can apply thermodynamics to the reversible changes. When a charge $Q$ is passed, the external work done is $QV$, where $V$ is the total e.m.f. in the circuit; the net heat absorbed at the junctions is $(\Pi_{T_1} - \Pi_{T_2})Q$; and the increase in internal energy is $-Q\int_{T_2}^{T_1}(\mu_{TA} - \mu_{TB})\,dT$, since this represents the energy required to heat the charge carriers from temperature $T_2$ to $T_1$ in metal A less that emitted in the reverse process in metal B. Hence from the first law of thermodynamics (dividing by $Q$),

$$V = \Pi_{T_1} - \Pi_{T_2} + \int_{T_2}^{T_1}(\mu_{TA} - \mu_{TB})\,dT. \qquad (11.51)$$

To detect this e.m.f., the circuit must be broken and a voltmeter introduced at, say XX, in Fig. 11.13. Although the voltmeter may contain metals different from A and B, the total e.m.f. in the circuit will be unaffected provided that both junctions to the instrument are at the same temperature. For, if the meter M has junctions at a temperature $T$,

$$V = (\Pi_{T_1} - \Pi_{T_2})_{A \to B} + (\Pi_T)_{B \to M} + (\Pi_T)_{M \to B} + \int_{T_2}^{T_1} (\mu_{TA} - \mu_{TB}) \, dT$$

$$= (\Pi_{T_1} - \Pi_{T_2}) + \int_{T_2}^{T_1} (\mu_{TA} - \mu_{TB}) \, dT,$$

which is the same as before. The same result holds if any number of metals at different temperatures are connected in the circuit, provided that each pair of junctions to a given metal are at the same temperature.

Since no temperature changes occur in the circuit, the external work $QV$ is also the free energy of the system, so that from the second law of thermodynamics

$$QV = U + T \frac{d(QV)}{dT},$$

where

$$U = Q \int_0^T (\mu_{TA} - \mu_{TB}) \, dT.$$

Hence

$$V = \int_0^T (\mu_{TA} - \mu_{TB}) \, dT + T \frac{dV}{dT},$$

and differentiation with respect to $T$ gives

$$\mu_{TA} - \mu_{TB} = -T \frac{d^2 V}{dT^2}. \tag{11.52}$$

Then elimination of $(\mu_{TA} - \mu_{TB})$ with the help of eqn (11.51) yields

$$\Pi_{T_1} - \Pi_{T_2} = [T(dV/dT)]_{T_2}^{T_1},$$

or

$$\frac{dV}{dT} = \frac{\Pi}{T}. \tag{11.53}$$

Although the Peltier and Thomson effects are reversible, the application of reversible thermodynamics to a thermoelectric circuit is open to a number of objections, since irreversible effects such as resistive heating and thermal conduction are always present. A better method is to use irreversible thermodynamics (see Adkins 1975) or to consider the change of entropy at a junction (see Cusack 1958) and to relate this to the net heat absorbed at that temperature. A more rigorous treatment on these lines confirms that eqns (11.52) and (11.53) are valid, and they have been tested experimentally over a large temperature range for many substances.

The quantity $dV/dT$ is called the 'thermoelectric power', and from eqn (11.52) we have

$$dV/dT = -\int \frac{(\mu_{TA} - \mu_{TB})}{T} dT + \text{constant}.$$

By the third law of thermodynamics (see, for example, Wilks 1961) $dV/dT \to 0$ as $T \to 0$, and we can therefore eliminate the constant by taking the integral from 0 to $T$ (to avoid infinities in the integrand this requires $\mu_T$ to vary at least as the first power of $T$). Then

$$\frac{dV}{dT} = \int_0^T \frac{\mu_{TB}}{T} dT - \int_0^T \frac{\mu_{TA}}{T} dT = S_B - S_A, \tag{11.54}$$

where the quantity

$$S = \int_0^T \frac{\mu_T}{T} dT \tag{11.55}$$

is known as the 'absolute thermoelectric power' and is a property of each substance separately. Our identification of the quantity $Q \int \mu_T \, dT$ as the change in the internal energy suggests that

$$\mu_T = (1/Q)(dU/dT) = C_e/Q = \gamma T/Q \tag{11.56}$$

where $C_e$ is the heat capacity of the electrons. Then

$$S = \int_0^T \left(\frac{\mu_T}{T}\right) dT = -\left(\frac{\gamma T}{F}\right), \tag{11.57}$$

where for 1 mol of electrons we have written $Q = -Ne = -F$, where $F$ is the Faraday constant.

The quantity $S$ is analogous to entropy per unit charge, since $\mu_T$ is the heat capacity per unit charge. The theory predicts that both $S$ and $\mu_T$ should be proportional to $T$, and at temperatures above 100 K this is largely true. Values of the Thomson heat $\mu_T$ at 273 K are listed in Table 11.3 together with the values calculated from the experimental electronic heat capacities in Table 11.2. The theoretical values are generally of the right order of magnitude, but the agreement is not exact. A serious difficulty is that for a number of cases the experimental values of $\mu_T$ are positive, and not negative as would be expected when the current is carried by negatively charged electrons. (See § 12.9.)

TABLE 11.3
*Values of the Thomson heat $\mu_T$ in $\mu V\ K^{-1}$*

| Metal | $\mu_T$ (experiment) | $\mu_T$ (theory) | Metal | $\mu_T$ (experiment) | $\mu_T$ (theory) |
|---|---|---|---|---|---|
| Li | +23·3 | −4·6 | Fe | −5·5 | −14·1 |
| Na | −5·1 | −3·9 | Co | −26·7 | −13·4 |
| K | −11·3 | −5·9 | Ni | −16·5 | −19·9 |
| Rb | −9·4 | −6·8 | Pd | −16·6 | −26·6 |
| Cs | −0·8 | −9·1 | Pt | −12·0 | −19·2 |
| Cu | +1·3 | −2·0 | Pb | −0·55 | −8·4 |
| Ag | +1·3 | −1·8 | | | |
| Au | +1·6 | −2·1 | | | |

The experimental values are measured near 273 K; the theoretical values are given by eqn (11.56), using the experimental values of $\gamma$ given in Table 11.2 and assuming $T = 273$ K.

## 11.9. Electron–electron interaction

An important effect which has been neglected in our treatment is the electrostatic repulsion of the conduction electrons. This tends to keep the electrons apart, and the chance of finding two electrons close together is less than it would be on our assumption that their motion is completely independent of each other; in other words, there is a *correlation* between their motions. There is a further effect due to the fact that the over-all wavefunction for the assembly of electrons must be antisymmetric, which is similar in nature to the 'exchange interaction' discussed in Chapter 15. These two effects contribute to the 'correlation energy' and are important in calculating the cohesive energy of a metal. The electrostatic repulsion is a long-range interaction which gives rise to 'plasma oscillations' (see § 10.10), for which the characteristic frequency in a metal is of the order of $10^{15}$ Hz. At low frequencies the dynamical properties of the electrons are not greatly altered, a fortunate circumstance which makes a simple treatment neglecting the correlation energy more accurate than might have been expected.

## References

BORELIUS, G., KEESOM, W. H., and JOHANSSON, C. H. (1928). *Proc. Acad. Sci. Amster.* **31**, 1046.

CORAK, W. S., GARFUNKEL, M. P., SATTERTHWAITE, C. B., and WEXLER, A. (1955). *Phys. Rev.* **98**, 1699.

CUSACK, N. (1958). *The electrical and magnetic properties of solids.* Longmans-Green, London.

DUYCKAERTS, G. (1939). *Physica* **6**, 817.

JAIN, S. C., and KRISHNAN, K. S. (1953). *Proc. R. Soc.* **A217**, 451

KITTEL, C. (1971). *Introduction to solid state physics*. Wiley, New York.
LANGMUIR, E. and KINGDOM, K. H. (1925). *Proc. R. Soc.* **A107**, 61.
MITCHELL, E. W. J. and MITCHELL, J. W. (1951). *Proc. R. Soc.* **A210**, 70.
O'DWYER, J. J. (1973). *Electrical conduction and breakdown in solid dielectrics*. Clarendon Press, Oxford.
RAMSEY, N. F. (1956). *Molecular beams*. Clarendon Press, Oxford.
WILKS, J. (1961). *The third law of thermodynamics*. Clarendon Press, Oxford.
ADKINS, C. J. (1975). *Equilibrium thermodynamics*. McGraw-Hill, London.

## Problems

11.1. Show that eqn (11.19) leads to a value for the temperature coefficient $(dW_F/dT)$ for tungsten $(W_F = 5·8 \text{ eV})$ at 3000 K of about $-0·6 \times 10^{-5} \text{ eV K}^{-1}$. $W_F$ will also vary because of thermal expansion, since the number of electrons per unit volume changes. If $\alpha$ is the linear coefficient of expansion, show that

$$dW_F/dT = -2\alpha W_F.$$

For tungsten at 3000 K, $\alpha$ is about $6 \times 10^{-6}$, from which $dW_F/dT = -7 \times 10^{-5} \text{ eV K}^{-1}$, showing that the effect of thermal expansion is considerably more important than the second term in eqn (11.19).

11.2. Show that eqn (11.57) leads to a formula of the type (11.46) for the thermoelectric e.m.f. in a circuit where the temperature difference between the two junctions is $t$. What predictions about the coefficients $a$, $b$ follow from eqn (11.57)?

11.3. It is found that the thermoelectric power $dV/dt$ of a copper–nickel thermocouple in the range 273–373 K can be expressed as $(20·4+0·0450 t)\mu \text{ V K}^{-1}$, where $t$ is the temperature in °C measured on a particular hydrogen gas thermometer. The Peltier coefficient is given by the expression $5560+32·6t+0·0448t^2 \mu \text{J C}^{-1}$. Verify that in this thermometer the hydrogen behaves as a perfect gas, and deduce the absolute temperature of the ice-point.

11.4. Show that for electrons obeying Maxwell–Boltzmann statistics the Thomson heat should be given by the formula

$$\mu_T = -\frac{3}{2}\frac{k}{|e|}$$

whose value is independent of temperature and equal to about $-130 \mu\text{V K}^{-1}$.

11.5. For an 'isotropic gas' of charged carriers the thermoelectric power $S$ is given by the formula

$$qS = \frac{2c_q}{3} - \frac{dW_F}{dT},$$

where $q$ is the charge and $c_q$ the 'heat capacity' of each carrier, irrespective of the statistics obeyed by the carriers. Show that for free electrons obeying Fermi–Dirac statistics this gives

$$-|e| S = \tfrac{1}{2}\pi^2 k^2 T/W_F.$$

# 12. The band theory of metals

## 12.1. The wave equation for free electrons

In the previous chapter the properties of free electrons in a metal were discussed using Fermi–Dirac statistics, taking account of the wave-like properties of electrons by using the de Broglie relation and the analogy with waves in a box. In a real solid the wave-like nature is important for yet another reason: the wavelength is comparable with the interatomic distance, giving rise to diffraction effects. Before considering these, we shall discuss the simpler problem of the wave equation for free electrons.

The wave equation for a free electron of total energy $W$ in an unbounded region where the potential is $V$ is

$$\frac{\hbar^2}{2m} \nabla^2 \psi + (W - V)\psi = 0. \tag{12.1}$$

For simplicity we consider first the one-dimensional case, for which the wave equation is

$$\frac{\hbar^2}{2m} \frac{\partial^2 \psi}{\partial z^2} + (W - V)\psi = 0, \tag{12.2}$$

and for convenience we further assume $V = 0$ everywhere. Then the solutions of this equation are of the form

$$\psi = A \exp jk_z z. \tag{12.3}$$

The momentum of the particle $p_z$ is given by

$$p_z = -j\hbar \int_{-\infty}^{+\infty} \psi^* \frac{\partial \psi}{\partial z} \, dz = -j\hbar(jk_z) \int_{-\infty}^{+\infty} \psi^* \psi \, dz = \hbar k_z \tag{12.4}$$

since the normalization of the wavefunction requires $\int_{-\infty}^{+\infty} \psi^* \psi \, dz = 1$. In order to satisfy eqn (12.2) we must have $(\hbar^2/2m)k_z^2 = W$, or

$$W = (\hbar^2/2m)k_z^2 = p_z^2/2m \tag{12.5}$$

so that $W$ corresponds to that part of the kinetic energy of the electron associated with its momentum $p_z$ in the $z$-direction.

The three-dimensional equation (12.1) also has a simple solution in Cartesian coordinates, corresponding to the product of three functions of the type (12.3). This solution is

$$\psi = A \exp(jk_xx)\exp(jk_yy)\exp(jk_zz)$$
$$= A \exp j(k_xx+k_yy+k_zz) = A \exp j(\mathbf{k} \cdot \mathbf{r}) \qquad (12.6)$$

since $(x, y, z)$ are the components of the vector $\mathbf{r}$, and we can similarly regard $(k_x, k_y, k_z)$ as the components of a vector $\mathbf{k}$, known as the wave vector. In order to satisfy eqn (12.1) we have

$$W = (\hbar^2/2m)(k_x^2+k_y^2+k_z^2) = (p_x^2+p_y^2+p_z^2)/2m = \mathbf{p}^2/2m, \qquad (12.7)$$

where $\mathbf{p}$ is the momentum vector, with components $(p_x, p_y, p_z)$. These components of $\mathbf{p}$ are just $\hbar$ times those of $\mathbf{k}$, that is, we have the de Broglie relation

$$\mathbf{k} = \mathbf{p}/\hbar \qquad (12.8)$$

which was used in § 11.1. The wavelength associated with the electron is $2\pi/k = h/p$.

In applying the free-electron model to a metal, we assume that the electrons move in a region of constant potential, with a sharp rise in the potential at the boundaries of the metal. Since the electrons do not have enough energy to surmount this barrier, they are confined within the metal (we neglect phenomena such as thermionic emission, which involve only an insignificant proportion of the electrons). Our model thus assumes a rectangular potential well, such as is shown in Fig. 12.1(a) for one dimension. Taking the floor of the well to be at $V = 0$, the solutions of the wave equation are of the form (12.6) inside the well. Outside the well, where the potential $V > W$, so that $(W - V)$ is negative, the solutions are real exponentials; the chance of finding an electron outside, which is proportional to

$$\psi^*\psi = \exp\left[\left\{\frac{8m(V-W)}{\hbar^2}\right\}^{\frac{1}{2}}z\right],$$

falls off very rapidly with distance. A proper solution of the problem requires that the wavefunctions and their derivatives be continuous at the boundary, but if we make the approximation of taking $V$ to be infinite this reduces to making $\psi$ vanish at the boundary. Then the problem is

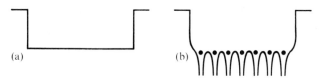

(a)                                              (b)

FIG. 12.1. (a) Rectangular potential well assumed in free-electron model of a metal. (b) Actual potential variation, showing sharp fall near each positively charged ion.

similar to that of electromagnetic waves in a perfectly conducting box; if the latter is rectangular with dimensions $(a, b, c)$, the allowed solutions (cf. Problem 9.12) are a set of standing waves where $\psi$ is a product of terms such as

$$\frac{\sin}{\cos}(\pi lx/a)\frac{\sin}{\cos}(\pi my/b)\frac{\sin}{\cos}(\pi nz/c),$$

and the wavelength is given by eqn (9.42)

$$\frac{1}{\lambda^2} = \left(\frac{l}{2a}\right)^2 + \left(\frac{m}{2b}\right)^2 + \left(\frac{n}{2c}\right)^2.$$

For electrons at the top of the Fermi distribution the wavelengths involved are of the order of a nanometre, which are very small compared with the dimensions of a metal of ordinary size. The spacing of the allowed wavelengths is therefore very close, and the number in a given wavelength range can be computed using the approximations adopted in the theory of heat radiation, as in § 11.1.

## 12.2. The energy-band approximation

At this stage we are still making the arbitrary assumption that in a metal some of the electrons are detached from their parent atoms and are merely bound to the metal as a whole by a potential well inside which they move quite freely. However, many solids are very good electrical insulators in which we must assume there are no such free electrons. There is also the intermediate class of solids, the semiconductors, which are much poorer conductors than metals and which generally possess a negative rather than a positive coefficient of resistivity. To understand why these different types of solids exist we must consider the interaction between the electrons and nuclei when they are closely packed in a solid, where the interatomic distance is of the same order as the atomic radius. The potential energy of an electron then varies rather as shown in Fig. 12.1(b), falling steeply when the electron approaches a positively charged nucleus. Obviously the motion of the electrons in such a potential is a very complicated problem and cannot be solved exactly. Approximate methods must be used, whose nature is illustrated by approaching the problem from two different standpoints.

In an isolated atom the electrons are tightly bound and have discrete, sharp energy levels. When two identical atoms are brought together, the energy levels of each atom, which are initially the same, are split into two, one higher and one lower than the corresponding levels of the separated atoms. The reason for this behaviour follows from a simple quantum-mechanical result. Suppose we have two electrons which have energies $W_1$, $W_2$ when they are on isolated atoms, with no mutual interaction; when the atoms approach one another, an energy of interaction $W_{12}$

arises between the two electrons. The new values $W$ of the energy are given by the solution of the determinant

$$\begin{vmatrix} W_1-W & W_{12} \\ W_{12} & W_2-W \end{vmatrix} = 0 \qquad (12.9)$$

and are

$$W = \tfrac{1}{2}(W_1+W_2)\pm\{\tfrac{1}{4}(W_1-W_2)^2+W_{12}^2\}^{\frac{1}{2}}. \qquad (12.10)$$

If $W_{12} \ll W_1-W_2$, the values of the energy are approximately

$$\begin{aligned} W_1' &= W_1+W_{12}^2/(W_1-W_2), \\ W_2' &= W_2-W_{12}^2/(W_1-W_2), \end{aligned} \qquad (12.11)$$

showing that the energy levels are now slightly further apart, the upper level (assumed to be $W_1$) being pushed up and the lower level $W_2$ being pushed down. The change in energy is small if $W_{12}$ is small compared with $W_1-W_2$, but increases sharply when this inequality no longer holds. In the opposite extreme where $W_{12} \gg (W_1-W_2)$ the energy levels approach the values

$$W' = \tfrac{1}{2}(W_1+W_2)\pm W_{12}\pm(W_1-W_2)^2/8W_{12}, \qquad (12.12)$$

and the splitting has the value $2W_{12}$ for electrons which have the same energy $(W_1 = W_2)$ in the unperturbed state. In general the perturbed wavefunctions are linear combinations of the wavefunctions $\psi_1$, $\psi_2$ for the isolated atoms, and in the limit $W_1 = W_2$ they are

$$2^{-\frac{1}{2}}(\psi_1\pm\psi_2). \qquad (12.13)$$

This means that we can no longer regard one electron as being on one atom and the other on the second atom; the electrons are shared between the two atoms.

If more atoms are brought together, further levels of slightly different energy are formed; for a solid of $N$ atoms, where $N$ is a very large number, the levels are so close together that they form an almost continuous band. The splittings depend on the interactions, which are small for atoms far apart and larger for atoms close together. The interaction arises from the electrostatic repulsion between each pair of electrons; the major contribution comes from the regions where the electron charge clouds are closest together, and in particular from the regions where the electron wavefunctions overlap. At a given distance the interactions are therefore greatest for electrons whose atomic wavefunctions spread out furthest from the parent nucleus, that is, for electrons with the highest energy, which are least closely bound to the parent atom. The over-all bandwidth depends on the degree of the overlap for electrons on adjacent atoms, since these have the largest interaction. Fig. 12.2

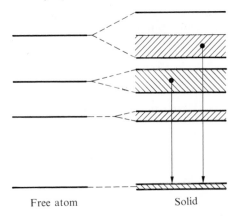

FIG. 12.2. Sharp energy levels in a free atom and the corresponding bands in a solid (the top band is shown only partly full). The arrows indicate allowed transitions giving X-ray emission bands when an electron has been ionized out of the lowest energy band.

is a rough diagram showing how the atomic levels develop into bands as the atoms are brought close together. The wavefunctions are linear combinations of atomic orbitals on every atom, so that the electrons must be regarded as being shared between all the atoms.

## 12.3. Electrons in a periodic potential

The problem may be approached from the opposite viewpoint, starting with electrons moving freely in the flat-bottomed potential well of Fig. 12.1(a), and investigating how the motion is modified when we allow for the variation in the potential energy near each atomic nucleus illustrated in Fig. 12.1(b). In a crystal the atoms form a regular array, and the potential has a corresponding periodic variation in three dimensions. The effect of the periodicity can be understood by considering a one-dimensional case, where a set of $N$ identical atoms form a linear chain, each atom a distance $a$ from its neighbours. We first obtain the boundary conditions. Normalization of the free-electron wavefunctions (12.3) requires

$$1 = A^2 \int_0^{Na} \psi \psi^* \, dz = A^2 \int_0^{Na} \exp(jk_z z)\exp(-jk_z z) \, dz = A^2(Na),$$

which gives $A = (Na)^{-\frac{1}{2}}$. We also need to make the wavefunctions orthogonal for different values of $k_z$, and this requires

$$0 = \int_0^{Na} \exp(jk_1 z)\exp(-jk_2 z) \, dz = \frac{1}{j(k_2-k_1)}[\exp j(k_1-k_2)z]_0^{Na},$$

which is satisfied provided that $(k_1 - k_2)Na$ is an integral multiple of $2\pi$. For this to hold for all values of $k_z$ we must have

$$k_z = 2\pi r/Na = r(G/N), \qquad (12.14)$$

where $r$ has integral values from 1 to $N$, $G = 2\pi/a$, and the number of allowed values of $k_z$ is just equal to the number of particles.

We now introduce the periodic potential, which must repeat in such a way that $V(z+a) \equiv V(z)$. Whatever the actual form of the variation of $V$ with $z$, it can be expanded in a Fourier series of the form

$$V(z) = V_0 + V_1 \cos Gz + V_2 \cos 2Gz + ..., \qquad (12.15)$$

where again $G = 2\pi/a$; the sine terms can be omitted (as here) by taking a suitable origin for $z$, since $V$ must be an even function of $z$ provided that $z = 0$ lies at the centre of an atom.

For two electrons of wave vectors $\mathbf{k}_1$, $\mathbf{k}_2$ the potential energy arising from the first term of (12.15) is

$$(Na)^{-1} \int_0^{Na} V_0 \exp(jk_1 z)\exp(-jk_2 z)\, dz, \qquad (12.16)$$

which vanishes unless $k_1 = k_2$, in which case it is equal to $V_0$. This corresponds to the steady potential at the bottom of the rectangular well in Fig. 12.1(a), and if we take this to be the zero of energy then $V_0 = 0$.

We can now find the effect of a term such as $V_s \cos sGz$ (where $s$ is a positive integer) in the periodic potential. It gives rise to an interaction energy between two electrons with wave vectors $\mathbf{k}_1$, $\mathbf{k}_2$ equal to

$$\tfrac{1}{2}(Na)^{-1} \int_0^{Na} \exp(jk_1 z)V_s\{\exp(jsGz)+\exp(-jsGz)\}\exp(-jk_2 z)\, dz,$$

which vanishes except when

$$k_1 - k_2 = \pm sG. \qquad (12.17)$$

Now the states $k_1$, $k_2$ have kinetic energies equal to $W_1 = (\hbar k_1)^2/2m$ and $W_2 = (\hbar k_2)^2/2m$, and the interaction energy $W_{12} = \tfrac{1}{2}V_s$ will have little effect if it is small compared with $W_1 - W_2$. On the other hand, it will have maximum effect for a pair of electrons for which $W_1 = W_2$, and this occurs for a pair for which $k_1 = -k_2$, provided that, from (12.17), $k_1 = -k_2 = \pm\tfrac{1}{2}sG$. For such a value of $k_z$, the interaction gives an additional energy which splits the levels by $\pm W_{12}$, as in eqn (12.12). The effect of this is shown in Fig. 12.3; the broken line represents the kinetic energy $W = (\hbar_z k_z)^2/2m$ in the unperturbed state. At the point $k_0 = \pm\tfrac{1}{2}G$ (that is, $k_0 = \pm\pi/a$) the periodic potential term $V_1$ causes a discontinuity in the

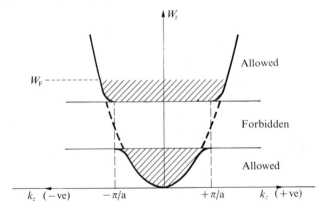

FIG. 12.3. Plot of $W_z$ against $k_z$ showing band structure due to periodic potential of lattice. - - -$W_z$ against $k_z$ in the absence of the periodic structure.

relation between $W$ and $k_z$, the curve for smaller values of $k_z$ being depressed and terminating at an energy $W = (\hbar k_0)^2/2m - \frac{1}{2}V_1$, and reappearing at a value $W = (\hbar k_0)^2/2m + \frac{1}{2}V_1$. Values of the energy between these two values are not allowed; the process is repeated at values of $k_z = \pm\frac{1}{2}sG$, whenever $s$ is an integer. We thus have a series of allowed energy bands alternating with forbidden bands, as in Fig. 12.2.

The reason why the potential energy has such a strong perturbing action at the points where the values of the wave-vector magnitude are $\pm k_0$ becomes clearer from a consideration of the wavefunctions. Since $W_1 = W_2$, the correct wavefunctions under the action of the perturbation are given by the linear combinations (12.13); these are here proportional to

$$\{\exp(jk_0z) \pm \exp(-jk_0z)\}, \qquad \text{or } \cos k_0z \text{ and } \sin k_0z.$$

It is easy to see that the former corresponds to the lower energy. The charge density is proportional to $\psi\psi^*$, which for $\psi = \cos k_0z$ varies as $\cos^2 k_0z$, whose maxima lie at the points where the potential energy is smallest. For the other wavefunction, the charge density varies as $\sin^2 k_0z$, which has its maxima at points half-way between atoms, where the potential energy is greatest (see Fig. 12.4). These wavefunctions correspond to standing waves; the electrons just oscillate forwards and backwards instead of moving on.

Our simple model also leads to a suitable modification to the simple wavefunctions $\exp(jk_zz)$. The effect of the perturbing potential $\frac{1}{2}V_s\{\exp(jsGz) + \exp(-jsGz)\}$ is, by simple perturbation theory, to mix in wavefunctions of the type $\exp\{j(k_z \pm sG)z\}$. The result is to produce a wavefunction of the form

$$a\exp(jk_z z)+\sum_s b_s^{\pm}\exp j(k_z\pm sG)z = \exp(jk_z z)\left\{a+\sum b_s^{\pm}\exp(\pm jsGz)\right\}.$$

We may write this in the simpler form

$$u(z)\exp(jk_z z),\qquad\qquad (12.18)$$

where $u(z)$ is a periodic function of $z$ such that $u(z+a)=u(z)$. Such a function clearly satisfies the property of translational symmetry in the lattice, and our perturbed wavefunction is of this form because $G=2\pi/a$ and $s$ is an integer.

Derivation of a general expression for the energy is clearly a complex problem, whose solution depends on the form of $V(z)$ in eqn (12.15). However, we can find readily how $W$ varies near a discontinuity such as that at $k_0=\tfrac{1}{2}G$, using the series expansion of eqn (12.12) with the condition that $W_1-W_2\ll W_{12}$. The interaction energy $W_{12}=\tfrac{1}{2}V_1$ acts between states for which the values of $k$ can be written as $k_0+k'$ and $k_0+k'-G=k'-k_0$, since $G=2k_0$. Provided that $k'\ll k_0$ for such a pair of states, we have

$$\tfrac{1}{2}(W_1+W_2)=\hbar^2 k_0^2/2m=W_0;\qquad (W_1-W_2)=\hbar^2(4k_0k')/2m.$$

Hence the values of the perturbed energy are

$$W=W_0\pm\tfrac{1}{2}V_1\pm\frac{\hbar^4 k_0^2 k'^2}{m^2 V_1}$$

$$=W_0\pm\tfrac{1}{2}V_1\pm\left(\frac{\hbar^2 k'^2}{2m}\right)\left(\frac{4W_0}{V_1}\right).\qquad (12.19)$$

This result shows that the energy varies with the square of $k'$ near a

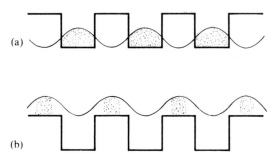

FIG. 12.4. Distribution of charge density for the perturbed wavefunctions at $k=k_0=\tfrac{1}{2}G$. In (a) the charge density is largest where the potential energy is low, reducing the over-all energy; in (b) the charge density is largest at points of high potential energy, increasing the over-all energy.

discontinuity; this corresponds to fitting the energy curve with a parabola, which is always the approximation valid for small values of $k'$.

In § 12.2 we tried to show how the energy bands in a solid are related to those in a free ion, while in this section we have derived them as a result of the periodic structure of the lattice. The use of a one-dimensional picture is not as unrealistic as might at first be supposed, since in a simple solid we can draw parallel planes through arrays of successive ions and consider the periodic variation of the potential as we move in the direction normal to the planes. Such planes are just the 'Bragg planes' which enter in the treatment of the diffraction of X-rays. Strong diffraction occurs when the X-ray wavelength satisfies the condition for interference between successive planes; for normal incidence this is $a = \frac{1}{2}\lambda$, which corresponds exactly to the relation $k_0 = 2\pi/\lambda = \frac{1}{2}G = \pi/a$, giving the wave vector at which the energy gap occurs.

## 12.4. Particle aspects

The foregoing treatment of electrons in a solid considers them as waves occupying the whole volume of the solid; these waves are the stationary states, solutions of the time-independent wave equation. We need to know how the electrons behave under the influence of a force (electric or magnetic); this is essentially a particle description, which must be related to the wave aspect of the electron. A free electron must be considered as a wave packet, where the group velocity corresponds to the particle velocity. The $z$ component of the group velocity is given by the relation

$$v_z = \frac{\partial}{\partial k_z}\left(\frac{W}{\hbar}\right) = \frac{1}{\hbar}\frac{\partial W}{\partial k_z},\tag{12.20}$$

which is analogous to the formula $d\omega/d\beta$ used for the group velocity in § 9.6; the energy $W$ corresponds to $\hbar\omega$ and $k_z$ to the phase constant $\beta$. A rigorous analysis shows that eqn (12.20) is valid for an electron moving in the periodic potential of a crystal lattice, and also that $\hbar\, dk_z/dt = F_z$, where $F_z$ is the component of an external force. Differentiation of eqn (12.20) then gives

$$\frac{dv_z}{dt} = \frac{d}{dt}\left(\frac{1}{\hbar}\frac{\partial W}{\partial k_z}\right) = \frac{1}{\hbar}\frac{\partial^2 W}{\partial k_z^2}\frac{dk_z}{dt} = \frac{1}{m^*}F_z,\tag{12.21}$$

where

$$\frac{1}{m^*} = \frac{1}{\hbar^2}\frac{\partial^2 W}{\partial k_z^2}.\tag{12.22}$$

This last equation defines the quantity $m^*$, which from eqn (12.21) clearly has the dimensions of mass, and is known as the 'effective mass'.

For a free particle the value of $W$ is given by eqn (12.5), and it is readily verified that eqns (12.20)–(12.22) satisfy (12.5) with $m^* = m$, so that the effective mass is equal to the true mass for a free particle. For an electron in a periodic potential the effective mass may depart markedly from the true mass. The advantage of the concept of effective mass is that the dynamic behaviour of an electron in a periodic potential can be treated as if it were a particle of mass $m^*$. The difference between $m^*$ and the true mass $m$ represents the effect on the motion of the electron which results from the electric potential of the ions forming the crystal lattice; when a force is applied to the electron, its change in momentum is different from that of a free electron, and $\mathbf{p} = \hbar\mathbf{k}$ is often referred to as the 'crystal momentum'. The fact that $F_z = m^*(\mathrm{d}v_z/\mathrm{d}t)$ and not $m(\mathrm{d}v_z/\mathrm{d}t)$ does not represent a breakdown of Newton's laws of motion, since the residual momentum is taken up by the lattice. Experiments to determine the ratio of current to momentum, similar to those of Kettering and Scott described in § 3.1, have been carried out by Scott (1951) and Brown and Barnett (1951). They find that even in the case of substances where the current is carried by 'positive holes' (see below), the ratio of current to net momentum has the same sign and is numerically the same as for free electrons.

The relation between the energy $W$ and $k_z$ does not differ greatly from that for a free electron except near the edges of an allowed band. At a point such as $k_0 = \pm\pi/a$ there is a discontinuity in the relation between $W$ and $k_z$, and differentiation of eqn (12.19) shows that at such points $\partial W/\partial k_z = \pm\partial W/\partial k'$ is zero since they correspond to $k' = 0$. This means that the electron velocity $v_z$ is zero at such points, a result which is exactly analogous to that for the group velocity of waves on a linear chain (§ 10.9), or in a filter at the edge of a stop band, or a waveguide at cut-off (§§ 9.1, 9.7).

From the shape of the curve of $W$ against $k_z$ (Fig. 12.3) it is clear that the effective mass $m^*$ depends on the value of $k_z$, and that it becomes negative near the top of an allowed band. From eqn (12.19) we can write

$$W = W_0 \pm \tfrac{1}{2}V_1 \pm \hbar^2(k_0 - k_z)^2/2m^*, \qquad (12.23a)$$

showing that the effective mass $m^*$ is constant near $k_z = k_0$, with a negative value just below $k_0$ and a positive value just above it. This is an important result in dealing with nearly full and nearly empty bands, which occur in semiconductors.

A full band is an allowed band in which all the states are occupied. It carries no electric current, since for every electron with a positive value of $k_z$ there is another with the value $-k_z$. Suppose we have a band which is full except for one state at the top. Since $\hbar(\mathrm{d}\mathbf{k}/\mathrm{d}t) = \mathbf{F}$, under the influence of an electric field the whole state distribution moves, giving a current $I$. But this must be the negative of the current corresponding to the state which is empty, so that $I = -(-ev) = +ev$, as for a positively charged particle. Again, for the empty state, $\hbar(dk_z/dt) = F_z = -eE_z$, so that the net effect of $E_z$ is $dI_z/dt = e(dv_z/dt) = (e/m^*)F_z = e^2E_z/(-m^*)$, where we have used eqn (12.21). Now for the full band lacking one electron the net value of $\mathbf{k}$ is equal and opposite in sign to the value of $\mathbf{k}$ for the electron needed to complete the band. Since the $W$–$\mathbf{k}$ curve is symmetrical (Fig. 12.3), $m^*$ has the same value for $\mathbf{k}$ as for $-\mathbf{k}$, and near the top of the band it is negative. Hence we can write $dI_z/dt = e^2E_z/|m^*|$, showing that the net effect of all the filled states is equivalent to that of a particle whose dynamic properties correspond to a positive charge and positive mass. Such a particle is called a 'positive hole'; its properties were first derived by Peierls in 1929, and it is analogous to the hole in the filled bands of electron states in Dirac's theory of the positron.

The great advantage of the concept of positive holes is that the current and momentum in a nearly filled band can be attributed to the presence of entities which behave like ordinary particles with positive charge and positive effective mass $|m^*|$. Near the top of such a band we can write (12.19) as

$$W = W_0' - \hbar^2 k'^2/2\,|m^*|. \tag{12.23b}$$

If we measure energy *downwards* from the top of the band $W_0'$, we can calculate the density of states for positive holes exactly as was done for free electrons in § 11.1. Furthermore, the chance of finding a state *not* occupied by an electron is

$$1 - f = \frac{1}{\exp\{(W_F - W)/kT\} + 1}.$$

This shows that for $W$ close to $W_0'$, and larger than $W_F$, $(1 - f)$ is close to unity, while for $W < W_F$ the chance of finding a hole decreases exponentially, and is zero when $T = 0$. These results for holes at the top of a band are just the mirror image of those for electrons at the bottom of a band, as we should expect from the fact that moving an electron *upwards* in energy into an unoccupied state corresponds exactly to moving a hole *downwards* into a previously occupied state.

*Three dimensions*

Treatment of a three-dimensional lattice corresponding to a real solid, even with simplified models of the potential variation, is very complex and will not be discussed here. The wavefunction of an electron associated with a wave vector $\mathbf{k}$ is of the form

$$\psi = u(\mathbf{r})\exp \mathrm{j}\mathbf{k}\cdot\mathbf{r}, \tag{12.24}$$

where

$$u(\mathbf{r} + \mathbf{a}_n) = u(\mathbf{r}).$$

Here $\mathbf{a}_n$ is a translation vector representing the repetitive property of the lattice; in particular, that the potential energy is periodic, obeying the rule that $V$ at the point $(\mathbf{r}+\mathbf{a}_n)$ is the same as at $\mathbf{r}$. From the analogy with X-rays it is clear that the values of $\mathbf{k}$ at which strong diffraction occurs will depend on the direction of $\mathbf{k}$; that is, the value of $\mathbf{k}$ at which there is a discontinuity in the energy (the boundary of a zone) is a function of direction. If we draw a vector $\mathbf{k}$ in '$\mathbf{k}$-space' whose length corresponds to this value, and repeat this process for all possible directions, the ends of the vectors will map out a three-dimensional figure, known as a 'Brillouin zone'. Its construction involves only geometry, and its symmetry is related to the symmetry of the crystal lattice.

Values of $\mathbf{k}$ whose vectors end on points inside the zone correspond to allowed energies; those which terminate at the zone boundary correspond to discontinuities in the energy. Higher zones corresponding to higher allowed energy bands exist, but it is possible to bring all wave vectors into the first zone (in the one-dimensional case this procedure corresponds to taking values of $k_z a$ only between $-\pi$ and $+\pi$). Values of $\mathbf{k}$ in the first zone then correspond to more than one allowed energy, but in general we are concerned only with the one band which is partly filled with electrons. At the absolute zero of temperature electrons fill this up to a certain energy, the Fermi energy, and it is therefore of interest to draw plots of constant $W$, or 'energy surfaces' in $\mathbf{k}$-space. Calculation of the energy surfaces is very complex: the results obtained by one method are shown in Fig. 12.5 for a simple cubic lattice. Energy surfaces well within the zone are spherical in shape, but this is by no means true near the zone boundary; in general at the boundary $\partial W/\partial \mathbf{k} = 0$ and the energy surfaces

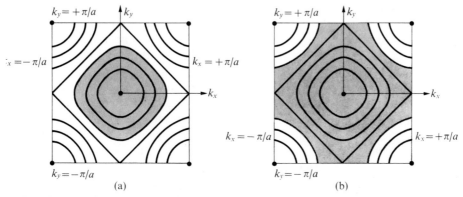

FIG. 12.5. Section through the constant-energy surfaces for a simple cubic lattice obtained by one method of calculation (the 'tight-binding' approximation). The energy surfaces end normally to the zone boundary except where two end at the same point. Shading indicates sections occupied by electrons, sufficient in number to reach the zone boundary in (b) but not in (a).

must end normal to the boundary. The boundary in **k**-space between the filled and empty states at 0 K follows the energy surface corresponding to the Fermi energy and is known as the 'Fermi surface'. Only electrons near the Fermi surface (see § 11.7) can take part in conduction processes, and many details of their behaviour are determined by the exact shape of the Fermi surface. An important property is the effective mass, which in general is a function of direction. Near a band edge the energy may be expanded in a power series in **k**, the lowest terms being quadratic; they can be reduced by a suitable choice of axes to the form

$$W = W_0 \pm \tfrac{1}{2}\hbar^2\left(\frac{k_x^2}{m_x^*}+\frac{k_y^2}{m_y^*}+\frac{k_x^2}{m_z^*}\right), \qquad (12.25)$$

where $k$ is measured from the value of $k_0$ at the band edge, as in eqn (12.23b). Here the upper sign must be taken for electrons near the bottom of a band, and $W_0$ is then the energy at the bottom; while the lower sign should be used for holes near the top of a band, and $W_0$ is then the energy at the top. Thus for holes the energy appears to be measured downwards from the top of the band (see above), a point to which we shall return in considering semiconductors (Chapter 17).

## 12.5. Conductors and insulators on the band theory

In older theories it was assumed that in an insulator all the electrons belonging to each atom are firmly bound to that atom, while in a conductor some of the outer electrons are detached from their parent atoms and able to move freely throughout the whole volume of the solid. On the band theory there is no such distinction between 'bound' and 'free' electrons; the electronic wavefunctions spread out through the whole volume of the solid, though for the states of lower energy (corresponding to the inner electrons of a single atom) the electronic density ($\psi\psi^*$ of the wavefunction) is concentrated near each nucleus. How then does the band theory explain the occurrence of both conductors and insulators?

At the absolute zero of temperature the electrons in a solid will have the lowest possible energy consistent with the Pauli exclusion principle, and they will fill the energy bands from the bottom upwards. The lowest energy bands will be fully occupied, but the highest occupied level may occur in the middle of an allowed band. The state of lowest energy is one in which as many electrons have positive values of **k** as have negative values, so that the net current is zero. To establish a current flow some electrons must be transferred from negative values of **k** to positive values, but because of the exclusion principle this is possible only if they make transitions to unoccupied states of higher energy, the energy being gained by acceleration through the application of an electric field. In weak fields this can occur only if adjacent levels are unoccupied, that is, if the top of the Fermi distribution comes in the middle of a band. If, on the other hand, the highest occupied band is completely full, an electron must gain sufficient energy from a movement in the applied electric field to raise it into the next higher band. This requires enormous electric fields, and for ordinary field strengths the substance is an insulator.

In the treatment of the one-dimensional array of $N$ atoms it was shown that $k_z$ was restricted to values which are integral multiples of $G/N$, and that there are just $N$ such values. In three dimensions we have a similar result, the number of translational levels (the number of allowed values of **k** which correspond to states which are physically not identical) again being just equal to the total number of atoms. In considering the Fermi–Dirac distribution just two electrons were allowed for each value of **k**, provided their spins were paired off. This clearly corresponds to s states on isolated atoms, and the treatment illustrated by Fig. 12.2 leads us to expect that there is a further multiplicity allowed for each value of **k**, corresponding to p, d states, etc. on isolated atoms. Each such state must have a different wavefunction (apart from the two-fold spin degeneracy), that is, it has a different function $u(r)$, such that near the nucleus of an atom $u(r)$ resembles the wavefunction for s, p, d electrons on an isolated atom. For each value of **k** we may then have two electrons in s states, six in p states, and ten in d states.

On this picture it is readily seen that the alkali metals, such as lithium, sodium, and potassium will be good conductors, for their atoms possess only one valence electron in an s state, while the energy band corresponding to this atomic state requires two electrons per atom to fill it. The Group II (or alkaline-earth) elements, magnesium, calcium etc., have two such electrons, which we should expect to fill the s band, making these substances insulators. In fact they are rather good electrical conductors; the reason for this is that in the solid the energy bands corresponding to the outer s and p electrons on the atom are sufficiently broad that they overlap appreciably. The state of lowest energy is then one where the electrons partly fill both the s band and the p band, up to the same level, and conduction is possible. This phenomenon of overlapping bands occurs for most of the elements, particularly those of high atomic number where the outermost electrons have rather extended wavefunctions (giving wide bands in the solid) and where the atomic energy levels are rather closely spaced. The progression can be seen in Group IV, where carbon in the form of diamond is a very good insulator while tin is a metal. The intermediate elements silicon and germanium are semiconductors, where the band gap is sufficiently small that some electrons are excited into the 'conduction band' already at room temperature. The number of conduction electrons increases exponentially as the temperature rises, so that the

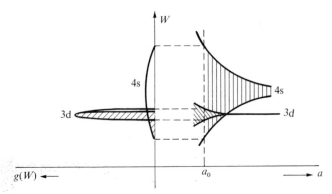

FIG. 12.6. Energy bands for nickel. To the right is shown the bandwidth of 4s and 3d states as a function of interatomic distance $a$ ($a_0$ is the value for solid nickel). To the left is shown g($W$), the shaded area indicating the filled parts of the bands.

conductivity rises with $T$ instead of falling, as in a pure metal. The influence of crystal structure on the energy bands is seen from the fact that carbon (as graphite) and the allotrope grey tin are both semiconductors. In Groups V and VI the light elements are insulators, but the heavy elements tend to be metallic, while in Groups VII and VIII all the elements are insulators. In transition elements the situation is more complex because there are energy bands corresponding to d-electron states. In copper the 3d band is completely filled and there is one electron per atom in the 4s band, making it a good conductor. In iron, cobalt, and nickel the 3d band is not completely filled; it is rather a narrow band, since d-electron wavefunctions do not spread so far out as s-electron functions, and interactions between d electrons on adjacent atoms are smaller than those between s electrons. The narrow 3d band is overlapped by a broad 4s band, and the 3d band (which can contain ten electrons per atom) gives an abnormally high value of the density of states g($W$), as shown in Fig. 12.6.

Information on energy bands in the solid state can be obtained directly from photoelectron spectroscopy, in which the kinetic energy is measured of the electrons emitted when electromagnetic waves (ultraviolet or X-ray) of known frequency are incident on the solid. Since the electrons can only escape if they come from within an extremely thin layer (~2 nm thick) at the surface, the latter must be specially cleaned to obtain reliable results. Results for gold are shown in Fig. 12.7; in the conduction band, the electron density stops sharply at the Fermi surface, and the broad 6s band is overlapped by the rather narrower 3d band, split by spin–orbit coupling. The inner levels $5f_{5/2}$, $5f_{7/2}$ give much narrower bands, less than 0·5 eV broad when examined with high resolution.

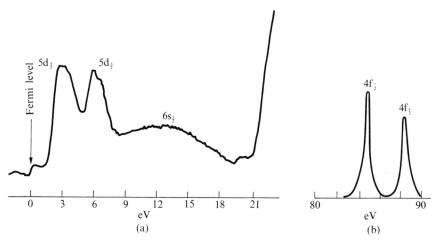

FIG. 12.7. Photoelectron spectrum of gold metal. (a) The conduction band, showing the broad 6s band overlapped by the narrower 5d bands, split by spin–orbit coupling. The sharp drop in electron density at the Fermi level can be seen at the left. (b) The much narrower 4f bands (width ~0·5 eV), whose ionization energies lie between 80 eV and 90 eV, again split by spin–orbit coupling.

## 12.6. Electronic heat capacity

The Pauli exclusion principle restricts the number of states allowed for a given value of $\mathbf{k}$ to two for s electrons, six for p electrons, and ten for d electrons, and this determines the number of states in a given range of wave vector from $\mathbf{k}$ to $\mathbf{k}+d\mathbf{k}$. One effect of the band structure is to alter the relation between $W$ and $\mathbf{k}$, so that $g(W)$, the corresponding number of allowed states between $W$ and $W+dW$, is not the same as in the simple free-electron theory. The presence of overlapping bands provides a further complication. Since only electrons near the Fermi energy $W_F$ can be thermally excited, we expect the electronic heat capacity to be determined by $\{g(W)\}_F$, the density of states at the Fermi energy. Theory shows that instead of eqn (11.38) the value of the mean energy $\bar{W}$ per electron becomes

$$\bar{W} = \bar{W}_0 + \frac{\pi^2(kT)^2}{6n}\{g(W)\}_F + ..., \qquad (12.26)$$

and the electronic heat capacity is given by the formula

$$C_V/R = \frac{\pi^2(kT)}{3n}\{g(W)\}_F... \qquad (12.27)$$

or

$$C_V = \gamma T. \qquad (12.28)$$

This formula shows that the electronic heat capacity gives the density of states at the Fermi energy, and the abnormally high values of $\gamma$ for the transition metals arise from the large density of states in the relatively narrow d bands. For the 'simpler' metals such as the alkali metals the reason why $\gamma$ is not equal to the free electron value (Table 11.2) can be ascribed to the effect of band structure on the relation between $W$ and $\mathbf{k}$. If $W$ varies as $k^2$ at the Fermi surface, we can write $W_F = \hbar^2 k_F^2 / 2m^*$, and the Fermi energy is then given by

$$W_F = (\hbar^2/2m^*)(3\pi^2 n)^{\frac{2}{3}}. \tag{12.29}$$

From eqn (11.11), $\{g(W)\}_F = 3n/2W_F$, so that a value of $m^*$ can be deduced from the electronic heat capacity. However, this assumes that the Fermi surface is spherical; this is seldom the case in practice, and in many cases the Fermi surface touches the zone boundary, so that, at best, $m^*$ can be regarded as some kind of average over that part of the Fermi surface which does not touch the zone boundary.

For an energy band which is completely filled the electronic heat capacity is obviously zero, since no electrons can be thermally excited except at temperature such that $kT$ approaches the value of the energy gap to the next allowed band. For the elements arsenic, antimony, and bismuth in Group V the highest bands are almost full, but with a small overlap to an almost empty band; the density of states is therefore rather low, giving rise to an exceptionally small electronic heat capacity, and to various peculiarities which have led to the name of 'semimetals' for these elements (see, for example, Saunders 1973).

## 12.7. Variation of electrical conductivity with temperature

The detailed calculation of a quantity such as the electric current using the quantum theory of transport phenomena is quite complex. In general the conductivity must be represented by a tensor quantity, but for a cubic metal the conductivity is independent of the direction of the current flow, so that it is a simple scalar quantity. For our purposes it is sufficient to use the formula

$$\sigma = n(e^2/m^*)\tau, \tag{12.30}$$

where $m^*$ represents an average value of the mass obtained by an integration over the Fermi surface. In a metal $m^*$ is independent of the temperature, and variation in the electrical conductivity must be attributed to a variation in the relaxation time $\tau$, as previously pointed out in § 11.7. The existence of a finite relaxation time is due to scattering of the conduction electrons through 'collisions', and in this section we consider the principal causes of scattering, together with the way in which the rates of scattering vary with temperature.

The reciprocal $\tau^{-1}$ of the relaxation time is given by the rate of scattering of the electrons and, provided that the rates of scattering

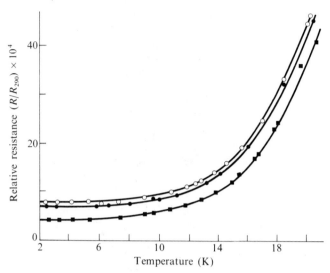

FIG. 12.8. Low-temperature resistance of three specimens of sodium. (MacDonald and Mendelssohn 1950.)

arising from different mechanisms are independent of one another, the total scattering rate is just the sum of the individual scattering rates. We can then write for the electrical resistivity

$$\rho = \frac{m^*}{ne^2}\left(\frac{1}{\tau_1} + \frac{1}{\tau_2} + ...\right), \tag{12.31}$$

a result which shows that the various contributions to the resistivity are additive. This is known as 'Matthiessen's rule', and it is used to find the resistivity of a 'perfect' crystal. Experimentally it is found that the electrical resistivity drops rapidly as the temperature is lowered (cf. Fig. 11.12); however, as $T$ tends to zero it reaches a value independent of temperature, but which varies with the amount of impurity and with the previous history of the specimen. For example, it can be reduced by annealing the specimen to remove lattice defects. The behaviour of three specimens of sodium metal is shown in Fig. 12.8, and is typical of many metals. On subtracting the residual resistivity $\rho_0$ from the resistivity at higher temperatures, results are obtained which are practically independent of the specimen used. This is assumed to be the resistivity of a perfect crystal with no impurities and no defects; it is of course zero at $T = 0$, as expected for a perfect lattice (§ 11.7).

We now consider the interaction between the electrons and the lattice vibrations, or phonons. The distortion of the lattice caused by such a vibration introduces a periodic variation in the lattice potential of the form $V_q \exp j(\omega t - \mathbf{q} \cdot \mathbf{r})$, where $\omega$ is the angular frequency of the vibrations and $\mathbf{q}$ its wave vector. This potential is oscillatory in time, and its effect on a conduction electron is somewhat different from that of a static potential, considered in § 12.3. The interaction of an electron with wavefunction exp $j\mathbf{k.r}$ with such a potential results in the generation of a new electron wave exp $j\mathbf{k'.r}$ such that $\mathbf{k'} = \mathbf{k+q}$. We say that the electron is 'scattered' from a state $\mathbf{k}$ to a state $\mathbf{k'}$ such that

$$\mathbf{k'-k=q}. \tag{12.32}$$

This result is analogous to eqn (12.17); in three dimensions the use of vectors makes it unnecessary to write $\pm\mathbf{q}$.

The state $\mathbf{k'}$ must not already be occupied by electrons, so that it cannot be inside the Fermi surface; in an 'elastic collision', where $\mathbf{k}$ and $\mathbf{k'}$ are electron states with the same energy, both must lie close to the Fermi surface (as in Fig. 12.9(a)). For a spherical Fermi surface, if the scattering rate from $\mathbf{k}$ to $\mathbf{k'}$ depends only on the angle $\alpha$ between $\mathbf{k}$ and $\mathbf{k'}$, the effect of scattering on the flow of charge carriers is to give a relaxation rate

$$\tau^{-1} = \int_0^\pi (1-\cos\alpha)f(\alpha)\sin\alpha\ d\alpha. \tag{12.33}$$

Here the factor $(1-\cos\alpha)$ represents the diminution in the component of velocity of the electron parallel to $\mathbf{k}$ when it is scattered from $\mathbf{k}$ to $\mathbf{k'}$, with a corresponding reduction in the contribution to the electric current. The presence of this factor means that the major part of the integral in (12.33) comes from scattering through large angles, and phonons with large values of $\mathbf{q}$ are therefore the most effective in scattering. At high temperatures the most abundant phonons are those with values of $\mathbf{q}$ close to the zone boundary, which in most metals is of the same order as $\mathbf{k_F}$, the

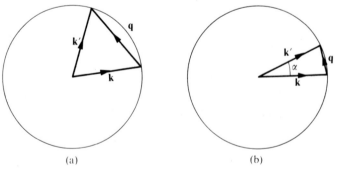

(a)                                          (b)

FIG. 12.9. Scattering of an electron with wave vector $\mathbf{k}$ into a state with wave vector $\mathbf{k'}$ by a phonon of wave vector $\mathbf{q}$. (a) At high temperatures the predominant phonons have high values of $\mathbf{q}$, and scattering is mostly through large angles. (b) At low temperatures only phonons with small values of $\mathbf{q}$ are excited, and scattering is limited to small values of the angle $\alpha$.

value of **k** at the Fermi surface. Scattering of electrons through large angles is therefore a rather probable process. By assuming that the predominant effect is due to phonons at the zone boundary, we can obtain a simple but rather general formula for the variation of $\tau^{-1}$ at high temperatures, using the fact that such phonons have frequencies corresponding to the Debye cut-off frequency given by $\hbar\omega_D = k\theta_D$, where $\theta_D$ is the Debye temperature.

The cross-section for scattering of electrons by the lattice vibrations is proportional to the mean-square displacement $\delta^2$ of an ion from its equilibrium position in the lattice. The mean-square displacement can be related to the vibrational energy, since, for an ion of mass $M$,

$$\text{mean kinetic energy} = \text{mean potential energy} = \tfrac{1}{2}M\omega^2\,\delta^2.$$

At high temperatures each of these has the value $\tfrac{1}{2}kT$ per vibrational mode, and for phonons of the Debye frequency

$$\delta^2 = kT/M\omega_D^2 = (\hbar^2/k)(T/M\theta_D^2). \tag{12.34}$$

Hence

$$\tau^{-1} \text{ should vary as } T/M\theta_D^2. \tag{12.35}$$

a result which is in general agreement with experimental measurements.

At lower temperatures the mean energy associated with a lattice vibration **q** is smaller than $kT$ by a factor

$$\frac{\hbar\omega_q/kT}{\exp(\hbar\omega_q/kT)-1} = \frac{x}{\exp x - 1}, \tag{12.36}$$

and the mean-square ionic displacement $\delta^2$ is smaller by the same factor. Because of the exponential factor we may assume that only lattice vibrations with values up to $\hbar\omega_q \sim kT$ are excited. By writing

$$\omega_q/\omega_D \sim q/q_D \sim T/\theta_D \tag{12.37}$$

we see that this restricts the phonons to those with wave vectors up to $\sim q_D(T/\theta_D)$. As a result, at low temperatures the electrons can only be scattered through rather small angles, as illustrated in Fig. 12.9(b); since we are concerned only with electrons at the Fermi surface, $f(\alpha)$ in eqn (12.33) is large only for values of $\alpha$ of the order of $q/k_F$. Using the relation

$$(1-\cos\alpha)\sin\alpha\,d\alpha = 8\sin^3\tfrac{1}{2}\alpha\,d(\sin\tfrac{1}{2}\alpha) \tag{12.38}$$

we can replace the integral in eqn (12.33) by an integral in which the variable is $\tfrac{1}{2}\alpha = q/2k_F$. On assuming that $f(\alpha)$ is proportional to $kT$ and independent of $\alpha$, we find that $\tau^{-1}$ varies as

$$kT \int_0^{(q_D/k_F)(T/\theta_D)} \frac{1}{2}\left(\frac{q}{k_F}\right)^3 d\left(\frac{q}{k_F}\right) = \frac{1}{8}\left(\frac{q_D}{k_F}\right)^4 kT\left(\frac{T}{\theta_D}\right)^4. \tag{12.39}$$

This predicts that at low temperatures the resistivity of a perfect crystal should be proportional to $T^5$, a result first obtained by Bloch.

Both the high and the low temperature results are contained in a semi-empirical formula, due to Gruneisen,

$$\rho = 4A\left(\frac{T}{\theta}\right)^5 \int_0^{\theta_D/T} \frac{e^x x^5}{(e^x-1)^2}\, dx = 4A\left(\frac{T}{\theta_D}\right)^5 J_5. \qquad (12.40)$$

Here the integral $J_5$ belongs to the same family as the integral $J_4$ which appears in the Debye theory of the lattice specific heat. At high temperatures where the approximation $e^x - 1 = x$ can be used, the integral $J_5$ reduces to

$$\int_0^{\theta_D/T} x^3\, dx = \tfrac{1}{4}(\theta_D/T)^4$$

and the resistivity becomes $\rho = A(T/\theta_D)$. At low temperatures the integration in $J_5$ can be taken to infinity, and it becomes a pure number, giving $\rho = 498A(T/\theta_D)^5$.

For the majority of real metals, apart from the alkali and the noble metals (copper, gold, silver), overlapping bands are present. Electrons in different bands, even with the same values of $W$, can have different effective masses. On a simple two-band approximation we may treat the two bands as carrying currents in parallel, and the conductivity is then

$$\sigma = \frac{n_a e^2 \tau_a}{m_a^*} + \frac{n_b e^2 \tau_b}{m_b^*}. \qquad (12.41)$$

If the total number of electrons is just sufficient to fill one band (if there were no overlap), then $n_a = n_b$, since $n_a$ is the number of holes in the nearly filled band and $n_b$ the number of electrons in the nearly empty band. If also $\tau_a \sim \tau_b$, eqn (12.41) shows that most of the current is carried by the particles with the smaller value of $m^*$; for overlapping s and d bands, as occurs in transition metals, this means that the current is predominantly carried by the s electrons.

In the previous discussion we considered only processes in which an s electron is scattered to another state in the s band, but it turns out that there is a much bigger probability of its being scattered into the d band. This reduces the mobility, and accounts for the rather high resistivity of the transition metals. Similarly, in divalent metals scattering from the s band to the overlapping p band gives a resistivity rather higher than in the monovalent metals.

## 12.8. Variation of thermal conductivity with temperature

In a solid heat is transported both by the lattice vibrations and by the conduction electrons. In a metal the phonons are scattered by collisions with electrons, and the heat current carried by the phonons should be smaller than in an insulator where this scattering process is absent. In general the thermal conductivity of metals is greater than that of insulators, apart from a few exceptional substances such as diamond (see Berman 1973). In practice the thermal transport in a metal is nearly all due to the electrons, the lattice conductivity being so small that it is difficult to measure experimentally except in superconductors, where thermal transport by the electrons is greatly reduced.

In the same approximation as eqn (12.30), the thermal conductivity can be written as

$$\frac{K}{T} = \frac{\pi^2}{3} \cdot \frac{nk^2}{m^*} \tau. \tag{12.42}$$

It follows that $K/\sigma T$ should be equal to the ideal Lorenz number $L_0$; in fact it can be shown that this should be true quite generally provided that the scattering of the electrons is elastic. In practice it always holds at temperatures approaching the Debye $\theta_D$, where the scattering of electrons by phonons occurs mainly through elastic collisions (as in Fig. 12.9(a)). Under these conditions $\tau^{-1}$ varies as $T$, and $K$ should be independent of temperature, as shown for copper in Fig. 11.12.

At lower temperatures 'inelastic' collisions become important for the thermal conductivity. The value of **k** for the electron changes little in a collision, but its energy is changed by amounts equal to the quantized energy of the phonons, that is, by amounts of the order of $kT$. Such collisions are effective in reducing the transport of energy, but have little effect on the transport of electric current, where a small change in energy matters little provided that **k**′ has the same direction as **k**. Thus in calculating the relaxation rate $\tau^{-1}$ for the thermal conductivity, the factor $(1-\cos\alpha)$ in eqn (12.33), which allows for the change in the component of electron velocity, is absent. Instead of eqn (12.39) we obtain an integral of the form

$$kT \int_0^{(q_D/k_F)(T/\theta_D)} \left(\frac{q}{k_F}\right) d\left(\frac{q}{k_F}\right) = \frac{1}{2}\left(\frac{q_D}{k_F}\right)^2 kT\left(\frac{T}{\theta_D}\right)^2, \tag{12.43}$$

so that at low temperatures $T/K$ should vary as $T^3$. In practice the thermal conductivity of most pure metals can be fitted to an expression of the form (see Rosenberg 1955)

$$T/K = aT^3 + b, \tag{12.44}$$

where the constant $b$ represents the effect of defect scattering. When the latter is subtracted to give the thermal resistivity $K^{-1}$ of a 'perfect' lattice, it is found that it can be well represented over a wide range of temperature by the function $(T/\theta_D)^3 J_3$, where

$$J_3 = \int_0^{\theta_D/T} \frac{e^x x^3}{(e^x - 1)^2} \, dx. \tag{12.45}$$

More complex expressions have been proposed; they all give the right asymptotic behaviour at high and at low temperatures, so that experimentally it is difficult to decide between them (see White and Woods 1959). These authors also conclude that the lattice conductivity may make an appreciable contribution to the thermal conductivity at low temperatures in metals such as titanium and zirconium, which have rather a high resistance.

The general behaviour of the electrical and thermal conductivity with temperature is illustrated in Fig. 12.10. At high temperatures $\tau^{-1}$ is the same for each, so that the value of the Lorenz number $K/\sigma T$ is close to the ideal value $L_0$. At low temperatures inelastic collisions become important in reducing $K$, so that $L$ falls below the ideal value, and for a 'perfect' crystal it should approach zero as $T^2$ when $T \rightarrow 0$. In an imperfect crystal, scattering from defects is an elastic process, and the contribution to $\tau^{-1}$ should be the same for the electrical and the thermal

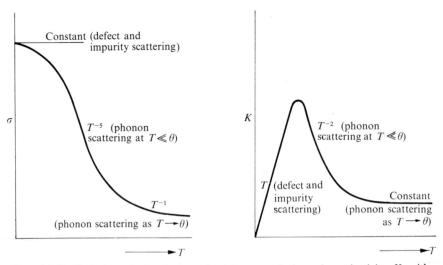

FIG. 12.10. Variation of electrical conductivity $\sigma$ and thermal conductivity $K$ with temperature in a metal. $\theta$ is the Debye temperature.

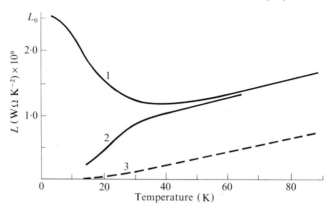

FIG. 12.11. Lorenz number for copper. (Berman and MacDonald 1952.)
1 experimental curve,
2 experimental curve for ideally pure copper, obtained by subtracting contributions to the electrical and thermal resistivity from impurities,
3 theoretical curve.

conductivity. Hence $L$ should again approach $L_0$ at very low temperatures, where defect scattering becomes the dominant process. The results of Berman and MacDonald for copper are shown in Fig. 12.11. $L$ rises towards $L_0$ (curve 1) at liquid-helium temperatures, but for the ideally pure metal (curve 2) it falls towards zero. The broken line represents calculations by Sondheimer; similar discrepancies between theory and experiments have been found for other metals.

## 12.9. Thermoelectric power

The presence of a temperature gradient in a metal produces a flow of charge carriers from the hot end to the cold end. This creates an electric current as well as a heat current, and formulae for the thermoelectric effects can be derived from general transport theory when both effects are included. The contributions from a given set of charge carriers to the heat current and the electric current depend respectively on their energy and their mobility, which do not necessarily vary together in the same way. Allowance for this is made in the following formula for the thermoelectric power $S$,

$$S = -\frac{\pi^2}{3}\cdot\frac{k^2 T}{|e|}\left(\frac{\mathrm{d}\ln\sigma_W}{\mathrm{d}W}\right)_{W=W_F}, \qquad (12.46)$$

where $\sigma_W$ is that part of the electrical conductivity arising from electrons of energy $W$. If we assume that $\sigma_W$ is proportional to $k^3$ (cf. Problem

12.4), we find that

$$S = -\frac{\pi^2}{3} \cdot \frac{k^2 T}{|e|} \left( \frac{3}{2W_F} + \frac{d \ln \tau}{dW} \right). \tag{12.47}$$

As can be seen from eqn (11.39), the first term in this equation is just the electronic specific heat per electron, in agreement with the assumption made in § 11.8. If $\tau$ increases with $W$, there is an extra term, reflecting the fact that the more mobile electrons carry the greater part of the electric current; and since they have larger values of $W$, for a given current they transport more heat.

In these equations the negative sign for $S$ appears because the electric current is in the opposite direction to the net flow of carriers when these are negatively charged. This immediately suggests that the thermoelectric power should have a positive sign when the current is carried by positive holes. The basis for this intuitive expectation can be seen by considering the shape of the $W-k$ curve near the top of a band. Electrons from the hot end of the metal have higher values of $W$, and hence also of $\mathbf{k}$, but because $(d^2 W/dk^2)$ is negative their velocity $\mathbf{v} = \hbar^{-1}(dW/d\mathbf{k})$ is lower than that of electrons from the cold end instead of being greater. The net effect is that electrons diffuse more rapidly from the cold end to the hot end, setting up an e.m.f. of the opposite sign to that we should expect from free electrons.

Quantitatively, if we write as in eqn (12.23b)

$$W_0' - W = \hbar^2 k'^2 / 2 |m^*| \tag{12.48}$$

and assume that $\sigma_W$ varies as $k'^3$ and as the power $\tau^s$ of the relaxation time, we have

$$\ln \sigma_W = \frac{3}{2} \ln(W_0' - W) + \frac{s}{2} \ln(W_0' - W),$$

and we obtain from eqn (12.46)

$$S = +\frac{\pi^2}{3} \cdot \frac{k^2 T}{|e|} \cdot \frac{(3+s)}{2(W_0' - W_F)}, \tag{12.49}$$

showing that the thermoelectric power is reversed in sign.

At low temperatures an effect known as 'phonon drag' becomes important. As a result of the 'collisions' between the electrons and the phonons the latter are pushed along in the direction of the electron flow, creating a drift of phonons which constitutes additional heat current. If such collisions are frequent compared with any other scattering processes for the phonons, the drift velocity of the phonons will equal that of the electrons, and the heat transported will be given by the sum of the lattice specific heat $C_L$ and the electronic specific heat. The former is much larger than

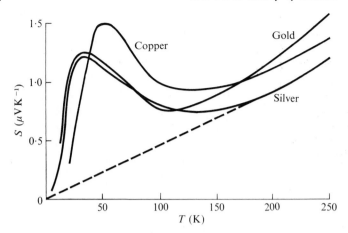

FIG. 12.12. Variation of the thermoelectric power $S$ for the metals copper, silver and gold. The large maxima at $T \sim 30$–$60$ K show the effect of 'phonon drag'. (After Pearson 1971.)

the latter, so that we should expect

$$|S| \sim C_L/n \, |e|.$$

This suggests that at low temperatures $S$ should vary as $T^3$, and be much larger than expected from eqn (12.47). Results for the metals copper, silver, and gold are shown in Fig. 12.12; at the higher temperatures the effect disappears because the main scattering processes for the phonons are collisions with other phonons rather than with electrons.

## 12.10. The anomalous skin effect

In Chapter 8 it was shown that at radio-frequencies the current flow in a good conductor is confined to a thin layer near the surface whose thickness is determined by the skin depth $\delta$ (see § 8.5). At 300 K in copper the skin depth at $10^{10}$ Hz is about $1 \, \mu$m, while the electron mean free path is only $0 \cdot 01 \, \mu$m. At liquid-helium temperatures the situation in a pure specimen of copper, free from defects, is reversed. The conductivity may be $10^4$ times greater, showing that the mean free path has increased to $100 \, \mu$m, while the skin depth should, from eqn (8.31), have fallen to about $0 \cdot 01 \, \mu$m. In the skin depth both the magnitude and the phase of the oscillating electric field are changing exponentially, and the velocity acquired by an electron through the action of the electric field can only be found by an integration, starting from the point at which the electron last made a collision. This means that the current density at a point is no longer directly related to the electric field at that point, and Ohm's law in its simple form is not valid. The current density depends on the integrated effect of the electric field over a volume whose linear dimensions are of the order of the mean free path. This is a 'non-local' situation which we shall meet again in discussing superconductivity (Chapter 13).

Under these conditions the effective value of the skin depth is altered; this is known as the 'anomalous skin effect'. Only those electrons moving almost tangentially to the surface complete a free path wholly within the skin depth, and the conductivity is thus effectively smaller. To estimate the 'anomalous' skin depth $\delta'$ which results, we assume that the fraction of the electrons which contribute to the current flow is reduced by a factor of the order of $\delta'/L$, where $L$ is the mean free path. The corresponding conductivity $\sigma'$ is thus given by $\sigma'/\sigma = \beta(\delta'/L)$, where $\beta$ is a number of the order of unity, and on using this value of $\sigma'$ in eqn (8.31) we obtain the result

$$\delta' = (2L/\beta\sigma\mu\omega)^{\frac{1}{3}}. \tag{12.50}$$

If the conductivity $\sigma$ can be represented by eqn (12.30) with $\tau = L/v_F$, then eqn (12.50) shows that $\delta'$ is independent of $\tau$ or $L$. In a single crystal it can be related indirectly to the curvature of the Fermi surface, since $\delta'$ depends on the number of carriers moving parallel to the metal surface in the direction of the electric field. The velocity of a carrier is parallel to a normal to the Fermi surface in **k**-space (see Problem 12.1), and the flatter the Fermi surface the larger the number of carriers moving close to a given direction.

### 12.11. The Hall effect

When a conductor carrying a current of density **J** parallel to the $y$-axis is placed in a magnetic flux density **B** parallel to the $z$-axis, a potential difference appears across the conductor in the direction of the $x$-axis. This effect was discovered by Hall in 1879. The force exerted by **B** on the moving current carriers tends to displace them in the $x$-direction, but this sets up a non-uniform charge distribution, giving rise to an electric field in the $x$-direction. In equilibrium this produces a force which just balances the magnetic force, so that

$$F_x = qE_x + qv_y B_z = 0.$$

If we can identify $v_y$ with the drift velocity of the carriers, then $J_y = nqv_y$, where $n$ is the number of carriers per unit volume. Then

$$E_x = -(J_y/nq)B_z = -R_H J_y B_z,$$

where $R_H$ is known as the Hall coefficient. Its magnitude should be

$$R_H = 1/nq, \tag{12.51}$$

a simple result which is important because it means that measurements of the Hall effect give both the value of $n$ and the sign of $q$.

TABLE 12.1

*Observed and calculated values of the Hall effect*

| Metal | Observed | Calculated, assuming x electrons per atom | |
|-------|----------|-------------------------------------------|---|
| Lithium | −17·0 | −13·1 | $(x = 1)$ |
| Sodium | −25·0 | −24·4 | $(x = 1)$ |
| Copper | −5·5 | −7·4 | $(x = 1)$ |
| Silver | −8·4 | −10·4 | $(x = 1)$ |
| Zinc | +4·1 | −4·6 | $(x = 2)$ |
| Cadmium | +6·0 | −6·5 | $(x = 2)$ |

$R_H$ (in units of $10^{-11}$ m$^3$ C$^{-1}$)

By combining the Hall effect with the conductivity $\sigma$, the mobility $u$ can be found, since $\sigma = nqu$. The mobility is directly related to an angle $\phi$, known as the 'Hall angle'. To maintain the current density $J_y$ an electric field strength $E_y$ must be present such that $J_y = \sigma E_y$. The resultant electric field thus makes an angle $\phi$ with the y-axis, given by

$$\tan \phi = E_x/E_y = -R_H J_y B_z/E_y = -(\sigma R_H)B_z = -uB_z. \qquad (12.52)$$

At first sight the appearance of the density of carriers in eqn (12.51) is surprising; it reflects the fact that the force exerted by **B** depends on the carrier velocity, and the larger the mobility the smaller the number of carriers for a given current density.

A comparison of the observed values of $R_H$ for various substances with those calculated from eqn (12.51) is given in Table 12.1. In the free-electron theory we expect $q = -e$, giving a negative value for $R_H$; this is observed for the monovalent metals, and the numerical value agrees fairly well with the assumption of one conduction electron per atom. For the divalent alkaline-earth metals, $R_H$ is found to be positive, suggesting the carriers are effectively positively charged. In such metals there are over-lapping s and p bands, the s band being nearly full so that it contains positive holes (see § 12.5). These carriers have the higher mobility (see end of § 12.7), and are predominant in deciding the sign of the Hall effect.

This result draws attention to the fact that the simple derivation of eqn (12.51) given above is not correct when carriers of different mobility are present. In a thin wire, it is only the *resultant* current which must flow parallel to the wire, and the relation $E_x + v_y B_z = 0$ cannot be used when there is more than one drift velocity present. A geometrical solution using the Hall angles can be constructed if just two sets of carriers are assumed, as in a model of a metal with two overlapping bands. Detailed evaluations, assuming that $\tau$ is constant over the Fermi surface, have confirmed that eqn (12.51) is correct if the Fermi surface is spherical or ellipsoidal.

If a velocity distribution of the Maxwellian type is assumed, an expression for $R_H$ larger by a factor $3\pi/8$ is obtained; this result is often used in semiconductors. Because of the relatively small carrier density, the values of $R_H$ in such materials are large (and very temperature-dependent). Rather large values are also observed for the semimetals; for bismuth $R_H$ corresponds to about $10^{-4}$ electrons per atom.

*Magnetoresistance*

If only one type of carrier is present, the electrical resistance is not changed by the application of magnetic flux, because any sideways motion induced by $B_z$ is exactly counterbalanced by the field strength $E_x$. Thus for free electrons there should be no transverse magnetoresistance effect, and obviously we do not expect a longitudinal effect since the drift velocity is parallel to the magnetic flux. In general most metals do show a resistivity which increases appreciably in high flux densities, since carriers of more than one type are deviated by the magnetic field. The effects are greater at low temperatures, where mean free paths are longer, and carriers follow curved paths for longer distances before being interrupted by collisions.

## 12.12. Conduction electrons in a strong magnetic field

The equation of motion of a free particle in a flux density $\mathbf{B}$ is

$$m\dot{\mathbf{v}} = \mathbf{F} = q(\mathbf{v} \wedge \mathbf{B}). \tag{12.53}$$

If $\mathbf{B}$ is constant and parallel to the $z$-axis, then $v_z$ is constant. In the $x,y$-plane the particle traces out a circle with angular velocity (eqn (4.73))

$$\omega_c = (q/m)B,$$

this circle being the projection of the helical path followed by the particle in three dimensions. No work is done on the particle, since the power expended is $\mathbf{v} \cdot \mathbf{F} = q\mathbf{v} \cdot (\mathbf{v} \wedge \mathbf{B}) = 0$, and its kinetic energy is constant.

In a solid the corresponding problem is much more complex, and we confine ourselves to a simplified illustration of the approach, using a free-electron model. The equation of motion is

$$\hbar\dot{\mathbf{k}} = q(\mathbf{v} \wedge \mathbf{B}); \tag{12.54}$$

if $\mathbf{B}$ is parallel to the $z$-axis, it follows that $k_z$ is a constant, and the movement of a representative point in $\mathbf{k}$-space is confined to a plane normal to $k_z$. For a particle at the Fermi surface the vector $\mathbf{v}$ is always normal to the surface (see Problem 12.1 for a special case), and $\dot{\mathbf{k}}$

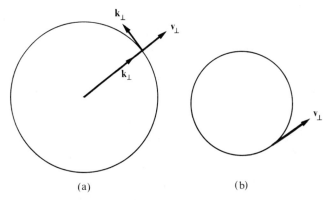

(a)                                             (b)

FIG. 12.13. Motion of a free electron in a magnetic flux density $B_z$. (a) The point in **k**-space describes a circular orbit in the plane normal to $B_z$ with angular velocity $\omega_c$. (b) In real space the particle trajectory is a helix whose projection on the $xy$-plane is a circle.

therefore lies in the surface, normal to **B**. The **k**-vector traces out a surface curve of constant energy which is just the intersection of the Fermi surface with the plane $k_z = $ constant. For the case of a free particle, we have $\hbar\mathbf{k} = \mathbf{p} = m\mathbf{v}$, so that

$$\hbar\dot{\mathbf{k}} = \hbar(q/m)(\mathbf{k} \wedge \mathbf{B}). \qquad (12.55)$$

If we write $k_\perp$ for the component of **k** normal to **B**, then $\dot{k}_\perp = (q/m)(k_\perp B_z)$, showing that $\dot{k}_\perp$ is normal to and proportional in magnitude to $k_\perp$ (Fig. 12.13). Hence $k_\perp$ performs a simple rotation about the $k_z$-axis with angular velocity $(q/m)B_z$, which is just equal to the cyclotron angular velocity of eqn (4.73). $k_\perp$ traces out a circle (a section of the Fermi surface, which is spherical for free electrons) the time taken being $2\pi/\omega_c$.

The kinetic energy can be written as

$$W = \frac{\hbar^2}{2m}(k_x^2 + k_y^2 + k_z^2) = \frac{\hbar^2}{2m}(k_\perp^2 + k_z^2). \qquad (12.56)$$

The component $k_z$ is unaffected by the presence of $B_z$ and the corresponding wavefunctions are of the form $\exp(jk_z z)$ as before. On the other hand, in the $x,y$-plane the energy levels turn out to be those of a two-dimensional harmonic oscillator, given by

$$W_\perp = \frac{\hbar^2}{2m}k_\perp^2 = (n+\gamma)\hbar\omega_c, \qquad (12.57)$$

where $\gamma$ is a phase constant and the quantum number $n$ is an integer. The allowed electric dipole transitions are $\Delta n = \pm 1$, so that they occur just at the classical cyclotron frequency.

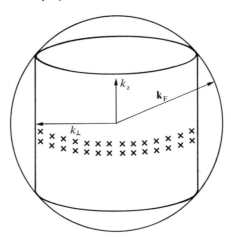

FIG. 12.14. In a flux density $B_z$ the representative points in **k**-space lie on the surfaces of coaxial cylinders, only one of which is shown here. The cross-sectional area $A_\perp = \pi k_\perp^2$ is proportional to $\omega_c$ and hence to $B_z$. Oscillations occur in the density of states at the Fermi surface as the radius $k_\perp$ of successive cylinders passes through the Fermi radius $k_F$.

The area of the circular orbit traced out in **k**-space is

$$A_\perp = \pi k_\perp^2 = \frac{2\pi m}{\hbar^2} W_\perp = \frac{2\pi m}{\hbar}(n+\gamma)\omega_c, \qquad (12.58)$$

showing that $A_\perp$ is proportional to $W_\perp$. Thus to each integral value of $n$ there corresponds a circle of given area in the $k_x k_y$-plane, and the allowed points in **k**-space lie on the circumference of this circle. Each time that $n$ increases by unity, the area $A_\perp$ increases by

$$\delta A_\perp = \frac{2\pi m}{\hbar}\omega_c.$$

We can use this result to find the number of allowed points in **k**-space for each value of $n$. For a cube of metal whose sides are of unit length, the number of points allowed in a two-dimensional system is $1/(2\pi)^2$ per unit area, and for each increment of area $\delta A_\perp$ the number of points is therefore

$$(2\pi m/\hbar)\omega_c \div (2\pi)^2 = (m/h)\omega_c = (qB_z)/h. \qquad (12.59)$$

These results are illustrated in Fig. 12.14. The allowed points lie on the surfaces of a set of concentric cylinders whose cross-sectional area is given by eqn (12.58); this replaces the simple cubic array of points which occurs when $B_z = 0$. Parallel to the $k_z$-axis the allowed points are equally spaced at intervals of $1/2\pi$. The surfaces of equal energy are spherical, corresponding to eqn (12.56), so that allowed points with the same energy occur at the circles where spheres and cylinders intersect.

We now consider the density of states. Each of the energy levels defined by eqn (12.57) is associated with a finite set of points defined by the value of $k_\perp^2$. Motion parallel to the $z$-axis gives points equally spaced in $k_z$, but since the associated energy is $W_z = (\hbar^2/2m)k_z^2$, the density of points $g(W_z)$ decreases as $W_z^{-\frac{1}{2}}$ (see Problem 12.2). The resultant density of states has the form shown in Fig. 12.15, being high at each value of $W$ which satisfies eqn (12.57) and then falling as $W$ increases until the next value of $W$ is reached, at which $n$ has increased by unity in this equation. The Fermi energy $W_F$ at $T = 0$ is defined by the fact that the number of states lying below it is just equal to the number of electrons which must be accommodated. As $B_z$ is increased, the vertical spacing between the energy levels in Fig. 12.15 increases in proportion to $B_z$, so that the density of states at the Fermi level oscillates as each crosses the Fermi level (which must also oscillate to fulfil the criterion for the total number of states below it). The result is a sharp change in the density whenever $W_F = (n+\gamma)\hbar\omega_c = (n+\gamma)\hbar(qB_z/m)$, and many physical properties such as the magnetic susceptibility and electrical resistance show a corresponding oscillation. This occurs at equally spaced intervals when plotted as a function of $B_z^{-1}$. To observe such effects, the sharp changes in the density of states must not be smeared over by the spread in the Fermi–Dirac distribution function around the Fermi energy; this requires that experiments must be made at low temperatures such that $kT \ll \hbar\omega_c$. For

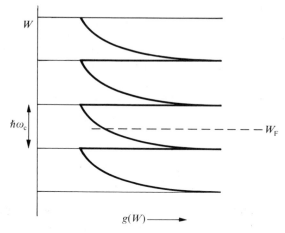

FIG. 12.15. Density of states in a magnetic flux density $B_z$. The variation is repeated at intervals of $\hbar\omega_c$, which depends linearly on $B_z$. As $B_z$ is altered, a sharp change in the density of states at the Fermi energy occurs whenever a cyclotron resonance level passes through $W_F$.

free electrons this means that a field of several tesla, and temperatures down to 1·5 K are needed.

It is important to realize that the separation of successive cyclotron resonance levels is very small compared to the Fermi energy. Even at a flux density $B_z = 10$ T the spacing for free electrons is only about $10^{-3}$ eV, and the oscillations are closely spaced as a function of $B_z^{-1}$; even at this rather high field they occur at intervals of less than $10^{-2}$ T. The changes in the density of states are correspondingly small, but are observable primarily because the density varies as $W_z^{-\frac{1}{2}}$, and so goes to infinity at each cyclotron resonance level. A further important condition is that the width of the levels ($\tau^{-1}$) must be less than the spacing, and for good resolution the criterion $\omega_c \tau \gg 1$ means that low temperatures and very pure samples are required.

The various phenomena observable in the presence of a magnetic field give important information about the Fermi surface. In the presence of band structure the cyclotron resonance frequency is given by

$$\omega_c = \frac{2\pi q B_z}{\hbar^2} \cdot \frac{dW}{dA_\perp}. \tag{12.60}$$

This result can also be expressed in terms of a 'cyclotron mass' $m_c$ defined by the relation $\omega_c = qB_z/m_c$, where

$$m_c = \frac{\hbar^2}{2\pi} \frac{dA_\perp}{dW}. \tag{12.61}$$

Direct determination of the cyclotron resonance frequency is discussed in Chapter 24. The full theory of the oscillatory effects in other properties is very complicated, and we mention only the effects in the resistivity, known as the de Haas–Shubnikov effect. The resistivity depends on the rate of scattering of the carriers, and at low temperatures the low-frequency phonons produce mainly elastic scattering to other parts of the Fermi surface, as in Fig. 12.9(b). Scattering can take place only to empty carrier states of nearly the same energy, which are most abundant when a cyclotron resonance level coincides with the Fermi level, at which point an increase in the resistivity occurs. For $k_z = 0$, $W = W_\perp$ and integration of eqn (12.60), combined with the quantization condition (12.57), gives

$$A_\perp = \frac{2\pi q B_z}{\hbar^2 \omega_c} W = \frac{2\pi q B_z}{\hbar} (n+\gamma). \tag{12.62}$$

This is the maximum cross-section of the Fermi surface, and it can be determined from the frequency of the oscillations plotted against $1/B_z$. In general oscillations are observed corresponding to either maximum or minimum values of the cross-section.

Oscillations in the magnetic susceptibility are known as the de Haas–van

Alphen effect. At fields too small for oscillations to be observable, there is a small temperature-independent susceptibilty, which will be discussed in § 14.11.

In this treatment we have omitted the electron spin, whose magnetic moment interacts with $B_z$ to give an additional energy $\pm\frac{1}{2}g_s\mu_B B_z$ in eqn (12.57). In the approximation $g_s = 2$ this is just $\pm(e\hbar/2m)B_z = \pm\frac{1}{2}\hbar\omega_c$ for a free electron, and the result is just a shift in the energy levels in Fig. 12.15. Cyclotron resonance is induced by an oscillatory electric field (see § 24.9); the spin vector remains fixed and the cyclotron frequency is not altered.

## References

BEATTIE, J. R. (1955). *Phil. Mag.* **46,** 235.
BERMAN, R. (1973). *Contemp. Phys.* **14,** 101.
—— and MACDONALD, D. K. C. (1952). *Proc. R. Soc. A,* **211,** 122.
BROWN, S. and BARNETT, S. J. (1951). *Phys. Rev.* **81,** 657.
MACDONALD, D. K. C. and MENDELSSOHN, K. (1950). *Proc. R. Soc. A,* **202,** 103.
PEARSON, W. B. (1971). *Solid St. Phys., USSR,* **3,** 1411.
ROSENBERG, H. M. (1955). *Phil. Trans. R. Soc. A* **247,** 441.
SAUNDERS, G. A. (1973). *Contemp. Phys.* **14,** 149.
SCOTT, G. G. (1951). *Phys. Rev.* **83,** 656.
WHITE, G. K. and WOODS, S. B. (1959). *Phil. Trans. R. Soc. A,* **251,** 273.

## General references

KITTEL, C. (1971). *Introduction to solid state physics.* Wiley, New York.
MACDONALD, D. K. C. (1962). *Thermoelectricity.* Wiley, New York.
ROSENBERG, H. M. (1963). *Low temperature solid state physics.* Clarendon Press, Oxford.
ZIMAN, J. M. (1970). *Electrons in metals.* Taylor and Francis, London.

## Problems

12.1. If $k_z =$ constant in eqn (12.25), show that at the point $k_x,k_y$ on the Fermi surface the slope of the tangent in the $k_x k_y$-plane is

$$dk_y/dk_x = -(k_x/k_y)(m_y/m_x)$$

and that the normal to the surface at this point is parallel to the electron velocity whose components are given by equations similar to (12.20).

12.2. Adapt the treatment of § 11.1 to the case of a one-dimensional system of unit length to show that the density of states (allowing for electron spin) is

$$g(W_z) = (1/2\pi)(m/2\hbar^2 W_z)^{\frac{1}{2}}.$$

Show similarly that for a square of unit side

$$g(W) = (1/2\pi)(2m/\hbar^2).$$

12.3. In a simple cubic lattice the energy surfaces given by the 'tight-binding' approximation are of the form

$$W = W_1 - W_2(\cos k_x a + \cos k_y a + \cos k_z a),$$

where $a$ is the atomic separation. Show that the width of the energy band is $6W_2$, and that near the bottom ($k_x a \to 0$, etc.) the energy is approximately.

$$W = (W_1 - 3W_2) + \tfrac{1}{2}W_2 k^2 a^2 + ...,$$

while near the top ($k_x a \to \pm\pi$, etc.) it is

$$W = (W_1 + 3W_2) - \tfrac{1}{2}W_2 k^2 a^2 + ...,$$

where $k^2 = k_x^2 + k_y^2 + k_z^2$. This shows that the energy surfaces are spheres about the centre of the zone, or the corners of the zone respectively (compare Fig. 12.5). Note that the effective mass $m^* = h^2/a^2 W_2$, and hence is inversely proportional to the bandwidth.

12.4. For a cubic crystal the electrical conductivity is given by

$$\sigma = \frac{e^2}{12\pi^3\hbar} \int \tau v_F \, dS_F,$$

where $v_F$ is the velocity at the Fermi surface, and the integral is taken over the Fermi surface. Show that if $\tau$ is constant over this surface, and $m^* v_F = \hbar k_F$, this reduces to

$$\sigma = (e^2/3\pi^2)(\tau k_F^3/m^*),$$

and that this reduces to eqn (12.30) on writing $n = (2/8\pi^3)\Omega_F$ where $\Omega_F$ is the volume of the Fermi sphere.

12.5. Show that if $2(2l+1)$ electrons of mass $m^*$ are allowed for each point in $\mathbf{k}$-space, the density of states $g(W)\,dW$ is still given by eqn (11.11), but that the value of $(W_F)_0$ becomes

$$(W_F)_0 = \frac{\hbar^2}{2m^*}\left(\frac{3\pi^2 n}{2l+1}\right)^{\frac{2}{3}},$$

where $n$ is the number of electrons per unit volume of the metal.

12.6. Using the treatment of § 8.4, find an expression for the complex refractive index $(n - jk)$ of a metal in the region where relaxation effects in the conductivity are important, that is, where the conductivity is complex as in eqn (10.51).

If the ordinary permittivity of the metal is neglected, show that

$$(n^2 - k^2)/2nk = -\omega\tau.$$

The measurements of Beattie (1955) show that for aluminium at room temperature at wavelengths between $6\,\mu$m and $12\,\mu$m, the quantity $(n^2 - k^2)/2nk$ is roughly equal to $-11/\lambda$, where $\lambda$ is the wavelength in micrometres. Show that this gives a value of $\tau$ of about $0.6 \times 10^{-14}$ s.

12.7. The resistivity of a sample of copper at 4 K is approximately $10^{-10}\,\Omega$ m. Assuming that $m^*/m = 1.5$, and $W_F = 4.7$ eV, show that the mean free path of electrons in copper at this temperature is about $7\,\mu$m, while the classical value of the skin depth (eqn (8.31)) at a wavelength of $0.03$ m is about $0.05\,\mu$m. Show also that at this wavelength and temperature the value of $\omega\tau$ is about $\frac{1}{4}$. (Assume one electron per atom for $n$.)

# 13. Superconductivity

## 13.1. Introduction

In 1911, three years after he had first liquefied helium, Kamerlingh Onnes discovered the phenomenon of superconductivity. This quite unexpected and fascinating property of a number of metals and alloys has given rise to a vast range of experimental and theoretical work. This chapter does not set out to cover all aspects of the subject, but concentrates on the fundamental properties and their explanation in terms of the quantum-mechanical treatment of electrons in metals discussed in the two preceding chapters.

On measuring the electrical resistance of mercury at temperatures of a few degrees Kelvin, Kamerlingh Onnes found that at 4·2 K it dropped abruptly to zero, as shown in Fig. 13.1. Subsequent investigations have shown that this sudden transition to perfect electrical conductivity is a characteristic of a number of metals and alloys. The transition is very sharp, being spread over less than a thousandth of a degree in good single crystals. It occurs at a fixed temperature $T_c$ which is a characteristic of the material; values of $T_c$ range from less than 1 K to over 20 K. Below the transition temperature the electrical resistivity is so close to zero that a persistent current in a superconducting loop has run for over 2 years. The absence of any detectable decay in its size has set an upper limit of $10^{-25}$ $\Omega$ m for the resistivity (Quinn and Ittner 1962), many orders of

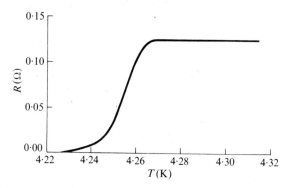

Fig. 13.1. The superconductive transition curve of mercury as measured by Kamerlingh Onnes in 1911; at 3 K the resistance was $10^{-7}$ of that at 300 K. Pure unstrained crystals give transitions spreading over less than $10^{-3}$ K.

magnitude lower than the residual resistivity at low temperatures of the purest copper ($\sim 10^{-12}\,\Omega$ m). Experimentally only an upper limit to the resistivity can be determined, but we shall assume that in the superconducting state it is zero.

A number of the elements which become superconducting are listed in Table 13.1, together with their transition temperatures $T_c$. In addition, the following elements of the transition groups are superconductors;

3d;  Ti, V;

4d;  Zr, Nb, Mo, Tc, Ru;        4f;  La (f.c.c.);

5d;  Hf, Ta, W, Re, Os, Ir;      5f;  Th, Pa, U($\alpha$).

A few other elements become superconducting under pressure. Though the superconducting elements tend to fall into groups in the periodic table, it is difficult to deduce any obvious pattern. It must depend on the crystal structure, since it is limited to the $\alpha$-forms of mercury and uranium and the face-centred cubic structure of lanthanum; the best-known indication of this is that white tin is a superconductor but grey tin is not. However, the known superconductors are not confined to any one type of crystal lattice, though their atomic volumes fall within a fairly small range. There is no correlation with the electrical conductivity immediately above the transition temperature; in fact superconductors tend to have a rather high resistivity at room temperature.

Kamerlingh Onnes realized that the property of zero electrical resistance would have an important application in the use of superconducting coils to produce very high magnetic fields, but in 1914 he discovered another important result, that superconductivity is destroyed by a magnetic field when the flux density is greater than a certain critical value $B_c$. The value of $B_c$ depends on the temperature and on the substance. For a large number of substances the variation can be expressed, with an

TABLE 13.1

*Values of the transition temperature $T_c$ and of the critical magnetic flux density $B_0$ at $T = 0$ K for a number of superconducting elements, grouped as they occur in the periodic table*

|       | Group II | | | Group III | | | Group IV | |
|       | $T_c$ (K) | $B_0$ (T) | | $T_c$ (K) | $B_0$ (T) | | $T_c$ (K) | $B_0$ (T) |
|-------|-----------|-----------|-----|-----------|-----------|-----------|-----------|-----------|
|       |           |           | Al  | 1·18      | 0·0105    |           |           |           |
| Zn    | 0·875     | 0·0053    | Ga  | 1·09      | 0·0051    |           |           |           |
| Cd    | 0·56      | 0·0030    | In  | 3·40      | 0·0293    | Sn (white) | 3·72     | 0·0309    |
| Hg($\alpha$) | 4·15 | 0·0412    | Tl  | 2·39      | 0·0171    | Pb        | 7·19      | 0·0803    |

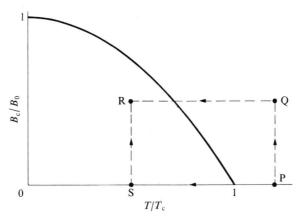

FIG. 13.2. Plot of eqn (13.1) in reduced coordinates. This equation is not exact; at $(T/T_c)^2 = \frac{1}{2}$, lead shows an upward deviation of about 2·2 per cent, and tin a downward deviation of about 3 per cent.

accuracy of a few per cent, by the relation

$$B_c = B_0\{1-(T/T_c)^2\} \tag{13.1}$$

This relation, which shows that $B_c$ falls to zero at the critical temperature, is illustrated in Fig. 13.2. The phase transition between the normal and superconducting state, which is very sharp at $B = 0$ for a pure sample, depends on the shape of the sample and its orientation when an external field is present. This is because of the presence of demagnetizing effects; the transition remains sharp provided that the magnetic flux density inside the specimen is uniform. The critical magnetic field may be produced by the superconducting current itself, so that at any temperature below $T_c$ there is an upper limit to the current which can be carried without causing a superconductor to revert to the normal state. The values of $B_0$, the critical flux density at $T = 0$, are also listed in Table 13.1.

## 13.2. The Meissner effect

Twenty-two years after the discovery of superconductivity, Meissner and Ochsenfeld (1933) found that a superconductor exhibits perfect diamagnetism as well as perfect electrical conductivity. This is illustrated in Fig. 13.3. In (a) the temperature is above $T_c$, and when a flux density **B** is applied, corresponding to moving from P to Q in Fig. 13.2, the magnetic flux penetrates the sample in the expected manner. If then (b) the sample is cooled through the transition temperature at constant field (that is, moving from Q to R in Fig. 13.2), the magnetic flux is completely expelled from the sample at the transition temperature. If, to avoid

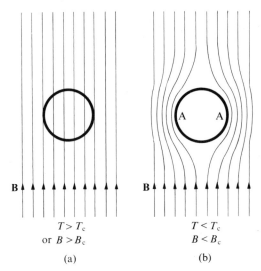

Fig. 13.3. The Meissner effect. In the normal state (a) the lines of magnetic flux penetrate the sample, but in the superconducting state (b) the flux density inside is zero.

problems with demagnetizing effects, we consider a sample in the form of a thin rod with **B** parallel to its axis, then we can write for the flux density inside the superconductor

$$\mathbf{B}_i = \mu_0\mathbf{H} + \mu_0\mathbf{M} = \mathbf{B} + \mu_0\mathbf{M},$$

and on putting this equal to zero we obtain

$$\mathbf{M}/\mathbf{B} = -1/\mu_0. \tag{13.2}$$

It must be emphasized that this is a new and fundamental property of superconductors and *not* a consequence of perfect electrical conductivity. For a conductor in which $\mathbf{E} = 0$, Maxwell's equation (8.3) curl $\mathbf{E} = -\partial\mathbf{B}/\partial t$ leads only to the result that **B** must remain fixed in time in the superconducting state. Thus on following the path PQR in Fig. 13.2, the flux in the metal in the normal state would remain trapped at the value which existed on entering the superconducting state. This means that the flux through a coil surrounding the metal would remain unaltered during the transition. In fact there is an abrupt change, consistent with the flux density in the superconducting state becoming zero. Fig. 13.3 shows that this results in a change in the flux distribution around the superconductor, and detailed measurements confirm that this is exactly that which would be expected if $\mathbf{B} = 0$ inside the superconductor.

Another way of proceeding from the point P to R in Fig. 13.2 is via the path PSR; the sample is first cooled in zero field to a point below the transition temperature, so that the flux density inside is zero. When an external field less than $B_c$ is applied, the flux density inside remains zero, and the same result is obtained whatever path is followed from P to R. Thus the Meissner effect suggests that there exists a unique equation of state for a superconductor, so that the superconducting transition can be treated using equilibrium thermodynamics. This would not be the case for a transition to a perfectly conducting state in which the resistance simply becomes zero, since the flux density would then remain fixed at the value which existed when the transition too place, and this depends on the way in which we go from P to R.

A second example of the Meissner effect is given by the hollow cylinder (or a hollow ring) of Fig. 13.4, in which the axis of the cylinder is parallel to the applied field. When the superconducting state is attained, either by reducing the temperature or the external field, the flux density **B** becomes zero within the body of the sample, but, in the space inside, **B** remains constant at the value it had when the transition took place. Diagrams (a) and (b) in Fig. 13.4 show that this is different, depending on whether the final state is reached by a path corresponding to PQR in Fig. 13.2 or by path PSR. These results were verified by Meissner and Ochsenfeld using a hollow lead cylinder.

Although a superconductor may be regarded as a perfect diamagnet, it would be wrong to assume that there is any volume distribution of magnetization inside the superconductor. The fact that **B** = 0 inside the body of the sample must be attributed to the presence of currents flowing

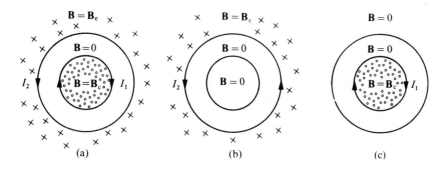

FIG. 13.4. The Meissner effect in a hollow cylinder whose axis is parallel to the magnetic flux density. (a) The cylinder is cooled into the superconducting state in a flux density $\mathbf{B} = \mathbf{B}_c$, and the flux density in the hole remains fixed at this value whatever the value of the external flux density $B_e$ becomes later. (b) The cylinder is cooled into the superconducting state in zero field; the flux density in the hole remains at **B** = 0, whatever the value of the external flux density $B_e$ becomes later. (c) The cylinder is cooled into the superconducting state in a flux density $\mathbf{B} = \mathbf{B}_c$, and the external field is subsequently reduced to zero.
The flux density in the body of the cylinder is always zero in the superconducting state.

at the surface of the sample. From the discussion of the infinite solenoid in § 4.6, we see that the situation represented in Fig. 13.4(b) could be ascribed to a current $I_2$ flowing round the outside surface of the cylinder, whose value per unit length is such that $I_2 = -B_e/\mu_0$, since this produces a field inside which just annuls the external field. Similarly, the situation in (a) can be ascribed to another current $I_1 = +B_e/\mu_0$ per unit length flowing round the inside surface.

If the hollow cylinder is cooled into the superconducting state in the presence of a flux density $\mathbf{B}_c$, and this external field is then removed, we reach the situation shown in Fig. 13.4(c). In the hollow space $\mathbf{B}$ remains equal to $\mathbf{B}_c$, while everywhere else $\mathbf{B} = 0$. The flux density $\mathbf{B}_c$ inside is just that which would be produced by the current $I_1$ per unit length, and the latter is known as the 'persistent current'. The fact that the flux density in the cylinder remains constant in the superconducting state, irrespective of changes in the external field, shows that a superconductor behaves as a perfect magnetic shield, as well as a perfect electric shield.

These surface currents are analogous to the 'magnetization currents' discussed in Chapter 4, except that they are real currents. It follows from that discussion that the discontinuity in the magnetization at the surface of an infinite cylindrical magnet may be attributed to a surface 'magnetization current', which conceptually can be an infinitely thin current sheet. With a real current this would imply an infinite current density, which intuitively seems to be unacceptable. In fact the current density decays exponentially as we go inwards and away from the surface (see § 13.4). The magnetic moment of the sample is exactly the same whether it arises from surface currents or an equivalent bulk magnetization; this enables us to use eqn (13.2) in the next section to calculate the magnetic energy.

### 13.3. Thermodynamics of the superconducting state

If the properties of a superconductor correspond to a unique equation of state, the normal–superconducting transition can be discussed using equilibrium thermodynamics. Since it takes place at constant temperature and constant pressure, the equilibrium condition is that the Gibbs function $G$ must be equal in the two phases at the transition point.

In order to find the effect on $G$ of applying a magnetic field to the superconductor, we use the relation

$$dG = -S\, dT - M\, dB, \qquad (13.3)$$

where $S$ is the entropy and $B$ is the flux density outside the sample. If $G_s$ is the thermodynamic potential of the superconductor at $B = 0$, and $G_s(B_c)$ the value when $B$ is increased to the transition field at constant temperature, then

$$G_s(B_c) = G_s - V \int_0^{B_c} M \, dB \qquad (13.4)$$

for a sample volume $V$. We consider again a long thin sample with its axis parallel to **B**, so that demagnetizing effects can be neglected. From eqn (13.2) we have

$$G_s(B_c) = G_s + VB_c^2/2\mu_0 \qquad (13.5)$$

and this must be equal to the thermodynamic potential of the normal state at the transition field. Hence

$$G_n = G_s + VB_c^2/2\mu_0. \qquad (13.6)$$

From this relation we can calculate the entropy difference between the normal and superconducting states. At constant $B$,

$$S_n - S_s = -\frac{d}{dT}(G_n - G_s) = -\frac{V}{\mu_0} B_c \left(\frac{dB_c}{dT}\right). \qquad (13.7)$$

From eqn (13.1) it is clear that $dB_c/dT$ is always negative, showing (see Fig. 13.5) that the entropy is lower in the superconducting state, that is, it is the state of greater order. Because of the entropy difference the transition occurs with a latent heat

$$L = T(S_n - S_s), \qquad (13.8)$$

and this is the heat that must be supplied to pass from the superconducting to the normal state at temperature $T$ and flux density $B_c$. The presence of a latent heat shows that we have a 'first-order' transition

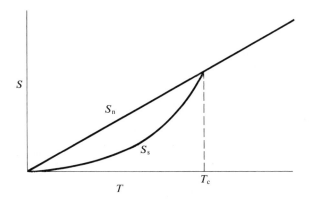

FIG. 13.5. The electronic entropy in the normal and superconducting states. At $T = T_c$ the entropy is $\gamma T_c$.

except at $T = T_c$, where $B_c = 0$. In this case there is only a jump in the heat capacity at the transition between the two phases, given by

$$C_n - C_s = T\frac{d}{dT}(S_n - S_s) = -\frac{VT}{\mu_0}\left\{B_c\frac{d^2B_c}{dT^2} + \left(\frac{dB_c}{dT}\right)^2\right\}. \qquad (13.9)$$

Such a jump in the heat capacity is characteristic of a transition of the second order. At $T = T_c$, where $B_c = 0$,

$$C_n - C_s = -\left(\frac{VT_c}{\mu_0}\right)\left(\frac{dB_c}{dT}\right)^2, \qquad (13.10)$$

showing that the heat capacity is greater in the superconducting state. However, $d^2B_c/dT^2$ is negative, and eqn (13.9) changes sign at lower temperatures (see Problem 13.1); the heat capacity becomes lower in the superconducting phase, the difference of course vanishing again at $T = 0$.

The heat capacity of superconducting gallium (in zero magnetic field)

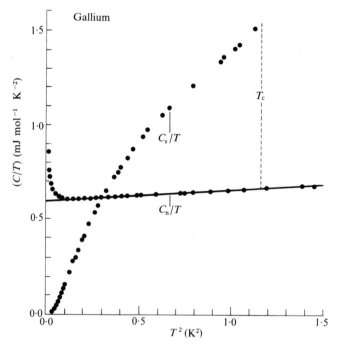

FIG. 13.6. Plot of $C_s/T$ and $C_n/T$ against $T^2$ for gallium. Apart from a small nuclear hyperfine contribution near $T = 0$, the normal heat capacity is well represented by the relation (in the units on the diagram)

$$C_n/T = 0.596 + 0.0568T^2.$$

At the lowest temperatures $C_s$ falls below $C_n$; the reverse is true as $T \to T_c$. The measurement of $C_n$ is made in an applied field of $0.02$ T. (After Phillips 1964.)

and of gallium in a magnetic field large enough to restore the normal state are shown in Fig. 13.6. The quantity plotted is $C/T$ against $T^2$ (cf. Fig. 11.10), and a straight line is obtained for $C_n/T$ except at the lowest temperatures, where an additional contribution from an electrical quadrupole splitting of the nuclear levels becomes significant. On the other hand, $C_s/T$ varies rapidly with temperature and lies well above $C_n/T$ at temperatures near $T_c$. Measurements on a range of substances agree well with the assumption that the transition to the superconducting state is reversible.

When $T$ is much smaller than $T_c$ the value of $C_s - C_n$ becomes negative, and the heat capacity in the superconducting state can be represented by a formula of the type

$$\frac{C_s}{\gamma T_c} \propto \exp\left(-\frac{\Delta}{kT}\right), \tag{13.11}$$

a result that is illustrated for Ga in Fig. 13.7. For this metal $\Delta = 1\cdot39kT_c$;

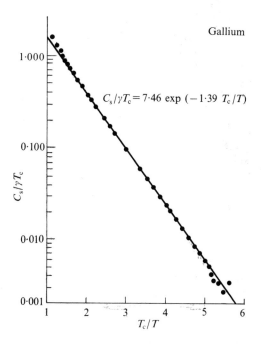

FIG. 13.7. Plot of $C_s/\gamma T_c$ (on logarithmic scale) against $T_c/T$ for gallium. The straight line corresponds to the exponential relation

$$C_s/\gamma T_c = 7\cdot46 \exp(-1\cdot39T_c/T).$$

The two-fluid theory predicts that $C_s$ should vary as $T^3$ when $(T/T_c) \ll 1$, but the exponential dependence suggests the presence of an energy gap. (After Phillips 1964).

similar results have been obtained for a number of superconducting elements, with a ratio of $\Delta$ to $kT_c$ of this order.

*The 'two-fluid' model*

A useful phenomenological model (or working model) of a superconductor was put forward by Gorter and Casimir (1934). It assumes that in the superconducting state there exist two types of charge carrier:

(1) 'superconducting' carriers, condensed into an ordered state whose entropy is zero—they are not scattered by collisions, and give rise to perfect electrical conductivity;

(2) 'normal' carriers which are scattered by collisions—they are 'shorted out' by the 'superconducting' carriers and do not affect the conductivity.

The proportion of superconducting and normal carriers is a function of temperature. At $T = 0$ all carriers are in the superconducting state, but as the temperature rises, the proportion falls, reaching zero at the transition temperature. The superconducting state is regarded as the 'ground state' of the system, and a finite amount of energy is required to excite a carrier into the normal state; the heat capacity represents the energy required to excite an increasing number as the temperature rises, and the exponential dependence in eqn (13.11) suggests that the energy 'gap' is somewhat greater than $kT_c$. The energy gap is not constant, but decreases as the temperature rises, becoming zero at $T_c$, above which the properties are those of a normal metal.

The important property of the condensed phase is that the carriers remain essentially at 0 K, with zero entropy; they carry the electrical current but their energy does not change with temperature, so that the Thomson heat is zero. They make no contribution to the transport of heat, so that in a superconductor heat is carried only by the normal carriers, whose number decreases to zero as $T \rightarrow 0$, and by the lattice. The phonons are not scattered by the condensed phase carriers, so that the lattice contribution to the thermal conductivity is increased. At low temperatures where no normal carriers are present to scatter the phonons, the scattering rate is determined solely by imperfections and crystal boundaries and is independent of temperature. The lattice conductivity is then simply proportional to the lattice specific heat that is, $K_s$ for the superconductor varies as $T^3$, while in the normal phase $K_n$ varies as $T$, following the electronic specific heat—see eqn (11.44). Hence $K_s/K_n$ varies as $T^2$, and may be as small as $10^{-5}$ if the rate of phonon scattering by imperfections is high. A superconducting wire may therefore be used as a 'heat switch' (Mendelssohn 1955), providing thermal contact only when the normal phase is restored by a local magnetic field.

## 13.4. The London equations

In the discussion of the Meissner effect in § 13.2, the fact that $\mathbf{B} = 0$ inside the sample was attributed to the presence of currents flowing at the surface. We now investigate how far such currents can be discussed in terms of the classical electromagnetic equations. In the absence of any dissipative forces, the current density $\mathbf{J}$ should obey eqn (3.9a), that is,

$$\dot{\mathbf{J}} = (nq^2/m)\mathbf{E}. \tag{13.12}$$

If $\mathbf{E}$ arises from a changing magnetic flux and grad $V = 0$, then from eqn (5.6) we can write

$$\dot{\mathbf{J}} = -(nq^2/m)\dot{\mathbf{A}}. \tag{13.13}$$

Provided that div $\mathbf{A} = 0$, we also have from eqn (4.54)

$$\nabla^2\mathbf{A} + \mu_0\mathbf{J} = 0, \tag{13.14}$$

so that

$$\nabla^2\dot{\mathbf{A}} - (\mu_0 nq^2/m)\dot{\mathbf{A}} = 0. \tag{13.15}$$

The solution of this for a semi-infinite solid bounded by the plane $z = 0$ is

$$\dot{\mathbf{A}} = \dot{\mathbf{A}}_0 \exp(-z/\lambda_L), \tag{13.16}$$

where

$$\lambda_L^{-2} = \mu_0 nq^2/m. \tag{13.17}$$

This result shows that any time variation of the field must fall off exponentially as we move into the superconductor, in a length of order $\lambda_L$, which for normal metals is of order $0\cdot01$–$0\cdot1$ $\mu$m, or about a hundred lattice spacings. This result is similar to the skin effect (§ 8.5), but at zero frequency the skin depth given by eqn (8.31) is indeterminate in a perfect conductor. All that the equations above tell us is that $\mathbf{B} = \operatorname{curl}\mathbf{A}$ must be fixed in time after the material has entered the superconducting state at the value it had just before entering this state. This expectation is inconsistent with the Meissner effect.

The brothers F. and H. London (1935) took the bold step of proposing that instead of eqn (13.13), involving the time differentials, one should write

$$(m/nq^2)\mathbf{J} + \mathbf{A} = \mu_0\lambda_L^2\mathbf{J} + \mathbf{A} = 0. \tag{13.18}$$

The condition div $\mathbf{A} = 0$, known as the London gauge, is here equivalent to div $\mathbf{J} = 0$, and ensures continuity in the current flow. On combining eqn (13.18) with eqn (13.14) we have

$$\nabla^2\mathbf{A} - \lambda_L^{-2}\mathbf{A} = 0, \tag{13.19}$$

whose solution is

$$\mathbf{A} = \mathbf{A}_0 \exp(-z/\lambda_L). \tag{13.20}$$

TABLE 13.2

*Some important length parameters in superconductivity (a few typical values are shown in Table 13.3)*

---

1. $\lambda$ is the penetration depth; see eqn (13.28).
2. $\lambda_L$ is the penetration depth on the London theory; see eqn (13.23).
3. $\lambda_L(0)$ is the value of $\lambda_L$ as $T \to 0$, and $(n_s/n) \to 1$; see eqn (13.17).
4. $\xi_0$ is the Pippard coherence length, and represents the minimum size of the wave packets of superconducting charge carriers in bulk material. It is the range within which electrons can interact to form Cooper pairs (see § 13.6). In a pure superconductor such as aluminium or tin, $\xi_0 \gg \lambda_L(0)$.
5. $L$ is the mean free path for scattering of electrons in the normal state.
6. $\xi$ is the Pippard coherence length in the presence of scattering; see eqn (13.25).

---

This shows that **A**, and hence also **B** = curl **A** and **J**, fall off exponentially inside a superconductor. Thus **B** = 0 only at depths large compared with $\lambda_L$ from the surface. The quantity $\lambda_L$ is known as the 'penetration depth'. It is the first of several important length parameters which we shall encounter; these which are listed in Table 13.2, and some values are given in Table 13.3.

The result obtained above shows that the London equation (13.18) corresponds to the property of perfect diamagnetism in the bulk of the sample. A second London equation corresponds to perfect electrical conductivity, and is given by eqn (13.12). Through eqn (13.13) it can be seen that this follows from differentiation of the first London equation (13.18) provided that grad $V = 0$.

For a free particle in a region of changing magnetic flux we have from eqn (5.7)

$$\mathbf{p}_0 = m\mathbf{v} + q\mathbf{A}, \tag{13.21}$$

where $m\mathbf{v}$ is the momentum when the magnetic vector potential has the value **A** and $\mathbf{p}_0$ is the momentum when **A** = 0. If **v** is the mean velocity of

TABLE 13.3

*Values of the London penetration depth $\lambda_L(0)$, the measured penetration depth $\lambda$, and the Pippard coherence length $\xi_0$*

|           | $\lambda_L(0)$ (nm) | $\lambda$ (nm) | $\xi_0$ (nm) |
|-----------|---------------------|----------------|--------------|
| Aluminium | 16                  | 50             | 1600         |
| Tin       | 34                  | 51             | 230          |
| Lead      | 37                  | 39             | 83           |

the carriers, then the current density $\mathbf{J} = nq\mathbf{v}$, and from eqn (13.18) we have

$$q(\mathbf{A} + \mu_0\lambda_L^2\mathbf{J}) = q\mathbf{A} + (m/nq)\mathbf{J} = q\mathbf{A} + m\mathbf{v} = 0. \qquad (13.22)$$

This result suggests that $\mathbf{p}_0$ is zero for the superconducting carriers, so that the condensation is into a state of zero momentum. It follows from the uncertainty principle that the wavefunctions of the superconducting carriers must be essentially unlimited in their spatial extension.

## 13.5. The Pippard coherence length

The value of $\lambda_L$ defined by eqn (13.17) involves the total number of electrons per unit volume (the total 'number density'), but in fact the current is all carried by the superconducting particles, whose density we denote by $n_s$. At $T = 0$, $n_s = n$ and eqn (13.17) thus gives only the limiting value $\lambda_L(0)$. At higher temperatures $n_s/n$ is less than unity on the two-fluid model, and we expect that the penetration depth, given by

$$\lambda_L = (m/\mu_0 n_s q^2)^{\frac{1}{2}}, \qquad (13.23)$$

should vary with temperature. Two types of experiments have been made to determine the penetration depth and its temperature dependence: (1) measurement of the reduction in the magnetic moments of thin films and colloidal particles; (2) measurement of the resonant frequency of a microwave cavity. The latter depends slightly on the depth to which radio-frequency currents penetrate and, although the frequency change is small, it can be measured accurately because of the very high quality factor $Q_F$ of the cavity; it has the advantage of giving results for the bulk metal. It was found that the effective depth was considerably greater than that calculated from eqn (13.23).

This discrepancy was resolved by Pippard (1953), who pointed out that the London equations are 'local equations' in which the current density $\mathbf{J}$ at a point is assumed to be determined entirely by the value of $\mathbf{A}$ at that point. We have already seen (§ 12.10) that this is not true in a normal metal when the mean free path $L$ of the electrons becomes greater than the skin depth; the value of $\mathbf{J}$ then depends on how $\mathbf{E}$ varies over a region with linear dimensions of the order of $L$. Pippard introduced the idea of a 'coherence length' $\xi_0$ for the wavefunctions of the superconducting carriers, such that a perturbation at one point influences every carrier within a distance $\xi_0$ of that point. This means we have a 'non-local' situation in which $\mathbf{J}$ depends on the values of $\mathbf{A}$ over a region of linear dimensions $\xi_0$.

To obtain a value of $\xi_0$, Pippard argued that only electrons with a range of energy of order $kT_c$ can be responsible for a phenomenon occurring below $T_c$ in temperature. Only electrons with the Fermi velocity $v_F$ are involved, so that the corresponding range of values of momentum $\delta p$ is of order $kT_c/v_F$. The uncertainty principle then gives a corresponding length

$$\delta x \geq \hbar/\delta p = \hbar v_F/kT_c,$$

which for aluminium, tin, and lead is of order 0·1 to 1 $\mu$m. If we write

$$\xi_0 = a(\hbar v_F/kT_c), \tag{13.24}$$

where $a$ is a numerical constant to be determined, it is found that for a number of pure metals, $a$ is about 0·15. Experiments on alloys show, however, that the coherence length $\xi$ depends on the rate of scattering of the carriers, being given by the relation

$$\frac{1}{\xi} = \frac{1}{\xi_0} + \frac{1}{L}, \tag{13.25}$$

where $L$ is the mean free path in the normal phase. In the absence of scattering, $L$ tends to infinity and $\xi$ becomes equal to $\xi_0$, the value for the pure metal, but for small values of $L$ the coherence length $\xi$ is simply determined by $L$.

To allow for the effect of the mean free path, Pippard suggested that instead of eqn (13.18) one should write

$$\mu_0\lambda_L^2\mathbf{J} + (\xi/\xi_0)\mathbf{A} = 0, \tag{13.26}$$

or

$$\mu_0\lambda^2\mathbf{J} + \mathbf{A} = 0, \tag{13.27}$$

where

$$\lambda/\lambda_L = (\xi_0/\xi)^{\frac{1}{2}}. \tag{13.28}$$

Combination of (13.27) with (13.14) gives, in place of (13.19),

$$\nabla^2\mathbf{A} - \lambda^{-2}\mathbf{A} = 0, \tag{13.29}$$

whose solution is

$$\mathbf{A} = \mathbf{A}_0 \exp(-z/\lambda). \tag{13.30}$$

The penetration depth is $\lambda$, given by eqn (13.28), but eqn (13.26) is still a 'local equation' which is valid when $\mathbf{A}$ changes only slowly over the distance $\xi$, that is, when $\xi \ll \lambda$.

In the opposite limit where $\lambda \ll \xi_0$ we assume instead that the appropriate value of $\mathbf{A}$ is that averaged over a volume limited in one dimension by $\lambda$, that is, a volume $\xi_0^2\lambda$ instead of $\xi_0^3$. $\mathbf{A}$ is then effectively reduced by a factor $(\lambda/\xi_0)$, and instead of eqn (13.26) we have

$$\mu_0\lambda_L^2\mathbf{J} + (\lambda/\xi_0)\mathbf{A} = 0. \tag{13.31}$$

It is easy to show that this leads to the result

$$\lambda/\lambda_L = (\xi_0/\lambda_L)^{\frac{1}{3}}. \tag{13.32}$$

These equations can be used to deduce a value of $\xi_0$ from the measured penetration depth λ, and give a value of about 0·15 for the constant $a$ in eqn (13.24).

## 13.6. The Bardeen–Cooper–Schrieffer (BCS) theory

Any theory of superconductivity must account for the two fundamental properties of an ideal superconductor: perfect diamagnetism and perfect electrical conductivity. In addition there is the striking similarity between the properties of different metals; the relation between critical field and temperature (see Fig. 13.2) is represented by eqn (13.1) with departures of not more than a few per cent. The difficulties are illustrated by the fact that the energy $kT_c$ is less than a millielectronvolt, and the magnetic energy $\frac{1}{2}B_c^2/\mu_0$ is $\sim 10^{-8}$ eV per atom, while the Fermi energy is several electronvolts.

The direct interactions between electrons arising from their electric charge are repulsive, even after allowing for the screening effects of other electrons. The interactions between the electrons and the fixed positive charges of the lattice give rise to the band structure discussed in Chapter 12. An important clue was the discovery in 1950 of the 'isotope effect'; using different isotopes of mercury it was found by Maxwell (1950) and by Reynolds, Serin, Wright, and Nesbitt (1950) that the transition temperature $T_c$ depended on the isotopic mass $M$, approximately as $M^{-\frac{1}{2}}$. Similar effects have been shown to exist for many other superconducting elements, with $T_c \propto M^a$. Though the exponent $a$ is not always $-\frac{1}{2}$, it is found that for the isotopes of a given element $B_0$ is accurately proportional to $T_c$, and that $B_c/B_0$ can be represented by exactly the same function of $(T/T_c)^2$.

The lattice structure of an element is independent of its isotopic constitution; so are its chemical properties, and static electronic properties such as the Fermi energy and the electronic heat capacity. The dynamical properties are different; the vibrational energy is $M\omega^2\delta^2$, and for a given quantum of energy the deformation $\delta$ varies as $M^{-\frac{1}{2}}$. Thus the isotope effect suggests that the interaction responsible for superconductivity involves deformation of the lattice. Such a deformation could result from the attraction between an electron and the nearby positive ions of the lattice, a possibility that was investigated by Fröhlich in 1950 independently of the discovery of the isotope effect, but it remained for the theory of Bardeen, Cooper, and Schrieffer (BCS) (1957) to show how the local electrical polarization of the lattice, associated with the deformation, would interact with another electron to produce an indirect force between the two electrons which is attractive in its nature.

It was first shown by Cooper (1956) that any force of attraction between a pair of electrons, no matter how weak, would produce an instability in the Fermi sea, in which the formation of some 'bound pairs' would lead to the formation of further bound pairs. This is a co-operative effect which results in the presence of a ground state that is separated by an energy gap from the allowed excited states of the system. The pairs of electrons are called 'Cooper pairs' and they are the key to the theory of superconductivity. It turns out that the range of the interaction binding a pair together is of the order of the Pippard coherence length $\xi_0$, about 1 $\mu$m. In an ordinary metal the number of conduction electrons per cubic metre is of order $10^{28}$, and the fraction of these which lie in the energy range $kT_c$ is about $kT_c/W_F$, or about $10^{-4}$. It follows that a volume of order $\xi_0^3$ contains about $10^6$ Cooper pairs. They must therefore have some special property to prevent collisions between pairs destroying the binding process.

It is found that the lowest energy for such a Cooper pair is achieved by binding together an electron with wave vector $+\mathbf{k}$ and an electron with $-\mathbf{k}$, and with their spins antiparallel, so that the net spin is zero, as in an atomic state $S = 0$. The net linear momentum of such a pair is zero, as suggested in the original London theory, but the pair has, of course, mass $2m$ (or $2m^*$) and charge $2e$. The same is true for all pairs; each has a total wave vector $+\mathbf{k}-\mathbf{k} = 0$, corresponding to a 'condensation' in $\mathbf{k}$-space (or momentum space). The wavefunction is coherent over macroscopic distances.

When a current of density $\mathbf{J}$ is flowing, this corresponds to a displacement of the Fermi sphere (see Fig. 11.11) by an amount $\delta\mathbf{k}$, such that $\mathbf{J} = -n(e/m)\hbar\,\delta\mathbf{k}$. A Cooper pair is then formed from electrons with wave vector $+\mathbf{k}+\delta\mathbf{k}$ and $-\mathbf{k}+\delta\mathbf{k}$, so that each pair has a net wave vector $\delta\mathbf{k}$. All pairs move together, with velocity corresponding to eqn (13.22). To account for a given 'supercurrent' density of this type, it is clear that $\hbar\,\delta k/m$ must be comparable with the *drift* velocity of the electrons in a normal current, so that $\delta\mathbf{k}$ will be very small compared with $\mathbf{k}_F$. The very long wavelength $2\pi/\delta k$ associated with each pair means that the scattering processes which are effective in scattering normal electrons with wave vector $\sim k_F$ are ineffective in scattering Cooper pairs, thus explaining the presence of persistent currents.

Since the electron energy is proportional to the square of the wave vector, a pair with $(+\mathbf{k}+\delta\mathbf{k}, -\mathbf{k}+\delta\mathbf{k})$ has an energy which is greater by $\delta W = \hbar^2(\delta k)^2/m$ than that of a pair $(+\mathbf{k}, -\mathbf{k})$. If this energy exceeds the binding energy of the pair, the pair will 'dissociate' and the property of superfluidity will disappear. Since $\delta\mathbf{k}$ is proportional to $\mathbf{J}$, this gives an upper limit to the current density.

The BCS theory leads to an expression in which the energy gap $\Delta_0$ is proportional to the Debye frequency $\omega_D$, which for a given force constant (that is, a given lattice) varies as $M^{-\frac{1}{2}}$, where $M$ is the ionic mass. The gap energy at $T = 0$ is related to the condensation energy at $T = 0$ by the formula

$$\frac{V}{2\mu_0} B_0^2 = \frac{g(W) \Delta_0^2}{4},$$                    (13.33)

and to the transition temperature $T_c$ by

$$2\Delta_0 = 3\cdot52kT_c.$$                    (13.34)

The density of states $g(W)$ is related to the electronic heat capacity constant $\gamma$ for the normal state through eqns (12.27)–(12.28) which with $R = nk$ gives, per mole,

$$\gamma = \tfrac{1}{3}\pi^2 k^2 g(W).$$                    (13.35)

Then elimination of $\Delta_0$ between (13.33) and (13.34) leads to the result

$$\frac{V}{2\mu_0} B_0^2 = 0\cdot235\gamma T_c^2,$$                    (13.36)

where the volume $V$ also refers (like $\gamma$) to 1 mol.

As the temperature rises, particles are excited across the gap, and the number of Cooper pairs decreases. The size of the gap $\Delta$ is correlated with the number of pairs, so that $\Delta$ also decreases, becoming zero at the transition temperature $T_c$, where the number of Cooper pairs vanishes. The BCS theory gives an explicit prediction for the way in which $\Delta$ varies with temperature, and since the condensation energy is simply proportional to $\Delta^2$ it follows from eqn (13.33) that $B_c/B_0 = \Delta/\Delta_0$. Fig. 13.8 shows how measurements of $B_c$ give a good fit to the theoretical prediction for pure samples of tin, indium, and mercury. In general the numerical value of the ratio $2\Delta_0/kT_c$, measured by various methods, lies between 2·5 and 4·5 (see Table 13.4), and is fairly close to 3·5 for a number of elements.

The BCS theory also leads to a relation between the Pippard coherence length and the energy gap

$$\xi_0 = \hbar v_F/\pi \Delta_0,$$                    (13.37)

TABLE 13.4

*Numerical parameters for aluminium, tin, and lead compared with the predictions of the BCS theory*

| | $2\Delta_0/kT_c$ From tunnelling measurements | $2\Delta_0/kT_c$ From electromagnetic absorption | $C_s/C_n$ at $T = T_c$ | $VB_0^2/2\mu_0\gamma T_c^2$ |
|---|---|---|---|---|
| Aluminium | 2·5–4·2 | 3·37 | 2·3–2·6 | 0·233 |
| Tin | 2·8–4·1 | 3·5 | 2·6 | 0·245 |
| Lead | 4·3–4·4 | 4·0–4·4 | 3·7 | 0·30 |
| BCS theory | { 3·52 eqn (13.34) | | 2·43 | 0·235 } eqn (13.36) |

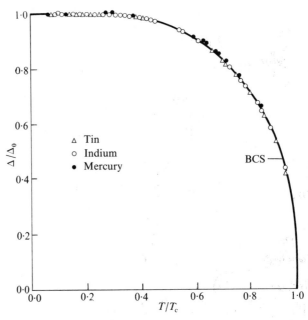

FIG. 13.8. Experimental values of $\Delta/\Delta_0$ as a function of $T/T_c$ for samples of tin, indium, and mercury, showing the good agreement with the BCS theory. The results are derived from measurements of $B_c$ as a function of temperature. (Finnemore and Mapother 1965.)

which may be combined with eqn (13.34) to give

$$\xi_0 = 2\hbar v_F/3.52\pi k T_c = 0.18\hbar v_F/k T_c, \tag{13.38}$$

which is in good agreement with eqn (13.24). A further prediction of the theory is that at $T = T_c$, $C_s/C_n = 2.43$ (see Table 13.4).

## 13.7. The energy gap

The energy gap in a superconductor is quite different in its origin from that in a normal metal. The band theory discussed in Chapter 12 shows that energy bands are a consequence of the static lattice structure. Discontinuities in the allowed energy occur at values of **k** determined by the spacing between the ions, that is, at the edges of the Brillouin zones. In a superconductor the energy gap is far smaller, and results from an attractive force between the electrons in which the lattice plays only an indirect role. It is present only at temperatures below $T_c$, and varies with the temperature when $T \leqslant T_c$. The gap occurs at the Fermi surface, as shown in Fig. 13.9, where the density of states is contrasted with that in a normal metal. On either side of the Fermi level there is an energy gap,

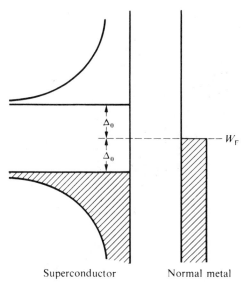

FIG. 13.9. The density of states near the Fermi energy $W_F$ in a superconductor, showing the energy gap $2\Delta_0$ at $T = 0$, and in a normal metal.

extending a distance $\Delta$ below and an equal distance $\Delta$ above $W_F$. At $T = 0$ all electrons are accommodated in states below the energy gap, and a minimum energy $2\Delta_0$ must be supplied to produce an excitation across the gap. The number of such excitations varies as $\exp[-(\text{energy gap})/2kT]$, just as in a semiconductor the number of electrons excited across the gap $W_g$ varies as $\exp(-W_g/2kT)$, see eqn (17.14). Thus, in a superconductor at the lowest temperatures, the heat capacity should have the form

$$C_s/\gamma T_c \propto \exp(-\Delta_0/kT).$$

At higher temperatures the behaviour of $C_s$ departs from a simple exponential because the gap energy itself varies with temperature, $\Delta$ being less than $\Delta_0$ and falling to zero at $T = T_c$.

The presence of an energy gap has been verified in a number of ways, of which we discuss only the use of electromagnetic radiation. It has been known for a long time that at optical frequencies superconductors exhibit no abnormal properties; the optical constants correspond to those for a normal metal with virtually no change at the transition temperature. This suggests that the quanta of optical radiation are large enough to excite carriers from the condensed phase into unoccupied electron states, while radio-frequency quanta are too small to reach such states. This is consistent with the idea of an energy gap in the electron states, which must obviously be expected to be of order $kT_c$. The corresponding frequency range is $10^5$–$10^6$ MHz, and much experimental work has been carried out in the microwave and far infrared regions.

The experimental results are expressed in terms of the a.c. conductivity, and normally involve comparisons between the superconducting state and the normal state, produced by imposing a magnetic field greater than the critical value, over a range of temperature. In the normal state the conductivity $\sigma_n$ is a real quantity and substantially independent of frequency. In the superconducting state it is necessary to use a complex conductivity

$$\sigma_s = \sigma_1 - j\sigma_2,$$

where both $\sigma_1$ and $\sigma_2$ vary with frequency. For an oscillatory current at frequency $\omega/2\pi$ differentiation of eqn (13.27) gives

$$\mathbf{E} = -\dot{\mathbf{A}} = \mu_0\lambda^2\dot{\mathbf{J}} = j\omega\mu_0\lambda^2\mathbf{J}, \tag{13.39}$$

from which $\sigma_1 = 0$ and $\sigma_2 = (\omega\mu_0\lambda^2)^{-1}$. However, at frequencies such that $\omega \geqslant \omega_g$, where $\hbar\omega_g$ is the energy gap, carriers can be excited across the energy gap into empty states, and energy is absorbed; $\sigma_1$ then becomes finite, rising from zero at $\omega/\omega_g = 1$ to $\sigma_n$ when $\omega/\omega_g$ is $\gg 1$. The behaviour of the ratios $\sigma_1/\sigma_n$ and $\sigma_2/\sigma_n$ is shown in Fig. 13.10.

In a superconducting film whose thickness is smaller than the penetration depth sufficient energy is transmitted to measure the transmission coefficient, which can be related to the conductivity by an extension of the theory of Chapter 8. This method has been extensively explored by

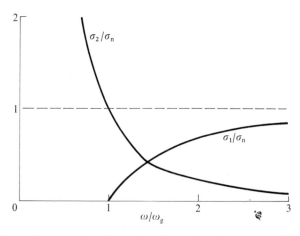

FIG. 13.10. Plot of $\sigma_1/\sigma_n$ and $\sigma_2/\sigma_n$ against the reduced frequency $\omega/\omega_g$, where $\hbar\omega_g =$ superconducting energy gap. From eqn (13.40), $\sigma_2/\sigma_n$ should vary as $\omega^{-1}$ when $(\omega/\omega_g) \ll 1$, while $\sigma_1$ becomes finite only when $\omega/\omega_g \geqslant 1$. Conductivity in superconducting state $= \sigma_1 - j\sigma_2$; conductivity in normal state $= \sigma_n$.

Tinkham; for example, the transmission through films of lead and tin, some 2 nm thick, was studied by Glover and Tinkham (1957) over the frequency range $\omega/2\pi = 0\cdot3$–40 $(kT_c/h)$. This and later work has shown that the results for $\sigma_2/\sigma_n$ can be fitted to a universal relation

$$\sigma_2/\sigma_n = a^{-1}(kT_c/\hbar\omega), \tag{13.40}$$

with values of $a$ of the order of $0\cdot2$–$0\cdot3$. This result provides a nice confirmation of the Pippard coherence theory. In such a thin film the carrier mean free path is limited by scattering at the surfaces, so that $L$ and $\xi$ are equal (see eqn (13.25)) and $\ll \lambda$. Then, using eqn (13.28) with $\xi = L$, we have

$$\sigma_2 = (\omega\mu_0\lambda^2)^{-1} = (nq^2/m)(L/\omega\xi_0),$$

while

$$\sigma_n = (nq^2/m)\tau = (nq^2/m)(L/v_F).$$

Hence

$$\sigma_2/\sigma_n = v_F/\omega\xi_0 = a^{-1}(kT_c/\hbar\omega),$$

where the last result is derived through the use of eqn (13.24). From eqn (13.38) we see that the BCS theory predicts that $a = 0\cdot18$, in reasonable agreement with the range of experimental values given above.

When $T \ll T_c$, $\sigma_1$ in the superconducting state should be close to zero; it starts to rise only when the frequency is large enough to cause excitation across the gap. The transmission through a thin film is a function of both $\sigma_1$ and $\sigma_2$; it has a maximum in the region just above $\omega/\omega_g = 1$, where both $\sigma_1/\sigma_n$ and $\sigma_2/\sigma_n$ are less than unity, and the experimental results lead to values of the energy gap in good agreement with the BCS theory.

### Tunnelling

If only a very thin insulating film (such as oxide) separates the metal and the superconductor in Fig. 13.9, electrons can tunnel from one to the other, provided that on the far side there exists an empty allowed state of the same energy. By applying a voltage which raises the potential energy in the metal, the top of the Fermi distribution can be raised to the top of the energy gap in the superconductor, and electrons can then tunnel from the metal into the superconductor. A voltage of opposite sign depresses the Fermi distribution, and when the top falls below the bottom of the energy gap, electrons can tunnel from the superconductor into the metal. The current–voltage characteristics obtained in this way give an accurate measure of the energy gap (see Table 13.4).

### Acoustic attenuation

High-frequency sound waves are low-frequency lattice vibrations which are scattered by normal electrons but not by Cooper pairs, so that their

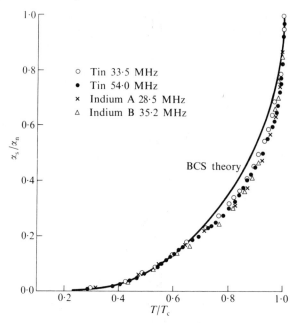

FIG. 13.11. Attenuation of 28–54 MHz sound waves in the superconducting state, relative to that in the normal state, as a function of $T/T_c$ for samples of tin and indium. The continuous line shows the prediction of the BCS theory. (Morse and Bohm 1957.)

attenuation is a measure of the fraction of normal electrons. The ratio of the attenuation in the superconducting state to that in the normal state is shown as a function of temperature in Fig. 13.11.

### 13.8. Type I and Type II superconductors

So far we have considered only the behaviour of superconducting materials in which the magnetic flux density in the bulk of the sample is zero up to the transition flux density $B_c$. The relation between the 'magnetization' and the applied field, given by eqn (13.2), follows the pattern shown in Fig. 13.12. Such materials are known as Type I superconductors, to distinguish them from a group of substances, mostly alloys, whose behaviour is illustrated in Fig. 13.13. As the applied field increases, all flux is excluded up to a critical flux density $B_{c1}$; above this field there is some flux penetration, but the substance remains superconducting until a second critical flux density $B_{c2}$ is reached. Above this field there is complete penetration and the substance behaves as a normal metal. The values of $B_{c1}$ and $B_{c2}$ lie respectively below and above the

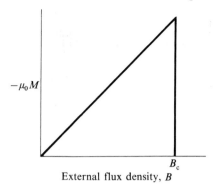

FIG. 13.12. Plot of the 'magnetization' against external flux density $B$ for a Type I superconductor.

'thermodynamic' $B_c$ which appears in eqn (13.6). The transition temperatures of a number of 'Type II superconductors', as these materials are
called, are given in Table 13.5. A number of the alloys are of considerable technical importance because they have high transition temperatures
and high critical fields; the value of $B_{c2}$ is over 20 T for niobium-tin
(Nb$_3$Sn). They are used in superconducting magnets, in which large
magnetic fields can be maintained at relatively low cost (see § 6.5).

The basic difference between Type I and Type II superconductors
arises from the relative sizes of the London penetration depth $\lambda$ and the
coherence length $\xi$. In a Type I material $\lambda$ is small compared with $\xi$, while
in a Type II material the reverse is true. The significance of this appears

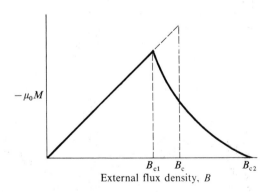

FIG. 13.13. Plot of the 'magnetization' $M$ against external flux density $B$ for a Type II
superconductor, showing how $M$ starts to decrease above $B_{c1}$, the critical field for flux
penetration. $M$ falls to zero only at a second critical field $B_{c2}$. The area under the magnetization curve $\int M \, dB$, gives the magnetic energy $B_c^2/2\mu_0$.

TABLE 13.5

*Transition temperatures $T_c$ and upper critical flux density $B_{c2}$ for some Type II superconductors*

|        | $T_c$ (K) | $B_{c2}$ (T) |
|--------|-----------|--------------|
| V      | 5·4       | 0·14         |
| Nb     | 9·5       | 0·19         |
| NbTi   | 16·0      |              |
| Nb₃Sn  | 18·1      | 22·5         |
| V₃Ga   | 14·0      | 19·5         |
| V₃Si   | 16·8      | 22·5         |

when we consider the energy involved in forming a boundary between normal and superconducting regions in a metal. There are two contributions to the energy:

(1) the magnetic energy, which rises from zero in the normal metal to a density of $B_c^2/2\mu_0$ in the superconducting region; the change occurs over a region determined by the flux penetration, whose thickness is of order $\lambda$;

(2) the electron energy, which is lower in the superconducting phase because of the negative condensation energy associated with the formation of Cooper pairs—in this case the change occurs over a region whose thickness is of order $\xi$.

The behaviour of the total energy density for a Type I material is illustrated in Fig. 13.14. The sum of the magnetic energy and the electron energy densities is greater in the boundary region, so that it is energetically unfavourable for a 'wall' to be present. The bulk of the material is therefore superconducting, the transition region being confined to the surface of the sample.

The consequences of the assumption that the coherence length $\xi$ is *small* compared with the penetration depth $\lambda$ were first investigated by Abrikosov (1957), using a theory proposed in 1950 by Ginzburg and Landau. The behaviour of the total energy density is shown in Fig. 13.15; the important difference is that a *lower* energy is associated with the boundary region, and it should therefore be energetically favourable for such boundaries to form. In fact this does not occur until the critical flux density $B_{c1}$ is reached; below this field the behaviour is the same as for a Type I material, with a complete Meissner effect where the flux density in the bulk of the sample is zero. In the region where the applied field has

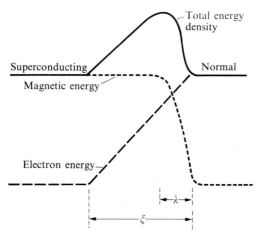

FIG. 13.14. Energy associated with a boundary in a Type I material between normal and superconducting regions. For the latter the magnetic energy is greater (eqn (13.6)), but falls in the region thickness $\lambda$, where magnetic flux penetrates. The electron energy is lower in the superconducting state through the formation of Cooper pairs, whose density falls over a distance whose thickness is of order $\xi$, the coherence length. The presence of a boundary involves extra total energy.

flux density greater than $B_{c1}$ but smaller than $B_{c2}$ there is a partial flux penetration, with a mixture of normal and superconducting regions; this persists up to the second critical flux density $B_{c2}$, beyond which there is complete penetration and no superconducting regions remain. In the presence of partial flux penetration, the volume from which it is excluded becomes smaller so that the energy involved in keeping the flux out is

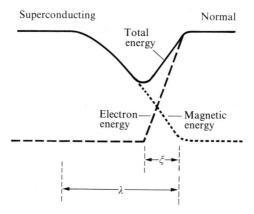

FIG. 13.15. Energy associated with a boundary in a Type II material, where the coherence length $\xi$ is smaller than the penetration depth $\lambda$. The formation of boundaries is favoured through the lower total energy in the boundary region.

reduced, and the critical flux density $B_{c2}$ can be greater than $B_c$. The theory leads to the result $B_{c2} = \sqrt{2}(\lambda/\xi)B_c$. In the region where $B$ lies between $B_{c1}$ and $B_{c2}$ the superconductor is said to be in a mixed or 'vortex' state (see below).

### Quantization of magnetic flux

In a superconducting ring such as the hollow cylinder shown in Fig. 13.4 there is a finite flux parallel to the axis in the centre. A supercurrent flows round the inner surface of the cylinder, its density decreasing to zero at distances inside the metal. This current consists of a flow of Cooper pairs, each with net wave vector $\delta\mathbf{k}$ and momentum $\hbar\delta\mathbf{k} = \mathbf{p}$. According to the Bohr–Sommerfeld quantization rule, the value of $\mathbf{p}$ must satisfy the relation

$$\int \mathbf{p} \cdot d\mathbf{s} = Nh, \tag{13.41}$$

where $h$ is Planck's constant and $N$ is an integer. From eqn (13.21) we must have

$$\mathbf{p} = m\mathbf{v} + q\mathbf{A},$$

and on relating $\mathbf{v}$ to $\mathbf{J}$ as in eqn (13.22) the quantization rule leads to the result

$$Nh = \int (m\mathbf{v} + q\mathbf{A}) \cdot d\mathbf{s} = \int (m/nq)(\mathbf{J} \cdot d\mathbf{s}) + \int q(\mathbf{A} \cdot d\mathbf{s}). \tag{13.42}$$

At points well into the metal $\mathbf{J} = 0$ so that the first integral vanishes. From eqn (4.50), $\int(\mathbf{A} \cdot d\mathbf{s}) = \Phi$, the magnetic flux threading the path. Since $q = -2e$ for a Cooper pair, it follows that the flux must be quantized in multiples of $|h/2e| = 2 \cdot 07 \times 10^{-15}$ Wb. The result $\mathbf{p}_0 = 0$ in the London theory corresponds to the absence of any units of quantized flux in the interior of a Type I superconductor.

The suggestion that the flux must be quantized has an important consequence for hollow superconducting bodies such as the hollow cylinder considered in § 13.2. When such a cylinder is cooled through the superconducting transition temperature in a finite magnetic field, the experiments of Meissner showed that the flux inside the cavity remained fixed at the value existing at the moment when the cylinder entered the superconducting state. We now see that this cannot be exactly true; the current flow round the inside surface of the cylinder must adjust itself to be such that the flux inside is an exact multiple of $h/2e$. Experiments to detect this have been carried out by Doll and Näbauer (1961) and by Deaver and Fairbanks (1961). Within the errors necessarily present in the determination of such small quantities it was shown that the flux is quantized in units of $h/2e$.

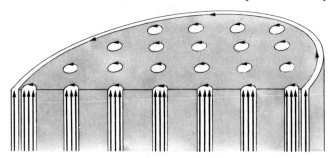

FIG. 13.16. The 'vortex state' in a superconductor between $B_{c1}$ and $B_{c2}$. At the centre of a vortex there is 'normal' metal carrying lines of magnetic flux, around which flow circulating currents out to a distance equal to the penetration depth $\lambda$. The total flux through each vortex is just equal to one quantum $\Phi_0$, and in a perfect crystal the vortices are equally separated in a triangular array.

An idealized model of a Type II superconductor in the mixed or 'vortex' state is shown in Fig. 13.16. It consists of tiny vortices of supercurrent in the form of Cooper pairs, circulating round regions of magnetic flux. Each region contains just one unit of quantized flux, spread over an area whose radius is about equal to the penetration depth. The current density and the number of Cooper pairs decreases towards the centre of each vortex, where the metal is essentially 'normal'. The correctness of this model of a Type II superconductor has been demonstrated in a beautiful experiment of Essmann and Träuble (1971), using an electron microscope. The pattern of vortices is triangular, in agreement with the theory which predicts this as having the minimum energy in a perfect crystal lattice; the pattern is distorted by lattice imperfections and grain boundaries. By counting the number of vortices and measuring the flux it was found that each vortex carried one quantum of magnetic flux, within the experimental error.

## 13.9. The Josephson effect

In a normal metal the electron wavefunctions are of the form $\exp j\mathbf{k}\cdot\mathbf{r} = \exp j\phi$; we may regard $\phi$ as a 'phase factor', but only its gradient in space is important since the wave vector $\mathbf{k}$ is related to the rate at which $\phi$ changes with distance. For a superconductor, the Ginzberg–Landau theory introduced a complex order parameter $\psi$ such that the density of superconducting carriers is given by $\psi\psi^* = n_s$. Here $\psi(r)$ has the form of a wavefunction $\psi = |\psi| \exp j\phi$ in which phase coherence is maintained over macroscopic distances. The phase factor is indeterminate in a bulk sample, but in going round a ring it must increase by a multiple of $2\pi$, a condition that corresponds to the phenomenon of flux quantization discussed above.

A remarkable effect was discovered by Josephson (1962) in considering the problem of phase coherence in a superconducting ring interrupted by a thin insulating layer (Fig. 13.17). The thickness of the layer (of the order of 1 nm) is such that electrons can tunnel through it, and with a normal metal the current through the layer is proportional to the voltage across it (the effective resistance is only a few ohms). The wavefunction in the insulating layer is real, but it must be continuous at the two boundaries between the layer and the metal. When the metal becomes superconducting, the presence of phase coherence round the ring combined with the boundary conditions leads to the result that the phases $\phi_1$, $\phi_2$ at the two surfaces of the insulator in contact with the metal must have a constant difference $\theta = \phi_1 - \phi_2$, and that a current of Cooper pairs traverses the layer given by the relation

$$I_s = I_c \sin \theta. \tag{13.43}$$

This current requires no voltage across the layer so long as $I_s$ does not exceed the critical value $I_c$ which is determined by the properties of the barrier (in particular, its thickness). If this critical value is exceeded, current flow must occur through some other process involving a finite voltage across the insulating layer. The current–voltage characteristic is shown in Fig. 13.18. The insulating layer is known as a 'weak link' and the device is called a 'Josephson junction'. Currents of the order of 1–100 $\mu$A or more can flow; the non-linear characteristic and the low dissipation make it important in a range of applications, of which we

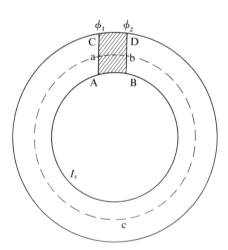

FIG. 13.17. A superconducting ring interrupted by a thin insulating layer ABCD. The current density inside the superconducting metal is zero, and the only contribution to the integral $\int \mathbf{J} \cdot \mathbf{ds}$ round the path abc comes from the part ab across the insulator.

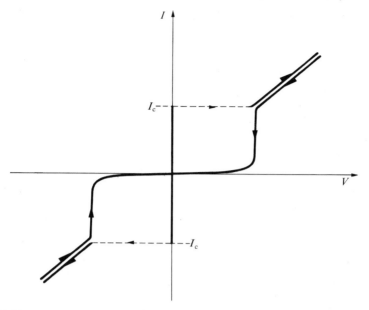

FIG. 13.18. D.c. current–voltage characteristic for a Josephson junction. When the current exceeds the critical current $I_c$ for Cooper pairs, the passage of normal electrons occurs and a voltage must be present across the junction.

mention below only two. Niobium metal ($T_c = 9.5$ K) is a convenient superconducting material for use at 4·2 K, the boiling-point of liquid helium. Various forms of weak link have been designed; a niobium screw with a conical end gives an easily adjustable point contact with a low capacity (Fig. 13.19).

*Superconducting quantum interference device (SQUID)*

We now return to the question of a superconducting ring, with an insulating layer as a weak link, in the presence of an external magnetic

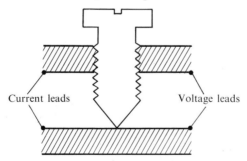

FIG. 13.19. A Josephson junction with a weak link at the point of contact between a niobium screw and niobium metal.

field. If the layer is so thick that no current can flow, the flux threading the ring is just that which would be expected from the applied field. In the opposite extreme, with zero thickness, a supercurrent flows in the ring which ensures that the flux threading it remains at zero if the applied field was zero when the ring became superconducting. In the presence of a weak link there is, however, a limit to the current of Cooper pairs which can flow through the link in the absence of any voltage. In general there will be a flux $\Phi$ through the ring, given by the relation

$$\Phi = \Phi_a + LI_s \tag{13.44}$$

where $\Phi_a$ = flux due to applied field, $L$ is the self-inductance of the ring, and $I_s$ the supercurrent producing a flux $LI_s$ (see eqn (5.8)). The value of $I_s$ is given by the Josephson relation (eqn (13.43)), whose maximum value $I_c$ is the critical current of Cooper pairs which can cross the link at zero voltage.

The phase difference $\theta$ across the link can be related to the flux, using eqn (13.42), since $\Phi = \int \mathbf{A} \cdot d\mathbf{s}$. If we write $\Phi_0 = h/|2e| = h/q$ for the unit of flux, and divide eqn (13.42) by $h$, we have

$$N = \frac{m}{nqh} \int \mathbf{J} \cdot d\mathbf{s} + \frac{q}{h} \int \mathbf{A} \cdot d\mathbf{s}$$

$$= \frac{m}{nqh} \int \mathbf{J} \cdot d\mathbf{s} + \frac{\Phi}{\Phi_0}. \tag{13.45}$$

Inside the superconductor $\mathbf{J} = 0$ and $\Phi$ must be an integral multiple of $\Phi_0$ for a complete superconducting ring. In a weak link $\mathbf{J}$ is finite, and a phase difference can occur across the link such that

$$\theta = 2\pi N - 2\pi(\Phi/\Phi_0). \tag{13.46}$$

Since $N$ is an integer,

$$\sin \theta = -\sin 2\pi(\Phi/\Phi_0),$$

and on combining this with eqns (13.43), (13.44) we have

$$\Phi = \Phi_a + LI_s = \Phi_a + LI_c \sin \theta$$

$$= \Phi_a - LI_c \sin 2\pi(\Phi/\Phi_0). \tag{13.47}$$

We can write this in a reduced form as

$$\frac{\Phi}{\Phi_0} = \frac{\Phi_a}{\Phi_0} - \frac{LI_c}{\Phi_0} \sin 2\pi\left(\frac{\Phi}{\Phi_0}\right), \tag{13.48}$$

whose solution is shown in Fig. 13.20. As we increase $\Phi_a$ from zero, $\Phi$ also increases slowly, but at the point $P_1$, where $I_s$ reaches the critical value $I_c$, there is an instability and $\Phi/\Phi_0$ increases by one unit, jumping to the point $Q_1$. Further increases in $\Phi_a/\Phi_0$ give rise to corresponding jumps

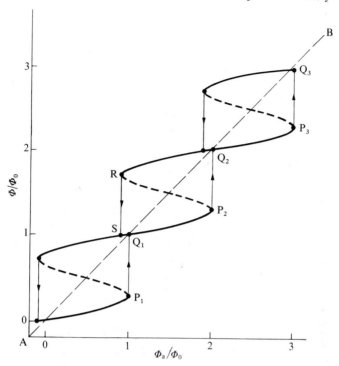

FIG. 13.20. Graphical representation of eqn (13.48) for a particular ratio $(LI_c/\Phi_0) \sim 0.8$. Essentially the graph of $\Phi/\Phi_0$ against $\Phi_a/\Phi_0$ is a sine curve about the line AB corresponding to $\Phi = \Phi_a$, which it crosses whenever $\sin(2\pi\Phi/\Phi_0) = 0$. Energy is dissipated when a path such as $P_2Q_2$ RS is traversed in the presence of an oscillatory flux $\Phi_a$.

of one unit in $\Phi/\Phi_0$. If the process is reversed, and $\Phi_a/\Phi_0$ is decreased, there are again discontinuous drops of one unit in $\Phi/\Phi_0$, but these occur at the points where $I_s = -I_c$. The discontinuities occur when $I_s = I_c$ because the ring then momentarily ceases to be superconducting as this limit is exceeded, so that flux can enter; $I_s$ then drops below the critical value, and the ring is again superconducting until the supercurrent again reaches a critical value.

If a loop of wire is placed inside the ring, a voltage impulse is induced in the loop each time a flux jump occurs. The voltage impulse can be amplified, and by counting the impulses the total flux change through a ring of known geometry can be measured. If the ring has a diameter of about 1 mm, a change in flux density of order $10^{-9}$ T is sufficient to change the flux by one quantum. In practice, a better system is to couple the loop to a tuned circuit oscillating at about 30 MHz. The oscillations build up in amplitude, exciting an oscillating supercurrent across the weak link; when this exceeds the critical amplitude, the SQUID traverses a path such as $P_2Q_2RS$ in Fig. 13.20, involving a dissipation of energy which causes the level of oscillation to drop abruptly. The advantage of this system is that the radio-frequency voltage across the tuned circuit can be about 100 times that across the weak link.

In the presence of a steady magnetic flux $\Phi_a$ the sum of the direct current $I_s = (\Phi - \Phi_a)/L$ and the amplitude of the oscillatory component $I'_s$ cannot exceed $I_c$, so that the level of oscillation must vary periodically as $\Phi_a$ increases, the repetition rate being just equal to one quantum of flux.

Because of its high sensitivity, the device must be carefully shielded from extraneous changes in magnetic flux. In the two-hole system shown in Fig. 13.21, the metal forms a continuous superconducting circuit inside which the total flux must be constant. If the flux $\Phi_1$ in one hole increases, there must be an equal decrease in the flux $\Phi_2$ in the second hole, while the net current through the weak link depends on $\Phi_1 - \Phi_2$. The second hole is coupled inductively to the radio-frequency oscillatory circuit, while the first contains a loop of wire which is in series with a search coil placed in the field to be measured. The whole circuit formed by these two coils is also superconducting, so that the total flux through it is constant. A flux change $\Phi_S$ in the search coil induces a screening current $I = -\Phi_S/L'$, where $L'$ is the total self-inductance of the two-coil circuit. The corresponding flux change in the SQUID is $\Phi_a = +MI = -(M/L')\Phi_S$, where $M$ is the mutual inductance between the loop and the SQUID; the system acts as a 'flux transformer' which changes the sensitivity by varying the ratio $M/L'$. Application of the SQUID to magnetometry is discussed by Swithenby (1974).

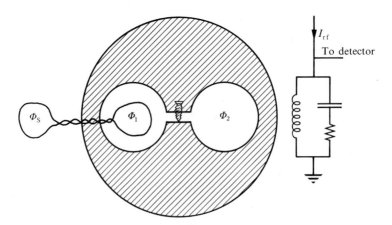

FIG. 13.21. A squid formed from a cylinder of niobium metal with two holes and a niobium screw; this gives a system which is mechanically robust and can stand many thermal cycles. Typical diameters for the holes are about 3 mm. The r.f. coil is inductively coupled to the right-hand cavity.

*The alternating-current Josephson effect*

The flow of Cooper pairs at zero voltage across a weak link is known as the direct-current Josephson effect. When a voltage is maintained across the junction, Josephson predicted that there would also be a flow of alternating current. This arises in the following way. On one side of the junction the wavefunction has a phase factor which oscillates in time, given by

$$\psi_1 = |\psi_1| \exp(-jW_1t/\hbar) = |\psi_1| \exp j\phi_1,$$

where $W_1$ is the energy. Similarly on the other side

$$\psi_2 = |\psi_2| \exp(-jW_2t/\hbar) = |\psi_2| \exp j\phi_2.$$

The current through the junction is therefore

$$I_s = I_c \sin \theta = I_c \sin(\phi_1 - \phi_2) = I_c \sin\{(W_2 - W_1)t/\hbar\}.$$

If the energy difference is $W_2 - W_1 = |2e| V$, we can write

$$I_s = I_c \sin \omega t \tag{13.49}$$

where

$$\hbar\omega = h\nu = |2e| V. \tag{13.50}$$

This is known as the alternating-current Josephson effect. Essentially the passage of a Cooper pair over the voltage drop $V$ results, not in the dissipation of energy, but the transformation of the energy $|2e| V$ into an electromagnetic quantum $h\nu$. The frequency $\nu$ is linearly proportional to the voltage $V$, and has the value 483·59 MHz for a voltage of 1 $\mu$V.

Power limitations stop the Josephson junction from becoming a useful radio-frequency generator, but it can be used as a detector. In the presence of microwave radiation, the d.c. characteristic (Fig. 13.22) shows a change in the current, with discontinuities at the points where the voltage is a multiple of $h\nu/2e$. The non-linear characteristic has been used for frequency multiplication up to about $10^6$ MHz, and for the determination of $h/e$ with great precision. High accuracy is readily available in the measurement of frequency (see § 22.5), and the small voltages across the junction are determined from comparison with a standard cell using Hamon's resistance ratio device (Problem 3.6). The accuracy of a few parts in $10^7$ (see Sanders and Wapstra 1972 (Part 9)) means that the method can be used to monitor changes in the voltage of standard cells, and to compare cells at different Standards laboratories.

## 13.10. Conclusion

In this chapter we have attempted to summarize the main experimental results which are basic to the understanding of the superconducting state.

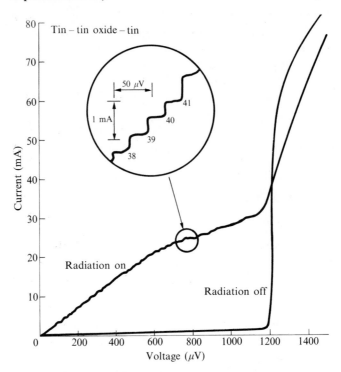

FIG. 13.22. Current–voltage characteristic at 1·2 K of a Josephson junction in the presence of 10 GHz microwave radiation. The steps occur at voltages which are integral multiples of $h\nu/2e$. (Parker, Langenberg, Denenstein, and Taylor 1969.)

Much important work has been omitted; for example, there is no discussion of the intermediate state which arises when the field is greater than the critical field over some parts of the surface of the sample (for the sphere of Fig. 13.3, the value of **B** is larger at the equator than at the poles). Necessarily the BCS theory has been presented in qualitative terms. Despite the simplifying assumptions it contains, the theory works remarkably well, as can be seen from the data in Table 13.4 for the Type I superconducting metals aluminium, tin, and lead.

## References

ABRIKOSOV, A. A. (1957). *Sov. Phys.* **5**, 1174.
BARDEEN, J., COOPER, L. N., and SCHRIEFFER, J. R. (1957). *Phys. Rev.* **108**, 1175.
COOPER, L. N. (1956). *Phys. Rev.* **104**, 1189.
DEAVER, B. S. and FAIRBANKS, W. M. (1961). *Phys. Rev. Lett.* **7**, 43.
DOLL, R. and NÄBAUER, M. (1961). *Phys. Rev. Lett.* **7**, 51.

ESSMANN, U. and TRÄUBLE, H. (1971). *Scient. Am.* **224**, 75.
FINNEMORE, D. K. and MAPOTHER, D. E. (1965). *Phys. Rev.* A**140**, 507.
GLOVER, R. E. and TINKHAM, M. (1957). *Phys. Rev.* **108**, 243.
GORTER, C. J. and CASIMIR, H. B. G. (1934). *Physica* **1**, 305.
JOSEPHSON, B. D. (1962). *Phys. Lett.* **1**, 251.
LONDON, F. and LONDON, H. (1935). *Proc. R. Soc.* A**149**, 71.
MAXWELL, E. (1950). *Phys. Rev.* **78**, 477.
MEISSNER, W. and OCHSENFELD, R. (1933). *Naturwiss.* **21**, 787.
MENDELSSOHN, K. (1955). *Prog. low Temp. Phys.* **1**, 185.
MORSE, R. W. and BOHM, H. V. (1957). *Phys. Rev.* **108**, 1093.
PARKER, W. H., LANGENBERG, D. N., DENENSTEIN, A., and TAYLOR, B. N. (1969). *Phys. Rev.* **177**, 639.
PHILLIPS, N. E. (1964). *Phys. Rev.* A**134**, 385.
PIPPARD, A. B. (1953). *Proc. R. Soc.* A**216**, 547.
QUINN, D. J. and ITTNER, W. B. (1962). *J. appl. Phys.* **33**, 748.
REYNOLDS, C. A., SERIN, B., WRIGHT, W. H., and NESBITT, L. B. (1950). *Phys. Rev.* **78**, 487.
SANDERS, J. H. and WAPSTRA, A. H. (1972). *Atomic masses and fundamental constants 4.* Plenum Press, London.
SWITHENBY, S. J. (1974). *Contemp. Phys.* **15**, 249.

## Problems

13.1. If the critical flux density $B_c$ for the superconducting transition is given by eqn (13.1), show that the discontinuity in the heat capacity $C_s - C_n$ should obey the relation

$$C_s - C_n = -\frac{B_0^2 T}{\mu_0 T_c^2}\left\{1 - 3\left(\frac{T}{T_c}\right)^2\right\}.$$

This suggests that $C_s - C_n$ should change sign at $T = T_c/\sqrt{3}$.

13.2. The reduced critical flux density $(B_c/B_0)$ can be expressed as a power series in $(T/T_c)^2$

$$B_c/B_0 = 1 - a_2(T/T_c)^2 + a_4(T/T_c)^4 + \dots.$$

By calculating the limiting value of $(C_s - C_n)/T$ as $T \to 0$ from eqn (13.9), show that

$$\gamma = 2a_2 \frac{V}{\mu_0} \frac{B_0^2}{T_c^2}$$

and that eqn (13.36) then leads to a value $a_2 = 1 \cdot 07$.

# 14. Paramagnetism

## 14.1. A general precession theorem

IN Chapter 6 the origin of paramagnetism was discussed, and it was shown to exist in substances containing permanent magnetic dipoles. Such dipole moments are associated with moving charges, being due either to the motion of electrons in their orbits about the atomic nucleus or the spin of the electron about its own axis. From observations of hyperfine structure in atomic spectra it was inferred that the nuclei of many types of atom also possess 'spin' due to rotation about an internal axis, and that a magnetic dipole moment is associated with this spin. In all these cases the magnetic moment is associated with some units of angular momentum, and the direction of the magnetic moment $\mathbf{m}$ is parallel to that of the angular-momentum vector $\mathbf{G}$, and proportional to it. Thus we may write (cf. eqn (6.2))

$$\mathbf{m} = \gamma \mathbf{G}, \tag{14.1}$$

where $\gamma$ is a constant whose reciprocal $1/\gamma$ is known as the gyromagnetic ratio. For an electron of charge $-e$ and mass $m_e$ moving in an orbit, $\gamma$ is equal to the classical value $-e/2m_e$; the magnetic moment associated with the intrinsic spin of the electron is anomalously large, the ratio being in this case very nearly equal to $-e/m_e$. The minus sign in each of these cases arises from the fact that the charge carried by the electron is negative, and shows that the magnetic moment is oppositely directed to the angular-momentum vector. In the case of the nucleus, the angular momentum is of the same order as that of an electron, being either a small half-integral or integral multiple of $\hbar$, but the magnetic moment is about 1000 times smaller, corresponding to the greater mass of the particles (proton and neutron) in the nucleus. The value of $\gamma$ is then $g_N(e/2m_p)$, where $m_p$ is the mass of the proton, and $g_N$ is a number which is of the order of unity but is not in general an exact integer or a simple fraction.

When an atom or nucleus with a permanent magnetic dipole moment $\mathbf{m}$ is placed in a steady magnetic flux density $\mathbf{B}$, a couple is exerted on it which may be written in vector form as $\mathbf{m} \wedge \mathbf{B}$. The angular momentum must therefore change (either in magnitude or direction) at a rate equal to this couple, that is

$$\dot{\mathbf{G}} = \mathbf{m} \wedge \mathbf{B}.$$

Since **m** is proportional to and parallel to **G**, we have

$$\dot{\mathbf{G}} = \gamma \mathbf{G} \wedge \mathbf{B}. \tag{14.2}$$

This is a vector equation whose solution is easily found by writing down its components referred to Cartesian coordinates. If the magnetic field is assumed to act along the $z$-axis the components are

$$\left.\begin{aligned}
\dot{G}_x &= \gamma B G_y, \\
\dot{G}_y &= -\gamma B G_x, \\
\dot{G}_z &= 0.
\end{aligned}\right\} \tag{14.3}$$

Integration of the last equation shows that the component $G_z$ along the $z$-axis is a constant. It follows that the angle $\alpha$ which **G** makes with **B** is constant, and we may write $G_z = G \cos \alpha$. The equations for the $x$ and $y$ components may be solved by differentiating one of them and eliminating either $G_x$ or $G_y$. We find

$$\ddot{G}_x = \gamma B \dot{G}_y = -(\gamma B)^2 G_x,$$

with an identical equation for $G_y$. The solution is of the form

$$G_x = A \cos(-\gamma B t + \epsilon),$$

and from eqn (14.3) we find

$$G_y = A \sin(-\gamma B t + \epsilon).$$

Thus it will be seen that the projection of $G$ on the $xy$ plane is of constant magnitude $A = G \sin \alpha$, but rotates with the angular velocity $-\gamma B$, which we may write as $\omega_L$. Thus our solution for the components of **G** is

$$\left.\begin{aligned}
G_x &= G \sin \alpha \cos(\omega_L t + \epsilon), \\
G_y &= G \sin \alpha \sin(\omega_L t + \epsilon), \\
G_z &= G \cos \alpha.
\end{aligned}\right\} \tag{14.4}$$

The motion is such that the magnitudes of both **m** and **G** remain constant, but their directions 'precess' at a constant angle about the direction of **B** as in Fig. 14.1. The angular velocity of the precession depends only on the gyromagnetic ratio and the size of the magnetic field. The direction of precession is that of a right-handed screw progressing along **B** if $\gamma$ is negative, and vice versa. The angle $\alpha$ depends on the initial conditions prevailing when the magnetic field was switched on.

Although we have chosen to solve eqn (14.2) by the use of a Cartesian coordinate system, we could have derived a certain amount of information about the motion by inspection of the vector equation (14.2) itself. Since the vector product of two vectors is a vector perpendicular to both,

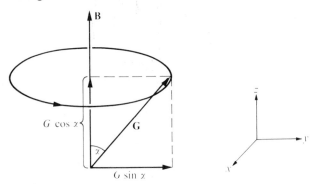

FIG. 14.1. Precession of **G** about **B**. The direction of precession shown is that for an electronic
momentum ($\gamma$ negative).

it follows that $\dot{\mathbf{G}}$ is normal to **G** and to **B**. Thus, in Fig. 14.1, if
instantaneously both **G** and **B** are in the plane of the paper, the motion of
**G** must be normal to the paper. This means that if the momentum vector
**G** is drawn from a fixed origin, then its tip must move out of the paper,
and since $\dot{\mathbf{G}}$ always remains normal to **G**, the tip must move in a circle
around **B**, that is, the angular momentum vector precesses around **B**.

This precessional motion, originally derived in a theorem due to
Larmor, is a quite general result, depending only on the connection
between angular momentum and magnetic dipole moment. In a quantum-
mechanical system, such as the atom, the angular momentum plays an
important role, as is well known from atomic theory. In the next section
we turn to consideration of the magnetic moments of single atoms,
making use of our general precession theorem. The chief difference we
shall find from the classical case considered above is that the angle $\alpha$ is
now fixed by the rules of quantization, only a small number of values
being possible, instead of any value.

## 14.2. The vector model of the atom

A complete understanding of the behaviour of magnetic moments in
atoms requires a thorough knowledge of quantum theory. Here we
assume the reader is acquainted with atomic structure as presented in
most elementary textbooks of atomic physics, and give only a résumé of
the position, mainly in terms of orbits rather than wavefunctions. The
results quoted will be those appropriate to the wave-mechanical theory.

The state of an electron in an atom is defined by four quantum
numbers $n$, $l$, $m_l$, and $m_s$, whose significance is as follows. The principal
quantum number $n$ has integral values from unity upwards, and the
energy of the electron is mainly determined by the value of $n$. On the

original Bohr theory the energy depended only on the value of $n$, its value being, for an atom with only one electron,

$$W_n = -(R_\infty Z^2/n^2), \tag{14.5}$$

where $R_\infty$ is a universal constant (Rydberg's number), and $Ze$ the charge on the nucleus. The same result is obtained by wave mechanics for a one-electron atom, but this result does not hold for atoms with several electrons owing to the electrostatic repulsion between the electrons. For most atoms it remains true that electrons with the lower values of $n$ have the lower energy, and the difference of energy for successive values of $n$ decreases as $n$ becomes larger (cf. Fig. 14.2).

The quantum number $l$ is defined by the value of the angular momentum which the electron possesses in its orbit around the nucleus; on the vector model the angular momentum $\mathbf{G} = l\hbar$, where $\hbar = h/2\pi$ and $h$ is Planck's constant. The allowed values of $l$ are integral, from 0 up to $(n-1)$ for an electron whose principal quantum number is $n$. Since the electron is a charged particle, its motion in an orbit is equivalent to a circulating current, and a magnetic dipole moment is associated with the orbit which has the same value as that expected on classical theory. This

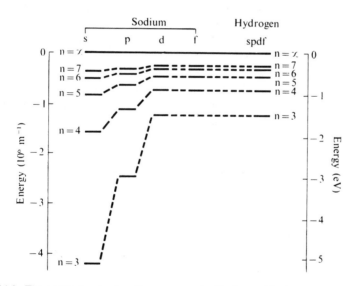

FIG. 14.2. The energy levels of sodium compared with those of hydrogen. For sodium the levels are those of the single electron outside the closed shells $1s^2$, $2s^2$, $2p^6$. For the higher values of $n$ the levels approach closely those of the hydrogen atom. This is because at large distances from the nucleus the electric field is that of the nuclear charge $+Ze$ surrounded by $(Z-1)$ electrons, and hence (by Gauss's theorem) is that of unit positive charge. Orbits with lower values of $n$ penetrate the closed electron shell and so feel a greater positive charge, giving a lower energy. This is most marked for the 'penetrating' s orbits.

dipole moment is

$$\mathbf{m} = (-e/2m_e)\mathbf{l}\hbar = -(e\hbar/2m_e)\mathbf{l} = -\mu_B\mathbf{l}.$$

Here the minus sign shows that $\mathbf{m}$ and $\mathbf{l}$ have opposite directions, owing to the negative charge of the electron. The quantity $\mu_B = e\hbar/2m_e$ is known as the Bohr magneton and is the atomic unit of magnetic dipole moment; its magnitude is $0.9274 \times 10^{-23}$ A m$^2$.

The precession theorem of § 14.1 shows that the dipole moment, and its associated angular momentum vector, precess at a constant angle about an applied magnetic field. The component $G_z$ of the angular momentum in this direction is therefore a constant. On quantum theory this component must be an integral multiple of $\hbar$; it is written $m_l\hbar$, where the 'orbital magnetic quantum number' $m_l$ can have all integral values, including 0, between $+l$ and $-l$. The components normal to this direction are not quantized, but the sum of their squares can be shown to be given by

$$G_x^2 + G_y^2 = \{l(l+1) - m_l^2\}\hbar^2.$$

Since $G_z^2 = m_l^2\hbar^2$, we see that the square of the angular momentum has the 'sharp' value $G^2 = l(l+1)\hbar^2$. It is a general feature of the quantum theory of angular momentum that one component has a value which is of the form $\hbar m_l$, corresponding to the stationary component in a field, while the square of the angular momentum is $l(l+1)\hbar^2$ and not $l^2\hbar^2$. These points are illustrated in Fig. 14.3.

Since the magnetic moment is proportional to the angular momentum, the moment associated with the orbit has a fixed component parallel to a magnetic field of magnitude $-m_l\mu_B$, and the square of the magnetic dipole moment has the value $l(l+1)\mu_B^2$.

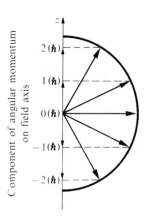

FIG. 14.3. Quantization of orbital angular momentum. The figure is drawn for the case of $l\hbar = 2\hbar$ and the radius of the circle is $\hbar\sqrt{(2 \times 3)} = \hbar\sqrt{6}$.

In addition to the orbital motion, the electron also possesses a spin about its own axis, with angular momentum $\mathbf{G}_s = s\hbar$, where the electronic spin quantum number $s$ always has the value $s = \frac{1}{2}$. The square of the angular momentum is $s(s+1)\hbar^2$. With the spinning charge of the electron is associated a magnetic moment

$$-g_s(e\hbar/2m_e)\mathbf{s} = -g_s\mu_B\mathbf{s}.$$

Here the coefficient $g_s$ is inserted because the ratio of the magnetic moment to the angular momentum differs from the classical value (corresponding to $g_s = 1$). For a long time it was thought that the value of $g_s$ for the electron spin was exactly 2; it has now been shown both experimentally and theoretically that the value is $2(1\cdot00115966)$, but for our purpose it is sufficient to omit the correction and assume that $g_s$ is 2 for electron spin. In an atom there are relativistic and diamagnetic corrections to both the orbital and spin magnetic moments (of the order of $10^{-6}$–$10^{-4}$), which we shall neglect. The minus sign in the expression for the magnetic moment shows that it is oppositely directed to the angular momentum vector, owing to the negative charge of the electron, as in the orbital case. In a magnetic field both the spin angular momentum and its magnetic moment precess about the direction of the field, as in Fig. 14.4, the steady component of the angular momentum in this direction having one of the values $\pm m_s\hbar = \pm\frac{1}{2}\hbar$, and the corresponding steady component of the magnetic moment having the values $\mp\frac{1}{2}g_s\mu_B \approx \mp\mu_B$. Note that these components amount to 1 $\mu_B$, though the spin is half-integral.

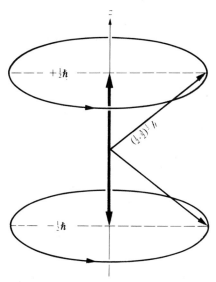

FIG. 14.4. Quantization of spin angular momentum $s = \frac{1}{2}$.

In an atom containing a number of electrons, the total angular momentum will be the vector sum of the individual momenta, both orbit and spin. In general this vector sum can be formed in a number of ways, with a number of different resultants. To know which of these is correct or, if several are allowed, which corresponds to the state of lowest energy (the ground or normal state of the atom), we need to know more about the mutual interactions between the various electrons. We shall see that these can be expressed in the form of a set of rules for coupling together the angular momenta in forming the vector resultant. These rules are subject to one overriding condition, expressed in the well-known Pauli principle: 'no two electrons in the same system can be in states with identical sets of quantum numbers'. When applied to an atom this means that no two electrons can have identical sets of values for $n$, $l$, $m_l$, and $m_s$. The Pauli principle sets a limit to the number of electrons in any one shell, and has the important result that the angular momentum of a filled shell containing $2(2l+1)$ electrons is zero: $L = S = 0$. Such a shell can have no resultant magnetic moment, so that the magnetic moment of an atom or ion must arise solely from its incompletely filled shells.

Our next problem is that of how to couple together the angular momenta in a partly filled subshell. This depends on the mutual interactions between the electrons, of which the two principal types are as follows:

1. *Mutual repulsion between the electrons, due to their electrical charge.* When this is treated by wave mechanics an unexpected result is found. The energy of the system contains two terms, one corresponding to the classical Coulomb interaction, the other known as an 'exchange energy', because it appears to be connected with an exchange of any pair of electrons between the states we assigned to them before including the effect of their mutual repulsion. These exchange forces have no analogue in classical theory, but play an important role in atomic theory. By means of the Pauli exclusion principle, their effect can be shown (see § 15.8) to correspond to a strong coupling between the electron spins, this coupling being such that within an atom the state with the spins parallel is more stable and has the lower energy. The energy of interaction of this coupling may be written in the form $W = -2\mathscr{J}_{ij}\mathbf{s}_i \cdot \mathbf{s}_j$, where $\mathbf{s}_i$, $\mathbf{s}_j$ are the spin vectors, and $\mathscr{J}_{ij}$ is called the 'exchange energy', being positive for any pair of electrons within a given atom but varying in magnitude, depending on their orbital quantum states. Thus with a number of electrons, the primary effect of the exchange forces is to couple together the various vectors $\mathbf{s}_i\hbar$, $\mathbf{s}_j\hbar$ to form a resultant $\mathbf{S}\hbar$, which in the state of lowest energy has the largest possible value consistent with the exclusion principle. The remaining orbital momenta are then coupled together by the electrostatic

forces to form a resultant $\mathbf{L}\hbar$, which in the state of lowest energy again has the largest possible value consistent with the exclusion principle. (These two rules are known as Hund's rules.) Other values of $\mathbf{S}\hbar$ and $\mathbf{L}\hbar$ are possible, but correspond to states of higher energy. This method of coupling the angular momenta is known as $L-S$ or Russell–Saunders coupling. In wave mechanics $S$ and $L$ are quantum numbers and the absolute magnitudes of the squares of the total angular momenta associated with them are $\{S(S+1)\}\hbar^2$ and $\{L(L+1)\}\hbar^2$.

2. *Magnetic coupling between the magnetic moments of orbit and spin (the 'spin–orbit' interaction).* The orbital motion of the electron in an atom means that it sees the nucleus as a moving charge, producing a flux density $\mathbf{B}$ which acts on the magnetic moment associated with the electron spin. This is the origin of the 'spin–orbit interaction', which can be derived as follows. In the usual frame of reference where the nucleus is fixed, there is an electric field of strength $\mathbf{E}$ at the electron, and in this frame the electron is moving past the nucleus with velocity $\mathbf{v}$. We need the value of $\mathbf{B}$ in a frame where the electron is fixed; in this frame the nucleus is moving with velocity $-\mathbf{v}$, and the electron experiences a magnetic flux density

$$\mathbf{B} = -(\mathbf{v} \wedge \mathbf{E})/c^2 = +(\mathbf{E} \wedge \mathbf{v})/c^2,$$

where $c$ is the velocity of light. This equation is derived from the special theory of relativity, and incidentally is a consequence of the experimental fact that the electric charge (unlike mass) is invariant of the velocity of the frame of reference.

In this field the electron precesses with an angular velocity $\boldsymbol{\omega}_s = -\gamma_s \mathbf{B}$, where $\gamma_s = -g_s(|e|/2m_e)$. This is, however, the precession velocity in the frame in which the electron is at rest, and from the theory of relativity Thomas showed that to transfer to the laboratory frame one must add a velocity of precession

$$\boldsymbol{\omega}_T = (\mathbf{f} \wedge \mathbf{v})/2c^2$$

where $\mathbf{f}$ is the acceleration of the electron in the laboratory frame. Clearly $\mathbf{f} = -e\mathbf{E}/m_e$, so that the net precession frequency in the laboratory frame is

$$\boldsymbol{\omega}_s + \boldsymbol{\omega}_T = (g_s - 1)\frac{e}{2m_e} \cdot \frac{\mathbf{E} \wedge \mathbf{v}}{c^2}.$$

Over all, the effect is equivalent to a magnetic flux density

$$\mathbf{B} = \frac{(g_s - 1)}{g_s} \cdot \frac{\mathbf{E} \wedge \mathbf{v}}{c^2}.$$

In a hydrogen-like atom with nuclear charge $Ze$ and a single electron the field strength $\mathbf{E}$ at distance $\mathbf{r}$ from the nucleus is $\mathbf{E} = \mathbf{r}(Ze/4\pi\epsilon_0 r^3)$. In an atom with many electrons the electric field is still radial to a good approximation (the 'central-field' approximation), but the field is reduced through the screening effect of the other electrons. We may write $\mathbf{E} = \mathbf{r}(e/4\pi\epsilon_0)\langle Z'/r^3 \rangle$, where $Z'e$ is the nuclear charge which gives the correct field at distance $r$ and $\langle Z'/r^3 \rangle$ represents the mean value averaged over the electron orbit. Hence

$$\mathbf{B} = \frac{(g_s - 1)}{g_s} \cdot \frac{e}{4\pi\epsilon_0 c^2} \left\langle \frac{Z'}{r^3} \right\rangle (\mathbf{r} \wedge \mathbf{v}).$$

But $m_e(\mathbf{r} \wedge \mathbf{v}) = \mathbf{l}\hbar$, the orbital angular momentum, and since $e\hbar/2m_e = \mu_B$, (the Bohr magneton), we have

$$\mathbf{B} = \frac{2(g_s - 1)}{g_s} \cdot \frac{\mu_0}{4\pi} \left\langle \frac{Z'}{r^3} \right\rangle \mu_B \mathbf{l}. \tag{14.6}$$

The interaction energy with the spin magnetic moment is then

$$-\mathbf{B} \cdot \mathbf{m}_s = g_s \mu_B (\mathbf{B} \cdot \mathbf{s}) = \zeta(\mathbf{l} \cdot \mathbf{s}). \tag{14.7}$$

The effect of this spin–orbit interaction tends to couple together the vectors $s\hbar$ and $l\hbar$ for each electron to form a resultant $j\hbar$; the various values of $j\hbar$ for the individual electrons would then be coupled together (vectorially) to form the total angular-momentum vector $\mathbf{J}\hbar$. However, the spin–orbit interaction is smaller in magnitude than the exchange interactions between the spins discussed in (1) above, except in the heaviest elements. We shall therefore confine ourselves to $L$–$S$ coupling, where the individual spins are coupled to form a resultant $\mathbf{S}\hbar$, and the individual orbital momenta to form a resultant $\mathbf{L}\hbar$. The spin–orbit interaction then couples $\mathbf{S}\hbar$ and $\mathbf{L}\hbar$ together with an energy

$$W = \lambda \mathbf{L} \cdot \mathbf{S}, \tag{14.8}$$

which is similar in form to eqn (14.7). It can be shown that the relation between the two constants is $\lambda = \pm\zeta/2S$, where the positive sign is required for a shell that is less than half-filled, and the negative sign for one that is more than half-filled; the spin–orbit coupling parameter $\lambda$ vanishes for a half-filled shell. This coupling of $\mathbf{S}\hbar$ and $\mathbf{L}\hbar$ gives a resultant vector $\mathbf{J}\hbar$, the square of whose angular momentum is $J(J+1)\hbar^2$. The number of possible values of $J$ is either $(2S+1)$ or $(2L+1)$, whichever is the smaller. $L$ has only integral values, and the values of $J$ are therefore integral or half-integral according to whether the value of $S$ is integral or half-integral. The latter depends on whether the number of electrons involved is even or odd.

The nomenclature used to describe atomic energy states is mainly derived from pre-quantum attempts to analyse atomic spectra, and therefore does not possess the simple logical sequence which quantum theory could give it. Single-electron states for which the orbital quantum number $l$ has the values 0, 1, 2, 3, 4, 5,... are called s, p, d, f, g, h,... states, and similarly the levels of a many-electron atom for which $L = 0$, 1, 2, 3, 4, 5,... are denoted by the symbols S, P, D, F, G, H,... . The value of $n$ for a single-electron state is given by the number preceding the symbol, that is, 1s, 2s, 2p, 3s, 3p, 3d, etc. The number of electrons with given values of $n$ is denoted by a superfix; thus, three electrons with $n = 2$, $l = 1$ appear as $2p^3$. The spectroscopic state of the whole atom is defined by the values of $S$, $L$, and $J$; the value of the spin multiplicity $2S + 1$ is given by a superfix preceding the symbol for $L$, and the value of $J$ by a following suffix. Thus the symbol $^4F_{\frac{3}{2}}$ means that the state has $S = \frac{3}{2}$, $L = 3$, $J = \frac{3}{2}$; the other possible values of $J$ in this case are $\frac{5}{2}, \frac{7}{2}, \frac{9}{2}$, ranging in all from $L - S$ to $L + S$.

The coupling scheme for a many-electron atom or ion may be illustrated by reference to the energy-level diagram for the $Cr^{3+}$ ion, shown in Fig. 14.5. The triply charged chromium ion has the configuration $1s^2 2s^2 2p^6 3s^2 3p^6 3d^3$ with three electrons in the partly filled 3d shell. By

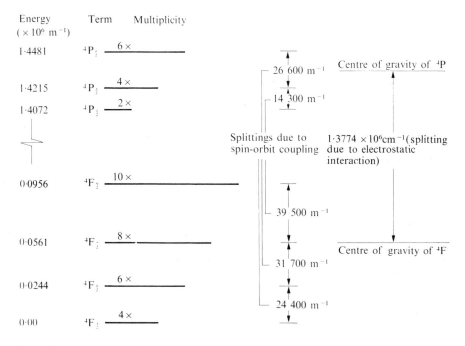

FIG. 14.5. Quartet energy levels of the free triply charged chromium ion, $Cr^{3+}$, $3d^3$. These are formed from the three electrons in the 3d shell: the splitting between the $^4F$ and $^4P$ terms is due to electrostatic repulsion between the electrons; the splittings between levels of different $J$ within each term are due to spin–orbit coupling. Doublet terms formed from $3d^3$ lie in the energy range $1\cdot4$–$3\cdot7 \times 10^6\,m^{-1}$. The next lowest levels are those belonging to the configuration $3d^2 4s$, and lie above $10^7\,m^{-1}$.

Hund's rules the energy is lowest when all three electrons have parallel spins, giving $S = \frac{3}{2}$. The electrons must then all have different values of $m_l$, by the Pauli principle, but since we have five possible values ($m_l = 2$, 1, 0, $-1$, $-2$) there are $(5!/3!2!) = 10$ possible arrangements. The largest possible value of $M_L = \sum m_l$ that we can have is $3 = 2+1+0$, and this belongs to an $L = 3$ state. This has $2L+1 = 7$ values of $M_L$, which therefore take up seven of the possible arrangements of electrons in the $m_l$ states; the other three belong to a state with $L = 1$. By Hund's rule, the $L = 3$ states will have lower energy than the $L = 1$ states. Both are shown in Fig. 14.5, the $^4$P ($L = 1$) states being higher in energy than the $^4$F ($L = 3$) states by about $1\cdot4 \times 10^6 \text{ m}^{-1}$. States of still higher energy (not shown) are formed by reversing one spin, giving $S = \frac{1}{2}$. Two electrons with opposite spin can now occupy the $m_l = 2$ state, so that the greatest possible value of $L_z = M_L$ is $5 = 2+2+1$, which belongs to a $^2$H state. Altogether six doublet states, $^2$H, $^2$G, $^2$F, $^2$D (twice), $^2$P are allowed; they have energies ranging from about $1\cdot4 \times 10^6 \text{ m}^{-1}$ to $3\cdot7 \times 10^6 \text{ m}^{-1}$. All other states are much higher in energy, the 3d$^2$4s configuration lying about $10^7 \text{ m}^{-1}$ above the ground state 3d$^3$, $^4$F.

The separation of the various quartet and doublet terms is determined by a combination of the exchange interaction and Coulomb interaction arising from the mutual repulsion of the electrons. The spin–orbit coupling splits the $^4$F term into four levels with values of $J$ ranging from $|L-S|$ to $|L+S|$, and the $^4$P term is similarly split, the only allowed values of $J$ in this case being $\frac{5}{2}$, $\frac{3}{2}$, and $\frac{1}{2}$. From the energy levels given in Fig. 14.5 it can be verified (see Problem 14.7) that the spin–orbit coupling constant $\lambda$ has the value $8\cdot7 \times 10^3 \text{ m}^{-1}$ for the ground states of Cr$^{3+}$. It can be seen that the splittings due to the 'magnetic' spin–orbit coupling are an order of magnitude smaller than those due to 'electrostatic' interactions.

## 14.3. Magnetic moments of free atoms

When we turn to consider the magnetic properties of atoms we find that the problem is simplified by the fact that these depend only on the partly filled electron shells, since completely filled shells (and of course empty shells) have $S$, $L$, and $J = 0$. The magnetic moment associated with each electron spin can be described by a vector parallel to and proportional to the angular-momentum vector $s\hbar$, and on forming the vector sum $S\hbar$ for a number of electrons the magnetic moments add in a similar way, so that the total magnetic moment of the spin is parallel to $S\hbar$ and has the same factor of proportionality to it. The same is true of the total orbital magnetic moment and the total orbital angular momentum $L\hbar$. When we come to make the vector addition of $S\hbar$ and $L\hbar$ the problem is not so simple because the factor of proportionality between the magnetic moment and the angular momentum is not the same for $S\hbar$ and $L\hbar$. The vector representing the total magnetic moment will not therefore be parallel to $\mathbf{J}\hbar$; this is illustrated by the vector diagram in Fig. 14.6. Here the magnetic-moment vector associated with $\mathbf{L}\hbar$ is drawn of the same length as $\mathbf{L}\hbar$, but on this scale the magnetic-moment vector associated with $\mathbf{S}\hbar$ must be drawn approximately twice as long as $\mathbf{S}\hbar$. The resultant magnetic-moment vector $\mathbf{m}$ is therefore at an angle to $\mathbf{J}\hbar$.

In considering this question further we must return to the discussion of the spin–orbit coupling between $\mathbf{L}\hbar$ and $\mathbf{S}\hbar$. This is primarily magnetic in origin, and arises from the magnetic moments of the orbit and spin. Each of these produces a magnetic field which interacts with the dipole moment of the other.

The interaction energy $\lambda\mathbf{L}.\mathbf{S}$ is equivalent to $-\mathbf{m}_{\mathbf{S}}.\mathbf{B}_L$ or to $-\mathbf{m}_L.\mathbf{B}_{\mathbf{S}}$, that is, to a flux density $\mathbf{B}_L = -\lambda\mathbf{L}/\gamma_S\hbar$ acting on the spin moment $\mathbf{m}_S = \gamma_S\hbar\mathbf{S}$, or to a flux density $\mathbf{B}_S = -\lambda\mathbf{S}/\gamma_L\hbar$ acting on the orbital moment $\mathbf{m}_L = \gamma_L\hbar\mathbf{L}$. Hence from eqn (14.2) the equations of motion will be

$$\left.\begin{aligned}\dot{\mathbf{L}} &= \gamma_L\mathbf{L}\wedge\mathbf{B}_S = \gamma_L\mathbf{L}\wedge(-\lambda\mathbf{S}/\gamma_L\hbar) = -(\lambda/\hbar)(\mathbf{L}\wedge\mathbf{S}),\\ \dot{\mathbf{S}} &= \gamma_S\mathbf{S}\wedge\mathbf{B}_L = \gamma_S\mathbf{S}\wedge(-\lambda\mathbf{L}/\gamma_S\hbar) = -(\lambda/\hbar)(\mathbf{S}\wedge\mathbf{L}).\end{aligned}\right\} \tag{14.9}$$

We note that the couples are equal and opposite, as they must be since no external couple acts on the system. Since $\mathbf{L}+\mathbf{S}=\mathbf{J}$, and

$$\mathbf{L}\wedge\mathbf{L}=\mathbf{S}\wedge\mathbf{S}=0,$$

we have

$$\left.\begin{aligned}\dot{\mathbf{L}} &= -(\lambda/\hbar)(\mathbf{L}\wedge\mathbf{S}+\mathbf{L}\wedge\mathbf{L}) = -(\lambda/\hbar)(\mathbf{L}\wedge\mathbf{J}),\\ \dot{\mathbf{S}} &= -(\lambda/\hbar)(\mathbf{S}\wedge\mathbf{L}+\mathbf{S}\wedge\mathbf{S}) = -(\lambda/\hbar)(\mathbf{S}\wedge\mathbf{J}),\end{aligned}\right\} \tag{14.10}$$

showing that $\mathbf{L}$ and $\mathbf{S}$ each precess about $\mathbf{J}$ with angular velocity $\lambda\mathbf{J}/\hbar$.

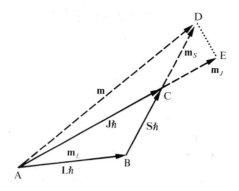

FIG. 14.6. Vector coupling of angular momentum vectors $\mathbf{L}\hbar$, $\mathbf{S}\hbar$ to form resultant $\mathbf{J}\hbar$ (represented by AB, BC, and AC), and of the associated magnetic moments $\mathbf{m}_L$, $\mathbf{m}_S$, and $\mathbf{m}$ (represented by AB, BD = $g_S$ BC, AD). AE is the projection $\mathbf{m}_J$ of $\mathbf{m}$ on $\mathbf{J}\hbar$.

The vector addition of $\mathbf{L}\hbar$ and $\mathbf{S}\hbar$ to form $\mathbf{J}\hbar$ is shown in Fig. 14.6. The magnetic moments $\mathbf{m}_L$, $\mathbf{m}_S$ are parallel and proportional to $\mathbf{L}\hbar$ and $\mathbf{S}\hbar$; in the figure they are represented by the lines AB (the same as $\mathbf{L}\hbar$), and BD, which is $g_S \times BC = g_S\mathbf{S}\hbar$. The resultant magnetic moment $\mathbf{m}$, given by AD, must precess around $\mathbf{J}\hbar$ at the same rate as $\mathbf{L}\hbar$ and $\mathbf{S}\hbar$. It therefore has a component $\mathbf{m}_J$, given by AE, and a precessing component ED which averages to zero over a time long compared with the precession period. When a small external field of flux density $\mathbf{B}$ is applied, its main interaction is with the component $\mathbf{m}_J$, which is a fixed vector when $\mathbf{B} = 0$. We may find the value of $\mathbf{m}_J$ in quantum mechanics by the following method. The energy of interaction with the flux density $\mathbf{B}$ is given by

$$W = -\mathbf{m}_L \cdot \mathbf{B} - \mathbf{m}_S \cdot \mathbf{B} = \mu_B(\mathbf{L} + g_S\mathbf{S}) \cdot \mathbf{B}. \tag{14.11}$$

If we write $\mathbf{m}_J = -g_J\mu_B\mathbf{J}$, where $g_J$ is a number yet to be found, its energy of interaction with $\mathbf{B}$ is

$$W = -\mathbf{m}_J \cdot \mathbf{B} = g_J\mu_B(\mathbf{J} \cdot \mathbf{B}), \tag{14.12}$$

which is identical with (14.11) if the time-averaged components are

$$\mathbf{L} + g_S\mathbf{S} = g_J\mathbf{J}. \tag{14.13}$$

On multiplying this equation by $\mathbf{J}$ we obtain

$$\mathbf{L} \cdot \mathbf{J} + g_S(\mathbf{S} \cdot \mathbf{J}) = g_J(\mathbf{J} \cdot \mathbf{J}) = g_J J(J+1), \tag{14.14}$$

and by squaring the quantities $\mathbf{S} = \mathbf{J} - \mathbf{L}$ and $\mathbf{L} = \mathbf{J} - \mathbf{S}$ we have

$$2(\mathbf{L} \cdot \mathbf{J}) = \mathbf{J} \cdot \mathbf{J} + \mathbf{L} \cdot \mathbf{L} - \mathbf{S} \cdot \mathbf{S} = J(J+1) + L(L+1) - S(S+1),$$
$$2(\mathbf{S} \cdot \mathbf{J}) = \mathbf{J} \cdot \mathbf{J} + \mathbf{S} \cdot \mathbf{S} - \mathbf{L} \cdot \mathbf{L} = J(J+1) + S(S+1 - L(L+1). \tag{14.15}$$

Hence

$$g_J = \frac{(g_S+1)J(J+1) - (g_S-1)\{L(L+1) - S(S+1)\}}{2J(J+1)}, \tag{14.16}$$

and if we make the approximation of writing $g_S = 2$,

$$g_J = \frac{3}{2} + \frac{S(S+1) - L(L+1)}{2J(J+1)}. \tag{14.17}$$

It is easy to see that if $S$ or $L$ is zero, so that $J = L$ or $J = S$, then $g$ is 1 or 2 respectively, corresponding to the cases of 'orbit only' and 'spin only'. The value of $g_J$ is known as the 'Landé g-factor'.

When an atom such as we have been considering is placed in an external magnetic field, the behaviour of the angular-momentum vectors in general will be rather complicated. The reason is that each of the magnetic moments associated with orbit and spin is acted on by the magnetic field due to the magnetic moment of the other as well as the external magnetic field. No simple description of the motion is possible when these fields are of the same order of magnitude, but when one is much larger than the other an approximate treatment is possible. We shall consider only the case when the external field is very small compared with the field due to the spin–orbit coupling. The vectors **L**, **S** then precess round **J** as in the case of zero external field, but **J** is no longer stationary in space, its motion being a precession round the external field of flux density **B**. The precession of **L**, **S** about **J** is at a much higher frequency than that of **J** about **B**, since the external field is small compared with that due to the spin–orbit coupling, and we may therefore picture the components of the magnetic moment precessing around **J** as averaging to zero, leaving only the steady component along **J**. This is acted on by the external field to give the precession of **J** about **B**, at an angular velocity $\omega = -g_J(-e/2m_e)\mathbf{B}$, where $g_J$ is the Landé g-factor. This is identical with the general result of § 14.1, if we take

$$\gamma = -g_J(e/2m_e) = -g_J\mu_B/\hbar.$$

The quantization rule for the projection of $\mathbf{J}\hbar$ on the flux density **B** is similar to the previous rules for other angular momentum vectors. The projection has the values $M_J\hbar$, where $M_J$ takes the values $J, J-1, J-2,...,$ $-(J-1), -J$. The component of the magnetic dipole moment of the atom parallel to the field has the value $-M_Jg_J\mu_B$, and the energy is

$$W_{M_J} = -\mathbf{m}_J\cdot\mathbf{B} = M_Jg_J\mu_BB. \tag{14.18}$$

Thus the $2J+1$ levels with different values of $M_J$ are split in energy by the application of a magnetic field, but have the same energy when $B=0$. In the latter case they are said to be 'degenerate', and the application of a field 'lifts the degeneracy'. This 'Zeeman splitting' is illustrated in Fig. 14.7 for the $^4F$ states of the $Cr^{3+}$ ion in a flux density $B=10$ T. The value of the spin–orbit coupling parameter $\lambda$ is about $8\cdot7\times10^3$ m$^{-1}$ for this ion (see Problem 14.7), and the frequency of precession of **L**, **S** about **J** is $\lambda J/h = 4\times10^6$ MHz in the lowest state $J=\frac{3}{2}$. In contrast the frequency of precession of **J** about **B** in the $J=\frac{3}{2}$ state is only abut $6\times10^4$ MHz in a field of $B=10$ T. Thus our assumption of a very fast precession of **L**, **S** about **J**, and a much slower precession of **J** about **B**, will be valid for all fields of ordinary magnitude. In Fig. 14.7 this corresponds to the fact that the Zeeman splittings between the states of different $M_J$ (but the same $J$) are very small compared with the separation between states of different $J$. It is only when this inequality holds that the energy of a Zeeman sublevel is linearly proportional to the applied field (eqn (14.18)); when the Zeeman energy ($\sim\mu_BB$) is comparable with $\lambda$ the behaviour of the energy levels is more complicated. In the opposite extreme when $\mu_BB \gg \lambda$, the coupling

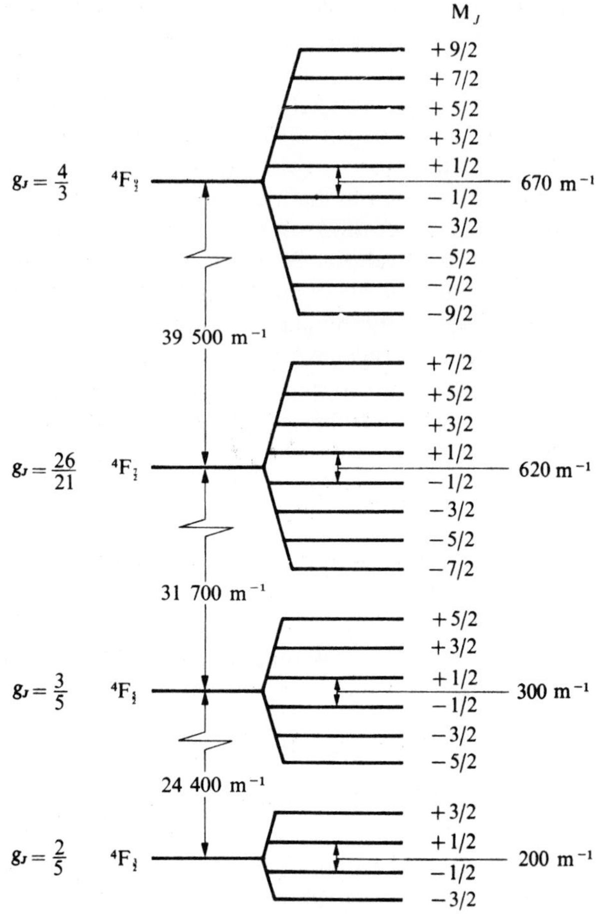

FIG. 14.7. Zeeman splitting of the ${}^4F$ states of the $Cr^{3+}$ ion (see Fig. 14.5) in a magnetic flux density $B = 10$ T. Note that the Zeeman splittings are very much smaller relative to the spin–orbit splittings (separation of states of different $J$) than the figure suggests.

between **L** and **S** is broken down, and each tends to precess independently about the external field; this is known as the Paschen–Back effect, and can be observed only in very high fields for light atoms where the spin–orbit coupling is small.

## 14.4. The measurement of atomic magnetic moments—the Stern–Gerlach experiment

The spatial quantization of angular momentum (that is, the fact that $M_J$ can have only a discrete number of values, and not a continuous range, as in classical theory) was directly demonstrated in the celebrated atomic beam experiment of Stern and Gerlach in 1922, which also made possible the direct measurement of the magnetic moment of an atom. Though this experiment has been succeeded by more refined and accurate methods, it remains an historical landmark; developments of this method, originally due to Rabi and his colleagues, have made experiments with atomic and molecular beams the basis of the extremely accurate knowledge we now possess about atomic magnetic moments, and their interaction with the magnetic moment of the nucleus.

A molecular or atomic beam is a beam of molecules or atoms moving with thermal velocities in a given direction. It is formed by heating the substance in an oven until its vapour pressure is about 1 Pa, the oven being in a highly evacuated enclosure. Atoms or molecules effuse through a narrow orifice (see Fig. 14.8), and if the pressure is so low that the mean free path is large compared with its dimensions no collisions occur in the orifice and all molecules will be moving in substantially the same direction. The angular diameter of the beam is then further limited by the slits $S_1$, $S_2$. The total path traversed by the beam in the apparatus may be up to 0·5 m, and the pressure must be so low ($<10^{-4}$ Pa) that very few collisions occur to scatter the molecules out of the beam.

In order to determine the magnetic moment of an atom, Stern and Gerlach deflected the beam in an inhomogeneous magnetic field. This was obtained from a magnet with one wedge-shaped pole piece, giving a field gradient $\partial B/\partial z$ in the $z$ direction, which is perpendicular to the path of the beam. The beam traversed the inhomogeneous field for a distance $l$ and then struck a detector, which in the early experiments was simply a target cooled in liquid air on which the molecules condensed. If $m_z$ is the component parallel to $\partial B/\partial z$ of the magnetic moment of the atom, then the force exerted on the atom in the $z$ direction is $m_z(\partial B/\partial z)$, and for

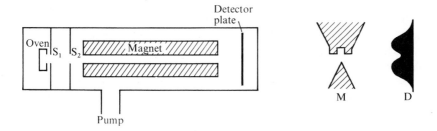

FIG. 14.8. Experimental arrangement for the Stern–Gerlach experiment.
M   end view of pole pieces producing inhomogeneous field,
D   density of trace on detector plate for atoms in doublet ground state, for example, Ag, $^2S_{\frac{1}{2}}$,
$S_1S_2$   narrow collimating slits.

a molecule of mass $M$ and velocity $v$ the deflection $s$ after traversing the field gradient is

$$s = \frac{\frac{1}{2}(l/v)^2 m_z (\partial B/\partial z)}{M} = \frac{l^2 m_z (\partial B/\partial z)}{2Mv^2}. \tag{14.19}$$

To obtain appreciable deflections (of the order of a millimetre) fields of about 1 T with gradients of about $1000\ \mathrm{T\ m}^{-1}$ are required. The deflection is inversely proportional to the thermal energy $\frac{1}{2}Mv^2$ of the molecules, and the traces are thus spread out owing to the distribution of velocities appropriate to the temperature of the oven. This limits the accuracy of this type of experiment, but with atoms such as sodium or silver, which are both in $^2S_{\frac{1}{2}}$ states, two distinct traces were obtained, with deflections appropriate to values of $m_z$ equal to $\pm 1\ \mu_B$. Thus both the existence of spatial quantization corresponding to $M_J = \pm\frac{1}{2}$ and the magnetic moment of nearly $1\ \mu_B$ associated with electron spin of $\frac{1}{2}\hbar$ were confirmed. Later modifications of these experiments gave fairly precise values of atomic magnetic moments, but much higher accuracy has been obtained by the magnetic-resonance method, outlined in Chapter 24.

## 14.5. Curie's law and the approach to saturation

A theoretical derivation of Curie's law due to Langevin was given in Chapter 6. This was based on a classical approach in its use of Boltzmann statistics, but used the idea of the existence of permanent magnetic moments of fixed values. This latter assumption is not in accordance with classical theory, for we should then expect a continuous range of magnetic moments from $-\infty$ to $+\infty$. It was shown independently by Bohr and by Miss van Leeuwen that if such a continuous range is assumed, the paramagnetic and diamagnetic contributions to the susceptibility of any system should be exactly equal and opposite, and thus, in a strictly classical calculation, the susceptibility would be zero. The quantum-mechanical approach outlined in § 14.2 shows that finite permanent magnetic dipoles do exist in atoms, and we must now examine how the Langevin calculation must be modified to take account of the fact that only a finite number of projections of the moment on an external field are allowed.

We assume that all the atoms are in the same spectroscopic state $L$, $S$, $J$, this being the ground state of the atom. It was shown in § 14.3 that the potential energy $W$ of such an atom in a magnetic field is $M_J g_J \mu_B B$, where $M_J$ is the magnetic quantum number, and $g_J$ the Landé factor appropriate to the spectroscopic state of the atoms. As in classical theory, the probability of an atom being in a state with an energy $W$ is proportional to $\exp(-W/kT)$, and for a given value of $M_J$ this is therefore proportional to $\exp(-M_J g_J \mu_B B/kT)$. Thus the fraction of all atoms in this state is

$\exp(-M_J g_J \mu_B B/kT)/\sum \exp(-M_J g_J \mu_B B/kT)$, where the summation is over all values of $M_J$. The component of the atomic magnetic moment parallel to **B** is $-M_J g_J \mu_B$, and the total magnetic moment of a system of $n$ atoms will therefore be

$$n\bar{m} = n \frac{\sum (-M_J g_J \mu_B)\exp(-M_J g_J \mu_B B/kT)}{\sum \exp(-M_J g_J \mu_B B/kT)}, \qquad (14.20)$$

where the summation in each case is over all values of $M_J$ from $+J$ to $-J$. This expression is rather clumsy to handle but it may be shown by an algebraic reduction that it reduces to the form (see Problem 14.1)

$$n\bar{m} = n g_J \mu_B J \left\{ \frac{2J+1}{2J} \coth\left(\frac{2J+1}{2J}y\right) - \frac{1}{2J} \coth\left(\frac{y}{2J}\right) \right\}, \qquad (14.21)$$

where $y = g_J \mu_B J B/kT$. The expression in brackets in eqn (14.21) is called the Brillouin function. When $J$ becomes very large it approaches as a limit the Langevin function $\{\coth y - (1/y)\}$, as we should expect from the fact that a summation over a large number of terms can be replaced by an integration, as used in the derivation in § 6.3.

At normal field strengths and ordinary temperatures, the value of $y$ is very small; at $B = 1$ T and $T = 290$ K, with $g_J = 2$ and $J = \frac{1}{2}$, $y$ is about $0 \cdot 002$. It is then possible to make a series expansion of either eqn (14.20) or (14.21). To the first order the former equation becomes

$$n\bar{m} = -n g_J \mu_B \sum_{-J}^{+J} M_J (1 - M_J g_J \mu_B B/kT) \bigg/ \sum_{-J}^{+J} (1 - M_J g_J \mu_B B/kT)$$

$$= \frac{n g_J^2 \mu_B^2 B}{(2J+1)kT} \sum_{-J}^{+J} M_J^2.$$

The summation amounts to $\frac{1}{3}J(J+1)(2J+1)$, and the susceptibility is thus

$$\chi = \frac{n\bar{m}}{H} = \frac{\mu_0 n g_J^2 \mu_B^2 J(J+1)}{3kT} = \frac{C}{T}, \qquad (14.22)$$

which is the same as the classical expression (eqn (6.13)) if we write

$$m^2 = g_J^2 \mu_B^2 J(J+1).$$

This is just the value of the square of the atomic magnetic moment expected on the quantum-mechanical theory, but at higher field strengths there is a difference between the Langevin and Brillouin functions. In particular, the limiting saturation moment reached at high field strengths and low temperatures (large values of $y$) is $g_J \mu_B J$, and not $g_J \mu_B \{J(J+1)\}^{\frac{1}{2}}$. This is because the greatest component of each moment parallel to **B** is $g_J \mu_B J$, and the actual magnetic moment always precesses at a finite angle to the field. The correctness of the Brillouin function has been verified in a number of experiments, representative results being those of Henry

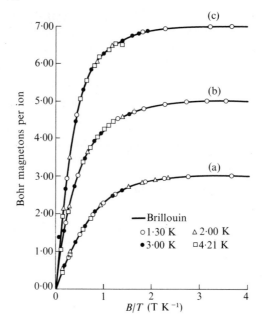

FIG. 14.9. Plot of average magnetic moment per ion m against $B/T$ for (a) potassium chromium alum $(KCr(SO_4)_2 \cdot 12H_2O)(J = S = \frac{3}{2})$, (b) iron ammonium alum $(NH_4Fe(SO_4)_2 \cdot 12H_2O)(J = S = \frac{5}{2})$, and (c) gadolinium sulphate octahydrate $(Gd_2(SO_4)_3 \cdot 8H_2O)(J = S = \frac{7}{2})$.

shown in Fig. 14.9. Note that the close approach to saturation is obtained by the combination of high field (5 T) and low temperature (4 K and lower).

## 14.6. Susceptibility of paramagnetic solids—the 4f group

The theory given above applies only to an assembly of free atoms, and the situation is rather different for matter in the aggregated state, because of the large forces exerted by the atoms on each other. These are mainly electrical in origin, and are generally far stronger than the interaction between the magnetic moment of an atom and an external magnetic field. We must therefore consider the effect of the interatomic forces first, and we shall find that, whereas most free atoms have permanent magnetic dipole moments, most bound atoms do not. This is due to the fact that the exchange forces between electrons in different atoms are nearly always of opposite sign to those between electrons in the same atom, and they therefore tend to make the electron spins line up antiparallel, giving no resultant spin whenever possible. Thus in the formation of a homopolar molecule such as $N_2$, the binding electrons from the nitrogen atoms are shared between the two atoms with their spins antiparallel; the orbits are also arranged so that the electrons have no resultant orbital angular momentum. The total angular momentum is therefore zero and the molecule has no permanent magnetic moment, though there will always be an induced negative moment when a magnetic field is applied, giving rise to diamagnetism. In heteropolar binding a molecule such as NaCl is formed of the two ions $Na^+$ and $Cl^-$, both of which have closed electron shells; thus again there is no resultant magnetic moment. Though this picture of molecule formation is oversimplified, and in general we have a mixture of homopolar and heteropolar binding, the general result of no permanent magnetic moment is still true. Thus the only common gases with permanent moments are NO and $NO_2$, which have an odd number of electrons, so that a resultant spin must remain (for NO there is also one unit of orbital angular momentum about the molecular axis); and $O_2$, where two electron spins are unpaired, giving oxygen gas a paramagnetism appropriate to $S = 1$, $g_J = 2$.

In the solid state most substances consist of ions with closed shells of electrons and are therefore diamagnetic. The main exceptions to this rule arise in compounds of the so-called 'transition elements', where an electron shell is in process of being filled. Such elements are marked by their possession of more than one chemical valency, and some (if not all) of their ions of different valency have unclosed shells, and hence a permanent magnetic moment. Since it is the electrons in the unclosed shell which determine the magnetic properties, the paramagnetism is typical of the ion, not the atom. Thus ions with the same electron configuration, even if formed from different atoms, have similar magnetic properties. These ions may be labelled by the spectroscopic description of the electron shell which is partly full; these are, 3d (iron group), 4d (palladium group), 4f (lanthanide group), 5d (platinum group), and 5f (actinide group). The titles in brackets are often used as being more descriptive, though less precise.

We consider first the 4f group, whose paramagnetism in the solid state is closest to that of an assembly of free ions. The spectroscopic states of the free ions of the 4f shell are shown in Table 14.1. It will be seen that they conform to Hund's rules, the values of first $S$ and then $L$ being the greatest possible consistent with the Pauli exclusion principle. The ground state has the smallest possible value of $J$ in the first half, and the largest value in the second half, as the spin–orbit coupling interaction changes sign when the shell is more than half full. The experimental values of $p^2$ have the following significance. If we assume that the susceptibility of a substance obeys Curie's law, we may write

$$\chi = \mu_0 nm^2/3kT = \mu_0 np^2\mu_B^2/3kT. \qquad (14.23)$$

Here $p$ is called the effective Bohr magneton number, and by comparison with eqn (14.22) we see that for an assembly of free ions

$$p^2 = g_J^2 J(J+1).$$

It is convenient to give the experimental results in terms of $p^2$, since this facilitates comparison with the theory, but it must be remembered that though we can always calculate a value of $p^2$ from the susceptibility at a given temperature, it has little significance if the susceptibility does not obey Curie's law. The latter can be established by measuring the susceptibility over a range of temperature. In this connection it must be emphasized that only measurements on 'magnetically dilute' salts are significant; by this phrase is meant salts where the paramagnetic ions are fairly far apart so that mutual interaction between them may be neglected (see Problem 14.2 and § 15.1). This condition is generally fulfilled for hydrated salts, and the values of $p^2$ in Tables 14.1, 14.2 are for salts where the effect of mutual interaction on the susceptibility is appreciable only at very low temperatures.

The average values of $p^2$ at room temperature measured experimentally and shown in the last column of Table 14.1 are in fair agreement with those calculated for an assembly of free ions with angular momentum $J$. The reason for this is that states of different $J$ are separated by energies corresponding to several thousand degrees Kelvin, so that (1)

TABLE 14.1

*Comparison of theoretical and measured values of $p^2$ for trivalent rare-earth ions*

| Number of electrons in 4f shell | Ion | Ground spectroscopic state | Theoretical values | | | | | Average experimental value of $p^2$ |
|---|---|---|---|---|---|---|---|---|
| | | | $S$ | $L$ | $J$ | $g_J$ | $p^2 = g_J^2 J(J+1)$ | |
| 0 | La$^{3+}$ | $^1S_0$ | 0 | 0 | 0 | — | 0 | 0 |
| 1 | Ce$^{3+}$ | $^2F_{\frac{5}{2}}$ | $\frac{1}{2}$ | 3 | $\frac{5}{2}$ | $\frac{6}{7}$ | 6·43 | 6 |
| 2 | Pr$^{3+}$ | $^3H_4$ | 1 | 5 | 4 | $\frac{4}{5}$ | 12·8 | 12 |
| 3 | Nd$^{3+}$ | $^4I_{\frac{9}{2}}$ | $\frac{3}{2}$ | 6 | $\frac{9}{2}$ | $\frac{8}{11}$ | 13·1 | 12 |
| 4 | Pm$^{3+}$ | $^5I_4$ | 2 | 6 | 4 | $\frac{3}{5}$ | 7·2 | — |
| 5 | Sm$^{3+}$ | $^6H_{\frac{5}{2}}$ | $\frac{5}{2}$ | 5 | $\frac{5}{2}$ | $\frac{2}{7}$ | 0·71 (2·5) | 2·4 |
| 6 | Eu$^{3+}$ | $^7F_0$ | 3 | 3 | 0 | — | 0 (12) | 12·6 |
| 7 | Gd$^{3+}$ | $^8S_{\frac{7}{2}}$ | $\frac{7}{2}$ | 0 | $\frac{7}{2}$ | 2 | 63 | 63 |
| 8 | Tb$^{3+}$ | $^7F_6$ | 3 | 3 | 6 | $\frac{3}{2}$ | 94·5 | 92 |
| 9 | Dy$^{3+}$ | $^6H_{\frac{15}{2}}$ | $\frac{5}{2}$ | 5 | $\frac{15}{2}$ | $\frac{4}{3}$ | 113 | 110 |
| 10 | Ho$^{3+}$ | $^5I_8$ | 2 | 6 | 8 | $\frac{5}{4}$ | 112 | 110 |
| 11 | Er$^{3+}$ | $^6I_{\frac{15}{2}}$ | $\frac{3}{2}$ | 6 | $\frac{15}{2}$ | $\frac{6}{5}$ | 92 | 90 |
| 12 | Tm$^{3+}$ | $^3H_6$ | 1 | 5 | 6 | $\frac{7}{6}$ | 57 | 52 |
| 13 | Yb$^{3+}$ | $^2F_{\frac{7}{2}}$ | $\frac{1}{2}$ | 3 | $\frac{7}{2}$ | $\frac{8}{7}$ | 20·6 | 19 |
| 14 | Lu$^{3+}$ | $^1S_0$ | 0 | 0 | 0 | — | 0 | 0 |

The values given in parentheses for Sm$^{3+}$ and Eu$^{3+}$ are those calculated by Van Vleck allowing for population of excited states with higher values of $J$, at $T = 293$ K.

only the ground state is occupied at ordinary temperatures, and (2) the precessing component of the magnetic moment represented by DE in Fig. 14.6 makes only a small contribution under these circumstances. The exceptions are the ions 4f⁵, 4f⁶ in which excited levels with higher values of *J* are exceptionally low in energy (see Problem 14.7). Van Vleck has shown that much better agreement is obtained when allowance is made for the presence of the excited levels, and his values, calculated for *T* = 273 K, are shown in parentheses.

For the ions Gd³⁺, Eu²⁺ whose spectroscopic state is 4f⁷, ⁸S₇⸍₂ the susceptibility follows Curie's law down to temperatures below 1 K, but for the other ions the susceptibility shows marked departures from this law at much higher temperatures. The reason for this is that we cannot neglect the influence of the charged ions which surround each paramagnetic ion in the solid state. In a magnetically dilute salt these immediate neighbours carry no permanent magnetic moment (they are diamagnetic ions such as F⁻, O²⁻, etc.), but they are electrically charged, and have an electrostatic interaction with the 4f electrons which are responsible for the paramagnetism. To a good approximation the 4f electrons can be regarded as

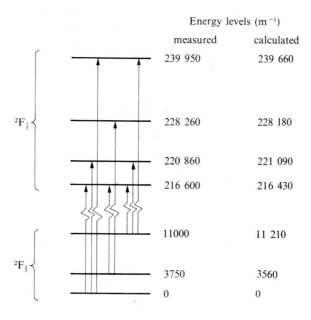

FIG. 14.10. Energy levels and optical transitions of the Ce³⁺ ion (4f¹, *L* = 3, *S* = ½) in LaCl₃. The splittings of both the *J* = 5⁄2 and *J* = 7⁄2 manifolds into series of doublet levels can be fitted within the experimental error by a crystal field of hexagonal symmetry and a spin–orbit coupling constant ζ = 6·26 × 10⁴ m⁻¹ in eqn (14.7). The over-all splitting of the ²F₅⁄₂ levels is about 160 K.

moving in an electric field set up by the neighbouring ions, known as the 'crystalline electric field'. The energy of interaction with this field is smaller, for ions of the 4f group, than the Coulomb, exchange, and spin–orbit interactions within the paramagnetic ion itself; it gives rise to a 'Stark' splitting of the $2J+1$ levels of the ion. This is similar in nature to the effect of an external electric field on the spectrum of an atom, first investigated in detail by Stark, but is considerably more complex because the electrostatic potential set up by the neighbours varies in a complicated way over the space occupied by the 4f electrons. The over-all splittings of the $2J+1$ levels of a 4f ion are generally of the order of a few hundred degrees Kelvin, as shown in Fig. 14.10 for $Ce^{3+}$, $4f^1$ in $LaCl_3$. The susceptibility is rather insensitive to such splittings, and approaches that of the free ion at temperatures where most of the levels are appreciably populated. At low temperatures where only the very lowest levels are populated, the susceptibility can be very different from that of the free ion, and in a single crystal may be highly anisotropic.

At first sight it may appear surprising that the crystalline electric field can have such a marked effect on the magnetic properties. The basic reason is that the wavefunctions corresponding to different values of the orbital magnetic quantum number $M_L$ have different angular dependencies, that is, for each value of $M_L$ the distribution of electronic charge has a different shape, and hence acquires a different electrostatic energy in

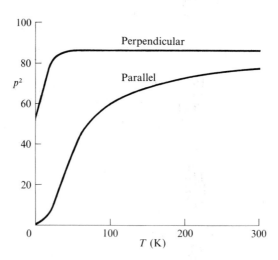

FIG. 14.11. The values of $p^2$ (parallel and perpendicular) for a single crystal of erbium ethylsulphate $(Er(C_2H_5SO_4)_3 \cdot 9H_2O)$. This forms hexagonal crystals, and the susceptibility is symmetrical about the hexagonal axis. The ground state of the $Er^{3+}$ ion is $^4I_{\frac{15}{2}}$, and this is split by the crystalline electric field into eight doublets lying at 0, 61, 108, 159, 249, 301, 375, and 438 K.

the crystalline electric field. Thus the primary interaction is associated with the electronic orbit, and the interaction is zero (except for a small residual effect due to a slight departure from pure $L-S$ coupling) for an ion such as $Eu^{2+}$ or $Gd^{3+}$ with a half-filled shell carrying no orbital angular momentum ($L = 0$). For the other ions the coupling together of $L$ and $S$ means that states of different $M_J$ have a different charge distribution, so that the crystalline electric field splits the $2J+1$ states which otherwise have the same energy in the absence of a magnetic field.

An important restriction on this splitting occurs for ions with an odd number of electrons, which have half-integral values of $S$ and hence also of $J$; in this case the states occur always in pairs which have the same charge distribution and differ only in the orientation of the magnetic moment. Each pair must thus retain the same energy in an electric field, though they can be split in a magnetic field. This result was proved in a theorem of Kramers and the double degeneracy of such states in an electric field is known as 'Kramers' degeneracy'. Typical examples are $Ce^{3+}$, $4f^1$ (Fig. 14.10), and $Er^{3+}$, $4f^{11}$. At low temperatures where just one such pair of states is occupied, the value of $p^2$ has a temperature variation of the form $A+BT$ (see Problem 14.3); in Fig. 14.11 the variation of $p^2$ for a compound of $Er^{3+}$ near $T = 0$ is of this form both for the perpendicular and for the parallel direction (in the latter case $g_{\parallel}^2$ is so small that $A$ appears to be zero in the figure). For a 'non-Kramers' ion, with an even number of electrons and an integral value of $J$, the ground state left by the crystal field may be a singlet state. This has no permanent

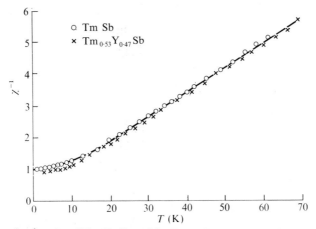

FIG. 14.12. Plot of $\chi^{-1}$ against $T$ for TmSb and for $Tm_{0.53}Y_{0.47}Sb$, in which about half of the paramagnetic $Tm^{3+}$ ions are replaced by diamagnetic $Y^{3+}$ ions. The continuous line is a theoretical fit using a cubic crystal field. The vertical axis is in units where $\chi^{-1} = 1$ when a field of 1 T produces a magnetic moment of 1 $\mu_B$ per terbium ion.

magnetic moment, but when a field is applied it acquires an induced moment which is proportional to the field strength in small fields, but independent of temperature. The susceptibility then becomes independent of temperature, a phenomenon first explained by Van Vleck. A clear example of this effect is the cubic compound TmSb, where the ions occupy a regular crystal lattice with the NaCl structure (it is not an alloy). The Sb antimony ion is not magnetic, but the ion $Tm^{3+}$, $4f^{12}$ has $J = 6$. The inverse susceptibility $\chi^{-1}$ is plotted in Fig. 14.12; it can be fitted accurately with a cubic crystal field whose potential is of the form given in eqn (14.24).

## 14.7. Susceptibility of paramagnetic solids—the 3d group

The spectroscopic ground states of the free ions of the 3d shell, shown in Table 14.2, again follow Hund's rules. An assembly of free ions should give a susceptibility corresponding to $p^2 = g_J^2 J(J+1)$, but a comparison of the values of this quantity with the experimental values for solids shows a striking disagreement. In fact the experimental values lie much closer to $p^2 = 4S(S+1)$, the value which would be expected if there were no orbital angular momentum and the magnetism were due entirely to the electron spin. This is clearly brought out in Fig. 14.13, in which average experimental values of $p^2$ are plotted together with the quantities $g_J^2 J(J+1)$ and $4S(S+1)$.

TABLE 14.2

| Number of electrons in 3d shell | Ion | Ground state | $S$ | $L$ | $J$ | $g_J^2 J(J+1)$ | $p^2$ (experimental) | $4S(S+1)$ |
|---|---|---|---|---|---|---|---|---|
| 0 | $K^+, Ca^{2+}, Sc^{3+},$ $Ti^{4+}, V^{5+}$ | $^1S_0$ | 0 | 0 | 0 | 0 | 0 | 0 |
| 1 | $Ti^{3+}, V^{4+}$ | $^2D_{\frac{3}{2}}$ | $\frac{1}{2}$ | 2 | $\frac{3}{2}$ | 2·4 | 2·9 | 3 |
| 2 | $V^{3+}$ | $^3F_2$ | 1 | 3 | 2 | 2·67 | 6·8 | 8 |
| 3 | $V^{2+}, Cr^{3+}$ | $^4F_{\frac{3}{2}}$ | $\frac{3}{2}$ | 3 | $\frac{3}{2}$ | 0·6 | 14·8 | 15 |
| 4 | $Cr^{2+}, Mn^{3+}$ | $^5D_0$ | 2 | 2 | 0 | 0 | (23·3) | 24 |
| 5 | $Mn^{2+}, Fe^{3+}$ | $^6S_{\frac{5}{2}}$ | $\frac{5}{2}$ | 0 | $\frac{5}{2}$ | 35 | 34·0 | 35 |
| 6 | $Fe^{2+}$ | $^5D_4$ | 2 | 2 | 4 | 45 | 28·7 | 24 |
| 7 | $Co^{2+}$ | $^4F_{\frac{9}{2}}$ | $\frac{3}{2}$ | 3 | $\frac{9}{2}$ | 44 | 24·0 | 15 |
| 8 | $Ni^{2+}$ | $^3F_4$ | 1 | 3 | 4 | 31·3 | 9·7 | 8 |
| 9 | $Cu^{2+}$ | $^2D_{\frac{5}{2}}$ | $\frac{1}{2}$ | 2 | $\frac{5}{2}$ | 12·6 | 3·35 | 3 |
| 10 | $Cu^+, Zn^{2+}$ | $^1S_0$ | 0 | 0 | 0 | 0 | 0 | 0 |

The values of $p^2$ (at 300 K) are for double sulphates of the type $M''M_2'(SO_4)_2 \cdot 6H_2O$ or $M'''M'(SO_4)_2 \cdot 12H_2O$ (where $M''$ = divalent paramagnetic ion, $M'''$ = trivalent paramagnetic ion, $M'$ = monovalent diamagnetic ion). In these salts the distance between nearest paramagnetic ions is at least 0·6 nm, and interaction between them is negligible. The value in parentheses is for $CrSO_4 \cdot 6H_2O$: no double sulphate of $Cr^{2+}$ has been measured.

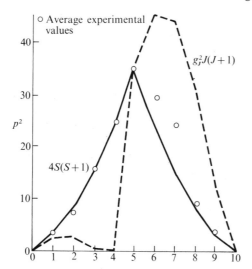

FIG. 14.13. Experimental (at 300 K) and calculated values of $p^2$ for the 3d group. In the second half of the group the orbital angular momentum is less effectively quenched than in the first half, so that the values of $p^2$ lie noticeably above the spin-only values.

This phenomenon, known as the 'quenching' of the orbital magnetism, is a result of the crystalline electric field. In the lanthanide group the 4f electrons, which are responsible for the paramagnetism, are fairly deep seated in the atom, but in the iron group the 3d electrons are in an outer shell which has a very much larger interaction with the crystalline electric field of neighbouring charged ions. On the other hand, the spin–orbit coupling in the 3d group is considerably smaller than in the 4f group. The result is that interaction between the orbit and the crystalline electric field is a good deal stronger than the spin–orbit coupling for 3d ions, so that the orbital momentum is primarily coupled to the crystal field, and it is no longer correct to regard $L$ and $S$ as coupled to form a resultant $J$. The quantitative expression of this situation is that the $2L+1$ orbital states are split in the crystal field, and have energies differing by about $10^6\ \mathrm{m}^{-1}$, which is much larger than the spin–orbit splittings (of the order of $10^4$–$10^5\ \mathrm{m}^{-1}$) between the states of different $J$ in the free ion. The simplest case to consider is that where the crystal field splitting of the orbital levels gives a singlet state as the lowest level. Such a state has no magnetic moment, and the orbital moment is completely 'quenched'. On the other hand, the electron spin has no direct interaction with the crystalline electric field, and remains free to orient itself in a magnetic field. Thus, in this case, the susceptibility would correspond exactly to the 'spin only' value of $p^2 = 4S(S+1)$ at all temperatures such that there is no appreciable population of an excited orbital state.

It can be seen from Fig. 14.13 that the values of $p^2$ do not follow exactly the 'spin only' values, particularly for ions with $d^6$ ($Fe^{2+}$) and $d^7$ ($Co^{2+}$) configurations. The basic reason for this is that the crystalline electric field does not always result in a singlet orbital state as the lowest state, but sometimes gives a group of low-lying orbital states which can make some contribution to the magnetic moment, though less than the full orbital contribution from a free ion. In principle we could calculate the splitting of the orbital levels, but in practice this is extremely difficult to do. However, the general features of the magnetic properties of salts of the 3d group are well understood, mainly through the work of Van Vleck, and we will now attempt to outline the main results of the crystal-field approach.

The size of an ion of the 3d group is such that six negatively charged ions can be packed round it; when these ions are identical, they are arranged in the form of an octahedron which is very nearly regular. In hydrated salts these six ions are commonly the oxygens of six water molecules, as shown in Fig. 14.14. In oxides such as MnO the octahedron is regular and formed by six $O^{2-}$ ions; in fluorides such as $KMnF_3$ there is a regular octahedron of $F^-$ ions. These ions are known as the 'ligand ions', and in general there is a small amount of sharing of electrons between them and the 3d ion. In the crystal field theory this is ignored, and the magnetic 3d electrons are assumed to be localized on the 3d ion, and to move in the electrostatic potential of the surrounding charged ligand ions. If the 3d ion is assumed to be at the point $(0, 0, 0)$, the ligand ions may be taken to lie at the points

$$(\pm a, 0, 0), \quad (0, \pm a, 0), \quad (0, 0, \pm a)$$

thus forming a regular octahedron. If we assign a charge $-2e$ to each oxygen ion, the electrostatic potential near the centre of the octahedron

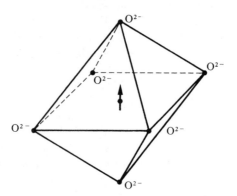

FIG. 14.14. Octahedron of oxygen ions round a paramagnetic 3d ion.

(see Problem 2.18) is

$$V = -\frac{12e}{4\pi\epsilon_0 a} - \frac{2e}{4\pi\epsilon_0}\left(\frac{35}{4a^5}\right)(x^4 + y^4 + z^4 - \tfrac{3}{5}r^4), \qquad (14.24)$$

and this will change the energy of an electron on the central ion by an amount $\int \psi^*(-eV)\psi \, d\tau$, where $\psi$ is the electronic wavefunction. This energy change is a quantitative expression of the fact that, since the electrons on the magnetic ion are negatively charged, they will have a lower energy in states where they avoid the negatively charged ligand ions as much as possible, and a higher energy when they do not, because of the electrostatic repulsion. Under the action of the potential (14.24), the correct $d$ orbitals are represented by real wavefunctions, which can be written as a radial function $f(r)$ times the following functions, which express the angular dependence in Cartesian coordinates instead of the spherical harmonics of § 2.2:

$$\left.\begin{aligned} C_{2,0} &= \tfrac{1}{2}(2z^2 - x^2 - y^2)/r^2 \\ \tfrac{1}{\sqrt{2}}(C_{2,2} + C_{2,-2}) &= \tfrac{\sqrt{3}}{2}(x^2 - y^2)/r^2 \end{aligned}\right\} \quad (d\gamma),$$

$$\left.\begin{aligned} -\tfrac{j}{\sqrt{2}}(C_{2,2} - C_{2,-2}) &= \sqrt{3}\, xy/r^2 \\ \tfrac{j}{\sqrt{2}}(C_{2,1} + C_{2,-1}) &= \sqrt{3}\, yz/r^2 \\ -\tfrac{1}{\sqrt{2}}(C_{2,1} - C_{2,-1}) &= \sqrt{3}\, zx/r^2 \end{aligned}\right\} \quad (d\epsilon). \qquad (14.25)$$

The last three (known as d$\epsilon$ states) are each zero along two of the cubic axes (see Fig. 14.15), so that the charge density (which is proportional to the square of the wavefunction) is also zero along two of the axes. This gives a lower energy for these three states (by symmetry each must have the same energy in a cubic field, since $x$, $y$, $z$ are all equivalent if the octahedron is regular) than for the other two (d$\gamma$) states which have a finite density along all three cubic axes. Hence we get a splitting of the D(d) state as shown in Fig. 14.16 for $d^1$. The splitting is similar, but inverted, for $d^9$, which is one electron short of a filled d shell. A filled shell has a spherical charge distribution, and the charge distribution for $d^9$ is equivalent to a filled shell plus a 'positive hole', for which the electrostatic energy in the crystal field has the opposite sign. A half-filled shell also has spherical symmetry, with $L = 0$; for this reason the ligand field plays virtually no role in affecting the paramagnetism of $d^5$ ions, whose ground state is $^6S_{\frac{5}{2}}$. However, a $d^6$ ion, with one electron outside the half-filled shell, has a similar splitting to $d^1$, while $d^4$ corresponds to a positive hole in a half-filled shell and behaves like $d^9$ (see Fig. 14.16).

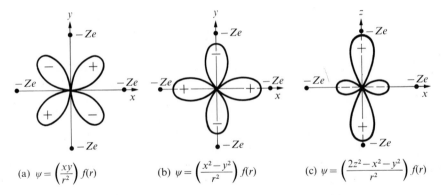

(a) $\psi = \left(\dfrac{xy}{r^2}\right) f(r)$     (b) $\psi = \left(\dfrac{x^2 - y^2}{r^2}\right) f(r)$     (c) $\psi = \left(\dfrac{2z^2 - x^2 - y^2}{r^2}\right) f(r)$

FIG. 14.15. Angular variation of the d orbitals of eqn (14.25). (a) is a $d\epsilon$ state (the other two $d\epsilon$ states are similar but differently oriented); (b) and (c) are $d\gamma$ states. $d\epsilon$ and $d\gamma$ functions have different symmetry properties: $d\gamma$ functions do not change sign on reflection in any of the cubic planes (that is, $x \rightarrow -x$, or $y \rightarrow -y$, or $z \rightarrow -z$), while the $d\epsilon$ functions change sign for two such reflections but not for the third.

Thus $d^1$, $d^4$, $d^6$, $d^9$ have a basically similar splitting pattern because each is equivalent to a single electron or single hole state, as far as the orbit is concerned (they do not all have the same spin).

Similarly, the remaining ions $d^2$, $d^3$, $d^7$, $d^8$ are orbitally equivalent to two-electron or two-hole states ($d^7 \equiv$ half-filled shell+2 electrons; $d^3 \equiv$ half-filled shell+2 holes; $d^8 \equiv$ filled shell+2 holes). They are all in F states, with $L = 3$, which are split by a cubic crystal field into a singlet and

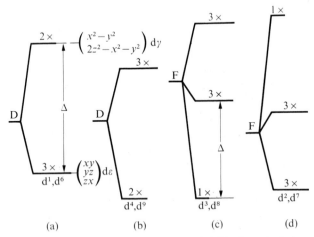

FIG. 14.16. Splittings of D and F states in a crystal field of octahedral symmetry. The over-all splittings lie generally in the range $3 \times 10^6 \, \mathrm{m}^{-1}$. The strong colours of many paramagnetic compounds of the 3d group are due to absorption bands in or near the visible region of the spectrum which arise from transitions between the ground state and excited states shown above, combined with vibrational effects.

two triplet levels, as shown in Fig. 14.16. In $d^3$ and $d^8$ the lowest level is a singlet (it corresponds to a wavefunction $xyz$, which has zero density along all three cubic axes), but in $d^2$ and $d^7$ the splitting pattern is inverted.

Using these crystal field splittings, we can distinguish between two separate cases.

1. When the orbital ground state is a singlet, it has no component of angular momentum along any axis, so that the magnetism is due primarily to the spin. The susceptibility follows Curie's law very exactly (for example, the susceptibility of a chrome alum such as $CrKSO_4 \cdot 12H_2O$ does not deviate from Curie's law by more than 2 per cent between room temperature and 2 K). There are, however, two residual effects of the spin–orbit coupling: (a) the effective g-value differs from the free spin value by an amount of order $\lambda/\Delta$, where $\lambda$ is the spin–orbit coupling and $\Delta$ the splitting between the ground orbital level and the excited orbital level shown in Fig. 14.16. The spin–orbit constant $\lambda$ is positive if the d shell is less than half-filled, and negative if it is more than half-filled. It is also larger for the ions at the end of the group because of the increased nuclear charge. Thus the effective value of g is about 1 per cent smaller than the free spin value for $Cr^{3+}$, $d^3$, but 10 per cent higher for $Ni^{2+}$, $d^8$. (b) Where the spin is 1 or more, the $2S+1$ spin states may be split by amounts of order $(\lambda^2/\Delta)$, which is usually of the order of $10-10^3 \, m^{-1}$ (roughly $0\cdot1 - 1$ K). This gives a heat capacity anomaly, of which a typical example is shown in Fig. 14.17.

2. When the ground state is not a singlet, there is a first-order contribution from the orbit to the paramagnetism, but less than that for

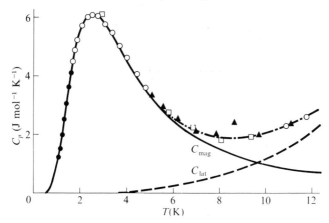

FIG. 14.17. Magnetic heat capacity anomaly of $\alpha$-nickel sulphate hexahydrate ($\alpha NiSO_4 \cdot 6H_2O$). The anomaly is associated with the spin triplet states, which lie at 0 K, 6·44 K, and 7·26 K respectively. (After Stout and Hadley 1964.)

the free ion. When the ground state is a triplet in Fig. 14.16, it behaves like a P state with $L = 1$, and an effective $g_L$ which is $-1$ for $d^1$ and $d^6$, and $-\frac{3}{2}$ for $d^2$ and $d^7$. Thus it can interact with the spin through the spin–orbit interaction, giving states with an effective $J$ of $S-1$, $S$, and $S+1$ if $S \geqslant 1$, or $\frac{1}{2}$ and $\frac{3}{2}$ if $S = \frac{1}{2}$. The splittings between these levels are of order $10^4\,\mathrm{m}^{-1}$ ($\sim 100$ K), so that Curie's law is not obeyed because excited states become occupied as the temperature is raised. This is particularly noticeable for cobalt ($Co^{2+}$, $d^7$) salts, as shown in Fig. 14.18. The ions $d^4$, $d^9$ are exceptional because the doublet orbital states left as their ground states by the octahedral field have effectively $g_L = 0$. Thus they behave rather like case 1. Fig. 14.18 shows a typical cupric salt, $Cu^{2+}$, $d^9$, where Curie's law is obeyed closely, but the effective g-value is more than 10 per cent higher than the free spin value, giving $p^2 = 3 \cdot 76$ instead of the value $4S(S+1) = 3$ we would expect for a single hole ($S = \frac{1}{2}$).

A striking effect in many single crystals of paramagnetic substances of the 4f group is the high anisotropy of the susceptibility; this arises because the surroundings of the paramagnetic ion in such crystals have less than cubic symmetry. For the regular octahedron in Fig. 14.14 there would be no anisotropy, but in the 3d group this octahedron is normally somewhat distorted; as a result the orbital contributions to the magnetism depend on the direction in which the external field is applied relative to the crystal axes. The anisotropy is large when the splitting of the lowest orbital levels is small, that is, when there are also considerable departures from Curie's law, as in cobalt salts.

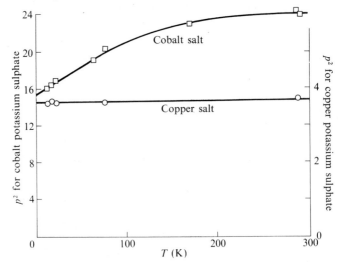

FIG. 14.18. Variation of $p^2$ with temperature for two iron-group salts. The change in $p^2$ with temperature for the cobalt salt is due to the presence of low-lying excited states (cf. Fig. 14.11).

## 14.8. Susceptibility of paramagnetic solids—strongly bonded compounds

Much less is known in detail about the magnetic properties of salts of the 4d and 5d groups, but in many cases it appears that the binding to the ligand ions is covalent rather than ionic in character. This is true also for a few salts of the iron group, notably the complex cyanides such as $K_3Fe(CN)_6$. Here, and in salts such as $K_2IrCl_6$, where the magnetic $Ir^{4+}$ ion has the configuration $5d^5$, the ligand ions (six CN groups in the former case, six $Cl^-$ ions in the latter) are again arranged in the form of a very nearly regular octahedron. We shall confine the discussion to this type of compound, as it affords an interesting comparison with the pure crystal field approach.

In a covalent bond the electrons are shared between the two ions concerned, in contrast with a purely ionic case where the electrons are localized on each ion. The latter is an oversimplification, and in practice there is always a small amount of covalent bonding, so that we distinguish only between weak bonding, and strong bonding. In a complex with octahedral symmetry, the $d\gamma$ states have maximum density along the cubic axes (towards the ligand ions), and can form $\sigma$-bonds with the ligand ions, while the $d\epsilon$ states can only form $\pi$-bonds. In the formation of a bond, a d-state on the magnetic ion is combined with the appropriate bonding state of the ligand ion, and the overlap of the electronic wavefunctions gives a splitting of the combined levels, in the same way as pointed out in § 12.2. The overlap is greater for the $d\gamma$ states (forming $\sigma$-bonds) than for the $d\epsilon$ states (forming $\pi$-bonds), giving the energy level diagram shown in Fig. 14.19. The lower bonding states are all filled with electrons, and behave as filled subshells. Thus the states available for magnetic electrons are the antibonding states, which are split in the same

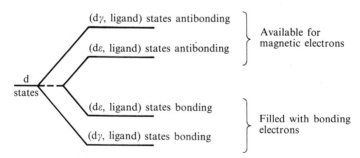

FIG. 14.19. Splitting of the d states on the bonding model. The lowest (bonding) states are filled with electrons, and only the antibonding states are available for the magnetic electrons. In the crystal-electric-field approach the bonding states play no role, and an octahedral field splitting (see Fig. 14.16) of the d state is obtained similar to that for the antibonding states above.

way as by an octahedral crystal field. In weakly bonded compounds this splitting is about $10^6 \, \text{m}^{-1}$, as mentioned in the previous section, but in the strongly bonded compounds it is very much larger, so that the latter behave as though subjected to a very much stronger crystalline electric field. However, the approach from the bonding viewpoint is more correct, since it allows the magnetic electron wavefunctions to spread out from the central ion on to the ligand ions, for which there is direct experimental evidence from measurements of the hyperfine interaction between the magnetic electrons and the nuclear moments of the ligand ions.

In this more general approach, allowing for bonding, the splitting between the $d\epsilon$ states and the $d\gamma$ antibonding states is ascribed to the 'ligand field', and it is interesting to contrast the two cases of small and large ligand field. Here the comparison is with the electrostatic and exchange energy which is responsible for $L$–$S$ coupling, and when the ligand field splitting is large compared with this exchange energy we must regard the $L$–$S$ coupling as broken. The way in which the $d\epsilon$ and $d\gamma$ antibonding states are occupied by the magnetic electrons in the two cases is determined by the competition between the exchange energy (which favours parallel orientation of the electron spins) and the ligand field splitting (which favours electrons going into the $d\epsilon$ states because of their lower energy).

We may represent each orbital state by a pair of square boxes, as in Fig. 14.20, into each of which we can put one electron, with spin up or

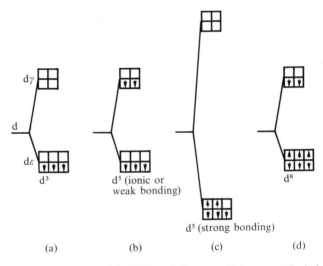

FIG. 14.20. Single-electron model of filling of d states split by an octahedral crystal field. The exchange energy favours parallel spin arrangements (subject to the exclusion principle); the crystal field splitting favours electrons in the $d\epsilon$ states.

down. With one electron, the state of lowest energy will obviously be when this electron is in the d$\epsilon$ states. When further electrons are added they will also go into the d$\epsilon$ states, but with spins parallel in order to make the exchange energy a minimum; up to three electrons can be accommodated in this way. The d$\epsilon$ shell is then half full, and behaves like a state with zero orbital momentum (corresponding to the singlet orbital ground state for d$^3$ in Fig. 14.16) and $S = \frac{3}{2}$. When more electrons are added, they cannot go into the d$\epsilon$ shell with parallel spin, because this would violate the exclusion principle. There is therefore a competition between the exchange energy, which prefers parallel spin, and the ligand field energy, which prefers the d$\epsilon$ states. In the hydrated salts, the latter is less important, and the fourth and fifth electrons go into the d$\gamma$ states with parallel spin, making d$^5$ a state with $L = 0$, $S = \frac{5}{2}$ (a half-filled d shell). In the more strongly bonded salts the ligand field splitting is so large that no electrons go into the d$\gamma$ states, since they have a lower energy by occupying the d$\epsilon$ states with antiparallel spin, as shown in Fig. 14.20(c). Thus a strongly bonded d$^5$ salt behaves as if it had one hole in the d$\epsilon$ shell; for example, $K_3Fe(CN)_6$, where the $Fe^{3+}$ ion has a d$^5$ configuration, has magnetic properties quite different from hydrated ferric compounds, showing considerable departures from Curie's law, a susceptibility close to that of a single spin, and strong anisotropy in single crystals. With six electrons, d$^6$, strongly bonded, the d$\epsilon$ shell is completed, and the ion has no permanent magnetic dipole moment ($K_3Co(CN)_6$ has only a small temperature-independent paramagnetism). Any further electrons would have to go into the d$\gamma$ states, but these are so high in energy that such ions are usually chemically unstable. However, when the ligand field is less strong, the d$\gamma$ states are occupied, as shown in Fig. 14.20(d) for a hydrated $Ni^{2+}$, d$^8$, ion. Here the d$\epsilon$ states are completely filled and the d$\gamma$ half-filled, giving again a ground state with no orbital momentum (cf. Fig. 14.16(c)). On this single electron picture we can see that the orbital momentum is effectively quenched whenever the two sets of d$\epsilon$ and d$\gamma$ states are each either empty, half-filled, or completely filled with electrons; the reader can verify that a half-filled subshell, with all electron spins parallel, can be achieved by only one possible arrangement of the electrons in the various boxes, and therefore corresponds to a singlet orbital state. On the other hand, when the d$\epsilon$ states are occupied by one or two electrons, or four or five electrons, there are three equivalent ways of arranging them, giving a triply degenerate orbital state. This corresponds to the triplet state which is lowest for d$^1$, d$^2$, d$^6$, d$^7$ in Fig. 14.16.

## 14.9. Electronic paramagnetism—a summary

The magnetic properties of insulating paramagnetic compounds are the result of competition between (1) electrostatic and exchange interactions within the ion which favour $L-S$ coupling; (2) spin–orbit interaction within the ion; and (3) interaction with the ligand field of neighbouring (diamagnetic) ions. In this rather complex situation the largest interaction must be considered first, and the different behaviour of the various transition groups arises from the change in the size of the ligand interaction (3) relative to (1) and (2). In broad outline, the situation may be summarized as follows.

(*a*) Electrons of the 4f group (and, to a lesser extent, of the 5f group) are 'inner' electrons whose interaction energy with the ligand field is of order $10^4 \, \text{m}^{-1}$. This is smaller than the spin–orbit coupling energy $(10^5 \, \text{m}^{-1})$, and splits the $(2J+1)$ levels by amounts up to $2$–$5 \times 10^4 \, \text{m}^{-1}$.

(*b*) Electrons of the 3d group are 'valence' electrons, and their interaction with the ligand field is of order $10^6 \, \text{m}^{-1}$, comparable with the $L-S$ coupling energy. In a first approximation the ligand field splits the $(2L+1)$ orbital levels, and the spin states are then affected through the spin–orbit coupling.

(*c*) Electrons of the 4d, 5d groups (and of the 3d group in some compounds) are valence electrons for which the ligand interaction is stronger than $L-S$ coupling. The $(2l+1)$ orbital states of each electron are strongly split by the ligand field, generating new single-electron states on which the effects of electrostatic and spin–orbit couplings within the atom are evaluated subsequently.

The ligand field is electrostatic in origin and interacts directly only with the orbitals, which have different charge distributions, and not with the electron spin. The latter interacts with a magnetic field, either that applied from outside or that generated through the orbital motion (spin–orbit coupling). By Kramers' theorem, every energy level of an ion with an odd number of electrons must be associated with at least two electron states which have the same distribution of electric charge and hence the same energy in an electric field. Such pairs of states have equal and opposite magnetic moments, and are split in energy by a magnetic field, either externally applied or originating in the fields of neighbouring dipoles. The dipoles also interact with each other through exchange interaction, which may result in spontaneous magnetization, even at room temperature (see Chapters 15 and 16), and the onset of an ordered magnetic state.

The susceptibility is a primary magnetic quantity for the paramagnetic state, but it is in fact rather a poor guide to the ligand field splittings. At temperatures where all states have roughly equal populations, the susceptibility behaviour approaches that given by Curie's law. It departs markedly from this only at temperatures where the populations are changing rapidly, that is, where $kT$ is of the same order as the energy splitting (see

Problem 14.4). Optical and infrared spectroscopy give much more accurate information about the actual energy levels, and the distinctive colours of transition-ion compounds are due to absorption bands at or near optical frequencies. Electron spin resonance gives very detailed information on the properties of the lowest levels, such as the ground state and others within about 10 K in energy. Splittings of this order can also be determined calorimetrically.

## 14.10. Nuclear moments and hyperfine structure

For nuclei, as for atoms, magnetic moments are associated with the presence of a quantized angular momentum or 'spin', whose value is $I\hbar$. For both proton and neutron the value of $I$ is $\frac{1}{2}$, and if a nucleus contains an odd number of nucleons, $I$ is half-integral. For an even number of nucleons $I$ is an integer, but if the numbers of protons and neutrons are each individually even, the value of $I$ is zero for the state of lowest energy of the nucleus. Nuclear magnetic moments are measured in terms of a unit called the 'nuclear magneton', equal to $\mu_N = e\hbar/2m_p$; this is analogous to the Bohr magneton, but the mass in the denominator is that of the proton instead of that of the electron. If the proton obeyed a similar wave equation to that for the electron, we would expect it to possess a moment of one nuclear magneton associated with its spin $\frac{1}{2}$, just as the electron has a moment of 1 $\mu_B$ and spin $\frac{1}{2}$. In fact the moment of the proton is $+2 \cdot 793$ nuclear magnetons and the neutron, although uncharged, has a moment of $-1 \cdot 913 \mu_N$. Here the significance of the plus and minus signs is that the magnetic moments are respectively parallel and antiparallel to the spin. Since the magnetic moment of neither neutron nor proton is an integral number of nuclear magnetons we should not expect the moments of more complicated nuclei to be simple integers. They do, however, follow the trend which the nuclear shell model would indicate (see Problem 14.6); in general we write the nuclear magnetic moment as $\mathbf{m}_N = g_N \mu_N \mathbf{I}$, where $g_N$ is the nuclear magnetogyric ratio.

Interactions between a nuclear magnet and its surroundings are small. In a magnetic field each of the $2m_I + 1$ states corresponding to different orientations of the nuclear moment takes up a different energy

$$W_{m_I} = -\mathbf{m}_N \cdot \mathbf{B} = -g_N \mu_N m_I B; \qquad (14.26)$$

in a field of 1 T the separation between successive levels corresponds to a frequency of order 10 MHz. In an atom or ion which has a permanent electronic magnetic moment, the latter sets up a magnetic field at the nucleus which may be as much as $10^3$ T, being generally larger in the heavier atoms. This magnetic field is partly due to the electronic orbit and partly to the spin, but for most purposes we need consider only the steady component of the electronic field of flux density $\mathbf{B}_e$, which is parallel to and proportional to the resultant electronic angular momentum vector $\mathbf{J}$. Thus we have an additional 'hyperfine' energy

$$W = -\mathbf{m}_N \cdot \mathbf{B}_e = A\mathbf{J}\cdot\mathbf{I}, \tag{14.27}$$

where $A$ is a constant whose order of magnitude can be estimated as follows. The electronic flux density $\mathbf{B}_e$ is $\sim \mathbf{m}_e \langle R_e^{-3}\rangle = -g_J\mu_B\mathbf{J}\langle R_e^{-3}\rangle$, where $\langle R_e^{-3}\rangle$ is the mean inverse cube of the distance of the electron from the nucleus; since $\mathbf{m}_N = g_N\mu_N\mathbf{I}$, $A$ is of order $g_J g_N\mu_B\mu_N\langle R_e^{-3}\rangle$. In frequency units $A/h$ generally lies in the range $10^2$–$10^4$ MHz, so that the hyperfine energy may approach $100\ \mathrm{m}^{-1} \equiv 1\cdot43$ K.

In addition to possessing a magnetic moment, a nucleus may have a non-spherical distribution of electric charge. Its electrostatic potential can then be expanded as in §2.3, giving an energy of interaction with the electrons of the form (see eqns (2.30)–(2.34))

$$W = \frac{1}{4\pi\epsilon_0}\iint \frac{\rho_e\rho_N\,d\tau_e\,d\tau_N}{|R_e - r_N|}$$

$$= \frac{1}{4\pi\epsilon_0}\left\{Ze\int\frac{\rho_e\,d\tau_e}{R_e} + \sum_{m=-2}^{+2}(-1)^{|m|}A_{2,m}B_{2,-m} + \dots\right\}. \tag{14.28}$$

Here the subscripts e, N refer to the electrons and nuclei respectively; $Ze = \int \rho_N\,d\tau_N$ is the nuclear charge, so that the first term is the Coulomb interaction due to a point charge at the nucleus, and the quantities in the second term are

$$A_{2,m} = \int \rho_N r_N^2 C_{2,m}(\theta_N, \phi_N)\,d\tau_N,$$

$$B_{2,-m} = \int (-1)^{|m|}\rho_e R_e^{-3}C_{2,-m}(\theta_e, \phi_e)\,d\tau_e.$$

This term represents the interaction between the electric quadrupole moments of the nucleus and of the electrons, whose nature is that of a tensor. The charge distribution is spherical for nuclei with $I = 0$ or $\tfrac{1}{2}$, and for electronic shells with $J = 0$ or $\tfrac{1}{2}$, so that the quadrupole interaction vanishes in either case. Since the nuclear charge is symmetric about the axis of nuclear precession, the nuclear terms can be expressed in terms of a single quantity

$$A_{2,0} = \int \rho_N \tfrac{1}{2} r_N^2(3\cos^2\theta_N - 1)\,d\tau_N = \tfrac{1}{2}eQ\left\{\frac{3m_I^2 - I(I+1)}{I(2I-1)}\right\}, \tag{14.29}$$

where

$$Q = \frac{1}{e}\int \rho_N r_N^2(3\cos^2\theta_N - 1)\,d\tau_N \tag{14.30}$$

is called the nuclear electric quadrupole moment and has the dimensions of an area of the same order as the (nuclear radius)$^2$. It is expressed in terms of the 'barn' $10^{-28}\,\mathrm{m}^2$. The sign of $Q$ is positive for a prolate spheroid and negative for an oblate spheroid, as illustrated in Fig. 14.21. The expression in parentheses in eqn (14.29) gives the variation of $A_{2,0}$ with the nuclear magnetic quantum number $m_I$, and it is easily verified that in the states $m_I = \pm I$, $A_{2,0} = \frac{1}{2}eQ$.

In the absence of an external magnetic field, the electronic and nuclear angular-momentum vectors $J$, $I$ are coupled together by the magnetic hyperfine energy (eqn (14.27)) to form a resultant vector $F$. Different values of $F$ correspond to different energies, since the angle between $J$ and $I$ is changed, and from the vector model it can be shown that

$$W_F = \tfrac{1}{2}A\{F(F+1)-J(J+1)-I(I+1)\} \tag{14.31}$$

so that the energies of successive states form an arithmetical progression (cf. Problem 14.7, for the corresponding case of spin–orbit coupling). This simple rule no longer holds when an electric quadrupole interaction is present, but $F$ is still a good quantum number, and the size of the quadrupole interaction can be deduced from the departures from eqn (14.31). It can be shown that

$$W_F = \tfrac{1}{2}AC + B_Q\frac{\tfrac{3}{4}C(C+1)-I(I+1)J(J+1)}{2I(2I-1)J(2J-1)}, \tag{14.32}$$

where

$$C = F(F+1)-I(I+1)-J(J+1); \qquad B_Q = 2eQB_{2,0}/4\pi\epsilon_0 = eQ(\partial^2 V/\partial z^2),$$

where $\partial^2 V/\partial z^2$ is the electric field gradient set up by the electrons at the nucleus.

In the solid state an assembly of nuclear dipoles behaves as a simple

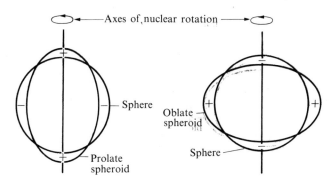

FIG. 14.21. Representation of non-spherical charge distribution in nucleus as combination of sphere and quadrupole.

paramagnetic substance, contributing an amount (cf. eqn (14.22))

$$\chi_N = \frac{\mu_0 n g_N^2 \mu_N^2 I(I+1)}{3kT},$$  (14.33)

which is only about $10^{-6}$ of that of any electronic paramagnetic substance, since the susceptibility depends on the square of the magnetic dipole moment. The nuclear contribution has been detected by static susceptibility measurements in solid hydrogen, where the nuclear paramagnetism just outweighs the electronic diamagnetism at about 1 K (see Problem 14.9).

The nuclear susceptibility follows Curie's law, eqn (14.33), only at temperatures such that $kT$ is large compared with any splittings of the nuclear levels. In substances without permanent electronic magnetic dipoles this means temperatures down to about $10^{-6}$ K, except where the nuclear levels are split through an electric quadrupole interaction with the electrostatic field gradient (the 'crystal field') set up by neighbouring ions. In such cases the gradient $(\partial^2 V/\partial z^2)$ is fixed, unlike in a free atom where it follows the precessing electronic angular-momentum vector for the orbit, which determines the orientation of the electronic charge cloud. Hence in a solid in which the local surroundings of a nucleus have symmetry about an axis (which we take to be the $z$-axis) the nuclear levels may be split according to the formula

$$W_{m_I} = \frac{1}{4\pi\epsilon_0} A_{2,0} B_{2,0} = eQ(\partial^2 V/\partial z^2) \frac{3m_I^2 - I(I+1)}{4I(2I-1)}.$$  (14.34)

The splittings range from a few kilohertz to over 2000 MHz for $^{127}$I in $I_2$ (solid iodine).

In substances containing ions with both electronic and nuclear magnetic dipoles the two contributions to the susceptibility are additive at temperatures such that $kT$ is large compared with any hyperfine structure splittings (in practice this usually means down to about 1 K). Such splittings arise from both nuclear magnetic dipole and electric quadrupole interactions in the same way as for free atoms, but the effects are more complicated because of the complex interaction of the electrons with the crystal or ligand field discussed in §§ 14.6–14.9. The hyperfine splittings usually correspond to temperatures in the range 0·001–1 K, and affect both the electronic and nuclear contributions to the susceptibility in this temperature range and below.

## 14.11. Magnetism of conduction electrons

This chapter has been concerned mainly with paramagnetism of insulating compounds in which the electrons are localized; the compounds can be regarded as assemblies of atomic ions in which the paramagnetism arises from ions of the transition groups. Many metals in these groups are strongly magnetic; they are discussed in Chapters 15 and 16, and here we confine ourselves to the interaction of a magnetic field with the conduction electrons in 'normal' metals.

A magnetic field alters the translational motion of the conduction electrons, through the Lorentz force, and this gives rise to a diamagnetic susceptibility which is constant at ordinary field strengths. A formula for this was first derived by Landau, and on using an effective mass $m^*$ this can be written as

$$\chi_d = -\frac{\mu_0 e^2}{6\pi m^*}\left(\frac{3n}{8\pi}\right)^{\frac{1}{3}},\tag{14.35}$$

where $n$ is the number of conduction electrons per unit volume.

The second effect of a magnetic field arises from its interaction with the electron spin, the two orientations of the spin dipole having energies $+\frac{1}{2}g_s\mu_B B$ and $-\frac{1}{2}g_s\mu_B B$ respectively. In the latter state, with the lower energy, the magnetic dipoles are parallel to the field, and the probability of occupation is a little larger than for the antiparallel orientation, giving a net positive susceptibility. For conduction electrons this must be calculated using the Fermi–Dirac distribution function, and since this varies very little with temperature, the paramagnetic spin susceptibility is practically independent of temperature. At the absolute zero an expression for its value can be derived rather simply.

At the absolute zero, two electrons with oppositely directed spins occupy each translational energy level up to a certain energy $W_F$, the top of the Fermi distribution. When a magnetic field is applied, an electron can only reverse its spin magnetic dipole from an antiparallel to a parallel orientation if the decrease in its magnetic energy $g_s\mu_B B$ is sufficient to supply the extra kinetic energy required to raise it to an empty translational energy level. This follows from the Pauli principle, which shows that two electrons with parallel spin cannot occupy the same energy level. The effect on the distribution of electrons in the energy band is shown in Fig. 14.22. This differs from Fig. 11.2 in that the band is drawn in two halves, one containing the electrons whose spin dipoles are parallel to the flux density **B**, the other those with their spin dipoles antiparallel. The two half-bands are then separated in energy by $g_s\mu_B B$, the potential-energy difference in the magnetic field. For the total energy, magnetic plus kinetic, of the whole system to be a minimum, the electrons must fill the two displaced half-bands up to the same level. Any deviation from this would require a transfer of electrons from one half-band to higher vacant levels in the other half-band, and so increase the energy.

The total magnetic moment of the system is $2x(\frac{1}{2}g_s\mu_B)$, where $x$ is the number of electrons transferred from the antiparallel to the parallel

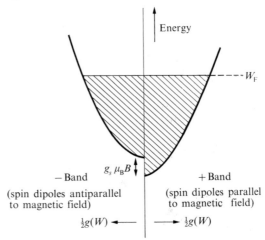

FIG. 14.22. Displacement of + and − bands of conduction electrons by an applied magnetic field. The displacement is equal to the difference of energy $g_s\mu_B B$ of a spin dipole parallel and antiparallel to the flux density $B$. The resultant magnetization is due to the excess of electrons in the + band.

orientation, since the excess in the latter is then $2x$ and the spin dipole moment is $\frac{1}{2}g_s\mu_B$. The value of $x$ can be found as follows. Suppose $w$ is the difference in kinetic energy between successive levels at the top of the Fermi distribution. To turn round the dipole on one electron requires that its kinetic energy increase by $w$, since we can take the electron from the highest filled level and transfer it to the next level, which is vacant. To turn round a second dipole requires an additional energy of $3w$, since the next two levels with parallel orientation are already filled. The third transfer requires $5w$ in energy, and the $x$th electron $(2x-1)w$, which may be taken as $2xw$ when $x$ is large. At equilibrium, $2xw$ must just equal $g_s\mu_B B$ in order that the two half-bands are filled to the same level.

Now the energy separation $w$ between successive kinetic energy levels is $[\frac{1}{2}\{g(W)\}_F]^{-1}$, and hence

$$2x = g_s\mu_B B/w = \tfrac{1}{2}g_s\mu_B B\{g(W)\}_F.$$

The susceptibility per unit volume is

$$\chi_P = 2x(\tfrac{1}{2}g_s\mu_B)/H = \tfrac{1}{4}\mu_0 g_s^2\mu_B^2\{g(W)\}_F, \tag{14.36}$$

and for free electrons, using the value of $\{g(W)\}_F$ from eqn (11.11) this can be written as

$$\chi_P = \frac{\mu_0(3ng_s^2\mu_B^2)}{8\,W_F} \tag{14.37}$$

This result was first obtained by Pauli, and the phenomenon is sometimes called 'Pauli paramagnetism'. For practical purposes $\chi_P$ is independent of temperature, the next term in a series expansion for $\chi_P$ being smaller by a factor of order $(kT/W_F)^2$, as in eqn (11.19). As in the case of the specific heat (§§ 11.6 and 12.6), the paramagnetic susceptibility depends on the density of states $\{g(W)\}_F$ at the Fermi surface, and is smaller by a factor $\sim(kT/W_F)$ than the corresponding quantity for particles obeying classical statistics. Similarly, instead of Curie's law (eqn (14.22)) we have an expression with $W_F$ in the denominator, instead of $kT$.

For free electrons, the value of $\chi_P$ given by eqn (14.37) with $g_s = 2$ is just 3 times the value of $\chi_d$ given by eqn (14.35) with $m^* = m$, as can be verified from eqn (11.7) for $W_F$ on substituting for $\mu_B$. Hence the net effect is a positive susceptibility. If $m^* > m$, the value of $\chi_d$ is diminished, while $\chi_P$ increases, since $W_F^{-1} \propto m^*$ by eqn (12.29). Transition metals show strong Pauli paramagnetism, corresponding to high values of $\{g(W)\}_F$. In addition, interaction effects between the electrons also cause an increase in $\chi_P$ (for a review see Van Vleck (1957)).

Static measurements of the susceptibility give the quantity $\chi = \chi_d + \chi_P + \chi_c$, where $\chi_c$ is the diamagnetic susceptibility of the electrons bound to the positive-ion cores. This can be estimated from values for neighbouring non-metallic elements, or from calculated values of $\sum (r^2)$ using eqn (6.7). Electron spin resonance can be used to determine $\chi_P$, and $\chi_d$ can then be found from the value of $(\chi - \chi_P - \chi_c)$. A comparison of theory and experiment is given in Table 14.3, based on Van Vleck (1957). Later measurements of $\chi_P$ give slightly different values, but a satisfactory comparison requires an experimental determination of $m^*/m$.

TABLE 14.3

*Experimental and theoretical values of the volume
susceptibility of lithium and sodium*

This table is in units of $10^{-6}$ e.m.u. cm$^{-3}$ (see Appendix D);
to convert to S.I. multiply by $4\pi$.

|  | Lithium | | Sodium | |
|---|---|---|---|---|
|  | Experiment | Theory | Experiment | Theory |
| $m^*/m$ |  | 1·46 |  | 0·985 |
| $\chi_P$ | 2·08±0·1 | 1·17† | 0·95±0·1 | 0·64† |
|  |  | 1·87‡ |  | 0·85‡ |
| $\chi$ | 1·89±0·05 |  | 0·70±0·03 |  |
| $\chi_c$ |  | −0·05 |  | −0·18 |
| $\chi_d$ | −0·14±0·15 | −0·19§ | −0·07±0·13 | −0·22§ |

(After Van Vleck 1957.)
† From eqn (14.37), using effective mass.
‡ Calculated by Pines, including interaction effects.
§ From eqn (14.35), using effective mass.

## References

Cooper, B. R. and Vogt, O. (1970). *Phys. Rev.* **B1**, 1211.
Stout, J. W. and Hadley, W. B. (1964). *J. chem. Phys.* **40**, 55.
Van Vleck, J. H. (1957). *Nuovo Cim.* **6**, 857.

## Problems

14.1. In statistical mechanics the partition function $Z$ is defined as

$$Z = \sum_i \exp(-W_i/kT),$$

where $W_i$ is the energy of the $i$th state. Show that for an assembly of non-interacting paramagnetic ions, each of angular momentum $J$, in a field of flux density $B$

$$Z = \frac{\sinh\{(2J+1)y/2J\}}{\sinh(y/2J)},$$

where $y = Jg_J\mu_B B/kT$.

The magnetization $M$ for an assembly of $n$ such atoms is given by the formula

$$M = nkT\frac{d}{dB}(\ln Z).$$

Using this formula, derive the Brillouin function of eqn (14.21).

14.2. Show that the energy of interaction of two magnetic dipoles $m$ a distance $r$ apart is of the order $\mu_0 m^2/4\pi r^3$.

In potassium chrome alum, each $Cr^{3+}$ ion carries a magnetic moment of $3\mu_B$, and the mean distance apart is about $0.78$ nm. Assuming that serious departures from Curie's law will occur when the interaction energy between two neighbouring dipoles is $\approx kT$, show that this temperature is approximately $0.01$ K. (In fact the levels of each $Cr^{3+}$ ion are split by about $0.2$ K through a high-order effect of the crystalline field, and this is more important than the magnetic dipole interaction between neighbouring ions; it also gives a specific heat anomaly at about $0.1$ K of the type shown in Fig. 14.17.)

14.3. For a $Cu^{2+}$ ion, $S = \frac{1}{2}$ and the energy levels of the ground state in a magnetic flux density $B$ are of the form

$$W = \pm\tfrac{1}{2}g\mu_B B - \tfrac{1}{2}\alpha B^2.$$

Show that in small fields $(g\mu_B B/kT \ll 1)$ the partition function

$$Z = 2 + \alpha B^2/kT + g^2\mu_B^2 B^2/4(kT)^2 + \ldots,$$

and hence that

$$\chi/\mu_0 = n\{g^2\mu_B^2/4kT + \alpha\} = np^2\mu_B^2/3kT.$$

This shows that the term $B^2$ in $W$ gives rise to a temperature-independent contribution to the susceptibility. Note that $p^2$ is then of the form $A + BT$. ($B$ is very small for $Cu^{2+}$, but $Co^{2+}$ ions obey this relation below $100$ K—see Fig. 14.18.)

14.4. A pair of singlet levels have separation $\delta$ when $B = 0$, and in a flux density

$B$ their energies become $W_1 = +\frac{1}{2}\Delta$ and $W_2 = -\frac{1}{2}\Delta$ respectively, where $\Delta = \{\delta^2 + (g\mu_B B)^2\}^{\frac{1}{2}}$. Show from the relation $m = -dW/dB$ that the magnetic moments of the two states are $\mp\frac{1}{2}(g^2\mu_B^2 B/\Delta)$.

Verify that in a small field at temperature $T$ the susceptibility is given by

$$\chi/\mu_0 = \frac{1}{2}(g^2\mu_B^2/\delta)\tanh(\delta/2kT)$$

and that at high temperatures when $\delta/kT \ll 1$ this approaches $g^2\mu_B^2/4kT$, while at low temperatures when $\delta/kT \gg 1$ it approaches the constant value $g^2\mu_B^2/2\delta$.

14.5. Show that for a system where the magnetic moment associated with orbital angular momentum $l\hbar$ is $g_l\mu_B l$ and that associated with spin $s\hbar$ is $g_s\mu_B s$ and $l$ and $s$ are coupled together to form a resultant $j$, the generalized Landé formula for the g-factor is

$$g = \frac{j(j+1)(g_l+g_s)+\{l(l+1)-s(s+1)\}(g_l-g_s)}{2j(j+1)}.$$

14.6. On the nuclear shell model the nuclear spin is due to the odd neutron or proton with spin $\frac{1}{2}$ moving in an orbit within the nucleus with angular momentum $l\hbar$. The observed nuclear spin $I$ is either $l+\frac{1}{2}$ or $l-\frac{1}{2}$. Apply the formula of the last question to calculate the magnetic moment, assuming that for a proton $g_l = 1$ and $g_s = 5\cdot586$, and for a neutron $g_l = 0$ and $g_s = -3\cdot826$. Show that the magnetic moment $m = g_N\mu_N I$ of a nucleus of spin $I$ is

(a) odd proton   $I = l+\frac{1}{2},$    $m = \mu_N(I+2\cdot293),$

   $I = l-\frac{1}{2},$    $m = \mu_N I(I-1\cdot293)/(I+1),$

(b) odd neutron  $I = l+\frac{1}{2},$    $m = -1\cdot913\mu_N = m_N,$

   $I = l-\frac{1}{2},$    $m = 1\cdot913\mu_N I/(I+1) = -m_N I/(I+1).$

These formulae are known as the Schmidt limits. Observed nuclear moments follow the trend given by these formulae but generally lie between these limits.

14.7. Use the treatment of § 14.3 to show that the spin–orbit coupling leads to an energy for a state of total angular momentum $J\hbar$ of

$$W_J = \frac{1}{2}\lambda\{J(J+1)-L(L+1)-S(S+1)\}$$

and that $W_J - W_{J-1} = \lambda J$ (this is known as the Landé interval rule).

Show from the splittings of the $^4F$ multiplet in Fig. 14.5 that $\lambda$ is about $8\cdot7\times10^3$ m$^{-1}$ for the $Cr^{3+}$ ion (slightly higher values are obtained from the $^4P$ states, but these are perturbed by doublet states which are not far away).

For the $Eu^{3+}$ ion, $4f^6$, $L = 3$, $S = 3$ the value of $\lambda$ is $2\cdot27\times10^4$ m$^{-1}$. Verify that (a) the ground state is $J = 0$; (b) the first excited state $J = 1$ lies at $2\cdot27\times10^4$ m$^{-1}$; (c) $g_J = \frac{3}{2}$ for all excited states.

14.8. Hydrogen molecules are of two types: (a) orthohydrogen, where the nuclear spin of $\frac{1}{2}\hbar$ of each proton is parallel to the other and the nuclear spin for the molecule is $I = 1$; (b) parahydrogen, where the two proton spins are antiparallel giving $I = 0$ for the molecule. Show that at high temperatures where the ratio of ortho- to parahydrogen molecules is $3:1$, the susceptibility due to the nuclear paramagnetism is identical with that of the same total number of hydrogen atoms with independent spin $I = \frac{1}{2}$. (Note that the equilibrium ratio of $3:1$ corresponds to the fact that there are three quantum states for $I = 1$, associated with three possible orientations of the spin, each with the same *a priori* probability as the single state for $I = 0$.)

14.9. Calculate the paramagnetic susceptibility of a mole of hydrogen at 1 K due to the nuclear moments, assuming that the ortho–para ratio is still $3:1$. Show that it is of the same order as the diamagnetic susceptibility due to the electrons, assuming that each of the two electrons is in an orbit for which the mean radius is the Bohr radius.

(*Answers:* $2 \cdot 2 \times 10^{-11}$ and $-2 \cdot 0 \times 10^{-11}$ S.I. units.)

14.10. The ground state of sodium is $^2S_{\frac{1}{2}}$, and the yellow D lines are due to transitions to the ground state from the two lowest excited states $^2P_{\frac{1}{2}}$ and $^2P_{\frac{3}{2}}$. Show that the Landé g-factors of these two states are $\frac{2}{3}$ and $\frac{4}{3}$ respectively.

Hence show that one D line will be split in a magnetic field of flux density $B$ into four components, with frequencies $D_1 \pm \frac{2}{3}\delta$, $D_1 \pm \frac{4}{3}\delta$; and the other D line into six components with frequencies $D_2 \pm \frac{1}{3}\delta$, $D_2 \pm \delta$, $D_2 \pm \frac{5}{3}\delta$, where $\delta = \mu_B B / h$, when viewed normal to the field. (Only the $\Delta M = \pm 1$ components are seen when viewed parallel to the field.)

14.11. The arrangement of parallel cylindrical conductors carrying equal and opposite currents of Problem 4.3 is used to give a large field gradient and deflect atoms in an atomic beam. Each cylinder carries a current of 1000 A and the axes of the cylinders are 0·01 m apart. A beam of atoms in the $^2S_{\frac{1}{2}}$ state from an oven at 900 K travels parallel to the cylinders along the line where the inhomogeneous field is a maximum. Calculate the separation between the two components of the beam after travelling a distance of 0·2 m.

(*Answer:* $\approx 0 \cdot 1$ mm.)

# 15. Ferromagnetism

## 15.1. Exchange interaction between paramagnetic ions

IN the discussion in §§ 14.5–14.9 of paramagnetism in the solid state, it was assumed that interactions between neighbouring paramagnetic ions could be neglected. Such interactions are of two types: (1) magnetic dipole–dipole interaction, arising from the magnetic field due to one dipole acting on another; (2) exchange interactions between the electrons in different paramagnetic ions, of the same nature as those between electrons within the same atom (giving rise to L–S coupling) or between the electrons of different atoms in chemical binding. Of these, exchange interaction greatly outweighs dipole–dipole interaction in ordinary substances. For example, the Curie point of nickel is 631 K (see Table 15.1). This is a rough indication of the temperature at which the interaction between neighbouring nickel ions (separation 0·25 nm) is of the order $kT$, whereas (Problem 14.2) the purely magnetic interaction of two atomic dipoles at this distance would be equivalent to $kT$ with $T$ less than 1 K. Exchange interaction decreases more rapidly than magnetic dipole interaction as the atomic separation is increased, though no simple law can be given for its rate of decrease. As an example, in paramagnetic salts of the 3d group the exchange interaction is more important than the magnetic dipole interaction until the separation between the paramagnetic ions is greater than about 0·6 nm, and then both are so small that they have an appreciable effect on the magnetic properties only well below 1 K.

The mechanism of exchange interaction, as originally proposed by Heisenberg in 1928, is one in which the forces involved are electrostatic in origin, but which, because of the constraints imposed by the Pauli exclusion principle, are formally equivalent to a very large coupling between the electron spins, of the type

$$W = -2\mathscr{J}\mathbf{s}_i \cdot \mathbf{s}_j. \tag{15.1}$$

The quantity $\mathscr{J}$ is known as the exchange energy. Though several types of indirect exchange interaction have since been suggested (see § 15.8), they all lead to a basic coupling between the spins of this form, dependent on the cosine of the angle between the two spin vectors. For two separate atoms with total spin vectors $\mathbf{S}_i$, $\mathbf{S}_j$ we may use the vector summations to show that the total interaction energy is

$$W = -2\mathscr{J} \sum_i \sum_j \mathbf{s}_i \cdot \mathbf{s}_j = -2\mathscr{J} \sum_i \mathbf{s}_i \cdot \sum_j \mathbf{s}_j = -2\mathscr{J} \mathbf{S}_i \cdot \sum_j \mathbf{s}_j = -2\mathscr{J} \mathbf{S}_i \cdot \mathbf{S}_j, \quad (15.2)$$

which depends only on the relative orientation of the two total spin vectors $\mathbf{S}_i$, $\mathbf{S}_j$. An immediate result of eqn (15.2) is that the exchange interaction vanishes for any closed shell of electrons, since then $\mathbf{S} = 0$. Thus we need consider only the partly filled shells which are responsible for permanent magnetic dipole moments in atoms and ions.

For an ion in which $\mathbf{J}$ is a good quantum number (such as ions of the 4f group), we must project $\mathbf{S}$ onto $\mathbf{J}$; the reason for this is that $\mathbf{J}$ is a constant of the motion, and hence so also is the projection of $\mathbf{S}$ onto $\mathbf{J}$. The components of $\mathbf{S}$ normal to $\mathbf{J}$ are precessing rapidly, so that their contribution to the scalar product $\mathbf{S}_i \cdot \mathbf{S}_j$ is zero on a time average. From the equivalences $\mathbf{L} + 2\mathbf{S} \equiv g\mathbf{J}$ (where $g$ is the Landé factor), $\mathbf{L} + \mathbf{S} \equiv \mathbf{J}$, we find at once that $\mathbf{S} \equiv (g-1)\mathbf{J}$. Thus, for a pair of such ions (assumed identical, with the same values of $\mathbf{J}$ and $g$) we have

$$W = -2\mathscr{J} \mathbf{S}_i \cdot \mathbf{S}_j = -2\mathscr{J}(g-1)^2 \mathbf{J}_i \cdot \mathbf{J}_j = -2\mathscr{J}' \mathbf{J}_i \cdot \mathbf{J}_j. \quad (15.3)$$

This gives a coupling of the angular-momentum vectors of the same form as eqn (15.2), but with a modified value of the apparent exchange energy.

In a solid, any given magnetic ion is surrounded by other magnetic ions, with each of which it will have an exchange interaction. The total interaction for each ion will therefore be a sum of terms such as (15.3) taken over all pairs of ions; the energy for atom $i$ is thus

$$W_i = -2\mathbf{J}_i \cdot \sum_j \mathscr{J}'_{ij} \mathbf{J}_j.$$

The magnetic dipole moment of each ion is proportional to the angular momentum $\mathbf{J}$, since $\mathbf{m} = g\mu_B \mathbf{J}$, so that the exchange energy can be expressed in terms of the dipole moments, giving

$$W_i = -2\left(\frac{\mathbf{m}_i}{g\mu_B}\right) \cdot \sum \mathscr{J}'_{ij}\left(\frac{\mathbf{m}_j}{g\mu_B}\right),$$

assuming again that all ions have the same Landé g-factor. In a ferromagnetic substance, or a paramagnetic substance subjected to an external magnetic field, each ion will have an average dipole moment in the direction of magnetization, together with fluctuating components in other directions whose time average is zero. In summing over the interaction with neighbouring ions, that part associated with the fluctuating components will tend to average out, since at any instant the contributions from different neighbours will be as often positive as negative. To a fair approximation we can therefore replace the vector sum over the neighbouring dipole moments by a sum over the average moment per neighbour $\bar{\mathbf{m}}_j$, and if we assume further that the only important interaction is with $z$ equidistant neighbours, each having the same interaction energy $\mathcal{J}'$, we can write

$$W = -2\left(\frac{\mathbf{m}}{g\mu_B}\right) \cdot \sum \mathcal{J}'\left(\frac{\bar{\mathbf{m}}_j}{g\mu_B}\right) = -2\left(\frac{\mathbf{m}}{g\mu_B}\right) \cdot \left(\frac{z\bar{\mathbf{m}}}{g\mu_B}\right)\mathcal{J}'$$

$$= -\left(\frac{2z\mathcal{J}'}{ng^2\mu_B^2}\right)\mathbf{m} \cdot \mathbf{M} = -\mathbf{m} \cdot \mathbf{B}_{int}. \tag{15.4}$$

Here we have dropped the subscript $i$, since we assume all ions are identical, and the energy is the same for each; and we have replaced the mean moment per ion by the magnetization $\mathbf{M} = n\bar{\mathbf{m}}$, where $n$ is the number of ions per unit volume. The result is an equation formally identical with the potential energy of a dipole $\mathbf{m}$ in a field

$$\mathbf{B}_{int} = (2z\mathcal{J}'/ng^2\mu_B^2)\mathbf{M} = \lambda\mathbf{M}.$$

We may therefore represent the effect of the exchange forces, to a good approximation, by an effective 'internal field' of flux density $\mathbf{B}_{int}$ which is proportional to the intensity of magnetization. This concept was first introduced by Weiss to account for the occurrence of spontaneously magnetized substances (ferromagnetics).

As a preliminary, we shall discuss the effect of this internal field in a paramagnetic substance. The total field acting on an ion is then $\mathbf{B}_0 + \mathbf{B}_{int} = \mathbf{B}_0 + \lambda\mathbf{M}$, where $\mathbf{B}_0$ is the flux density of the external field. So long as the magnetization is small compared with the saturation value we may assume that Curie's law $\chi = C/T$ still holds if we replace $\mathbf{B}$ in our earlier theory by $\mathbf{B}_0 + \lambda\mathbf{M}$. Then we have

$$\mathbf{M} = \frac{(C/T)\mathbf{B}}{\mu_0} = \frac{C(\mathbf{B}_0 + \lambda\mathbf{M})}{\mu_0 T}$$

and hence

$$\chi = \frac{\mu_0 M}{B_0} = \frac{C}{(T - \lambda C/\mu_0)} = \frac{C}{(T - \theta)}. \tag{15.5}$$

This is known as the Curie–Weiss law, and represents the behaviour of paramagnetic substances at temperatures $T > \theta$ with fair accuracy; $\theta = \lambda C/\mu_0$ is often called the 'Weiss' constant.

The form of eqn (15.5) shows that some radical change in the magnetic properties is to be expected at the temperature $\theta$, and we may interpret the infinite susceptibility which is predicted by eqn (15.5) at this point in the following way. Since $\chi = \mu_0 M/B_0$, and the maximum value of $M$ is finite, being limited to the saturation moment obtainable when all the dipoles are aligned parallel to one another, we must assume $B_0 = 0$; in other words, the substance is magnetized even in the absence of an external field. This 'spontaneous magnetization', due to the internal field, is a characteristic of ferromagnetism, and the temperature $\theta$ is the boundary between paramagnetic behaviour at $T > \theta$ and ferromagnetic behaviour when $T < \theta$. The temperature below which spontaneous magnetization appears is known as the Curie point, and the experimental values for a number of substances are given in Table 15.1. The 'ferromagnetic Curie temperature' is defined as that below which spontaneous magnetization sets in, and it often differs by 10–20 K from the value of $\theta$ determined in the paramagnetic region by fitting the observed susceptibility to eqn (15.5). The latter value is sometimes called the 'paramagnetic Curie temperature'. On our simple theory there is no difference between the two Curie temperatures.

Since the Curie constant $C = \mu_0 ng^2 \mu_B^2 J(J+1)/3k$, the value of the Weiss constant in eqn (15.5) is

$$\theta = \frac{\lambda C}{\mu_0} = \frac{(2z\mathscr{J}'/ng^2\mu_B^2)\{\mu_0 ng^2\mu_B^2 J(J+1)/3k\}}{\mu_0}$$

$$= \frac{2z\mathscr{J}'J(J+1)}{3k} \tag{15.6}$$

and on simple theory this is also the value of the Curie temperature $T_C$. More sophisticated methods of calculation produce a somewhat different value of the numerical constant, and Rushbrooke and Wood (1958) show that the results can be fitted remarkably well by the empirical formula

$$T_C = \frac{5\mathscr{J}'}{96k}(z-1)\{11J(J+1)-1\}. \tag{15.7}$$

This predicts somewhat lower values for the Curie point than eqn (15.6),

TABLE 15.1

*Saturation moment and Curie point of some ferromagnetic materials*

| Substance | Saturation moment at 0 K | Curie point (K) |
|---|---|---|
| | Bohr magnetons per atom | |
| Iron | 2·22 | 1043 |
| Cobalt (<670 K) | 1·715 | |
| (>670 K) | (1·76) | 1394 |
| Nickel | 0·605 | 631 |
| MnBi | 3·52 | 630 |
| MnAs | 3·40 | 318 |
| Fe$_2$O$_3$ | 1·20 (per atom of Fe) | 893 |

*Note:* Cobalt has a phase transition at about 670 K, being hexagonal in structure below that temperature, and face-centred cubic above. The values in brackets are obtained by extrapolation.

and conversely, gives higher estimates of the exchange interaction. For example, nickel has its Curie point at 631 K; its crystal structure is face-centred cubic (f.c.c.), for which the number of nearest neighbours is 12, which we take to be the value of $z$. If we make the further assumption that $J = S = \frac{1}{2}$, then we find that $\mathscr{J}'/k$ is 105 K from eqn (15.6), and 150 K from (15.7). Thus the exchange energy (there is no difference between $\mathscr{J}'$ and $\mathscr{J}$ when we are dealing with spin-only magnetism), is about $10^{-2}$ eV. The magnitude of this interaction can perhaps be appreciated best by expressing it in terms of the internal flux density $\mathbf{B}_{int}$ of eqn (15.4), which is found to be of order $10^3$ T. This is over 100 times larger than any field which can easily be produced in the laboratory, so that external fields would be expected to have little effect on the spontaneous magnetization below the Curie point.

Eqns (15.6) and (15.7) show that the sign of $\theta$ and $T_C$ is the same as that of $\mathscr{J}'$ (and hence also of $\mathscr{J}$, so long as we are dealing with identical ions). Thus a positive value of the exchange energy is required to give a vanishing denominator in the Curie–Weiss law (eqn (15.5)), and a co-operative state in which the electron spins are parallel to each other. This ferromagnetic state is a direct consequence of the fact that the exchange coupling (eqn (15.1)) gives a lower energy for any pair of electrons when their spins are parallel, provided the exchange energy $\mathscr{J}$ is positive. If it is negative, the state of lower energy is one with antiparallel spins; the Weiss constant is also negative, and the denominator of the Curie–Weiss law does not vanish at any real temperature. Nevertheless a co-operative state does then occur, but one in which the basic arrangement is of antiparallel spins. This phenomenon is called 'antiferromagnetism', and is discussed in Chapter 16.

## 15.2. The Weiss theory of spontaneous magnetization

Since the internal field in a ferromagnetic substance is so large, the magnetization will approach the saturation value even at ordinary temperatures. The assumption that the magnetization is small and proportional to the effective field, used in deriving eqn (15.5) for the susceptibility above the Curie point, thus cannot be used below the Curie point. If we retain the concept of an internal field, the magnetization may be calculated using the Brillouin function (see eqn (14.21)) which may be written in the form

$$M/M_s = \phi(y). \tag{15.8}$$

Here $M_s$ is the saturation magnetization per unit volume, and equals $ngJ\mu_B$, where $n$ is the number of atomic dipoles per unit volume. The argument of the Brillouin function may be written as

$$y = gJ\mu_B B/kT = M_s B/nkT$$

and $B$ must be taken as the sum of the external flux density $B_0$ and the internal field $\lambda M$. Hence we have

$$y = M_s(B_0+\lambda M)/nkT, \tag{15.9}$$

which may be solved for $M$, giving

$$M/M_s = y(nkT/\lambda M_s^2)-(B_0/\lambda M_s). \tag{15.10}$$

The value of the magnetization under any given conditions of $B_0$ and $T$ may be found by eliminating the parameter $y$ between the two equations (15.8) and (15.10). This cannot be done analytically, but the general behaviour of the magnetization can be found from a graphical solution. We begin by equating $B_0$ to zero, and finding the value of the spontaneous magnetization $M_0$ in zero field. To obtain a graphical solution we then plot the two functions $M_0/M_s = \phi(y)$ (from eqn (15.8)) and $M_0/M_s = y(nkT/\lambda M_s^2)$ (from eqn (15.10)) against $y$, as in Fig. 15.1. The second function gives a straight line which passes through the origin and intersects the curve for $\phi(y)$ at this point. Thus one possible value of the magnetization is always zero. If the temperature $T$ is sufficiently high, the slope of the line $M_0/M_s = y(nkT/\lambda M_s^2)$ is so great that this is the only point of intersection, and the substance must therefore be unmagnetized in zero external field. This corresponds to the paramagnetic behaviour above the Curie point, discussed in the last section. As the temperature $T$ falls, the slope of the line given by eqn (15.10) decreases, until at a certain temperature $T_C$ it is tangential to the curve (a) at the origin. For small values of $y$,

$$\phi(y) = M/M_s = y(J+1)/3J,$$

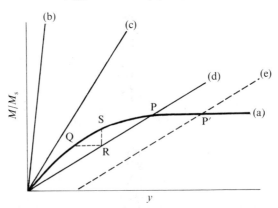

FIG. 15.1. Graphical solution of eqns (15.8) and (15.10) for spontaneous magnetization. (a) is the Brillouin function $\phi(y)$ (eqn (15.8)); (b), (c), (d) are the straight lines $M/M_s = y(nkT/\lambda M_s^2)$ for temperatures $T>T_C$, $T=T_C$, and $T<T_C$ respectively, where $T_C$ is the Curie point (all with $B_0 = 0$); (e) is the function in eqn (15.10); an external flux density $B_0$ is applied, with the temperature the same as for (d).

and on equating this to the value of $M/M_s$ given by eqn (15.10) with $B_0 = 0$, the value of $T_C$ is found to be

$$T_C = \frac{\lambda M_s^2}{nk}\left(\frac{J+1}{3J}\right) = \frac{\lambda n g^2 \mu_B^2 J(J+1)}{3k} = \frac{\lambda C}{\mu_0} = \theta, \qquad (15.6a)$$

where $\theta$ is the Weiss constant defined by eqn (15.6). At still lower temperatures, the slope of the line is less than the initial slope of $\phi(y)$, and there will be two points of intersection, and two possible values of the magnetization, one zero and the other finite. It is easy to show that the former is unstable and the latter stable. For, if we imagine the magnetization at any instant to correspond to the point Q on $\phi(y)$, then the internal field produced by the magnetization corresponds to the point R, and this field will produce the greater magnetization corresponding to the point S on $\phi(y)$. Thus the magnetization will increase until the point P is reached where the two curves intersect. Above P, the two curves cross and any further increase in the magnetization would produce an internal field insufficient to sustain the increased magnetization. It thus appears that the state of spontaneous magnetization corresponding to the point P is stable, while the unmagnetized state is unstable.

Since the value of the spontaneous magnetization is determined by the intersection with $\phi(y)$ of the line corresponding to eqn (15.10) (with $B_0 = 0$), and the slope of this line depends on the temperature, it is obvious that the whole of the curve $\phi(y)$ will be traced out as we lower the temperature from the Curie point to the absolute zero. From eqn (15.6a), $\lambda M_s^2/nk = 3\theta J/(J+1)$, and hence we may express eqn (15.10) (with $B_0 = 0$) in the form

$$\frac{M_0}{M_s} = y\left(\frac{T}{\theta}\right)\left(\frac{J+1}{3J}\right).$$

Elimination of $y$ between this equation and eqn (15.8) shows that a 'reduced equation' may be found of the form

$$M_0/M_s = f(T/\theta), \qquad (15.11)$$

where the function $f(T/\theta)$ is the same for all substances with the same value of $J$. This function is plotted in Fig. 15.2 (broken line) for the special case of $J = \frac{1}{2}$; the curves for other values of $J$ lie slightly inside this curve at intermediate values of $(T/\theta)$. The experimental determination of $M_0/M_s$ and the verification of this 'law of corresponding states' will be discussed in § 15.6.

When a considerable external magnetic field of flux density $B_0$ is applied the effect on the magnetization is found by a graphical solution of eqns (15.8) and (15.10), where the term in $B_0$ is retained in the latter equation. The straight line corresponding to a plot of $M/M_s$ against $y$ is

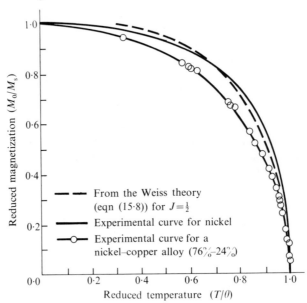

FIG. 15.2. Reduced equation of state for a ferromagnetic substance. (After Oliver and Sucksmith 1953.)

now displaced to the right compared to that for $B_0 = 0$ at the same temperature. The intersection with $M/M_s = \phi(y)$ occurs at the point P′ in Fig. 15.1, and the magnetization is slightly increased over that corresponding to P, the value for zero external field. At temperatures well below the Curie point $M_0$ is already close to $M_s$ and $\phi(y)$ increases only very slowly, so that the effect of $B_0$ is small. At temperatures near the Curie point P is on the steeper part of the curve for $\phi(y)$ near the origin and the increase in $M$ produced by an external field is more noticeable.

The theory outlined above is similar to the original theory of Weiss, except that the Brillouin function has been substituted for the Langevin function. Its great success lies in the explanation of the presence of spontaneous magnetization in a ferromagnetic substance, but there are also difficulties. The fact that the unmagnetized state is unstable appears to be contrary to experience, since it is well known that a piece of iron can be demagnetized by dropping it. Moreover, in a single crystal the magnetization can be restored by applying an external field of less than $10^{-4}$ T, although the internal field is about $10^3$ T! We also require some explanation of the hysteresis curve. To overcome these difficulties Weiss introduced the concept of domains of magnetization within the specimen. Each domain contains some $10^{17}$–$10^{21}$ atoms, and a piece of unmagnetized iron contains many domains all spontaneously magnetized, but the directions of magnetization of different domains are oriented at random. The theory of spontaneous magnetization applies to a single domain, but the magnetization of the whole specimen depends on whether the domains themselves are aligned towards the field or whether they are randomly oriented. This theory, which was conceived before the nature of the exchange interaction which causes the spontaneous magnetization was known, is remarkably successful in explaining the main features of ferromagnetic substances. The existence of domains has been confirmed by the experiments of Bitter, briefly described in the next section, where we shall first consider what factors determine the size and shape of the domains.

## 15.3. Ferromagnetic domains

A considerable advance in the understanding of ferromagnetism occurred when it became possible to obtain single crystals of iron, cobalt, and nickel sufficiently large for their magnetization curves to be measured. In each case it was found that the crystals are anisotropic, that is, the magnetization depends on the direction the field makes with the crystal axes. Fig. 15.3 shows the curves for iron, which forms body-centred cubic (b.c.c.) crystals. The $M-B_0$ curve is found to rise more steeply when $B_0$ is

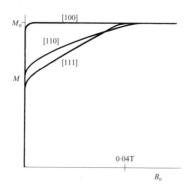

Fig. 15.3. Magnetization curves for a single crystal of iron. The directions of easy magnetization are the cube edges (for example, [100]). When the field is not along a cube edge, the initial process is of magnetization along the cube edges in directions nearest to that of the field; hence the curve for [110] breaks off roughly at $M_0/\sqrt{2}$, and that for [111] at $M_0/\sqrt{3}$, since further magnetization requires domain rotation against the anisotropy energy.

parallel to the edge of the unit cube [100] than any other direction, such as a face diagonal [110] or a body diagonal [111]. The energy of magnetization is $\int B_0 \, dM$, and is represented by the area between the magnetization curve and the $M$-axis ($B_0 = 0$). This energy is least when the single crystal of iron is magnetized along the [100] direction (or its equivalents, [010] and [001]), and these are known as directions of easy magnetization. In the case of nickel, with a f.c.c. structure, the directions of easy magnetization are the body diagonals, while for cobalt, with a hexagonal structure at room temperature, there is only one direction of easy magnetization, the hexagonal crystal axis.

The excess energy required to magnetize the substance in a hard direction is known as the anisotropy energy. It is clear that the anisotropy energy cannot arise from the exchange interaction, for the latter depends only on the mutual orientation of the dipoles and not on the angle which they make with the crystal axes. Its origin is thought to be similar to that of paramagnetic anisotropy (see end of § 14.7), arising from the combined effect of spin–orbit coupling and the electric field of the neighbouring charged ions. The anisotropy energy has the same symmetry properties as the crystal, and is smallest for crystals of high symmetry. Thus it is less for iron or nickel, which are both cubic, than for cobalt, which has only axial symmetry. The anisotropy energy also causes a change of length on magnetization (magnetostriction).

In zero field the specimen, whether it is a single crystal or an aggregate of crystals, will be in equilibrium when its potential energy is a minimum. In an unstrained crystal the important contributions are the exchange energy, the anisotropy energy, and the magnetostatic energy (the energy stored in the magnetic field). If the crystal consisted of one single domain, as in Fig. 15.4(a), the 'free poles' at the ends would give rise to a large external magnetic field and to a large magnetostatic energy. This is reduced by having two domains oppositely magnetized as in Fig. 15.4(b), when the two poles partially cancel one another. If there are no free poles on any surface the magnetostatic energy is reduced still further. For this to be the case, the flux density **B** at the surface of the crystal must always be parallel to the surface, and the normal component of **B** must be continuous across the boundary between two domains. If the two domains are magnetized in perpendicular directions the wall between them must run at an angle of 45° to each direction of magnetization, and Fig. 15.4(c)

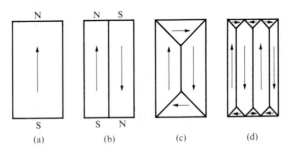

FIG. 15.4. Idealized domain structures in a single crystal, where the directions of easy magnetization are along the edges of a cube. (a) Single domain; external lines of field run from north to south pole and give large external field. (b) Double domain, where external lines of field run mostly between adjacent north and south poles, and the energy stored in external field is much reduced. (c) Arrangement with no free poles and no external field; the domains with perpendicular magnetization at top and bottom are called 'domains of flux closure'. (d) As in (c), but with further subdivision into smaller domains.

shows a possible arrangement. The little surface domains which produce a closed circuit of **B** are called domains of closure, and are generally much smaller than the inner domains. The size of the domains depends very much on the size and shape of the crystal, and this depends on the previous history of the substance. This fits in with the fact that the hysteresis curve is very sensitive to the composition and state of the specimen, since the domain structure must determine the shape of this curve.

The configuration in Fig. 15.4(d) is an alternative to that of Fig. 15.4(c), and we might expect the domains always to be very small in size and large in number; but energy is required to form the boundary between two domains, since the magnetization on either side is in opposite directions. The boundary between two domains is known as a 'Bloch wall'. It has a finite thickness, extending over a number of atoms whose spins change gradually in direction as we proceed through the wall (Fig. 15.5).

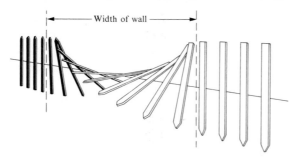

FIG. 15.5. Variation of spin orientation in a Bloch wall.

From eqn (15.3) the exchange energy between neighbouring identical spins is approximately $\Delta W_e = -2\mathcal{J}'J^2 \cos \psi$, where $\psi$ is the angle between the directions of the spin momentum vectors. Therefore the total exchange energy in going through the wall is

$$W_e = -\sum_{i>j} 2\mathcal{J}'J^2 \cos \psi_{ij}.$$

If the wall thickness extends over many atoms, and the angle between neighbouring spins is small, we may write $\cos \psi_{ij} \approx 1 - \psi_{ij}^2/2$, and the total increase in exchange energy because the spins are not exactly parallel is

$$W_e \approx \mathcal{J}'J^2 \sum \psi_{ij}^2.$$

For a wall which forms the boundary between two domains where the spins are antiparallel, the total change in angle in going through the wall is $\sum \psi_{ij} = \pi$. If there is a line of $n'$ atoms in the thickness of the wall, and $\psi_{ij}$ is the same for all adjacent pairs of atoms, $n'\psi_{ij} = \pi$ and

$$W_e \approx n'\mathcal{J}'J^2(\pi/n')^2 = \pi^2 \mathcal{J}'J^2/n'.$$

This equation shows that the exchange energy is reduced by making $n'$ large, and it would seem that the wall should be infinitely thick. This would increase the anisotropy energy $W_a$, however, since a number of spins in the wall are pointing at an angle to the direction of easy magnetization, and this number increases with the wall thickness. Thus $W_a$ is proportional to $n'$, and the total energy per unit area of wall in the substance is

$$W_e + W_a = \pi^2 \mathcal{J}' J^2 / n' a^2 + K n' a, \tag{15.12}$$

where $K$ is a constant roughly equal to the anisotropy energy per unit volume. $a$ is the lattice constant of the substance, so that, for a simple cubic crystal, there are $1/a^2$ atoms per unit area of wall, and $n'a$ is the thickness of the wall.

The form of eqn (15.12) shows that there will be a minimum value of the total energy for some value of $n'$, which by differentiation is found to be $n' = (\pi^2 \mathcal{J}' J^2 / K a^3)^{\frac{1}{2}}$. For Ni, $J = \frac{1}{2}$, $\mathcal{J}$ is about $10^{-21}$ J, $K$ is about $10^4$ J m$^{-3}$, and $a^3$ is about $10^{-29}$ m$^3$. Hence $n'$ is of the order of 100 atoms, and the thickness of the wall is a few tens of nanometres. Substitution of the optimum value of $n'$ in eqn (15.12) gives the expression $2\pi(\mathcal{J}' K J^2 / a)^{\frac{1}{2}}$ for the wall energy per unit area, whose order of magnitude is found to be about $10^{-3}$ J m$^{-2}$.

As the domain width decreases in the flux closure arrangement shown in Fig. 15.4(d), the number of walls per unit area of the crystal surface increases, with a corresponding increase in the energy. The energy required to form a wall therefore tends to keep the domains small in number, and large in size. When $K$ is large, particles of about 1 $\mu$m diameter are found to consist of a single domain, because the energy required to form a wall is more than the reduction in the magnetostatic energy which would result from the subdivision into domains. In large crystals another factor which enters into the determination of domain size is that the domains of closure in Fig. 15.4(d) may require to be magnetized in a hard direction, thereby increasing the anisotropy energy. The volume occupied by the domains of closure decreases as the width of the domains decreases, and the anisotropy energy therefore tends to reduce the domain size, while the wall energy tends to increase it. The optimum domain size is determined by a compromise between these two effects.

The most striking evidence for the existence of domains is provided by the Bitter patterns which are obtained when finely powdered iron or cobalt, or colloidal magnetite, is spread on the surface of the crystal. The surface must be very carefully prepared and electrolytically polished to remove irregularities. The particles deposit themselves along the domain boundaries since here there are strong local inhomogeneous magnetic fields which attract the particles. A typical Bitter pattern is shown in Fig. 15.6; the 'fir-tree' effect is obtained when the surface makes a small angle

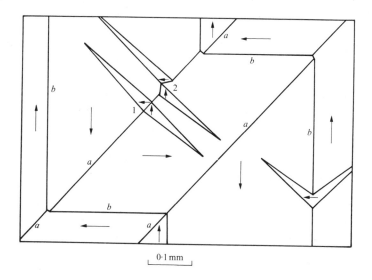

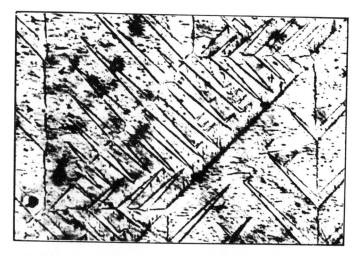

FIG. 15.6. Domain patterns on a demagnetized single crystal of silicon–iron (the surface is very nearly a (100) crystal plane). The magnetization is normal to the fine scratches visible on the surface, and is directed as shown in the key diagram above. Domain walls labelled *a* form the boundary between domains magnetized in directions differing by 90°, and those labelled *b* are boundaries between domains differing by 180°. The 'fir-tree' closure domains arise because the surface is not exactly a crystal plane. Two types of closure domain (labelled 1 and 2) can be seen on the 90° wall. (Photograph by L. F. Bates and A. Hart.)

of 2° or 3° with the true (100) crystal plane. The branches of the tree are the domains of closure which close the flux circuit over the primary domains below. On looking through a microscope the patterns can be seen to change as a magnetic field is applied. The direction of magnetization in a domain is found by making a tiny scratch on the surface with a fine glass fibre. If the scratch is parallel to the magnetization the pattern is unchanged, but if it is normal to it the pattern is distorted. This is because a scratch parallel to the field behaves as a long narrow cavity, with no free poles at the ends; a scratch perpendicular to the field will have induced poles on its sides, and there will be a strong field in the cavity so that the pattern is distorted. Experiments of this type, and others, in which the scattering of beams of electrons or polarized neutrons have been used to investigate domain structure, show that the theory outlined above is correct in its main features.

The changes in the domain structure which occur when a magnetic field is applied, and the correspondence between these changes and the various parts of the magnetization curve, have already been outlined in § 6.4. The initial portions of the magnetization curve are associated with movements of the Bloch walls, which are reversible in small fields but irreversible after larger fields have been applied. Where there are strains or inclusions of impurities the energy depends on the position of the wall, as can be seen from considering the effect of a small particle embedded in the material. Such a particle will be a small domain magnetized in one of its own easy directions of magnetization, which do not in general coincide with those of the surrounding material, or it may be a particle of a non-ferromagnetic substance. In the latter case there will be free poles at its surface, as in Fig. 15.7(a), and the field of these poles gives extra

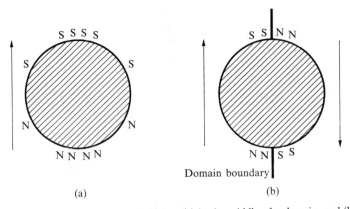

(a)                                              (b)

FIG. 15.7. Effect of a non-magnetic inclusion, (a) in the middle of a domain, and (b) when intersected by a domain boundary.

magnetostatic energy. If a Bloch wall intersects the particle, as in Fig. 15.7(b), this energy will be reduced, just as in the case of free poles on the surface of a ferromagnetic substance in Fig. 15.4(a) and (b). This gives a minimum of energy when a wall intersects as many inclusions as possible. In a small external field the wall is displaced slightly away from the minimum energy, but returns when the field is removed; this gives a reversible wall movement. In larger fields the wall may be shifted to a more distant position where the energy curve has passed through a maximum and then diminished; on removing the field the wall cannot cross the energy maximum and so is unable to return to its initial position. The displacement is then irreversible. The more free the material is from strains and inclusions, the greater the size of reversible wall movements, and the lower the field required to produce a movement, thus giving a large initial permeability, and a 'soft' magnetic material. With large strains and many inclusions the smaller is the possibility of boundary movement, and the higher the coercive force.

## 15.4. The gyromagnetic effect

It was pointed out in § 14.1 that the magnetic moment of an atom is proportional to the total electronic angular momentum of the atom. For a macroscopic system, the total magnetic moment $\mathbf{M}$ and the total electronic angular momentum $\mathbf{G}_e$ are formed by similar vector addition of the individual components, and they should therefore be related in the same way. Thus

$$\mathbf{M}/\mathbf{G}_e = \gamma = -g'e/2m_e,$$

where $g'$ is an effective Landé factor. It follows from this that if we could measure in some way the change in electronic angular momentum associated with the change in magnetization of a specimen, the value of $g'$ would be determined. Since $g'$ differs by a factor of 2 according to whether the magnetic moments are associated with orbital or spin angular momentum, this affords a method of verifying the assumption that ferromagnetism in the 3d group is associated with the electronic spins.

Since no external couple is exerted on a specimen by the act of changing its magnetization, the total angular momentum of the system must remain unaltered. The change $\Delta G_e$ in the electronic angular momentum must be accompanied by an equal and opposite change

$$\Delta G_{\text{lat}} = -\Delta G_e$$

in the angular momentum of the 'lattice', defined as the rest of the specimen, apart from the electrons responsible for the magnetization. It is this latter change in angular momentum which is observed, but it is very small. In a cubic centimetre of nickel there are some $10^{23}$ electrons whose individual momenta can be changed by $\hbar \approx 10^{-34}$ J s by reversal of the spin. The total angular momentum thus imparted to the lattice is only about $10^{-11}$ J s.

A variety of experimental methods have been used to determine $\gamma$, but only a short account will be given here (more details are given by the authors to whom references are made in this section). The methods fall into two classes. In one, an unmagnetized specimen is set into rotation and the resultant magnetization is measured. This is the Barnett effect, and typical experiments are those of Barnett (1944). Even with a large specimen, the magnetic moment induced is very small owing to the limited velocities of rotation which can be employed. In the second class, the magnetization is changed by a known amount and the change in angular momentum is determined; this is known as the Einstein–de Haas effect, though first suggested by Richardson. This method has the advantage that resonance can be used to enhance the effect. A ferromagnetic rod is suspended inside a long solenoid supplied with alternating current whose period is equal to the torsional oscillation period of the suspended rod. If $\mathscr{I}$ is the moment of inertia of the rod, $b$ the damping constant, $c$ the torsion constant of the suspension, and $M \sin \omega t$ the magnetic moment of the specimen at any instant, the equation of motion is

$$\mathscr{I}\frac{d^2\theta}{dt^2} + b\frac{d\theta}{dt} + c\theta = \frac{dG}{dt} = \frac{1}{\gamma}\omega M \cos \omega t.$$

At resonance, the amplitude of the angle of rotation is $(1/\gamma)(M/b)$; $b$ is found from the logarithmic decrement, and $M$ must be measured independently.

This method of measuring $\gamma$ was employed by Scott (1951) using a modification of the apparatus built for the determination of $e/m$ of the carriers of electric current (see § 3.1). The specimen, in the form of a rod, is suspended as a torsional pendulum. A coil is wound on the rod, and by reversing a current in this coil the magnetization of the rod can be reversed. The change in magnetization is measured by a null magnetometer placed half-way between the rod and a standard coil carrying a steady current which is simultaneously reversed with that in the specimen. This steady current is adjusted until a balance is obtained. The magnetometer is fitted with a mirror, and the light reflected from it falls on a pair of photocells each followed by an amplifier; by means of detection of the difference in the two outputs, a null magnetometer of great sensitivity is produced. A correction was made for the non-uniformity of magnetization of the rod, and the earth's field was neutralized by a system of Helmholtz coils. The period of oscillation of the rod was 26 s, and its rotation was observed by reflections of a beam of light from a mirror mounted immediately above the specimen. The procedure used was to reverse the magnetizing current at a moment when the specimen passed through the centre of its swing. The direction of reversal was chosen so that for 60 current reversals the amplitude was increased, and then for 60 reversals it was decreased. With small damping, the progressive change in amplitude was very nearly linear, and the amplitude change for one reversal was obtained from the two slopes of the plot for 120 reversals.

A number of experiments of high precision have been carried out using both the Einstein–de Haas and Barnett effects, and a mean of the results obtained between 1944 and 1960 is given in a survey by Meyer and Asch (1961), who show also that there is good agreement with results of ferromagnetic resonance experiments using microwave radiation (see § 24.7). The results are shown in Table 15.2, and are expressed in terms of two quantities g and g′, obtained from ferromagnetic resonance and gyromagnetic experiments respectively. When we have a mixture of orbit and spin, the magnetic moment and angular momentum may be written as $M = M_L + M_S = (e/2m_e)(G_L + 2G_S)$, $G = G_L + G_S$, and the ratio is

$$\gamma = \frac{M}{G} = \frac{e}{2m_e}\left(\frac{G_L + 2G_S}{G_L + G_S}\right) = g'\left(\frac{e}{2m_e}\right), \tag{15.13}$$

The 'spectroscopic splitting factor' g measured in a ferromagnetic resonance experiment has been shown by Kittel and Van Vleck to be defined by

$$\frac{M}{G_S} = \frac{e}{2m_e}\left(\frac{G_L + 2G_S}{G_S}\right) = g\left(\frac{e}{2m_e}\right), \tag{15.14}$$

from which it follows that

$$1/g + 1/g' = 1. \tag{15.15}$$

The results in Table 15.2 show that this relation is fulfilled within the experimental error for iron and nickel, and Meyer and Asch show that this is true also for a wide range of alloys of the 3d group. The fact that g, g′ are so close to 2 shows that the magnetism of the ferromagnetic

TABLE 15.2

*Some values of the quantities g′ and g*

| Substance | g′ | g | $1/g + 1/g'$ |
|---|---|---|---|
| Iron | 1·928±0·004 | 2·094±0·003 | 0·996±0·004 |
| Cobalt | 1·854±0·004 | — | — |
| Nickel | 1·840±0·008 | 2·185±0·010 | 1·001±0·009 |

The quantity g′ is derived from gyromagnetic (magnetomechanical) experiments, the quantity g from ferromagnetic resonance experiments; the values quoted are the means of a number of experimental results, given by Meyer and Asch (1961).

metals of this group is almost entirely due to spin. This result is similar to that found for the paramagnetism of salts of 3d group ions (Chapter 14), and there is little doubt that it is due essentially to the same cause, 'quenching' of the orbital moment by electrostatic interaction with the neighbouring (ligand) ions.

## 15.5. Thermal effects in ferromagnetism

When a substance is magnetized, with all the electron spins pointing in one direction, it is in a state of greater order than when it is unmagnetized, with the spins pointing in random directions. The magnetized state is therefore one of lower entropy than the unmagnetized state, and in passing from the former to the latter there will be an increase in the entropy of the spin system. If the transition is accomplished by heating a ferromagnetic substance through its Curie point, the entropy change appears as an anomaly in the heat capacity. If it is accomplished by the sudden (adiabatic) removal of a magnetic field, the entropy change appears as a fall in the temperature of the substance; this is known as the magnetocaloric effect. Both this effect and the heat capacity anomaly have been used to obtain information about the ferromagnetic state.

The heat capacity of a substance is $C = T(dS/dT)$, where the entropy change $dS$ is given by the relation

$$T\,dS = dU - \frac{1}{\rho}B\,dM;\qquad(15.16)$$

$dU$ is the change in the internal energy, and $-B\,dM$ is the increase in the magnetic potential energy when the magnetization is increased by $dM$ at constant flux density $B$. The density $\rho$ appears because $M$ is the magnetization per unit volume, while the specific heat (and other thermal quantities) are per unit mass. In a ferromagnet, $B = B_0 + \lambda M$, and the heat capacity is thus

$$C = T\frac{dS}{dT} = \frac{dU}{dT} - \left(\frac{B_0 + \lambda M}{\rho}\right)\left(\frac{dM}{dT}\right)$$

$$= C_M - \left(\frac{B_0 + \lambda M}{\rho}\right)\left(\frac{dM}{dT}\right).\qquad(15.17)$$

Below the Curie point any external field $B_0$ is very small in comparison with the internal field $\lambda M$, so that we can write

$$C = C_M - \frac{1}{2}\left(\frac{\lambda}{\rho}\right)\frac{d(M^2)}{dT}.\qquad(15.18)$$

Here $C_M$ is the heat capacity of the substance at constant magnetization, while the second term arises from the change in magnetization with temperature. Since $M$ falls with increasing $T$, it gives a positive contribution to the heat capacity (as the temperature rises, the degree of order in the magnetic system decreases, and the entropy associated with the magnetization increases). Fig. 15.2 shows that the rate of change of $M$ with temperature is greatest just below the Curie point, and the anomalous heat capacity arising from the magnetic properties should be greatest at this point, followed by a sharp drop above the Curie point where $M$ becomes zero.

An experimental curve showing the variation of $C$ with $T$ for nickel is given in Fig. 15.8. The anomalous heat capacity is appreciable only near the Curie point, where it appears as a sharp discontinuity. In order to obtain a value for $\lambda$ from the heat capacity anomaly, $C_M$ must be estimated and subtracted from the measured heat capacity so that only the magnetic contribution remains. Measurements are made at constant pressure, so that we can write $C_M = C_V + (C_p - C_V) + C_e$. $C_V$ is obtained by extrapolation, using the Debye formula, from measurements at low temperatures; $(C_p - C_V)$ may be found from the expansion coefficient and compressibility using a standard thermodynamical formula; $C_e$ is the electronic heat capacity. This is abnormally large in a ferromagnetic metal and difficult to estimate since it is associated with a high electron density in the 3d band (see § 12.6). $dM^2/dT$ must be found by plotting $M^2$ as a function of temperature, and then $\lambda$ is obtained. This is not a very accurate method of finding $\lambda$ and the value does not agree too well with

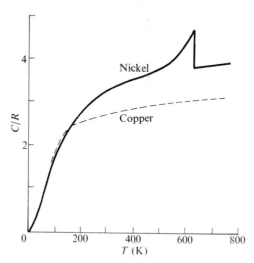

FIG. 15.8. The molar heats $C/R$ of copper and nickel, showing the excess in the ferromagnetic metal, and the sharp drop at the Curie point.

the value obtained from the magnetization curve, probably because of errors in $C_e$. However, the general form of the heat capacity curve is not incompatible with theory and this also applies to iron and cobalt, although the measurements on these metals are less certain (for experimental details, see Grew (1934)). The general form of the heat capacity anomaly ('lambda type') is typical of a 'co-operative' transition from an ordered to a disordered state.

If a field is applied to a magnetic substance, there is in general an increase in magnetization, and this results in a state of greater order than in zero field. In other words the entropy of the system has decreased, and the loss of (magnetic) potential energy of the dipoles in turning towards the field appears as heat of magnetization. If the field is switched off isothermally, heat is absorbed. If the field is switched off adiabatically the entropy of the system must remain constant; the increase in entropy due to increased disorder of the dipoles is then compensated by a decrease in the entropy associated with thermal agitation, and there is therefore a fall in temperature. This is the basis of the 'magnetic cooling' method for obtaining temperatures below 1 K using paramagnetic substances. This 'magnetocaloric' effect also has applications to ferromagnetics. Since $dS = 0$ in a reversible adiabatic process, we have from eqn (15.17)

$$dT = \left(\frac{B_0 + \lambda M}{\rho C_M}\right) dM. \tag{15.19}$$

Above the Curie point saturation effects are negligible and $M/B_0$ is a constant at a given temperature, so that in a finite change of the magnetization we have

$$\Delta T = \left(\frac{B_0/M + \lambda}{2\rho C_M}\right) \Delta(M^2). \tag{15.20}$$

Below the Curie point we can neglect $B_0$ in comparison with $\lambda M$, and we obtain

$$\Delta T = \left(\frac{\lambda}{2\rho C_M}\right) \Delta(M^2). \tag{15.21}$$

If the external field is initially zero, so that the magnetization of each domain has the spontaneous value $M_0$, the temperature rise on applying a field is

$$\Delta T = \left(\frac{\lambda}{2\rho C_M}\right)(M^2 - M_0^2). \tag{15.22}$$

If $\Delta T$ is plotted as a function of $M^2$, a curve of the form shown in Fig. 15.9 is obtained. It becomes a straight line in the region where the external field is large enough to change the magnetization of the domains, with a curved tail at lower fields where the magnetization of the substance

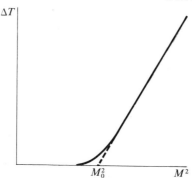

FIG. 15.9. Curve showing the variation of $\Delta T$ with $M^2$ in the magnetocaloric effect. $\Delta T = (\lambda/2\rho C_{\mathrm{M}})(M^2-M_0^2)$.

is mainly due to wall movements or the rotation of domains. Extrapolation of the straight portion to the axis $\Delta T = 0$ gives $M_0^2$ from the intercept.

The magnetocaloric effect may be used for a number of purposes, such as investigation of the hysteresis curve, one of the most important being the determination of the spontaneous magnetization $M_0$ near the Curie point. A good description of experimental technique is given by Oliver and Sucksmith (1953) in work on a copper–nickel alloy (24%Cu; 76%Ni).

## 15.6. Measurement of the spontaneous magnetization $M_0$ as a function of temperature

In § 15.2 it was shown that the spontaneous magnetization of a single domain should obey an equation of state which depends only slightly on $J$ (see Fig. 15.2). To test this relation it is necessary to determine the value of $M_0$ for a single domain at zero field over a wide range of temperature. Since in practice any specimen consists of a number of domains randomly oriented, so that (apart from remanence) the net magnetization will be zero, it follows that the spontaneous magnetization of a single domain cannot be directly measured. If we apply a sufficiently strong field, however, the various domains will rotate until they point in the direction of the external field, and the resultant magnetic moment will be close to the spontaneous magnetization of the individual domains. It will slightly exceed it, since the magnetization under these conditions corresponds, not to the point P in Fig. 15.1, but to the point P′, the stable state in the presence of a magnetic field. In order to find the value corresponding to P, we must make measurements of $M$ for a range of values of the external field, and then extrapolate back to zero field. Since the fields which are applied are small compared with the internal field, the point P′ is never far from P, and the extrapolation required is not very great at temperatures well below the Curie point. Near the Curie point the magnetocaloric effect is used as described in the previous section.

In a number of magnetic materials the nucleus of the magnetic ion possesses a nuclear magnetic dipole moment $\mathbf{m}_N$ which interacts with the magnetic flux density $\mathbf{B}_e$ of the electrons (see § 14.10). ($\mathbf{B}_e$ is the actual magnetic flux density at the nucleus generated by the magnetic electrons, and is nothing to do with the effective molecular field of flux density $\mathbf{B}_{int}$ introduced by Weiss to explain ferromagnetism.) The interaction energy

$$W = -\mathbf{m}_N \cdot \mathbf{B}_e$$

gives a hyperfine splitting of the nuclear levels, from observation of which $B_e$ can be found if the nuclear moment is known. In a ferromagnetic substance $B_e$ is parallel to the magnetization, and its time-average value is proportional to the average magnetic moment on each ion (apart from some small corrections). Thus $B_e$ is proportional to the magnetization, and observation of the hyperfine structure separation as a function of temperature gives a convenient and accurate method of determining the saturation magnetization curve. This can be done in zero external field, since it is not necessary to line up the domains, and many of the difficulties of direct measurement of the bulk magnetization are avoided.

The magnitude of the electronic flux density $B_e$ lies generally between $10\,\mathrm{T}$ and $10^3\,\mathrm{T}$. The nuclear levels have energy $W_{m_I} = -g_N\mu_N m_I B_e$, where $m_I$ is the nuclear magnetic quantum number, and are equally separated by an amount corresponding to a frequency of $10^2$–$10^4\,\mathrm{MHz}$. Two methods are available for measuring this separation over a range of temperature. One of these is the Mössbauer effect, in which a low-energy $\gamma$-ray is emitted from a nucleus in an excited state and absorbed by a nucleus in the ground state. A $\gamma$-ray photon of energy $h\nu$ carries momentum $h\nu/c$, so that the emitting or absorbing nucleus is given a recoil momentum, and hence takes up energy which reduces the photon energy. If the nuclei are in a solid the recoil momentum is generally taken up by the solid as a whole, and the energy taken from the $\gamma$-ray is negligible. Thus in a solid, unlike a gas (see Problem 15.3), there is no spread in energy of the photon due to the varying amounts of energy taken up by the recoil.

Only those $\gamma$-rays which are extremely narrow are of use, since the width of the $\gamma$-ray must be smaller than the hyperfine splitting. For magnetic purposes, the $14\cdot4\,\mathrm{keV}$ transition between the excited state ($I = \frac{3}{2}$) and the ground state ($I = \frac{1}{2}$) of the isotope $^{57}\mathrm{Fe}$ has been especially useful. This gives a linewidth of about $3\,\mathrm{MHz}$, and the hyperfine levels and structure of the Mössbauer $\gamma$-ray are shown in Fig. 15.10. The structure is exactly analogous to the Zeeman effect in an atomic transition. The splittings are a very small fraction of the $\gamma$-ray frequency, and

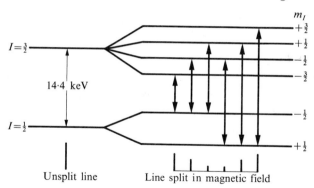

FIG. 15.10. Hyperfine splitting of the nuclear states of $^{57}$Fe in a magnetic field. The ground state $I = \frac{1}{2}$ has $g_N = +0.18$, and the excited state $I = \frac{3}{2}$ has $g_N = -0.010$; the allowed transitions are those for which $\Delta m_I = 0, \pm 1$. Note the gross disparity in scale; the hyperfine splittings are about $10^{-11}$–$10^{-12}$ of the $\gamma$-ray frequency. In some substances there is also an electric quadrupole interaction in the $I = \frac{3}{2}$ state.

their analysis is made by means of the Doppler effect produced by a relative motion of the source and absorber. It is convenient to use a source with no hyperfine structure, such as $^{57}$Fe (derived from the nuclear decay of $^{57}$Co) in stainless steel, which is non-magnetic, since this gives a single emission line, as shown on the left of Fig. 15.10. For this to be absorbed by a $^{57}$Fe nucleus in a magnetic substance, where six transitions are allowed with slightly different frequencies as on the right of Fig. 15.10, a Doppler shift is needed of the correct velocity to bring one of the transitions to the same frequency as the single line on the left. Thus the entire hyperfine pattern can be scanned by systematically changing the relative velocity of source and absorber (the velocity required is of order a few millimetres per second). Of course source and absorber can be interchanged, and the choice is determined by experimental convenience.

The hyperfine field in metallic iron has been determined as a function of temperature by means of the Mössbauer effect; the results (see Fig. 15.11) of Nagle, Frauenfelder, Taylor, Cochran, and Matthias (1960) show close agreement with the saturation curve determined by conventional means. At sufficiently low temperatures for the magnetization to reach the saturation value, the hyperfine field is 33 T. On applying an external field parallel to the magnetization Hanna, Heberle, Perlow, Preston, and Vincent (1960) found that the net field at the $^{57}$Fe nucleus was reduced, showing that the hyperfine field was in the opposite sense to the external field, and hence also to the magnetization.

The hyperfine splitting of the ground nuclear levels in a ferromagnetic substance has also been measured by the method of nuclear magnetic resonance (see § 24.5). For $^{57}$Fe this gives a direct observation of transitions between the states $m_I = +\frac{1}{2}$ and $-\frac{1}{2}$ of the ground state $I = \frac{1}{2}$, at a

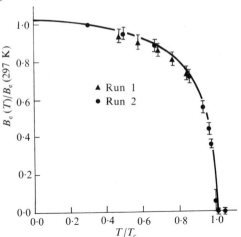

Fig. 15.11. The hyperfine magnetic field at a $^{57}$Fe nucleus in metallic iron, relative to that at room temperature, plotted against the reduced temperature $T/T_0$. The experimental points are measured by the Mössbauer effect, the solid line indicates the relative saturation magnetization as determined by a bulk measurement (see Nagle *et al.* 1960).

frequency such that

$$h\nu = W_{-\frac{1}{2}} - W_{+\frac{1}{2}} = g_N \mu_N B_e, \qquad (15.23)$$

where $g_N$ is the value for the ground state $I = \frac{1}{2}$. This gives a more precise measurement of $B_e$ than the Mössbauer method (in which the minimum linewidth is determined by the lifetime of the excited nuclear state). In iron at 0 K, the resonance frequency of 46·648 MHz has been determined to 1 kHz by Riedi (1973).

## 15.7. Spin waves

The molecular field theory introduced by Weiss is remarkably successful in giving a simple interpretation of the phenomenon of ferromagnetism. When combined with the Brillouin function it predicts a unique curve for the reduced magnetization as a function of reduced temperature for each value of **J**; the prediction is in good but not exact agreement with experiment (see Fig. 15.2). In this section we outline a more sophisticated approach, first formulated by Bloch (1932), in which the collective excitations of the assembly of magnetic dipole moments are analysed in terms of wave-like motions known as 'spin waves'. Formulae for the heat capacity and magnetization are obtained in the form of expansions which are valid for the co-operative state at low temperatures.

At 0 K, where the magnetization has the saturation value, all the spins are rigorously parallel, but this is obviously not so at a non-zero temperature where the magnetization is smaller. However, it would be incorrect to regard the reduction in magnetization as due to the reversal of any given individual spin, because any such deviation would be passed on to neighbouring spins through the exchange interaction in a time of order $\hbar/\mathcal{J}'$, so that it would not remain localized on any given atom. In fact the average deviation of each spin from exact parallelism is small, and can be analysed in terms of sinusoidal spatial variations throughout the crystal, known as 'spin waves'. In a spin wave of wave vector $\mathbf{k}_s$, the angle between adjacent spins a distance $\mathbf{r}_0$ apart is $\mathbf{k}_s \cdot \mathbf{r}_0$, and since the exchange energy varies with the cosine of this angle, the extra energy required to excite a wave is proportional to $\mathcal{J}'(1-\cos \mathbf{k}_s \cdot \mathbf{r}_0) \sim \frac{1}{2}\mathcal{J}'(\mathbf{k}_s \cdot \mathbf{r}_0)^2 = \frac{1}{2}\mathcal{J}'k_s^2 r_0^2 \cos^2 \theta_{k,r_0}$, where $\theta_{k,r_0}$ is the angle between $\mathbf{k}_s$ and $\mathbf{r}_0$. This must be summed over all neighbours, giving

$$W_k = \hbar\omega_k = Dk_s^2. \tag{15.24}$$

This is the dispersion relation connecting the frequency and wavelength of a spin wave. $D$ is proportional to the exchange energy $\mathcal{J}'$, and in a cubic lattice with $z$ equidistant neighbours with angular momentum $J$,

$$D = \frac{1}{3}\mathcal{J}'zJr_0^2. \tag{15.25}$$

The energy required to excite a spin wave is proportional to $k_s^2$, and at low temperatures only long waves will be excited; in the limit at $T = 0$ the only wave present will be $k_s = 0$, which corresponds to all the dipoles being parallel. As the temperature rises more spin waves of shorter wavelengths will be excited, and the energy required for this shows up as an additional term in the heat capacity. This is quite distinct from the abnormal electronic heat capacity discussed above, and would be present in a ferromagnetic insulator. It is purely magnetic in origin, and constitutes the low-temperature tail of the magnetic heat capacity anomaly discussed in § 15.5. Van Kranendonk and Van Vleck (1958) have shown that a spin wave behaves formally like a harmonic oscillator, its mean energy being

$$\bar{W}_k = \frac{\hbar\omega_k}{\exp(\hbar\omega_k/kT)-1} = \bar{n}\hbar\omega_k, \tag{15.26}$$

where $\bar{n} = \{\exp(\hbar\omega_k/kT)-1\}^{-1}$ is known as the 'occupation number'. The number of spin waves of wave vector $k_s$ (we have used $k_s$ to avoid confusion with Boltzmann's constant $k$ which occurs in eqn (15.26) is $g(k_s)\,dk_s = (V/2\pi^2)k_s^2\,dk_s$, (cf. eqn (11.5)). Hence the internal energy at temperature $T$ is

$$U = \int \bar{n}(\hbar\omega_k) g(k_s)\, dk_s$$

$$= \frac{V}{2\pi^2} \int \frac{(Dk_s^2)k_s^2\, dk_s}{\exp(Dk_s^2/kT)-1}$$

$$= \frac{VD}{4\pi^2}\left(\frac{kT}{D}\right)^{\frac{5}{2}} \int \frac{x^{\frac{3}{2}}\, dx}{\exp x - 1},$$

where we have made the substitution $x = (Dk_s^2/kT)$. The integral can be taken to infinity at low temperatures, and is therefore just a numerical constant. Differentiation with respect to temperature gives

$$C_V = dU/dT = c(kT/D)^{\frac{3}{2}}, \tag{15.27}$$

where $c$ is a numerical constant whose value depends somewhat on the crystal structure; for a simple cubic lattice ($z = 6$) we have (per mole)

$$C_V = 0.113 k V (kT/D)^{\frac{3}{2}} = 0.113 k (V/r_0^3)(kT/2\mathcal{J}'J)^{\frac{3}{2}}$$

$$= 0.113 R (kT/2\mathcal{J}'J)^{\frac{3}{2}}. \tag{15.28}$$

For the ordinary ferromagnetic metals this magnetic heat capacity is small and difficult to measure in the presence of the abnormally large electronic heat capacity, and the lattice heat capacity. It has been detected in some non-conducting ferrimagnetic compounds (see § 16.3).

The departure of the magnetization from the absolute saturation value $M_s$ can be computed similarly, using the fact that each spin wave reduces the magnetic moment by an amount $g\mu_B\bar{n}$. Thus

$$M_s - M_0 = g\mu_B \sum \bar{n} = \frac{g\mu_B V}{2\pi^2} \int \frac{k_s^2\, dk_s}{\exp(Dk_s^2/kT)-1}$$

$$= \frac{g\mu_B V}{4\pi^2}\left(\frac{kT}{D}\right)^{\frac{3}{2}} \int_0^{\infty} \frac{x^{\frac{1}{2}}\, dx}{\exp x - 1}.$$

On substituting for $D$ and using the fact that $g\mu_B(V/r_0^3) = M_s$, in a simple cubic lattice, we obtain the relation

$$M_0/M_s = 1 - a(kT/2\mathcal{J}'J)^{\frac{3}{2}}, \tag{15.29}$$

where $a = 0.059/J$ in a simple cubic lattice. This result is the first term of a power series where the next terms contain $T^{\frac{5}{2}}$, $T^{\frac{7}{2}}$, and $T^4$. The terms in $T^{\frac{5}{2}}$ and $T^{\frac{7}{2}}$ have been verified in a special case where they are unusually large (see Gossard, Jaccarino, and Remeika 1961), using nuclear magnetic resonance of the $^{53}$Cr nucleus in $CrBr_3$, where the nuclear resonance frequency is accurately proportional to the magnetization. In gadolinium metal, where the spontaneous magnetization has been measured by Elliott, Legvold, and Spedding (1953) by the bulk magnetization method, the $T^{\frac{3}{2}}$ law holds closely almost up to the Curie point, a result that has been explained as due to a near cancellation of the higher terms (Goodings 1962). In iron, Riedi (1973) has shown that nuclear magnetic resonance measurements are consistent with a variation in the magnetization which can be fitted accurately by a single $T^{\frac{3}{2}}$ term up to 32 K, and up to 400 K with one extra term in $T^{\frac{7}{2}}$. These experimental results confirm the validity of the spin-wave approach; the existence of spin waves is confirmed in magnetic resonance experiments using thin ferromagnetic films (§ 24.7).

## 15.8. Mechanisms of exchange interaction

In this chapter we have assumed that ferromagnetism is due to exchange interaction; certainly no other interaction of the correct form is known whose magnitude is large enough to produce a co-operative magnetic state at room temperatures. However, attempts to calculate the direct interaction from first principles have proved disappointing both as regards sign and size, and it appears that indirect mechanisms of exchange interaction play a predominant role.

The original treatment of Heisenberg and Dirac deals with the interaction between two electrons on the same atom. If the two electrons did not influence one another the solution of the wave equation would be a simple product of the two solutions for a single electron, of the form

$$\psi_{\mathrm{I}} = \phi_k^{(1)} \phi_m^{(2)}.$$

The physical interpretation of this is that electron (1) is in orbital $k$, and electron (2) is in orbital $m$. Since the two electrons are equivalent, the energy is unchanged if the two electrons are interchanged, giving another solution

$$\psi_{\mathrm{II}} = \phi_k^{(2)} \phi_m^{(1)}.$$

In general, any linear combination of these two solutions is allowed, the correct combination being determined when we include the electrostatic energy $e^2/r_{12}$ (where $r_{12}$ is the distance between the two electrons) of repulsion between the two negatively charged electrons. The correct solutions are then the symmetric and antisymmetric combinations

$$\psi_{\mathrm{sym}} = (2)^{-\frac{1}{2}}(\psi_{\mathrm{I}} + \psi_{\mathrm{II}}),$$
$$\psi_{\mathrm{ant}} = (2)^{-\frac{1}{2}}(\psi_{\mathrm{I}} - \psi_{\mathrm{II}}).$$

These two solutions no longer have the same energy, because the symmetrical solution allows the wavefunction to have a large amplitude if the two electrons are at the same point, while the antisymmetric wavefunction then vanishes because $\psi_{\mathrm{I}} = \psi_{\mathrm{II}}$. Thus the electrostatic repulsive energy between the two electrons is larger in the first case than in the second.

So far it has not been necessary to include the electron spin. With two electrons, the spin states (two for each electron, hence four in all) are divided between the triplet states ($S = 1$, $M_s = 1$, $0$, $-1$) and the singlet state $S = 0$, according to whether the two individual spins are parallel or antiparallel. The triplet states are symmetrical with respect to interchange of the two electrons, the singlet state is antisymmetrical. Since only states whose over-all symmetry is antisymmetrical are allowed in nature, the spin triplet states can only be combined with the orbital $\psi_{\mathrm{ant}}$, and the singlet spin state with the orbital $\psi_{\mathrm{sym}}$. Thus the difference in electrostatic energy between the symmetrical and antisymmetrical orbital states carries with it a corresponding energy difference between the spin singlet and triplet. This is formally similar to the introduction of a coupling energy between the electron spins of the type assumed in § 15.1.

For two electrons within the same atom, the exchange energy is always positive, so that the state of lower energy is with the spins parallel. This coupling is 'ferromagnetic' in nature, and is the justification for Hund's rule (§ 14.2) which makes the ground state of the atom the one with maximum multiplicity in the spin. In the phenomenon of ferromagnetism proper, however, we are concerned with exchange interaction between electrons on different atoms, and the electrostatic energy involved contains terms arising both from the repulsive forces between the two nuclei and between the two electrons and from the attractive forces between an electron on one atom and the nucleus of the other atom. (The much larger energy of attraction between an electron and the nucleus of its own atom has already been included in the wave equation for each electron.) In simple molecules like $H_2$ the over-all exchange term is easily shown to be negative, in agreement with the ground state of the molecule being a singlet, but with more complex ions such as the 3d group, opinions have differed as to whether the net effect (which obviously varies with interatomic distance) would be positive or negative at the ionic separations typical of 3d-group metals. Since improved wavefunctions have become available from electronic computers, attempts have been made to carry out calculations which might be reasonably realistic. Stuart and Marshall (1960) obtained a positive energy, though two orders of magnitude too small, but Freeman, Nesbet, and Watson (1962) find a negative energy. Thus 'direct exchange', due to direct overlap of the electronic wavefunctions, appears incapable of accounting for ferromagnetism. A negative exchange energy would give rise to 'antiferromagnetism' (see Chapter 16), where neighbouring spins are arranged antiparallel rather than parallel.

A striking feature of ferromagnetism is that it is observed practically only in metals; very few ferromagnetic insulators have been discovered (one example is $CrBr_3$ whose Curie temperature is 33·84 K). This qualitative difference suggests that the conduction electrons play an important role, and measurements of electrical resistivity show there must be an interaction mechanism with the magnetic dipoles which scatters the conduction electrons. For example, the resistivity of gadolinium metal at room temperature is over 100 times that of copper, and in general appreciable changes in the resistivity are observed near the Curie point in ferromagnetic substances. We can go further and question whether the electrons responsible for the observed magnetic dipole moments are 'localized' on given ions, as in an insulator, or whether they are itinerant electrons which should be treated by a 'collective electron' model using Fermi–Dirac statistics, as in Chapters 11 and 12.

The consequences of the latter model have been investigated mainly by Bloch, Slater, Stoner, and Wohlfarth. The electrons occupy certain energy bands, and the exchange interaction is introduced as an internal field $\lambda \mathbf{M}$, proportional to the magnetization, as in the Weiss treatment. This gives a difference in energy between spin dipoles pointing parallel and antiparallel to the internal field, which can be represented by dividing the band into two halves, as in Fig. 14.22. The important difference is that the effective field is now proportional to the magnetization. This depends on the number of electrons transferred from one half-band to the other, which thus determines the energy separation between the two half-bands which causes the transfer. This is a self-balancing situation in which a transfer occurs and an internal field is created only if this results in a lower energy for the magnetized state than for the unmagnetized state. The number of electrons transferred for a given energy difference is proportional to the density of states $g(W)$, and it turns out that a decrease of energy results only if the density of states is sufficiently large. This condition is satisfied for d-electron bands but not for s electrons. Thus sodium does not become ferromagnetic, and remains a Pauli paramagnet (§ 14.11), but transition metals with overlapping d and s bands may have co-operative magnetic states.

This collective model has the advantage that the saturation magnetization is determined by the net transfer of spins between the two half-bands, so that there is no reason why the net dipole moment per atom should correspond exactly to an integral multiple of a Bohr magneton. For example, the saturation moment of nickel is about 0·60 $\mu_B$ per atom. The 3d band is nearly full of electrons, and the saturation moment can be fitted by assuming that in the ferromagnetic state the half-band of lower energy is full as in Fig. 15.12, while that of higher energy has about 0·6 holes per atom. The collective model also gives a neat explanation of the change in magnetic properties which occurs on alloying with non-magnetic metals. For example, the copper atom has one more electron than nickel; when copper is alloyed into nickel, the extra electrons fill up

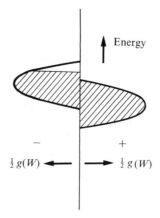

FIG. 15.12. Energy bands in nickel. The difference in energy of the two half-bands with opposite spin orientation is due to exchange interaction. The half-band of lower energy is full, while that with higher energy contains about 0·6 holes per atom.

the holes in the upper half-band, so that the average magnetic moment per nickel atom decreases steadily as the percentage of copper increases. Alloys with more than about 60 per cent of copper are not ferromagnetic, suggesting that at this point the 3d bands are full.

The collective model assumes that all the d electrons are 'itinerant' conduction electrons. The main evidence against this comes from neutron diffraction, where experiments indicate that the spatial distribution of the magnetic electrons in iron, cobalt, and nickel are similar to those in the free atoms. Furthermore, the excess magnetic heat capacity in iron corresponds to an entropy of $R \ln 3$, that is, to a spin $S = 1$ on the iron atom. Obviously some d electrons must be itinerant to account for the electronic specific heat and various transport properties, but it has been suggested (Stearns 1973) that the fraction is only about 5 per cent. This corresponds to about half an electron per atom out of a d band which can accommodate ten electrons when full.

Whatever the exact degree of localization, there is general agreement that the itinerant electrons play a major role in the following way. As a result of direct exchange interaction with a local magnetic moment, the itinerant electrons in the vicinity become spin-polarized. The effect resembles the 'Pauli paramagnetism' discussed in § 14.11, but the polarization arises from a 'molecular field' (cf. § 15.1) and not from the true magnetic field of the local moment. The spin polarization of the itinerant electrons interacts with the spins on neighbouring local moments, producing an indirect exchange interaction between any pair of magnetic ions. This is a second-order interaction, but it proves to be more important than the direct exchange interaction between ions because it is a longer-range effect. For two ions separated by a distance $R$, a calculation involving a number of simplifying assumptions such as a spherical Fermi surface gives an indirect exchange interaction which falls off as

$$x^{-4}(x \cos x - \sin x),$$

where $x = 2k_F R$. The important features of this result are (1) the indirect exchange interaction is oscillatory and changes sign with distance, and (2) it decreases more slowly with distance than the direct exchange interaction, which drops exponentially with $R$. The order of magnitude is $\mathcal{J}_{mi}^2/W_F$, where $\mathcal{J}_{mi}$ is the exchange interaction between the spin of a localized moment $m$ and that of an itinerant electron $i$, and $W_F$ is the Fermi energy.

A mechanism of this type was first proposed for the interaction between nuclear spins in a metal, the hyperfine interaction between the nuclear spin and the itinerant electron spin appearing instead of the exchange interaction $\mathcal{J}_{mi}$; the oscillatory nature of the interaction, and its variation with distance were deduced by Rudermann and Kittel (1954). The corresponding effect for electron moments was proposed by Kasuya (1956) and de Gennes (1958) as the most important interaction in the lanthanide metals (see § 16.4), where the magnetic 4f electrons are strongly localized; it was extended to other systems by Yosida (1957).

### Exchange interaction in insulators

In electrical insulators no conduction electrons exist to provide an exchange mechanism of the type just considered, but exchange interactions of considerable magnitude are found in compounds where the inter-ionic distance is so large that direct exchange between electrons localized on the magnetic ions must be negligible. As an example we consider a simple compound such as an oxide MO, where M is a dipositive ion of the 3d group (for example, $Mn^{2+}$, $Fe^{2+}$, $Co^{2+}$, $Ni^{2+}$). These have a f.c.c. structure similar to that of NaCl, a typical plane of atoms being shown in Fig. 15.13. Here the circles are drawn in proportion to the ionic radii, and it can be seen that the much larger anions separate almost completely even the nearest-neighbour cations. However, neutron diffraction results suggest that the stronger exchange interactions are between ions on next nearest neighbour sites such as A, B rather than between nearest neighbour sites such as A, C. Direct overlap between the wavefunctions of magnetic electrons on the cation sites is very small, but a purely ionic model where the magnetic electrons are localized on the cations is an oversimplification. Some degree of covalent binding is always present, whereby the wavefunctions of the magnetic electrons spread over onto the adjacent anions. Direct overlap of the magnetic electrons from

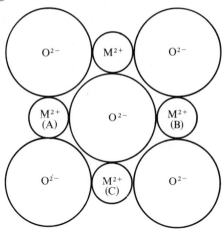

FIG. 15.13. A plane of atoms in an oxide MO, where $M^{2+}$ is a dipositive ion of the 3d group. The ions are drawn approximately to size, showing how the small cations ($M^{2+}$) are well separated by the large anions ($O^{2-}$).

neighbouring cations can therefore take place on the intervening anions. Of course the degree of overlap depends on the amount of covalent bonding, and so also does the size of this 'indirect' exchange interaction; it is very much smaller for ions of the 4f group, which take almost no part in covalent bonding, than for ions of the 3d group; for example, the transition temperature of the lanthanide oxides is below 10 K, while those of the 3d group oxides are over 100 K.

Although this solves the problem of interaction between ions at relatively large distances, there are two difficulties. The first is that the potential energy due to the electrostatic repulsion between the magnetic electrons in the overlap region leads to a ferromagnetic interaction (as for electrons within the same atom, where it results in Hund's rules—see § 14.2), whereas the vast majority of insulating magnetic compounds have an antiferromagnetic interaction. Second, the interacting magnetic electrons form a partly filled band, which should make the substance an electrical conductor. However, the energy band is narrow, and the fall in kinetic energy of an electron (cf. Fig. 12.2) on transferring from a localized state (which corresponds in energy to the centre of the band) to a conduction state at the bottom of the band is only ~1 eV. On the other hand, because of the electrostatic repulsion between the electrons, their potential energy is least when they are uniformly distributed, giving each ion the same number of electrons. This potential energy is part of the 'correlation energy' mentioned briefly in § 11.9. For s electrons, and to a smaller extent for p electrons, the correlation energy is small because their wavefunctions are extended and the charge density within the atom is small; however, the extended wavefunctions give a larger overlap and greater bandwidth in the solid (cf. Fig. 12.6). Thus such electrons have a lower energy over all when they are non-localized, and become conduction electrons. For d electrons the bandwidth is smaller and the correlation energy greater, making the latter correspondingly more important; in compounds the d-electron bandwidth is not more than about 1 eV, while about 10 eV is required to transfer an electron from one ion to another (that is, to create a pair of ions in $d^{n-1}$, $d^{n+1}$ states from a pair both in $d^n$ states). Thus it is energetically favourable for the electrons to remain localized and the substance is an electric insulator.

As a result of these conflicting energy considerations, the localization is not, however, absolutely complete. If $b$ is the reduction in kinetic energy which would result from moving from site to site, while $U$ is the potential energy required to overcome the electrostatic repulsion, the equilibrium state is one where the chance of such a movement is of order $b/U$, and the net reduction in kinetic energy is of order $b^2/U$. Through the exclusion principle this possibility of movement to adjacent sites is restricted nearly always to electrons with antiparallel spin, which can therefore acquire a lower energy than those with parallel spin. This is equivalent to an antiferromagnetic exchange energy of order $b^2/U$, which is of the order of 0·01–0·1 eV (a few hundred degrees Kelvin).

The first explanation of how exchange interaction could arise between ions at the rather large interionic distances found in compounds was put forward by Kramers (1934), and a number of subsequent attempts were made to arrive at more explicit interpretations. The theory outlined above is due to Anderson (1963), and though difficult to explain in simple terms, appears to be the most satisfactory in its general approach.

## References

ANDERSON, P. W. (1963). *Adv. Solid St. Phys.* **14,** 99.
BARNETT, S. J. (1944). *Phys. Rev.* **66,** 224.
BLOCH, F. (1932). *Z. Phys.* **74,** 295.
DE GENNES, P. G. (1958). *C.r. hebd. Séanc. Acad. Sci., Paris* **247,** 1836.
ELLIOTT, J. F., LEGVOLD, S., and SPEDDING, F. H. (1953). *Phys. Rev.* **91,** 28.
FREEMAN, A. J., NESBET, R. K., and WATSON, R. E. (1962). *Phys. Rev.* **125,** 1978.
GOODINGS, D. A. (1962). *Phys. Rev.* **127,** 1532.
GOSSARD, A. C., JACCARINO, V., and REMEIKA, J. P. (1961). *Phys. Rev. Lett.* **7,** 122.
GREW, K. E. (1934). *Proc. R. Soc.* **A145,** 509.
HANNA, S. S., HEBERLE, J., PERLOW, G. J., PRESTON, R. S., and VINCENT, D. H. (1960). *Phys. Rev. Lett.* **4,** 513.
KASUYA, T. (1956). *Prog. theor. Phys.* **16,** 45.
KRAMERS, H. A. (1934). *Physica* **1,** 182.

MEYER, A. J. P. and ASCH, G. (1961). *J. appl. Phys.* **32**, 330.

NAGLE, D. E., FRAUENFELDER, R. D., TAYLOR, R. D., COCHRAN, D. R. F., and MATTHIAS, B. T. (1960). *Phys. Rev. Lett.* **5**, 364.

OLIVER, D. J. and SUCKSMITH, W. (1953). *Proc. R. Soc.* **A219**, 1.

RIEDI, P. C. (1973). *Phys. Rev.* **B8**, 5243.

RUDERMANN, M. A. and KITTEL, C. (1954). *Phys. Rev.* **96**, 99.

RUSHBROOKE, G. S. and WOOD, P. J. (1958). *Molec. Phys.* **1**, 257.

SCOTT, G. G. (1951). *Phys. Rev.* **82**, 542.

STEARNS, M. B. (1973). *Phys. Rev.* **B8**, 4283.

STUART, R. and MARSHALL, W. (1960). *Phys. Rev.* **120**, 353.

VAN KRANENDONK, J. and VAN VLECK, J. H. (1958). *Rev. mod. Phys.* **30**, 1.

YOSIDA, K. (1957). *Phys. Rev.* **106**, 893.

## Problems

15.1. Show that for a substance consisting of atoms or ions in the $^2S_{\frac{1}{2}}$ state, the Brillouin function becomes

$$M/M_s = \tanh y, \quad \text{where} \quad y = \mu_B B/kT.$$

Show that for such a ferromagnetic substance at low temperatures, where y is large, the Weiss internal field treatment of § 15.2 leads to the formula

$$M_0/M_s = 1 - 2 \exp(-2\lambda M_0 M_s/nkT)$$

for the spontaneous magnetization $M_0$ in zero field. Note that this does not give a simple power law such as in eqn (15.29).

15.2 Using the result of Problem 5.11, show that the magnetostatic energy of a small spherical particle of nickel, of radius b and magnetized to saturation ($M_s = 5 \cdot 1 \times 10^5$ A m$^{-1}$), is approximately $2 \times 10^5 b^3$ J (b in metres).

From the results of § 15.3, the energy required to form a Bloch wall increases with $b^2$ (for nickel the wall energy is about $10^{-3}$ J m$^{-2}$). Hence show that for particles whose radius is less than about $10^{-8}$ m, the reduction in magnetostatic energy obtained by division into two domains is less than the energy required to form a wall.

15.3. The nucleus of an atom of mass M moving with velocity v emits a γ-ray of energy hν in the forward direction. Show by considering the change in momentum and energy of the atom that the γ-ray energy is increased by a fraction v/c, provided that the γ-ray energy is small compared with the rest mass of the atom ($h\nu/Mc^2 \ll 1$). Note that this is the same as the classical Doppler shift.

# 16. Antiferromagnetism and ferrimagnetism

## 16.1. Antiferromagnetism

In a paramagnetic substance the dipoles are free to orient themselves at random, and there is a correspondingly high entropy; if there are $2J+1$ levels having the same energy in the ground state, the entropy is $R \ln(2J+1)$. The substance would obey Curie's law down to 0 K if the ground state contains two or more levels with the same energy in the absence of an external field, but this would be a violation of the third law of thermodynamics, by which the entropy in a substance in thermodynamic equilibrium must be zero at 0 K. In practice there is always some mutual interaction between the dipoles (either through exchange or magnetic dipolar interaction), such that the internal energy $U$ of the system is lower when the dipoles are oriented in an orderly array than when they are randomly oriented. Thus, at the absolute zero, where the free energy $F = U - TS$ is equal to $U$, the equilibrium state of lowest *free energy* will be the ordered state with the lowest *internal energy*. At a sufficiently high temperature, on the other hand, the paramagnetic state, with its higher entropy corresponding to the random orientation of the dipoles, will have the lower free energy because of the second term in $F = U - TS$, and will thus be the equilibrium state. As the temperature falls, any substance where the dipoles still have some freedom of orientation (this excludes some paramagnetic substances which have a singlet ground state and a temperature-independent susceptibility) will make a transition from a disordered phase into an ordered phase. The ferromagnetic state discussed in the last chapter, in which all the dipoles are oriented parallel to one another at 0 K, is the state of lowest energy when the exchange energy $\mathscr{J}'$ has a positive sign. However, ferromagnetism is exhibited by relatively few substances, out of a large number which contain transition group ions.

It was suggested by Néel (1936) that in many substances the exchange interaction is large but negative, resulting in an ordered state where neighbouring dipoles are aligned in an antiparallel arrangement. Such an arrangement for a simple cubic lattice is shown in Fig. 16.1; the dipoles at adjacent corners of each cubic cell point in opposite directions. Another simple case is the body-centred cubic (b.c.c.) lattice, with an ion at the centre of each cube as well as at the corners; here all the ions at the

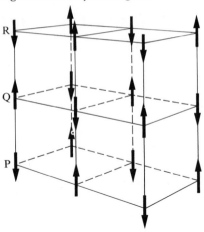

FIG. 16.1. Antiferromagnetic arrangement of dipoles in a simple cubic lattice.

corners have their dipoles parallel to each other, but antiparallel to the ions at the centres. In each case a given dipole is surrounded by a number of equidistant dipoles all pointing in the opposite direction, while the next nearest neighbours point in the same direction again. The system may be thought of as consisting of two interlocking sublattices, one of which is spontaneously magnetized in one direction, while the other is spontaneously magnetized in the opposite direction. As in ferromagnetism, this spontaneous magnetization of the sublattices sets in only below a certain transition temperature, in this case known as the 'Néel temperature'. Above this temperature the dipoles are randomly oriented, and the substance is paramagnetic, obeying a Curie–Weiss law with the Weiss constant of opposite sign to that found in ferromagnetism, as we should expect from the reversed sign of the exchange energy. The onset of spontaneous magnetization in the sublattices as the substance is cooled through the Néel temperature is accompanied by a heat capacity anomaly of the co-operative type, as illustrated in Fig. 16.2. The substance as a whole exhibits no spontaneous magnetization in zero field, since the two sublattices are equally and oppositely polarized. When an external field is applied, a small magnetization occurs giving a positive susceptibility; the general behaviour of the susceptibility can be explained quite well on a molecular-field model, as outlined below.

## 16.2. The molecular field and two-sublattice model

Let the two sublattices be denoted by A and B. Then a dipole in lattice A is subject to an external field of flux density $\mathbf{B}_0$ and an internal field proportional to the magnetization of sublattice B; we write this internal

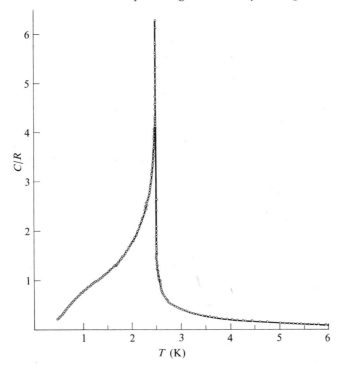

FIG. 16.2. Heat capacity of $GdVO_4$. The lattice heat capacity is almost negligible, unlike the
case of nickel (see Fig. 15.8). (Cashion, Cooke, Hoel, Martin, and Wells 1970.)

field as $-\lambda \mathbf{M_B}$, where the minus sign appears because of the reversed sign
of the exchange interaction. The effective fields acting on dipoles in
lattices A and B are therefore

$$\mathbf{B_A} = \mathbf{B_0} - \lambda \mathbf{M_B},$$
$$\mathbf{B_B} = \mathbf{B_0} - \lambda \mathbf{M_A}. \tag{16.1}$$

At high temperatures where the dipoles are randomly oriented, we
assume that the magnetization of each sublattice is proportional to the
field acting on it. Then we have

$$M_A = \tfrac{1}{2}(\chi_0/\mu_0)B_A, \qquad M_B = \tfrac{1}{2}(\chi_0/\mu_0)B_B, \tag{16.2}$$

where the factor $\tfrac{1}{2}$ appears because only half the dipoles are in each
sublattice, and $\chi_0$ is the value which the susceptibility would have in the
absence of any exchange interaction. The total magnetization is then

$$M = M_A + M_B = (\chi_0/\mu_0)\{B_0 - \tfrac{1}{2}\lambda(M_A + M_B)\}$$
$$= (\chi_0/\mu_0)(B_0 - \tfrac{1}{2}\lambda M). \tag{16.3}$$

This equation has the simple solution

$$\frac{1}{\chi} = \frac{B_0}{\mu_0 M} = \frac{1}{\chi_0} + \frac{\lambda}{2\mu_0}, \tag{16.4}$$

a result which shows that the inverse susceptibility $\chi^{-1}$ should be increased (that is, the susceptibility is smaller) in the presence of antiferromagnetic exchange interaction. This is the reverse of ferromagnetism, where a similar treatment shows that the susceptibility is increased by ferromagnetic exchange. Experimentally the result expressed in eqn (16.4) is illustrated in Fig. 16.3 for an alloy in which the value of $\lambda$ can be changed by altering the concentration of magnetic ions.

If we assume that $\chi_0$ obeys Curie's law so that $\chi_0 = C/T$, we find that

$$\chi = \frac{C}{T + (\lambda C/2\mu_0)} = \frac{C}{T + \theta}. \tag{16.5}$$

The equation for the susceptibility above the Néel point is similar to eqn (15.5) for ferromagnetic exchange, except that the Weiss constant $\theta$ has the opposite sign.

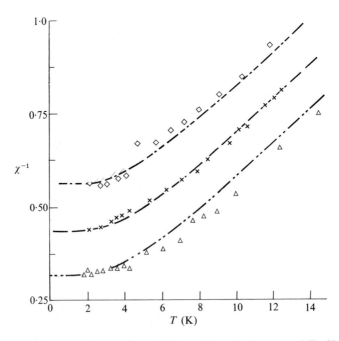

FIG. 16.3. Inverse susceptibility $\chi^{-1}$ per $Tb^{3+}$ ion of the mixed compound $Tb_x Y_{1-x} Sb$. The values of $x$ are 0·05, 0·22, and 0·40, proceeding upwards. As the concentration of $Tb^{3+}$ ions rises, $\lambda$ increases, and the curve is displaced upwards. (After Cooper and Vogt 1970).

At low temperatures we cannot assume that the magnetization is linearly proportional to the effective field, since a large spontaneous magnetization will be created in each sublattice by the internal field. Instead of eqns (16.2) we must use modified forms of eqn (15.8):

$$M_A = \tfrac{1}{2} n g \mu_B J \; \phi(y_A), \Big\}$$
$$M_B = \tfrac{1}{2} n g \mu_B J \; \phi(y_B). \Big\} \qquad (16.6)$$

where $y = (g\mu_B/kT) \times$(effective flux density acting on each sublattice), and $\phi(y)$ is the Brillouin function given by eqn (14.21). The factor $\tfrac{1}{2}$ again appears because only half the dipoles are on each sublattice, and we have assumed that the dipole moment is $g\mu_B J$.

When $B_0 = 0$ the magnetization of each sublattice is given by the solution of eqn (16.6) with

$$B_A = -\lambda M_B = -\lambda(-M_0) = +\lambda M_0,$$
$$B_B = -\lambda M_A = -\lambda(+M_0) = -\lambda M_0,$$

since the two sublattices A and B will have equal and opposite magnetization, $+M_0$ and $-M_0$ respectively. The equation may be solved graphically, as in ferromagnetism. The spontaneous magnetization is given by the points A and B in Fig. 16.4, which correspond to the stable condition, while the other possible solution $M_A = M_B = 0$ is unstable. The value of $M_0$ at any temperature is given by the root of the transcendental equation

$$M_0 = \tfrac{1}{2} n g \mu_B J \; \phi\left(\frac{g\mu_B J}{kT} \lambda M_0\right).$$

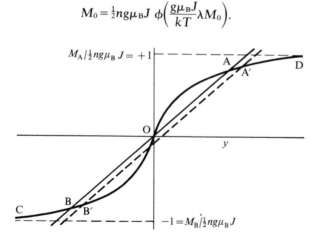

FIG. 16.4. Graphical solution of equation for spontaneous magnetization of an antiferromagnetic substance. Curve CBOAD is the Brillouin function $\phi(y)$; BOA is the straight-line relation between $M$ and $y$ when the external field is zero, and B'OA' is a similar line for the case when an external field is applied parallel to the direction of the spontaneous magnetization.

As the temperature rises the line BOA becomes steeper, and the points A, B move back towards the origin; the spontaneous magnetization disappears at the Néel temperature $T_N$ where the line AB is tangential to the Brillouin function at the origin. Since $\phi(y) = y(J+1)/3J$ for small values of y, we have, when $T = T_N$,

$$M_0 = \tfrac{1}{2}ng\mu_B J\left\{\left(\frac{g\mu_B J}{kT_N}\right)\left(\frac{J+1}{3J}\right)\lambda M_0\right\},$$

or

$$T_N = \tfrac{1}{2}\lambda ng^2\mu_B^2 J(J+1)/3k = \lambda C/2\mu_0 = \theta. \qquad (16.7)$$

Thus on this simple theory, due to Van Vleck (1941), the Néel point $T_N$ should have the same value as the Weiss constant $\theta$. The values of both $\theta$ and $T_N$ are given in Table 16.1 for a number of substances now established as being antiferromagnetic. It will be seen that in general $\theta$ and $T_N$ are different, and this can be accounted for by an extension of the theory given above where interactions with next nearest neighbours belonging to the same sublattice are included (Van Vleck 1951; see Problem 16.1). In addition other types of arrangement of the dipoles, where not all the nearest neighbours are antiparallel, are possible.

When an external field is applied at a temperature below the Néel point, a positive magnetization results whose magnitude can be estimated from the theory given above. In general, the effect of applying a field is to change the magnetization of each sublattice slightly, so that we may write

$$\left.\begin{array}{l}\mathbf{M}_A = \mathbf{M}_0 + \delta\mathbf{M}, \\ \mathbf{M}_B = -\mathbf{M}_0 + \delta\mathbf{M},\end{array}\right\} \qquad (16.8)$$

where these must be taken as vector equations if the external magnetic field is applied at an arbitrary angle to the direction of the spontaneous

TABLE 16.1

| Substance | Néel temperature $T_N$ (K) | $\theta/T_N$ | $\dfrac{\chi(T=0)}{\chi(T=T_N)}$ |
|---|---|---|---|
| Cr | 311 | — | — |
| $\alpha$-Mn | 100 | — | — |
| $MnF_2$ | 67 | 1·2 | 0·76 |
| $FeF_2$ | 78 | 1·5 | 0·72 |
| $CoF_2$ | 50 | — | — |
| $NiF_2$ | 73 | — | — |
| MnO | 116 | 4–5 | 0·69 |
| FeO | 198 | 3 | 0·75 |
| $CuCl_2 \cdot 2H_2O$ | 4·3 | — | — |
| $NiCl_2 \cdot 6H_2O$ | 5·3 | — | — |

magnetization $M_0$. If $B_0$ is parallel to $M_0$, so also will be $\delta M$, and, if we return to the graphical solution of our transcendental equation, we see that the magnetizations of the sublattices will be given by the intersection of the dotted line $B'A'$ in Fig. 16.4 with the Brillouin curve. For the effective fields become, using eqn (16.1),

$$\mathbf{B}_A = \mathbf{B}_0 - \lambda(-\mathbf{M}_0 + \delta\mathbf{M}) = \lambda\mathbf{M}_0 + (\mathbf{B}_0 - \lambda\delta\mathbf{M}),$$
$$\mathbf{B}_B = \mathbf{B}_0 - \lambda(+\mathbf{M}_0 + \delta\mathbf{M}) = -\lambda\mathbf{M}_0 + (\mathbf{B}_0 - \lambda\delta\mathbf{M}).$$

The resulting net magnetization $2\delta M$ will depend on the slope of the Brillouin function (for small fields) at the point $M_0$. As $M_0$ increases, this slope decreases, reaching zero at saturation. It follows that the susceptibility $\chi_\parallel$ (in the direction parallel to $M_0$) decreases to zero as the temperature falls to zero, as illustrated in Fig. 16.5. The exact shape of the curve depends only slightly on the value of $J$, as in the case of the saturation curve for a ferromagnetic substance.

If the external field is applied perpendicularly to $M_0$, we can evaluate the susceptibility more easily. In this case the magnetizations $M_A$, $M_B$ of each of the sublattices turn through an angle $\alpha$ towards $B_0$, as shown in Fig. 16.6. The effective field on a dipole now has the two components

$$B_x = B_0 - \lambda M_x, \qquad B_z = \lambda M_z,$$

and hence

$$\tan\alpha = M_x/M_z = B_x/B_z = (B_0 - \lambda M_x)/\lambda M_z.$$

This leads immediately to the result $2M_x = B_0/\lambda$, and hence the susceptibility in the perpendicular direction is

$$\chi_\perp = \mu_0(2M_x)/B_0 = \mu_0/\lambda = C/2T_N, \tag{16.9}$$

since $M_x$ is the component of each sublattice magnetization parallel to $B_0$. Thus $\chi_\perp$ should be constant below the Néel point and equal to the value

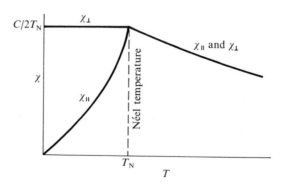

FIG. 16.5. Variation of $\chi_\parallel$ and $\chi_\perp$ on the simple theory of antiferromagnetism.

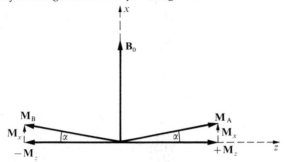

FIG. 16.6. Effect of applying a flux density $\mathbf{B}_0$ perpendicular to the spontaneous magnetization in an antiferromagnetic substance.

at the Néel point, as shown in Fig. 16.5. For a powder specimen consisting of microcrystals with random orientation, we have

$$\chi = \tfrac{1}{3}(\chi_{\|}+2\chi_{\perp}),\tag{16.10}$$

and we should expect

$$\frac{\chi(T=0)}{\chi(T=T_{N})} = \frac{2}{3}.\tag{16.11}$$

The values of this ratio for a number of powdered antiferromagnetics are also given in Table 16.1.

A more direct check on the theory is obtained from measurements on a single crystal, such as $GdVO_4$ (Fig. 16.7). Here the only ion with a dipole

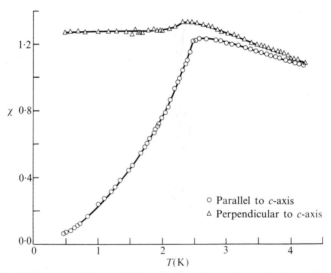

FIG. 16.7. Molar susceptibility of $GdVO_4$. (Cashion, Cooke, Hoel, Martin, and Wells 1970.)

moment is $Gd^{3+}$, $^8S_{\frac{7}{2}}$, and the susceptibility shows little anisotropy above the Néel temperature. Below this temperature $\chi_{\parallel}$ falls rapidly and approaches zero as $T \to 0$, while $\chi_{\perp}$ remains almost constant, as predicted by theory.

This model with two sublattices is valid for many antiferromagnetic substances, but in some cases there are more (in a f.c.c. lattice there are generally four). As in ferromagnetism, the exchange interaction itself gives no preference to any particular orientation of the spins relative to the crystal axes; this arises from the anisotropy energy. In a simple tetragonal crystal such as $MnF_2$ or $GdVO_4$, the spins are aligned along the tetragonal axis in a simple two sublattice antiparallel arrangement, but much more complicated arrangements are possible in which the vector sum of the dipole moments is zero—the distinctive feature of an antiferromagnetic.

## 16.3. Ferrimagnetism

The technical importance of magnetic materials in electrical industry has increased continuously, the ideal substance being one with a large magnetic moment at room temperature, which is also an electrical insulator. Ferromagnetic metals and alloys have been widely exploited, but their high electrical conductivity is a serious handicap in radio-frequency applications because of the eddy-current losses. For this reason a number of magnetic oxides ('ferrites', of which magnetite, $Fe_3O_4$, is the most famous as the original 'lodestone') have become of great technical interest because of their low electrical conductivity. They show spontaneous magnetization, remanence, and other properties similar to ordinary ferromagnetic materials, but the spontaneous moment does not correspond to the value expected for full parallel alignment of the dipoles.

In 1948 Néel put forward a theory for such materials; he suggested that they contain two sublattices in which the magnetizations are oppositely directed, but which give a net moment because the two sublattice moments are unequal. For this phenomenon he coined the word 'ferrimagnetism'. It can arise from a number of arrangements, of which the simplest are illustrated in Fig. 16.8. In (a) all the dipoles are equal in magnitude, but there are more on one sublattice than on the other; the

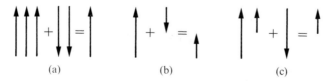

(a)                    (b)                    (c)

FIG. 16.8. Three possible arrangements of the dipole moments in a ferrimagnetic material. (a) unequal numbers of identical moments on the two sublattices, (b) unequal moments on the two sublattices, (c) two equal moments and one unequal.

most notable example is yttrium iron garnet (YIG). The simple arrangement (b) with ions of unequal moments occurs rather rarely. The arrangement (c), with two equal and opposite moments, and a third moment on one sublattice is typical of ferrites such as $MnFe_2O_4$.

### Ferrites

These have the typical formula $M^{2+}Fe_2^{3+}O_4^{2-}$ (equivalent to $MO,Fe_2O_3$), where $M^{2+}$ is a dipositive ion, commonly $Mn^{2+}$, $Fe^{2+}$, $Co^{2+}$, $Ni^{2+}$, $Cu^{2+}$, $Zn^{2+}$, or $Mg^{2+}$; other tripositive ions such as $Cr^{3+}$ can replace $Fe^{3+}$. The crystallographic structure is cubic and similar to the mineral spinel ($MgAl_2O_4$), and the unit cell, with eight formula units, equivalent to $M_8Fe_{16}O_{32}$, contains eight cation sites with tetrahedral coordination (to four oxygen ions) and sixteen cation sites with octahedral coordination (to six oxygen ions), known as the A and B sites respectively. The division of cations between these sites is not unique, the limiting cases being:

|                      | A sites (8)       | B sites (16)              |
| -------------------- | ----------------- | ------------------------- |
| 'Normal' structure   | $8M^{2+}$         | $16Fe^{3+}$               |
| 'Inverse' structure  | $8Fe^{3+}$        | $8M^{2+}+8Fe^{3+}$        |

Intermediate arrangements are also found, and we consider only the inverse structure. Each $Fe^{3+}$ ion is in a $^6S_{\frac{5}{2}}$ state with a moment of 5 $\mu_B$; however, the moments on the A and B sites are antiparallel. If $m_M$ is the moment of the $M^{2+}$ ion, the net saturation moment at 0 K for the unit $MFe_2O_4$ will be

$$m = m_M+(m_{Fe})_B-(m_{Fe})_A = m_M+5\mu_B-5\mu_B = m_M. \qquad (16.12)$$

The moments calculated thus (assuming that the M ion has a 'spin only' moment) are compared with the observed moments in Table 16.2. The agreement is satisfactory; some orbital moment would be expected in the $Fe^{2+}$, $Co^{2+}$, $Ni^{2+}$, $Cu^{2+}$ ions, and in magnesium ferrite the structure is not completely inverse.

TABLE 16.2

| M  | $m_M$ (spin only) | Observed moment | $T_N$ (K) |
| -- | ----------------- | --------------- | --------- |
| Mn | 5                 | 4·4–5·0         | 573       |
| Fe | 4                 | 4·0–4·2         | 858       |
| Co | 3                 | 3·3–3·9         | 793       |
| Ni | 2                 | 2·2–2·4         | 858       |
| Cu | 1                 | 1·3–1·4         | 728       |
| Mg | 0                 | 0·9–1·1         | —         |

The magnetic moments are in Bohr magnetons per unit $MFe_2O_4$.

Néel suggested that all the interactions in the ferrites are antiferromagnetic in sign, but that the A–B interaction is considerably stronger than the A–A or B–B interactions. Thus in the inverse structure the dominating A–B interaction makes the spins within each group parallel, despite their mutual antiferromagnetic interaction. This is supported by the fact that $ZnFe_2O_4$, which has the normal structure, has no net moment. Here the A sites are entirely occupied by zinc ions, with no moment, so the A–B interactions are zero. The ferric ions on the B sites are then aligned antiparallel through the antiferromagnetic B–B interaction, in equal numbers, so that the compound is antiferromagnetic. Its Néel temperature (9 K) is quite low, as would be expected if the B–B interactions are weak.

### Garnets

These have the typical formula $M_3Fe_5O_{12}$ (of which two units are equivalent to $5Fe_2O_3.3M_2O_3$), where both the M cation and the Fe are tripositive ions; the $M^{3+}$ ion is commonly yttrium or a member of the 4f group. The crystallographic structure is cubic and similar to the mineral garnet, though this has cations of other valencies. The unit cell is complex, containing eight units of $M_3Fe_5O_{12}$; for simplicity we shall discuss mainly YIG, where the $Y^{3+}$ ion has a closed shell and carries no magnetic moment. The ferric ions occupy two types of site; in each unit $Y_3Fe_5O_{12}$ two $Fe^{3+}$ ions occupy 'a' sites, coordinated to six oxygen ions, and three $Fe^{3+}$ ions occupy 'd' sites, coordinated to four oxygen ions. The magnetic moments of the two 'a' ions are antiparallel to those of the three 'd' ions, giving the arrangement shown in Fig. 16.8(a); the net moment per unit $Y_3Fe_5O_{12}$ is thus that of one $Fe^{3+}$ ion, or $5\,\mu_B$ (the best experimental value is $4\cdot96\,\mu_B$). The Néel temperature is 545 K.

Amongst other ferrimagnetic materials we mention only $BaFe_{12}O_{19}$ (equivalent to $BaO.6Fe_2O_3$). This has a hexagonal structure, with a number of inequivalent sites for the ferric ions. Of the twelve ferric ions per formula unit, the moments of eight are antiparallel to the remaining four, giving a net moment of $4\times5 = 20\,\mu_B$; the Néel temperature is about 820 K. Barium ferrite, as it is frequently called, has a high value of $(BH)_{max}$ and is used as a permanent magnet material (cf. Chapter 6). Being hexagonal, it has a high anisotropy energy; it is used in the form of pressed oriented fine particles.

### Discussion

Néel's theory of ferrimagnetism had considerable success in explaining the anomalous behaviour of the susceptibility above the Néel point. Using a molecular field approximation with three constants representing the

A–B, A–A, and B–B interactions he deduced the relation

$$\frac{1}{\chi} = \frac{T}{C} + \frac{1}{\chi_0} - \frac{\sigma}{T-\theta} \qquad (16.13)$$

for a substance where all the magnetic ions have the same moment, such as YIG, or $MFe_2O_4$ when M carries no moment. Here $C$ is the usual Curie constant, but the other parameters are functions of the molecular field constants and the numbers of ions in each sublattice. The general behaviour of the inverse susceptibility $\chi^{-1}$ given by Néel's relation, fitted to experiments on YIG, is shown in Fig. 16.9. The theory also explains qualitatively the complex behaviour of the spontaneous magnetization curve below the Néel temperature. The magnetization does not always increase monotonically as the temperature falls, and in ferrimagnetic compounds containing more than one type of magnetic ion whose spontaneous magnetization varies in different ways with temperature a 'compensation point' may be observed, where the magnetization of the two sublattices is equal and opposite.

The magnetization curve of gadolinium iron garnet, $(Gd_3Fe_5O_{12})$ is shown in Fig. 16.10, together with that of $Y_3Fe_5O_{12}$. The latter is not unlike that of a ferromagnetic, but at low temperatures the former has a much higher magnetization, falling to zero at the compensation point at about 295 K. At 0 K we would expect each $Gd^{3+}$ ion to have a moment of $7\ \mu_B$; if these are mutually parallel, but opposed to the net ferric moment, we would expect an over-all moment for the unit $Gd_3Fe_5O_{12}$ of $(3\times7-\{3\times5-2\times5\})\mu_B = (21-5)\mu_B = 16\mu_B$; this is close to the observed moment. As the temperature rises the magnetization of the gadolinium

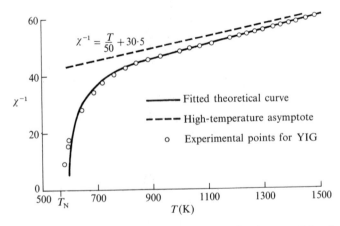

FIG. 16.9. Inverse magnetic susceptibility of the ferrimagnetic substance yttrium iron garnet (YIG), which has three and two $Fe^{3+}$ ions on the two sublattices (arrangement (a) in Fig. 16.8).

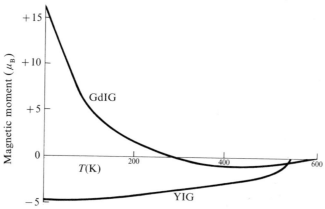

FIG. 16.10. Variation of the spontaneous moment with temperature for $Gd_3Fe_5O_{12}$ (GdIG) and $Y_3Fe_5O_{12}$(YIG) in Bohr magnetons per formula unit.

ions, which are subjected to a comparatively weak interaction with the ferric ions, falls much more rapidly than that of the iron lattice with its strong mutual interactions between the ferric ions. In fact the behaviour of the gadolinium ions is not far from that of paramagnetic ions with $S = \frac{7}{2}$, subjected to an internal field generated by the iron lattice. The Néel temperature of $Gd_3Fe_5O_{12}$ (564 K) is not appreciably different from that of $Y_3Fe_5O_{12}$ (545 K), as would be expected on this basis.

Apart from their technical importance, ferrimagnetic materials have played a major role in advancing our understanding of magnetic problems; for this purpose the garnets are more favoured than the ferrites, since the structure is unique and there are no uncertainties concerning the sites occupied by the magnetic ions. The absence of conduction electrons is a great asset, not only technically but also scientifically. On the one hand, we are dealing with localized magnetic moments, so the theory rests on a much firmer foundation; on the other many important experiments can be carried out to check the theory which would otherwise be impossible. An obvious example is magnetic resonance experiments (cf. Chapter 24) in the frequency range $10^4$–$10^7$ MHz, which have been a very fruitful field both for ferrimagnetics and for antiferromagnetics. The absence of conduction electrons plays a less direct but no less important role in measurements of the magnetic specific-heat contribution predicted by spin-wave theory. A term proportional to $T^{\frac{3}{2}}$ was first confirmed by Kouvel (1956) using $Fe_3O_4$ (it has also been measured in compounds such as YIG), whereas in the ordinary ferromagnetic metals it is obscured by the electronic specific heat. Another experimental achievement is the optical demonstration of the presence of domains, using the rotation of the plane of polarized light propagated parallel to the direction of magnetization (the Faraday effect); when a thin crystal of YIG is placed under a polarizing microscope the domains are visible as light and dark regions whose motion can be observed under the action of an applied field.

## 16.4. The lanthanide ('rare-earth') metals

Measurements of the susceptibilities of the lanthanide metals at high temperatures give generally a Curie–Weiss law where the size of the Curie constant agrees well with that expected for the tripositive ions. There are two notable exceptions to this: europium metal and ytterbium metal, which are cubic in structure with an ionic size indicating the presence of dipositive ions. In addition, cerium tends to show a phase transition at low temperatures to a cubic structure with $Ce^{4+}$ ions, dependent on the thermal history of the specimen. The $Ce^{4+}$ and $Yb^{2+}$ ions have closed shells and no magnetic moment, so they are not of interest here; the $Eu^{2+}$ ion has a half-filled shell, ground state $S = \frac{7}{2}$, but the magnetic behaviour of the metal shows unexpected complications. We shall therefore restrict ourselves to the metals containing tripositive ions, data for which are given in Table 16.3

No lanthanide metal shows a co-operative state above 300 K, so that exchange interactions are small compared with the spin–orbit coupling.

TABLE 16.3

*Magnetic data for the lanthanide metals containing $Ln^{3+}$ ions*

|    | $gJ$ | $(g-1)^2 J(J+1)$ | $T_N$ (K) | $T_C$ (K) |
|----|------|------------------|-----------|-----------|
| La | 0    | 0                | —         | —         |
| Ce | 2·14 | 0·18             | 12·5      | —         |
| Pr | 3·2  | 0·8              | ?         | —         |
| Nd | 3·17 | 1·84             | 7·5, 19   | —         |
| Sm | 0·71 | 4·5              | 14        | —         |
| Gd | 7    | 15·75            | —         | 293       |
| Tb | 9    | 10·5             | 230       | 220       |
| Dy | 10   | 7·1              | 176       | 88        |
| Ho | 10   | 4·5              | 130       | 19        |
| Er | 9    | 2·55             | 85        | 20        |
| Tm | 7    | 1·17             | 57        | 32        |
| Lu | 0    | 0                | —         | —         |

The value of $gJ$ (from Table 14.1) gives the moment per ion at 0 K assuming no crystal field effects are present. Promethium has no radioactively stable isotopes; europium metal becomes antiferromagnetic below 91 K, the ion being $Eu^{2+}$, $^8S_{\frac{7}{2}}$; ytterbium metal contains $Yb^{2+}$ ions, with a filled 4f shell like $Lu^{3+}$.

We may therefore regard the spin and orbit as coupled to give a resultant angular momentum $J$, as in the paramagnetic salts (cf. § 14.6). On this basis the saturation moment per ion at 0 K should be $gJ\,\mu_B$, where $g$ is the Landé factor appropriate to the ground state $J$ of the free tripositive ion. Values of $gJ$ are given in column 2 of Table 16.3, and are generally substantiated by the magnetic evidence for gadolinium and the heavier metals.

In considering the exchange interaction, we have to project the spin vector $\mathbf{S}$ onto the total angular momentum vector $\mathbf{J}$, as pointed out in § 15.1. The exchange interaction $-2\mathcal{J}\mathbf{S}_i.\mathbf{S}_j$ between the spins thus becomes equivalent to a coupling $-2\mathcal{J}'\mathbf{J}_i.\mathbf{J}_j$ between the total angular momenta, with $\mathcal{J}' = (g-1)^2\mathcal{J}$, as given by eqn (15.3). If in fact $\mathcal{J}$ were the same for all the lanthanon metals, we should expect the Curie points to vary as $(g-1)^2J(J+1)$, from eqn (15.6). This quantity is largest for $Gd^{3+}$, with a half-filled shell, and this metal shows co-operative effects at a higher temperature (293 K) than any other lanthanide metal. Reference to Table 16.3 shows that the temperature at which co-operative effects appear varies qualitatively in accordance with this relation in the second half of the group. However, these metals show more than one ordered phase, being antiferromagnetic at higher temperatures and ferromagnetic at lower temperatures. This effect appears to require a reversal in sign of the exchange interaction as the temperature falls, but it turns out that the magnetic moments are arranged in very complex patterns which change with temperature.

In the lanthanons the 4f electrons belong to an inner shell, and their wavefunctions are less extended than those of d electrons. For this reason direct exchange, involving overlap of f-electron wavefunctions on adjacent ions, is unlikely to be important. Instead, the Rudermann–Kittel–Kasuya–Yosida (RKKY) indirect exchange mechanism, discussed in § 15.8, is regarded as dominant, giving a picture of the metals as consisting of ions with well-localized moments due to the 4f shells, imbedded in a 'sea' of conduction electrons formed from the valence electrons, $5d^16s^2$. The conduction electrons are polarized by exchange interaction with the 4f shells, and serve as the medium whereby the orientation of the moment on one ion can influence that on neighbouring ions. The conduction electrons contribute little to the magnetic moment, but the oscillatory nature of the RKKY interaction (see § 15.8) plays a major role in producing wave-like forms in the magnetic ordering observed in the ordered state.

Electrostatic interaction between the 4f electrons on a given ion and the charge density on neighbouring ions provides a 'crystal field' interaction in the metals of the same order as that in salts of the lanthanide group. Direct evidence for this comes from Schottky-type anomalies in the specific heats of the first members of the group. It is difficult to obtain reliable values of the crystal field splittings of the $2J+1$ levels from this excess 'magnetic' specific heat, but the overall splitting is about a few hundred degrees Kelvin for cerium, praseodymium, neodymium, and samarium. Since the values of $(g-1)^2 J(J+1)$ are rather low for these elements, the exchange interaction is less important than the crystal field and a co-operative phase sets in only below 20 K.

The normal crystal structure of the lanthanon metals is hexagonal, apart from those containing $Ce^{4+}$, $Eu^{2+}$, or $Yb^{2+}$ ions, which are cubic, but the structure changes slightly at gadolinium. In this and the heavier metals the crystal field is smaller than in the lighter metals, while the value of $(g-1)^2 J(J+1)$ tends to be larger; thus exchange interactions preponderate over crystal field splittings, and the co-operative magnetic phase sets in at temperatures where crystal field effects are relatively less important, though they give rise to a large anisotropy energy. In the ordered phase gadolinium appears to be a normal ferromagnet, though the direction of the magnetic moment in the crystal varies with temperature. The behaviour of the other lanthanon metals is much more complicated. In the heavy metals the moments of all ions in each layer normal to the hexagonal axis lie parallel to one another, but successive layers have different orientations so that the over-all magnetization is zero below $T_N$ and above $T_C$. The moments of the layers in some cases lie on the surface of a cone (Fig. 16.11); the turn angle between successive layers depends on the oscillatory exchange interaction, which thus determines the period at which the structure repeats along the $c$-axis. The angle $\theta$ which the moments make with the $c$-axis is determined by a number of competing

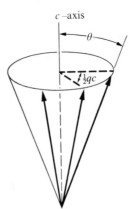

Fig. 16.11. A conical spin arrangement. The moments of successive layers (distance $\frac{1}{2}c$ apart) perpendicular to the $c$-axis are indicated by arrows. The periodicity can be represented by a wave vector $\mathbf{q}$ parallel to the $c$-axis, and the turn angle between layers is $\frac{1}{2}qc$. When $q = 0$ the alignment is ferromagnetic.

interactions (exchange, crystal field, magnetostriction); the equilibrium state is that with the lowest free energy, and varies with temperature. At the lowest temperatures all the heavy metals enter a ferromagnetic phase; the saturation magnetization per ion is a little larger than $g\mu_B J$ for gadolinium, terbium, dysprosium, and holmium, the excess being attributed to polarization of the conduction electrons. In erbium, thulium, and the lighter metals the moments are all lower than the free-ion value, showing the importance of crystal field effects.

Detailed information about the complicated magnetic structures has become available only through the use of neutron diffraction, particularly on single crystals. In some cases neutron diffraction also suggests that the exchange interaction is itself anisotropic, the energy depending on the orientation of the moments as well as the distance between them.

## References

CASHION, J. D., COOKE, A. H., HOEL, L. A., MARTIN, D. M., and WELLS, M. R. (1970). *Colloques Internationaux du C.N.R.S.* No. 180, p. 417.
COOPER, B R., and VOGT, O. (1970). *Phys. Rev.* **B1**, 1218.
KOUVEL, J. S. (1956). *Phys. Rev.* **102**, 1489.
NÉEL, L. (1936). *Ann. Phys.* (Fr.) **5**, 256.
VAN VLECK, J. H. (1941). *J. chem. Phys.* **9**, 85.
—— (1951). *J. Phys. Radium* **12**, 262.

## Problems

16.1. The theory of antiferromagnetism can be extended by assuming that the molecular field acting on each sublattice contains a term due to exchange interaction with ions on the same sublattice as well as a term due to ions on the other sublattice. Show that in the paramagnetic state the equations

$$M_A = (C/2\mu_0 T)(B_0 - \lambda M_B - \lambda' M_A),$$
$$M_B = (C/2\mu_0 T)(B_0 - \lambda M_A - \lambda' M_B)$$

lead to a Curie–Weiss law for the susceptibility (eqn (16.5)) with

$$\theta = (C/2\mu_0)(\lambda + \lambda').$$

16.2. The Néel temperature $T_N$ can be found by putting $B_0 = 0$ in the preceding question, and finding the condition that the pair of equations for $M_A$, $M_B$ still have a solution. ($T_N$ is the temperature at which a vanishingly small magnetization can exist when $B_0 = 0$, and the Brillouin function is approximated by Curie's law.) Show that this gives $T_N = (C/2\mu_0)(\lambda - \lambda')$.

16.3. Show that eqns (16.1)–(16.3) lead to the result

$$(M_A - M_B)\left(1 - \frac{\chi_0 \lambda}{2\mu_0}\right) = 0.$$

Hence either $M_A = M_B$, as in the paramagnetic state, or the second bracket vanishes, determining the Néel point at which a finite sublattice magnetization can occur.

# 17. Semiconductors

## 17.1. Intrinsic and extrinsic conductivity

A SUBSTANCE in which the number of electrons is just sufficient to fill the lowest energy bands at 0 K is an insulator at very low temperatures. At a non-zero temperature a few electrons may have sufficient energy to be excited into the lowest unoccupied band (the 'conduction' band), leaving holes in the highest 'occupied' band (the 'valence' band). This gives a small electrical conductivity whose magnitude depends on the temperature and on the width of the energy gap $W_g$ between the full and empty bands. We shall find in § 17.3 that the number of electrons excited into the conduction band is proportional to $\exp(-W_g/2kT)$, and if the gap is not more than about 1 eV, which corresponds to a value of $kT$ with $T \approx 12\,000$ K, there will be a measurable conductivity at room temperature. This phenomenon is known as 'intrinsic conductivity' and is a characteristic of pure semiconductors; it has been observed in pure silicon, germanium, indium antimonide (InSb), and some other substances.

For each electron in the conduction band, there will be a corresponding 'hole' in the filled band. Both electrons and holes contribute to the conductivity $\sigma$, so that

$$\sigma = |e|(nu_e + pu_h), \tag{17.1}$$

where $n$ and $p$ are the numbers per unit volume of electrons and holes respectively. The mobilities $u_e$, $u_h$ are taken as positive numbers and no sign is attached to $|e|$, though the holes and electrons drift in opposite directions under the influence of an electric field. For intrinsic conductivity $n = p$, since electrons and holes occur only in pairs. The equilibrium concentration rises rapidly as the temperature rises, but the mobilities vary much less rapidly with temperature; hence the increase in $n$ is the dominant factor and the conductivity rises as the temperature increases. This is the hall-mark of a semiconductor, and one of the features (together with its much smaller conductivity) which distinguishes it from a metal. Another difference is that electrons in the conduction band are in excited states, and have only a finite lifetime. An electron from the conduction band can drop down into the top of the valence band, recombining with a hole and releasing an energy $W_g$; conversely an electron–hole pair can be created by lifting an electron from the valence band to the conduction band. Both processes occur repeatedly, giving a dynamic equilibrium concentration which is a function of temperature.

The properties of a semiconductor are generally profoundly modified by the presence of an impurity, or some other cause of irregularity in the lattice. If these are present in not too great a concentration, they produce discrete energy levels. The reason for this is that the levels only spread out into bands when the impurity atoms are sufficiently close for their wavefunctions to overlap, and at low concentrations the impurity atoms are so far apart that any such overlap is negligible. These discrete energy levels are important when they lie in the forbidden band, and particularly so if they lie close to the conduction or the valence band, as illustrated in Fig. 17.1. In the former case electrons may occupy the impurity level D at low temperatures, and are then localized on the impurity atom and unable to partake in electrical conduction. As the temperature rises, these electrons are excited into the empty band. They behave then as conduction electrons, with negative charges; the material is known as an n-type semiconductor, and the impurity levels from which the electrons come are known as 'donor' levels. In the second case, the impurity levels A which lie just above the valence band will be unoccupied at 0 K, but as the temperature rises electrons are excited from the valence band into these levels, which are therefore known as 'acceptor' levels. This process leaves holes in the valence band, which behave as positively charged carriers, and the material is known as a p-type semiconductor.

If the impurity level lies close to a conduction or valence band, the temperature at which appreciable numbers of electrons or holes may be excited is relatively low, and the 'extrinsic conductivity' due to this cause may outweigh any intrinsic conductivity, even with small concentrations of impurities. In pure germanium at room temperature, for example, the number of intrinsic electrons in the conduction band is less than $10^{20}$ per cubic metre, whereas the number of germanium atoms per cubic metre is $4 \cdot 5 \times 10^{28}$. If an impurity atom which is easily ionized at room temperature is present to a concentration of one part in $10^8$ ($4 \cdot 5 \times 10^{20}$ impurity atoms

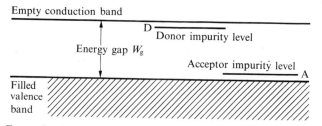

Fig. 17.1. Energy bands in a semiconductor, showing the gap between the valence band and the conduction band. At 0 K the valence band is full and the conduction band is empty, so that the substance behaves as an insulator. The discrete levels D, A are due to the presence of impurities in low concentration, where the impurity atoms are too far apart for their electronic wavefunctions to overlap.

per cubic metre), it can give rise to an extrinsic conductivity which exceeds the intrinsic conductivity.

The extrinsic conductivity increases as the temperature rises until all the donor impurity atoms are fully ionized, or all the acceptor levels fully occupied. The number of extrinsic conduction electrons, or holes, then becomes substantially constant; the conductivity becomes constant, or may fall with temperature because of a decrease in the mobility. This is known as the 'exhaustion range'.

Extrinsic and intrinsic conductivity may of course be present simultaneously, but the former will depend on the impurity content while the latter is a property of the pure material. In each case conduction depends on excitation into higher levels, and the charge carriers have a finite lifetime. All substances would be expected to show intrinsic conductivity at a sufficiently high temperature; 'insulators' are substances with such large energy gaps that appreciable conductivity sets in only at temperatures outside the normal laboratory range, and which may be above the melting-point of the substance.

## 17.2. Elementary and compound semiconductors

A number of elements are known to be semiconductors in their normal allotropic form; the principal ones are silicon, germanium, boron, selenium, and tellurium. By far the most important are silicon and germanium; they are used in many solid-state devices and their properties have been extensively investigated.

The elements carbon, silicon, germanium, tin, and lead belong to Group IV of the periodic table. Silicon, germanium, and the allotrope grey tin crystallize in the diamond structure (see Fig. 17.2) in which each

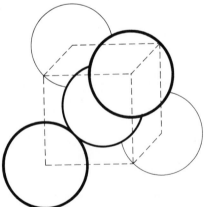

FIG. 17.2. The diamond structure, consisting ↑ four atoms centred on alternate corners of a simple cubic lattice, bonded to one at the cei ↑e of the cube. The structure is repeated so that every atom is in id ↑tical surroundings.

TABLE 17.1
*Values of the energy gap (eV) at 0 K*

| Group IV elements | | III–V compounds | | II–VI compounds | |
|---|---|---|---|---|---|
| Diamond | ~5·3 | BN | ~10 | | |
| Silicon | 1·21 | AlP | 3 | | |
| Germanium | 0·78 | GaAs | 1·35 | ZnSe | ~5 |
| Grey tin | ~0·08 | InSb | 0·24 | CdTe | 1·6 |
| | | | | PbS | ~0·4 |

atom has four equidistant neighbours arranged in the form of a regular tetrahedron. Each atom forms four covalent bonds with these neighbours, donating one electron to each bond whose spin is paired off with that of the corresponding electron donated by the neighbour. These electrons can be regarded as being in a filled valence band, above which is an energy gap to the next band which is empty and forms a possible conduction band. This picture represents the position at 0 K, where the substances behave as insulators. The energy gaps are listed in Table 17.1. At a non-zero temperature some electrons may be excited from the valence band into the conduction band, making the substance an intrinsic semiconductor. On a localized electron model this corresponds to taking an electron out of a bond to behave like a 'free electron'. This leaves a 'hole' in one bond; an electron can migrate from an adjacent bond to fill this hole, thereby transferring the hole to another bond. In this way the hole can be pictured as moving in a random way through the crystal, and being mobile like the 'free' electron, though not necessarily with the same mobility.

Intrinsic conductivity is exhibited only by very pure specimens, and the properties of silicon and germanium are drastically modified by small amounts of impurity. If an impurity from Group V of the periodic table, such as phosphorus, is introduced, it enters 'substitutionally', occupying the place of a silicon or germanium atom. Like the atom it replaces, it forms four covalent bonds with its four immediate neighbours. This uses up four of its valence electrons, leaving one in excess, which experiences an electrostatic attraction to the phosphorus because of the extra positive charge of its parent ion. This system of a singly charged ion and excess electron resembles a hydrogen atom, but one in which the potential due to the positively charged nucleus is modified by the presence of the surrounding silicon or germanium ions. These are electrically polarized by the excess positive charge of the phosphorus nucleus, and their effect on the electrostatic potential can be crudely approximated by the introduction of a relative permittivity, making the potential $V = e/4\pi\epsilon_r\epsilon_0 r$. Obviously this device of using a relative permittivity is only realistic at distances large compared with the interatomic distance, where the electron orbit is so large that it embraces many atoms, and whose effect then resembles that of a continuous medium. For an electron of effective mass $m^*$ in an orbit of principal quantum number $n$ the energy is then found to be (see Problem 17.1)

$$W = -\frac{m^*}{m}\frac{R}{\epsilon_r^2 n^2} = -\frac{m^*}{m\epsilon_r^2 n^2} \times 13\cdot5\ \text{eV},\qquad(17.2)$$

where $13\cdot5$ eV is the ionization potential of an electron in the $n = 1$ state of a free hydrogen atom. The value of $\epsilon_r$ for germanium is 16, and using this and a value of $m^*/m = 0\cdot2$ (an average of values obtained from other evidence), we find $W \sim 0\cdot01$ eV for the lowest state $n = 1$. This is fairly close to that actually observed for Group V donors in germanium (for phosphorus the observed value is $0\cdot012$ eV). The corresponding orbit radius is over 4 nm, which is quite large compared with the interatomic distance of $0\cdot245$ nm. Since $0\cdot01$ eV is equivalent to a temperature of only 120 K, it is evident that such donor impurities will release nearly all their electrons into the conduction band at room temperature, a process similar to that of ionization of free hydrogen atoms at the very high temperatures in the interior of stars. For Group V donors in silicon the corresponding binding energy is about $0\cdot04$ eV (see Problem 17.1).

If a Group III element is added as an impurity we have a different situation. The impurity atom now has one electron too few to fill the four bonds which it should make on replacing a silicon or germanium atom, and we are therefore left with a hole in one bond. If this hole moves away from its parent impurity, all four bonds to the impurity atom become filled and it has one net negative charge. Since the hole is effectively a positive charge, it has an electrostatic attraction to the negatively charged impurity ion, and we have an 'inside-out' hydrogen atom consisting of a negatively charged 'nucleus' with the positive hole in orbit around it. This gives an 'acceptor' level just above the valence band, the height above the top of the valence band being again about $0\cdot01$ eV. At 0 K the valence band is full of electrons, and the hole occupies the impurity level; it is then localized on the impurity ion, forming a neutral atom in a bound state. At a finite temperature an electron may be excited from the valence band into the impurity level; this leaves a hole in the valence band which is free to move, and corresponds to ionization of the 'inside out' impurity atom.

This analysis shows that at 0 K the hole is in the highest level (the acceptor level) and as the temperature rises more and more holes are excited in the lower levels (the valence band). This behaviour is similar to that of electrons being excited from donor levels into the conduction band, except that for the holes energy must be measured downwards instead of upwards. In § 17.3 we shall find that holes obey similar equations to electrons provided we measure energy downwards from the top of the filled band, an example of which has already occurred in eqn (12.23b).

A wide range of compounds show semiconducting properties. Of these, only a few which illustrate general classes, and for which sufficient information exists to make their properties reasonably well understood, can be mentioned here. Following the Group IV compounds germanium and silicon, it is natural to discuss first the Group III–Group V compounds, taking as example indium antimonide, the most studied of such materials. The two elements, indium and antimony, come in the periodic table immediately before and after tin. They form a compound in which each atom is surrounded by four equidistant neighbours at the apices of a regular tetrahedron, as in the diamond structure, but with the difference that each of the four nearest neighbours is of the other type (this is known as the zinc-blende structure, after one form of the compound ZnS—see Fig. 17.3). These four bonds are mainly covalent in character, and link lattice sites which may be regarded as occupied by $In^-$ and $Sb^+$ ions in regular alternation. Each of these ions has the same electron configuration as tin, the Group IV element, and forms covalent bonds in a similar fashion. However, the fact that we now have ions of alternate negative and positive charge gives rise to some ionic binding. Indium antimonide can be prepared in a sufficiently pure state to behave as an intrinsic semiconductor, with an energy gap of about $0.24$ eV at 0 K, decreasing to $0.18$ eV at room temperature.

The next binary compounds in sequence are the II–VI and I–VII compounds; these grow progressively more ionic in character, with larger energy gaps. This is illustrated in the sequence formed from atoms in the

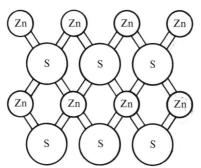

FIG. 17.3. The zinc-blende structure (two-dimensional representation). It is similar to the diamond structure, except that zinc and sulphur ions alternate.

seventh row of the periodic table:

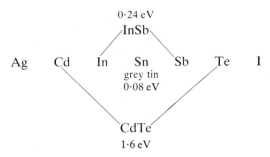

Silver iodide (AgI) is a good insulator, and clearly has a large energy gap. Though not as pure a polar compound as NaCl, we may regard it as consisting of $Ag^+$ and $I^-$ ions, with closed shells of electrons. With a II–VI compound such as CdTe we have the dilemma of whether to regard it as a polar compound, consisting of $Cd^{2+}$ and $Te^{2-}$ ions, or a covalent compound formed of $Cd^{2-}$ and $Te^{2+}$ ions. However, the crystal structure resembles that of zinc-blende, suggesting a covalent compound; on the other hand, the group of salts PbS, PbSe, PbTe have the NaCl structure, suggesting a polar compound. These difficulties illustrate the reserve with which such extreme classification should be regarded. In fact the group of lead salts have smaller energy gaps than CdTe, and their electrical properties are more typical of semiconductors.

The energy gaps of a number of substances are given in Table 17.1. General (but not invariable) rules are that the energy gap diminishes the heavier the atoms involved (that is, reading downwards in the table), but increases on moving from covalent to polar compounds (that is, from left to right). Much less information is available about the polar semiconductors, owing to the difficulty of preparing them in the pure state. Apart from foreign atoms, which act as acceptors or donors according to their group in the periodic table, such crystals may be non-stoichiometric. For example, PbS may have an excess of lead, producing donor levels, or of sulphur, producing acceptor levels.

The energy gap varies considerably with temperature, and most of the above are rounded values. In the case of silicon, germanium, and indium antimonide the values given in the table are those for 0 K; at room temperature they are approximately $1 \cdot 12$ eV, $0 \cdot 66$ eV, and $0 \cdot 18$ eV respectively.

## 17.3. Electron distribution and the Fermi level

Under conditions of thermal equilibrium the number of electrons with energy between $W$ and $W+dW$ can be calculated by means of statistical mechanics, using of course the Fermi–Dirac statistics appropriate to particles of half-integral spin. This number is

$$dn = f(W)g(W)\,dW, \tag{17.3}$$

where $f(W)$ is the Fermi–Dirac function (written as $f$ in eqn (11.15))

$$f(W) = \frac{1}{\exp\{(W-W_F)/kT\}+1} \tag{17.4}$$

and $g(W)$ is the density of states. $W_F$ is the Fermi level defined as the energy at which the function $f(W) = \frac{1}{2}$. In a metal, $W_F \gg kT$ at ordinary temperatures, and the only electrons which are thermally excited or can take part in conduction processes are those very close to the Fermi level.

In a semiconductor, it is not so obvious where the Fermi level lies with respect to the conduction and valence bands. We shall consider first an intrinsic semiconductor, where at low temperatures only a few electrons are excited into the conduction band. In the limit of extremely few electrons the chance of an electron occupying a given state is very low, and the restrictions imposed by the exclusion principle play little role. We are thus in a situation where the classical Maxwell–Boltzmann statistics are a good approximation, so that we can write

$$f_c(W) = \exp\{-(W-W_F)/kT\} \quad \text{very nearly.} \tag{17.5}$$

This is the approximation of eqn (17.4) in the limit where $W-W_F \gg kT$, so that we can neglect the second term in the denominator; this approximation is appropriate for electrons in the conduction band, which is empty of electrons at 0 K. At this temperature the valence band is full, and thus corresponds to energies well below the Fermi level. In the region where $W_F-W \gg kT$ the first term in the denominator of eqn (17.4) is now very small, and we can write

$$1-f_v(W) = \exp\{-(W_F-W)/kT\} \quad \text{very nearly.} \tag{17.6}$$

The quantity $1-f_v(W)$ is the chance of finding a hole in the valence band with energy $W$ (cf. § 12.4).

We must now consider the density of states $g(W)$. This is zero at the edge of a band and we assume that it varies as the square-root of the distance from the edge of the band. That is, we can modify eqn (11.8) and write (using $C' = Cm^{-\frac{3}{2}}$)

$$g_c(W) = C'(m_c^*)^{\frac{3}{2}}(W-W_c)^{\frac{1}{2}} \tag{17.7}$$

and

$$g_v(W) = C'(m_h^*)^{\frac{3}{2}}(W_v-W)^{\frac{1}{2}}, \tag{17.8}$$

where $W_c$, $W_v$ are the energies at the bottom of the conduction band and the top of the valence band respectively.

The total number of electrons in the conduction band is thus

$$n = \int_{W_c}^{\infty} f_c(W) g_c(W) \, dW$$

$$= C'(m_e^*)^{\frac{3}{2}} \int_{W_c}^{\infty} (W - W_c)^{\frac{1}{2}} \exp\{-(W - W_F)/kT\} \, dW.$$

On writing $y = (W - W_c)/kT$, this integral becomes

$$n = C'(m_e^* kT)^{\frac{3}{2}} \exp\{-(W_c - W_F)/kT\} \int_0^{\infty} y^{\frac{1}{2}} \exp(-y) \, dy$$

$$= (\pi^{\frac{1}{2}}/2) C'(m_e^* kT)^{\frac{3}{2}} \exp\{-(W_c - W_F)/kT\}$$

$$= N_c \exp\{-(W_c - W_F)/kT\}. \tag{17.9}$$

This result is the same as if we had a number $N_c$ of states at energy $W_c$; thus $N_c$ is the effective density of states at the bottom of the conduction band, and on substituting for $C' = C_m^{-\frac{3}{2}}$ as in eqn (11.9) we find

$$N_c = 2(2\pi m_e^* kT/h^2)^{\frac{3}{2}}. \tag{17.10}$$

Similarly for the number of holes in the valence band

$$p = \int_{-\infty}^{W_v} \{1 - f_v(W)\} g_v(W) \, dW$$

$$= N_v \exp\{-(W_F - W_v)/kT\}, \tag{17.11}$$

where

$$N_v = 2(2\pi m_h^* kT/h^2)^{\frac{3}{2}}. \tag{17.12}$$

For the product $np$ we find

$$np = N_c N_v \exp\{-(W_c - W_v)/kT\} = N_c N_v \exp(-W_g/kT), \tag{17.13}$$

and since in an intrinsic conductor $n = p = n_i$ we obtain

$$n_i = n = p = (N_c N_v)^{\frac{1}{2}} \exp(-W_g/2kT), \tag{17.14}$$

where $W_g = W_c - W_v$ is the width of the energy gap between the valence and conduction bands.

To find the position of the Fermi level $W_F$ we must equate the formulae for $n$ and $p$, which yields

$$\frac{N_v}{N_c} = \exp\left(\frac{2W_F - W_c - W_v}{kT}\right),$$

whence, since $N_v/N_c = (m_h^*/m_e^*)^{\frac{3}{2}}$,

$$W_F = \tfrac{1}{2}(W_c + W_v) + \tfrac{3}{4}kT \ln(m_h^*/m_e^*). \qquad (17.15)$$

This result shows that for most intrinsic semiconductors, where $m_h^*$, $m_e^*$ are nearly equal, the Fermi level lies in the middle of the energy gap, as shown in Fig. 17.4. In some cases, such as indium antimonide, where $m_h^*/m_e^* \approx 20$, the level varies markedly in position with temperature, and at room temperature is shifted well towards the bottom of the conduction band.

When impurities are present, and conduction is partly intrinsic and partly extrinsic, the position is a good deal more complicated. There is, however, one important general result which holds provided the numbers of electrons in the conduction band and holes in the valence band are small compared with the density of states. In that case the relations (17.9)–(17.13) are still valid, since they do not depend on any supposition

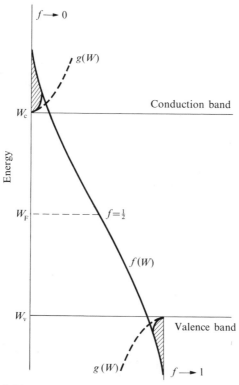

FIG. 17.4. The Fermi–Dirac distribution of electrons and holes in an intrinsic semiconductor; the figure is drawn for a case where $m_h^* = m_e^*$, so that the Fermi level $W_F$ is in the centre of the forbidden band. The shaded areas indicate the numbers of electrons in the conduction band, and holes in the valence band.

about the position of the Fermi level. Thus we have in eqn (17.13) an important relation between the numbers of electrons and holes, and in the light of eqn (17.14) we have also $np = n_i^2$, where $n_i$ is the number of intrinsic electrons which would exist at the same temperature. The ratio of numbers of electrons and holes depends on the position of the Fermi level, for which we shall quote some results only for extreme cases.

When donors or acceptors (but not both together) are present, which produce discrete levels lying close to the conduction or valence band respectively, the conductivity at low temperatures is dominated by the ionization of the impurity levels. The Fermi level at 0 K then lies between the donor level and the conduction band, or between the acceptor level and the top of the valence band (see Fig. 17.5). As the temperature rises the Fermi level shifts because of a term similar to the second term in eqn (17.15), but involving $\ln(N_d/N_c)$ or $\ln(N_a/N_v)$, where $N_d$ and $N_a$ are the number of donor and acceptor levels per unit volume respectively. Thus for donors we have

$$W_F = \tfrac{1}{2}(W_d + W_c) + \tfrac{1}{2}kT \ln(N_d/N_c) \quad (n \ll N_d), \tag{17.16}$$

showing that as the temperature rises the Fermi level will rise if $N_d > N_c$, or fall if $N_d < N_c$.

In the exhaustion range the donor levels are fully ionized and $n = N_d$; in this case the Fermi level is given approximately by

$$W_F = W_c + kT \ln(N_d/N_c) \quad (n = N_d). \tag{17.17}$$

If $N_d < N_c$ the Fermi level lies below the conduction band, the number of electrons excited into the conduction band is small compared with the number of available states and they obey the classical statistics. This condition is called 'non-degenerate'. On the other hand, if $N_d > N_c$ the

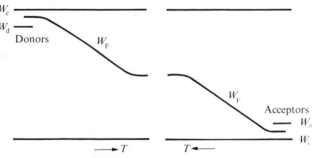

FIG. 17.5. Variation of Fermi level when *either* donors *or* acceptors are present. At very low temperatures the Fermi level lies midway between the impurity level and the conduction or valence band. For small impurity concentrations ($N_d < N_c$ or $N_a < N_v$) the level moves towards the centre of the forbidden band with rising temperature. At higher temperatures the electron distribution is dominated by the intrinsic contribution, and $W_F$ is at the centre of the forbidden gap if $m_h^* = m_e^*$.

Fermi level lies in the conduction band, and since $n = N_d$, the number of conduction electrons is greater than the number of available states; this condition is called 'degenerate'. This situation is similar to that in a metal, where the exclusion principle limits the number of electrons in a given energy range, and the electron distribution must be treated by Fermi–Dirac statistics instead of classical statistics.

Similar results are obtained for acceptor impurities, provided we count energy as increasing downwards from the top of the valence band rather than upwards from the bottom of the conduction band (cf. Fig. 17.4).

The importance of the Fermi level lies in the fact that its value is equal to that of the thermodynamic potential $G = U - TS + |e| V$ of the electrons. If the electron distributions in two substances are in thermal equilibrium with each other, then the values of the thermodynamic potential in the two substances are equal, and hence so also are the Fermi levels. Thus the position of the Fermi level plays an important role in discussing the properties of junctions (cf. § 11.3).

We end this section by giving some numerical values, assuming $m^* = m$. For intrinsic material at room temperature, the values of $n_i$ are roughly $10^{20}$ m$^{-3}$ for germanium and $10^{16}$ m$^{-3}$ for silicon. The latter contains $5 \times 10^{28}$ atoms per cubic metre, and if one atom in $5 \times 10^6$ is replaced by a donor atom which is fully ionized at room temperature, the density of electrons in the conduction band becomes $n = N_d \sim 10^{22}$ m$^{-3}$. The density of holes is $p = n_i^2/n \sim 10^{10}$ m$^{-3}$. In this material electrons are the 'majority carriers' and holes the 'minority carriers'. In each case the density is small compared with $N_c = N_v = 2 \cdot 5 \times 10^{25}$ m$^{-3}$, so that classical statistics are a valid approximation.

## 17.4. Optical properties

Semiconductors such as germanium and silicon look very much like metals; they are opaque to visible light and have a high reflectivity. This is because the quantum carried by a visible photon, which corresponds to an energy roughly between 1·5 eV and 4 eV, is sufficient to excite an electron from the valence band right across the forbidden energy gap into the conduction band. If the absorption coefficient is measured at lower frequencies, a sharp drop in absorption would be expected when the photon energy becomes smaller than the energy gap $W_g$, that is, at frequencies such that $h\nu < W_g$. The change in the absorption coefficient can be quite dramatic, from $10^6$ m$^{-1}$ or $10^7$ m$^{-1}$ at frequencies higher than the absorption edge, down to 10 m$^{-1}$ at frequencies beyond the edge, as illustrated in Fig. 17.6. The absorption beyond the edge depends on the purity of the specimen, since impurities produce levels in the forbidden band from or to which electrons can still be excited by photons of lower energy.

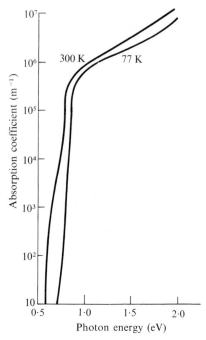

FIG. 17.6. Absorption coefficient for electromagnetic radiation in pure germanium, showing the steep rise at the band edge. The curves for two different temperatures show that the energy gap is slightly higher at the lower temperature.

An optical determination of the position of the absorption edge gives in principle a direct measurement of the energy gap, whose accuracy is limited by the fact that the drop in absorption is spread out over a small but finite range of frequency. Careful analysis of the experimental results in the light of a detailed theory of how the absorption coefficient should vary with frequency in the vicinity of the absorption edge has given quite accurate measurements of the energy gap, and shows directly how it varies with temperature. In the interpretation allowance must be made for the presence of selection rules analogous to those discussed in § 10.9 for optical absorption and lattice waves. The simple absorption of a photon can only excite an electron from the valence band into a state in the conduction band with the same crystal momentum $\hbar\mathbf{k}$, so that we have the selection rule $\Delta\mathbf{k} = 0$. If the maximum of the valence band and the minimum of the conduction band both occur at the same value of $\mathbf{k}$, then the lowest absorption frequency gives the energy gap. This is the case in indium antimonide, where the valence-band maximum and conduction band minimum both occur at $\mathbf{k} = 0$. In germanium and silicon, however, the conduction band has only a subsidiary minimum at $\mathbf{k} = 0$, and a deeper minimum occurs at a finite value of $\mathbf{k}$, as shown in Fig. 17.7. Thus

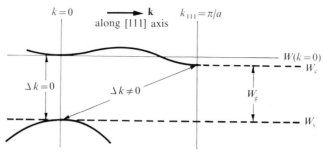

FIG. 17.7. Shape of the band edges against crystal momentum **k** for germanium. The momentum of a photon is negligible, so that there can be no net transfer of momentum on absorption. Either $\Delta\mathbf{k} = 0$ for the electron, or the difference in momentum when $\Delta\mathbf{k} \neq 0$ must be taken up by the creation or destruction of a phonon. For germanium $W_g = 0.75$ eV at 0 K but $W(\mathbf{k}=0) - W_c \approx 0.14$ eV, so that the $\Delta k \neq 0$ transitions give structure on the absorption edge.

for these substances transitions in the vicinity of $\mathbf{k} = 0$ do not determine the minimum value of $W_g$, defined as the difference in energy between the highest point in the valence band and the lowest point in the conduction band. (This is the energy required for thermal excitation of electrons across the gap, when no selection rules are involved.) In fact optical transitions such as that marked $\Delta\mathbf{k} \neq 0$ in Fig. 17.7 are allowed through the simultaneous creation or destruction of a phonon. Such transitions are called indirect or phonon-assisted transitions, while transitions in which $\Delta\mathbf{k} = 0$ are known as direct transitions.

*Excitons*

When an electron is excited from the valence band into the conduction band by a direct transition, a hole is created in the valence band whose momentum must be equal and opposite to that of the electron in the conduction band in order to make $\Delta\mathbf{k} = 0$. The electron and hole therefore move apart in opposite directions. In the vicinity of $\mathbf{k} = 0$, they move apart rather slowly, and their mutual Coulomb attraction begins to play a role; finally at $\mathbf{k} = 0$ itself the electron and hole stay together. Under these conditions their behaviour resembles that of an electron and proton in a hydrogen atom; a better comparison is with an electron and donor impurity atom, as discussed in § 17.2. Electron and hole may move in discrete orbits about the mutual centre of mass, giving rise to a series of energy levels

$$W = -\frac{m_r}{m}\frac{R}{\epsilon_r^2 n^2},\tag{17.18}$$

where $m_r$ is the reduced mass given by the relation

$$\frac{1}{m_r} = \frac{1}{m_e^*} + \frac{1}{m_h^*}.\tag{17.19}$$

In eqn (17.18) $W = 0$ corresponds to separation of the electron and hole by such a large distance that their mutual attraction is negligible (that is, to 'ionization' of the electron–hole 'atom') and thus corresponds to the bottom of the conduction band at $\mathbf{k} = 0$. A lower energy is obtained when the electron and hole are together, so that the levels of eqn (17.18) lie just below the conduction band (like those of a donor impurity and bound electron), as shown in Fig. 17.8. The energy is also of the same order; if $m_e^* = m_h^* = 2m_r$, the energy levels are just half those given by eqn (17.2), and lie therefore very close to the conduction band.

The electron–hole bound pair is known as an 'exciton' and has been identified through its hydrogen-like spectrum in some semiconductors with large energy gaps, together with cuprous oxide and germanium. In germanium excitons are associated both with the direct and indirect transitions. In the former case a sharp line spectrum would be expected at frequencies just short of the energy gap at $\mathbf{k} = 0$; in the indirect transitions the electron–hole pair can be formed with finite momentum and possess kinetic energy, so that the exciton levels are not sharp but broadened into a band.

*The spectra of impurity atoms*

When an impurity atom is substituted for a 'host' atom, it produces discrete levels in the energy gap, as discussed in § 17.2. For shallow donors (donors with a low ionization energy) the levels obey a relation

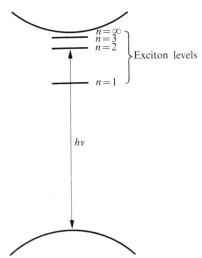

FIG. 17.8. Exciton levels lying just below the conduction band in the vicinity of $\mathbf{k} = 0$. The quantum of energy shown is that required to excite an electron from the valence band to the $n = 2$ level, and is slightly smaller than that corresponding to the absorption edge, which requires excitation to the bottom of the conduction band ($n = \infty$).

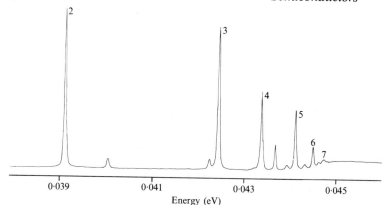

Fig. 17.9. Absorption spectrum at T = 20 K of phosphorus as a donor impurity in silicon. The sharp lines are due to transitions from the ground state to excited states of the donor electron bound to the impurity atom, ending at about 0·045 eV in the photo-ionization continuum. Note that in the highest quantum states the radii of the electron orbits are over 100 nm. (Courtesy Dr. R. A. Stradling).

close to that given by eqn (17.2), and transitions can be induced between them. Fig. 17.9 shows the spectrum of phosphorus in silicon, with a number of discrete lines, ending in the continuum corresponding to excitation of an electron from the donor atom into the conduction band.

*Photoconductivity*

When electromagnetic radiation excites electrons into the conduction band (or creates holes in the valence band), the electrical conductivity of a semiconductor changes appreciably; this phenomenon is known as photoconductivity. For small intensities of illumination the increase in conductivity is roughly proportional to the intensity, and forms an important method of detecting radiation of wavelengths shorter than the absorption edge. Cadmium sulphide gives a flat response between 0·5 $\mu$m and 0·65 $\mu$m, while lead sulphide has its peak response at about 2 $\mu$m and indium antimonide at 6 $\mu$m. Photoconductive detectors have a high sensitivity, and a short response time (from 1 $\mu$s to 100 $\mu$s) because the excess carriers quickly disappear through recombination, etc. This makes it possible to modulate the incoming radiation (using, for example, a mechanical chopping device) to produce an a.c. signal which can be amplified and detected.

The use of a semiconductor doped with suitable impurities makes it possible to construct detectors which are sensitive to longer wavelengths, the photoconductive effect then being due to electrons excited into the conduction band from donor impurity levels (or holes created by electrons being lifted into acceptor levels from the valence band). If the smallest separation of such discrete levels from the adjacent bands is about 0·01 eV, suitably doped samples are sensitive to wavelengths up to about 100 $\mu$m. Such detectors must of course be cooled to a temperature where thermal ionization of the impurity levels is unimportant.

*Emission of radiation*

Electrons excited into the conduction band may lose their energy in a collision, or they may make optical transitions in the reverse direction to those considered above in absorption processes. This results in the emission of electromagnetic radiation which lies in the visible region when a semiconductor with a suitable gap is used. Gallium arsenide (GaAs), and alloys of GaAs and gallium phosphide (GaP) are used for this purpose. A p–n junction (§ 17.7) is biased to inject a large density of electrons into the conduction band of p-type material, where they recombine with the holes to emit radiation with high efficiency. This process is the basis of the light-emitting diode (LED), and laser action can also be obtained under suitable circumstances.

More detailed discussions of the optical properties of semiconductors, and their applications, are given by Houghton and Smith (1966).

## 17.5. Transport properties

In addition to the energy gap $W_g$, the most important quantities we need for a semiconductor are the numbers of charge carriers of either sign, their effective masses, and their mobilities. The best method of measuring the effective mass is by means of cyclotron resonance, which is discussed in Chapter 24. Determination of $W_g$ by optical methods is possible only on very pure specimens, and in many cases it can only be deduced indirectly, using eqn (17.13). To bring out the temperature-dependence explicitly we rewrite this in the form

$$np = (2 \cdot 33 \times 10^{43})(m_e^* m_h^*/m^2)^{\frac{3}{2}} T^3 \exp(-W_g/kT), \qquad (17.20)$$

where the units are (metre)$^{-6}$. At first sight the simplest method of finding the number of charged particles would be measurement of the Hall coefficient $R_H$, which from eqn (12.51) is inversely proportional to the number of carriers. Since this number is so much smaller in a semiconductor than in a metal, the Hall effect is much larger ($\approx 0 \cdot 1$–1 m$^3$ C$^{-1}$ for pure silicon and germanium at room temperature) and correspondingly easier to measure. However, we have the difficulty that the Hall coefficient changes sign according to whether the current is carried by electrons or positive holes; in particular, in an intrinsic semiconductor, with equal numbers of each, we would expect the Hall coefficient to vanish. In practice this does not happen. because the electrons and holes have different mobilities, and so carry different fractions of the current. With two sets of carriers the expression for the Hall Coefficient is

$$R_H = -\frac{B}{|e|}\frac{(nb^2-p)}{(nb+p)^2},\tag{17.21}$$

where $b = \mu_e/\mu_h$ is the ratio of the mobilities of electrons and holes, and $B$ is a coefficient not far from unity. The presence of $B$ arises from the fact that different averages over the distributions of velocity and relaxation time are involved in calculating the conductivity and the Hall effect. For a metal or a degenerate semiconductor, $B = 1$; for a non-degenerate semiconductor the value of B depends on how the scattering cross-section varies with carrier velocity.

Measurement of the Hall coefficient $R_H$ and the conductivity $\sigma$ makes it possible to determine both $n$ and $p$, using eqns (17.1) and (17.21), provided that the value of $b$ is known. The direct measurement of mobility is discussed below, but this is possible only with certain substances such as silicon and germanium. In other cases we can proceed only by making some assumptions about $b$. In an intrinsic semiconductor, $n = p = n_i$, and we have the relations ($n_i$ per cubic metre)

$$n_i = (4{\cdot}8\times10^{21})(m_e^* m_h^*/m^2)^{\frac{3}{4}}T^{\frac{3}{2}}\exp(-W_g/2kT)\tag{17.22}$$

and

$$R_H = -\frac{B}{n_i|e|}\frac{(b-1)}{(b+1)}.\tag{17.23}$$

Although the mobilities $u_e$ and $u_h$ both vary with temperature, their ratio $b$ does not vary rapidly in comparison with $n_i$, which is dominated by the exponential factor in eqn (17.22). Thus a plot of $\ln(R_H T^{\frac{3}{2}})$ against $T^{-1}$ should be a straight line, and this is found to hold for silicon and germanium. The slope of the straight line yields a value of $W_g$, but since $W_g$ is itself temperature dependent, care must be exercised in the interpretation (see Problem 17.2).

Another approximate method of finding $W_g$ involves only measurement of the conductivity. For an intrinsic semiconductor (or an impurity semiconductor at high temperatures where the conductivity is dominated by the intrinsic electrons), we have

$$\sigma = |e|n_i(u_e+u_h),\tag{17.24}$$

and from measurement of the conductivity over a range of temperature we can find the variation of $n_i$ and hence determine the energy gap provided we known how the mobilities vary with temperature. When the mobility is determined by scattering processes due to lattice vibrations (phonons) the mobility (see below) should vary as $T^{-\frac{3}{2}}$. Since from eqn (17.22) $n_i$ varies as $T^{\frac{3}{2}}\exp(-W_g/2kT)$, we should expect that the conductivity would follow the law

$$\sigma = \sigma_0\exp(-W_g/2kT),\tag{17.25}$$

so that the energy gap can be found from a plot of $\ln \sigma$ against $T^{-1}$. This method was used in early work, but such plots show a slight curvature, indicating that the mobility does not follow a $T^{-\frac{3}{2}}$ law exactly.

Determination of the energy gap from the optical absorption edge needs careful experimentation and interpretation, but it is the most satisfactory method and the only one which gives directly the energy gap at a given temperature. Use of the Hall effect or the conductivity depends on assumptions about the mobility, whose experimental determination will now be discussed.

### Measurement of mobility

The mobility of electrons and holes in semiconductors can be measured directly by a method due originally to Shockley and Haynes. The specimen is in the form of a narrow rectangular bar, about 0·5 mm square and a few centimetres long. A steady voltage is applied between the ends to give a field of order $10^3 \, \text{V m}^{-1}$ along the bar; two electrodes A, B are applied to the specimen, as shown in Fig. 17.10. A short voltage pulse of duration about 1 $\mu$s is applied to electrode A; if the semiconductor is n-type, and A is made positive in the pulse, electrons are withdrawn from the semiconductor by the electrode; some of these may come from the valence band, creating an excess of holes in the semiconductor. To preserve electrical neutrality of the specimen, electrons enter at any terminals which are negative with respect to A. One of these is B, which is connected (through an amplifier) to the Y-plates of an oscilloscope; this registers a voltage because of the flow of such electrons through the resistance R. This pulse does not quite coincide in time with the pulse at A, but the time delay is that required by an electromagnetic wave set up by the disturbance at A to travel to B, which is of the order of $10^{-4} \, \mu$s and quite negligible. The holes injected at A are swept by the field towards B,

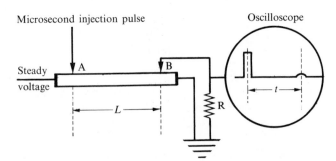

FIG. 17.10. Apparatus for direct measurement of mobility. The steady voltage applied at extreme left produces a uniform field in the semiconductor bar under whose influence a microsecond pulse of minority carriers drift from A to B. The mean drift velocity is $L/t$, where $t$ is measured on the oscilloscope.

and arrive at time $t = L/v$ later, where $L$ is the distance between electrodes A, B and $v$ is the drift velocity in the steady field. On arrival at B they appear as a second voltage pulse on the oscilloscope, and the time interval $t$ between the two pulses can be determined from the oscilloscope trace by calibrating the time-base. The voltage $V$ between the electrodes A, B due to the steady field is measured independently. Since the drift velocity $v = uE = uV/L$, the time $t = L/v = L^2/uV$ and the mobility is found from the relation

$$u = L^2/Vt. \tag{17.26}$$

In a typical experiment the value of $t$ is about 30 $\mu$s. The second pulse at B due to the arrival of holes is broader and smaller than the first for two reasons: (1) diffusion of the holes in random directions; for accurate results the holes must be swept from A to B in a time $t$ short enough to make diffusion effects small, and the voltage pulse applied at A must be short compared with $t$; (2) holes are lost by recombination with electrons within the specimen.

To measure the mobility of electrons, all that would seem necessary at first sight would be to use n-type material and apply a negative pulse at A. This would inject electrons, creating a local excess, which is, however, dissipated in an extremely short time (see Problem 17.3), restoring the equilibrium concentration of electrons everywhere in the semiconductor. In this case only the first pulse is observed at B, and no second pulse. When a positive pulse is applied at A, with an n-type semiconductor, holes are created, and although electrical neutrality is restored by an inflow of electrons, we have now a non-equilibrium distribution with an excess of holes and a corresponding excess of electrons. Equilibrium is restored only when the holes flow out at B, or are annihilated within the semiconductor by recombination with electrons. The net result is that we can measure directly the mobility only of 'minority carriers'; that is, of holes in an n-type semiconductor, or electrons in a p-type semiconductor. By observing the spread in the second pulse at B the diffusion constant can be determined, and by observing its size as a function of $L$ or of electric field strength $E$ the rate of recombination can be found.

The quantity measured directly in such experiments is called the 'drift mobility', but in many semiconductors it cannot be so determined because of rapid diffusion of the carriers. In that case the mobility, when only one type of carrier is present, can be found from the conductivity and Hall effect, since then $|R_H| = B/(n|e|)$ and $|R_H\sigma| = Bu$. The quantity $|R_H\sigma|$ is often written $u_H$ and called the 'Hall mobility' to distinguish it from the drift mobility.

*Variation of mobility with temperature*

The mobility of electrons or holes in semiconductors is limited by scattering processes which are basically similar to those in metals, but the temperature-dependence of the mobility is very different for two reasons. In a metal only electrons at the Fermi surface contribute to the conduction current, and their velocity $v$ is substantially independent of temperature; hence we do not need to take into account any velocity-dependence in a scattering cross-section. Similarly the fact that the scattering cross-section determines the free path $l$, while the mobility depends on the relaxation time $\tau = l/v$, does not of itself introduce any temperature-dependence. In a semiconductor, however, the average kinetic energy of the charge carriers is $\approx kT$, and the velocity variation as $T^{\frac{1}{2}}$ plays an important role.

Measurements of mobility at various temperatures and different impurity concentrations show that it is a function of both. In a pure material the charge carriers are scattered by the lattice vibrations (phonons); at all but the lowest temperatures (when impurity scattering dominates in any case) the scattering cross-section is proportional to the mean-square amplitude of the thermal fluctuations, and hence is proportional to $kT$. This gives a mean free path $l$ proportional to $T^{-1}$ which is the same for all carrier velocities, and the relaxation time $\tau = l/v$ and hence also the mobility should vary as $T^{-\frac{3}{2}}$. Table 17.2 shows that this law is not very well obeyed, except for electrons in germanium. The discrepancies may be due either to scattering by short-wavelength lattice vibrations where adjacent atoms vibrate in antiphase ('optical modes'), or to the complicated band structure (see Fig. 17.7), both of which allow scattering processes with a large change in electron wave vector. The observed mobilities in silicon and germanium at room temperature are found from Table 17.2 to lie between about $0\cdot05$ m$^2$ V$^{-1}$ s$^{-1}$ and $0\cdot4$ m$^2$ V$^{-1}$ s$^{-1}$, and are thus considerably higher than those in metals (for copper the value is about $0.004$ m$^2$ V$^{-1}$ s$^{-1}$). The high mobility is partly due to low values of the effective mass of electrons and holes in semiconductors. In InSb and GaAs where the electron masses (see Table 24.2) are particularly small, the mobilities are exceptionally high ($\sim 8$ m$^2$ V$^{-1}$ s$^{-1}$ for electrons in InSb).

The scattering cross-section due to neutral impurities is inversely proportional to carrier velocity, giving a relaxation time independent of

TABLE 17.2

*Mobilities in elemental semiconductors*

|    | Electrons | Holes |
|----|-----------|-------|
| Si | $(40\times10^4)T^{-2\cdot6}$ | $(2\cdot5\times10^4)T^{-2\cdot3}$ |
| Ge | $(0\cdot35\times10^4)T^{-1\cdot6}$ | $(9\cdot1\times10^4)T^{-2\cdot3}$ |

The units are m$^2$ V$^{-1}$ s$^{-1}$, and the values are taken from Ziman (1960).

velocity and hence also of temperature. Charged (that is, ionized) impurities will scatter carriers by a process analogous to Rutherford scattering of alpha particles; the cross-section is inversely proportional to the square of the carrier energy and hence varies as $T^{-2}$, so that the mean free path varies as $T^2$. Since the mean velocity varies as $T^{\frac{1}{2}}$, the relaxation time varies as $T^{\frac{3}{2}}$. A special case of scattering by charged particles is the mutual scattering of electrons and holes.

TABLE 17.3

*Summary of dependence of electron (hole) scattering on velocity v and temperature T*

| Scattering mechanism | Cross-section $\sigma$ | Free path $l \propto \sigma^{-1}$ | Relaxation time $\tau = l/v$ |
|---|---|---|---|
| Phonons (at ordinary temperatures) | $T$ | $T^{-1}$ | $T^{-\frac{3}{2}}$ |
| Neutral impurities | $v^{-1}$ | $v$ | constant |
| Ionized impurities | $v^{-4}$ | $v^4$ | $v^3 \equiv T^{\frac{3}{2}}$ |

The velocity- and temperature-dependence of these scattering mechanisms are summarized in Table 17.3. In a first approximation the rates of scattering by different processes are additive, that is, we can write (cf. eqn (12.31).

$$\frac{1}{\tau} = \sum_i \frac{1}{\tau_i}.$$  (17.27)

At low temperatures the lattice scattering decreases as the lattice vibrations die away, and the mobility is eventually dominated by the impurity scattering.

*Recombination and diffusion*

At any given temperature there is an equilibrium concentration of electrons and holes in a semiconductor, the two concentrations being equal in intrinsic material and generally unequal in extrinsic material. In the experiment of Shockley and Haynes we have seen that an excess of minority carriers can be injected at a contact, and to preserve electrical neutrality there will be a corresponding injection of majority carriers, possibly at another electrode. Any abnormal charge distribution caused thereby vanishes in about $10^{-5}\,\mu\text{s}$ (see Problem 17.3), so that we can write $\Delta p = \Delta n$, where $\Delta p$, $\Delta n$ are the local excesses in the number of holes and electrons respectively per unit volume. If we did not have this equality, a space charge $\rho = |e|(\Delta p - \Delta n)$ would be set up, which by Poisson's equation

$$\text{div } \mathbf{E} = \rho/\epsilon = (e/\epsilon)(\Delta p - \Delta n)$$

would give rise to strong electric fields. These would cause current to flow which would neutralize the space charge in the time given above. Obviously this current flow consists mainly of the more numerous majority carriers, and the controlling factor is the departure from the equilibrium value of the number of minority carriers. The fact that $\Delta p = \Delta n$ (in practice the equality is not exact, but departures from it are very small and can be neglected for present purposes) means that the *fractional* change in the number of minority carriers may be appreciably greater than the *fractional* change in majority carriers. A number of important devices described in later chapters depend on changes in the minority carrier concentration; such changes represent departures from equilibrium, and the mechanisms by which they decay play an important role in the design of such devices.

As mentioned in § 17.1, the charge carriers in a semiconductor have a finite lifetime, but this varies widely with the purity of the crystal. Simple recombination of an electron and a hole can only take place if certain restrictions on momentum and energy are satisfied, and measurements on very pure germanium show that this process would give a lifetime greater than $10^{-2}$ s. The observed lifetimes are generally much shorter, owing to the presence of chemical impurities which provide extra levels, known as 'traps'. In an n-type material, for example, electrons may drop from the conduction band into such traps; holes may then collide with these electrons to give recombination. The energy and momentum considerations involved in this 'indirect' process are much less restrictive than for the direct process of recombination, and the lifetime of the carriers is correspondingly shorter. These processes take place in the body of the semiconductor, but discrete levels at the surface (see § 17.6) may also act as traps which promote recombination through 'indirect' processes. A further loss of minority carriers also occurs at the electrodes.

The chance of a hole and an electron recombining is proportional to the concentration of each species; hence the rate of annihilation is $-apn$, where $a$ is a constant which depends on the mechanism involved. In thermal equilibrium the loss by recombination is balanced by the creation of new carriers through thermal excitation into the conduction band; if we denote the rate of creation by $c$, then clearly $c = ap_0 n_0$ where $p_0$, $n_0$ are the equilibrium concentrations. When a departure from equilibrium takes place we have

$$dp/dt = dn/dt = c - apn = a(p_0 n_0 - pn).$$

As mentioned above, the *fractional* change in the majority carrier concentration is much smaller than the *fractional* change in the minority carrier concentration, and to a first approximation the former may be neglected. Hence taking, for example, the holes in an n-type material, we may write

$$dp/dt \sim an_0(p_0-p) = -\Delta p/\tau_h, \tag{17.28}$$

showing that the rate of decay of the minority carrier concentration is simply proportional to the excess of minority carriers. The quantity $\tau_h$ is the 'recombination lifetime' of the minority carriers.

Recombination at the surface or extraction at an electrode causes a local diminution in the excess of minority carriers; this is counteracted by the movement of carriers from regions where they are more numerous. This is a process of diffusion, and for small field strengths diffusion currents are much larger than conduction currents. For simplicity, we consider a case where the minority carrier density $p$ (taking again holes in n-type material) varies only in one dimension. Then the number crossing unit area per second is $-D_h(\partial p/\partial x)$, and the net rate of increase in a thickness $dx$ is $d\{-D_h(\partial p/\partial x)\} = -D_h(\partial^2 p/\partial x^2)\,dx$, where $D_h$ is the diffusion coefficient for the minority carriers. In the steady state this equals the rate of loss by recombination, which in thickness $dx$ is $\{(p_0-p)/\tau_h\}\,dx$; hence we obtain the relation

$$\frac{\partial^2 p}{\partial x^2} = \frac{p-p_0}{D_h\tau_h}. \tag{17.29}$$

This has a solution (writing $\Delta p$ for $(p-p_0)$)

$$\Delta p = (\Delta p)_0 \exp(-x/L_h), \tag{17.30}$$

where $(\Delta p)_0$ is the excess concentration at $x = 0$, and $L_h = (D_h\tau_h)^{\frac{1}{2}}$ is a measure of the mean distance a minority carrier will move under the action of diffusion before it is lost by recombination. It is known as the 'diffusion length' and is an important quantity in transistor design. A typical value for germanium is 1 mm for both holes in n-type material and electrons in p-type material; typical values in silicon are smaller by a factor of about 3.

### Hot carriers and the Gunn effect

As the electric field in a bulk semiconductor rises, the carrier drift velocity at first increases linearly with the field, as in a metal. At higher field strengths the additional energy acquired by the carriers through acceleration by the field becomes comparable with or greater than their thermal energy. The relaxation rate is a function of carrier energy, so that the mobility changes. In most cases the mobility increases, but in gallium arsenide (GaAs) the drift velocity decreases as shown in Fig. 17.11. The explanation for this effect, due to Gunn, is based on the band structure. The main minimum in the conduction band is at $\mathbf{k} = 0$ (Fig. 17.12), but at certain values of $\mathbf{k}$ along the [001] axes subsidiary minima occur with an

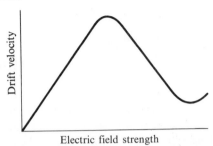

FIG. 17.11. Electron drift velocity against field strength for gallium arsenide. The velocities are about $10^5 \, \text{m s}^{-1}$ and the minimum occurs at a field strength of about $10^6 \, \text{V m}^{-1}$.

energy about $0 \cdot 36 \, \text{eV}$ greater than the central minimum. In the presence of an electric field sufficiently large to raise the carrier energy to $0 \cdot 36 \, \text{eV}$, electrons are scattered by optical phonons into the subsidiary minima. In these minima the effective mass $m^*$ is much higher than it is in the central minimum, so that carriers transferred into the subsidiary minima have a lower mobility. As the electric field increases, a greater proportion of the electrons are scattered into the subsidiary minima, and the mean drift velocity decreases. The velocity–field plot of Fig. 17.11 leads to a negative resistance effect, but the current–voltage characteristic is complicated by instabilities which lead to rapid current oscillations.

## 17.6. Metal–semiconductor junctions

To investigate the electrical properties of semiconductors it is necessary to make electrical connections to them. The behaviour of a junction between a metal electrode and a semiconductor depends on the nature and geometry of the connection, as well as on the properties of the metal and semiconductor, making a full treatment very complex. The most important property of such a junction is that the current flow for a given

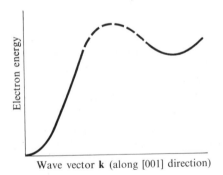

FIG. 17.12. Plot of electron energy against **k**-vector for the [001] axes in gallium arsenide, showing the subsidiary minimum at about $0 \cdot 36 \, \text{eV}$.

voltage is quite different in opposite directions, so that it acts as a rectifier.

When contact is made between a metal and a semiconductor, a potential difference is set up between the two in a similar manner to that between two metals (the contact potential). For an n-type semiconductor whose Fermi level is above that of the metal, electrons pass from the semiconductor to the metal until the two Fermi levels are equal. This process is illustrated in Fig. 17.13. The excess negative charge on the metal repels electrons near the surface of the semiconductor, creating a layer which is depleted of conduction electrons and so has a higher resistance than the bulk of the semiconductor. This layer is known as the 'barrier layer' and is a region of positive space charge because it contains the ionized donor impurities without the compensating charge of the negative conduction electrons. By Poisson's equation (eqn (2.1)) the potential will vary through the space-charge layer, so that there will be a potential difference $V_0$ between the position of the bottom of the conduction band at the surface and in the bulk semiconductor; the energy bands are therefore distorted near the surface, as shown in Fig. 17.13.

In equilibrium there will be no net current flow across the barrier, but this is a dynamic equilibrium between a current $-I_d$ of electrons flowing out of the metal into the semiconductor, and an equal current of electrons (and, to a much lesser extent, holes) leaving the semiconductor for the metal. The latter may be written in the form $I_0 \exp(-|e| V_0/kT)$, since the fraction of the electrons which have sufficient energy to surmount the barrier $V_0$ is proportional to $\exp(-|e| V_0/kT)$. If a voltage $V$ is now

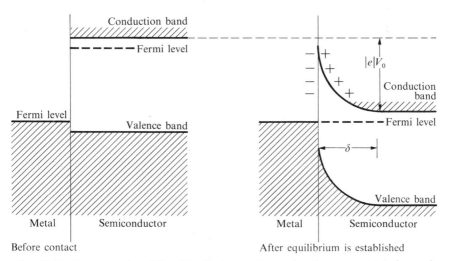

FIG. 17.13. Energy bands and Fermi levels at a metal to semiconductor contact, before and after equilibrium is established.

applied which makes the metal positive with respect to the semiconductor, this extra voltage appears almost entirely across the barrier since this has a much larger resistance than either the metal or the bulk semiconductor. The current leaving the semiconductor then becomes $I_0 \exp\{-|e|(V_0-V)/kT\}$, because the barrier height is reduced from $V_0$ to $V_0-V$; on the other hand the current leaving the metal is still $-I_d$, since the barrier which the electrons in the metal have to surmount is unaltered (see Fig. 17.14). Since

$$I_d = I_0 \exp(-|e| V_0/kT),$$

the net current flow is

$$I = I_d\{\exp(|e| V/kT)-1\}. \tag{17.31}$$

If $V$ is positive and greater than $(kT/|e|)$, large forward currents can flow, while if $V$ is negative $I$ approaches $-I_d$; the current–voltage characteristic therefore has the form shown in Fig. 17.15(a), and the junction is an efficient rectifier.

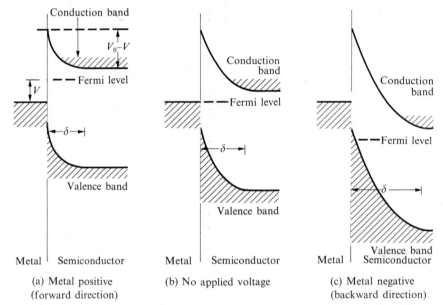

(a) Metal positive        (b) No applied voltage        (c) Metal negative
(forward direction)                                      (backward direction)

FIG. 17.14. Effect of applied voltage $V$ at metal–semiconductor junction. In (a) a forward voltage $V$ is applied, reducing the barrier height to $V_0-V$, and giving a large current flow of electrons from the semiconductor into the metal; in (c) a reverse voltage is applied, increasing the barrier and reducing the current flow. Note that there is a change with voltage in the effective thickness of the barrier

$$\left(\delta = \left\{\frac{2\epsilon(V_0-V)}{eN_d}\right\}^{\frac{1}{2}}\right),$$

and that the Fermi level in the metal is depressed when the metal is made positive because of the negative sign of the electronic charge.

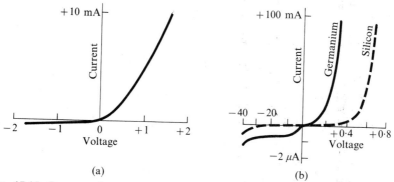

Fig. 17.15. Current–voltage characteristics of (a) silicon–tungsten diode for centimetre wavelengths, and (b) p–n junction. Note differences in scales.

This treatment gives a satisfactory qualitative treatment of the rectifying properties, though agreement with experiment is by no means exact. A satisfactory feature is that it gives the right sign for the forward direction, that is, that the direction of easy flow of electrons (for an n-type semiconductor) is from semiconductor to metal. In an early theory the flow of electrons through the barrier was ascribed to the tunnel effect, so that the easy flow was from metal (where the electron density is high) to semiconductor. However, the tunnel effect is only appreciable when the barrier thickness is comparable with the electron wavelength in the metal, that is, less than a nanometre, whereas the theory of a barrier due to a depletion layer gives a thickness of order 10 nm. A simple version of a theory for the barrier thickness put forward by Schottky is as follows.

We assume that all conduction electrons are removed from the barrier layer, so that the charge density $\rho = eN_d$, where $N_d$ is the number of donors per unit volume (for simplicity we take these all to carry unit charge, and all to be positively ionized, that is, we are in the exhaustion region). Let the surface of the semiconductor be the plane $x = 0$, and the barrier extend from the surface to the plane $x = \delta$, so that the space charge becomes zero for $x \geqslant \delta$. Then the potential is constant for $x \geqslant \delta$, and the electric field vanishes at $x = \delta$. In the barrier, Poisson's equation reduces to

$$\nabla^2 V = d^2V/dx^2 = -eN_d/\epsilon \qquad (17.32)$$

Integration, using the boundary conditions $dV/dx = 0$ (no electric field) at $x = \delta$, and $V = 0$ at $x = 0$, gives

$$V = -(eN_d/2\epsilon)\{(x-\delta)^2 - \delta^2\}. \qquad (17.33)$$

Hence the difference of potential at the surface $x = 0$ from that in the interior of the semiconductor $(x \geqslant \delta)$ is

$$V_0 = eN_d\delta^2/2\epsilon. \qquad (17.34)$$

If we take $\epsilon \approx 12$ (as in silicon) and $N_d = 10^{24}\,m^{-3}$, $\delta$ is found to be about 30 nm if $V_0$ is about 1 V.

A difficulty in the Schottky theory is that $V_0$ should be equal to the difference in the work functions of metal and semiconductor, and hence $I_d$, which is proportional to $\exp(-eV_0/kT)$, should depend on the metal used, whereas experimentally it does not. To overcome this difficulty Bardeen put forward the idea of 'surface states'. At the surface of the semiconductor the atomic arrangement is different because there are no atoms on the one side with which to form bonds. Thus the surface atoms have different energy levels, and these levels are discrete because bands are only formed from the levels of atoms which are identical. There may also be impurity atoms absorbed on the surface. Those surface levels which lie below the Fermi level of the semiconductor will be filled by electrons from the conduction band, giving a negative charge on the surface which repels electrons near the surface. This gives a depletion layer just inside the surface, which acts as the barrier. If $N_s$ is the number of surface states per unit area, and $N_d$ the number of donors per unit volume, then the thickness $\delta$ of the barrier will be determined by the relation $N_s = N_d\,\delta$ if we assume that all the conduction electrons for a distance $\delta$ are drawn into the surface states. Application of Poisson's equation to the barrier layer yields eqn (17.34) as before, but on writing $\delta = N_s/N_d$ we have

$$V_0 = 2N_s^2/2\epsilon N_d. \qquad (17.35)$$

This equation shows that $V_0$ depends on $N_s$ and $N_d$ for the semiconductor and is independent of what metal is used to make the contact; in fact the barrier layer exists in the absence of any contact.

The discussion above has been on the basis of an n-type semiconductor, but similar arguments apply to a p-type semiconductor if holes are substituted for electrons (thereby reversing the direction of easy current flow). Thus any metal–semiconductor junction will act as a rectifier. However, in order to use it two junctions must be made to complete a circuit, and since the forward direction will be in the opposite sense at the two junctions, no rectifying action will result unless the two junctions are different in nature. At microwave frequencies one junction must be very small in cross-section, since otherwise the capacitance between the metal and bulk semiconductor across the barrier layer acts as a short-circuit, the current flowing as displacement current through this capacitance instead of real current through the barrier. This small contact is made by a thin metal whisker (usually tungsten) pressed against the semiconductor (silicon); the other contact is soldered and of large area, so that it offers little resistance (and large capacitance) to the flow of current in what would otherwise be its 'backward' direction (see Fig. 18.16).

Such fine contacts have a relatively high resistance, since the current has to spread out from a fine point through the interior of the semi-conductor, and they cannot carry more than a few milliamperes of current, as in Fig. 17.15(a). At low frequencies metal–semiconductor junctions with a larger area can be used, the metal and semiconductor being separated by an insulating oxide layer sufficiently thin that electrons can readily tunnel through it. Such 'Schottky' diodes' can carry large currents, comparable with those in the p–n junction described below. They have a low forward resistance because electrons crossing the barrier are majority carriers with relatively high velocities, arriving in the metal with the extra energy equivalent to the difference in work functions.

## 17.7. Theory of the p–n junction

Single crystals of a semiconductor (silicon or germanium) can be prepared in which one end is doped to make it p-type, and the other end n-type, the change from p-type to n-type taking place in a region whose thickness is $\sim 10$ nm. Such a unit is called a p–n junction. The p-type material is made by doping with a concentration $N_a$ of acceptors, the n-type by doping with a concentration $N_d$ of donors; obviously there will be a narrow region at the junction where the doping concentration varies from one extreme to the other, but provided this region is small in width compared with the thickness of the barrier region estimated below, we can regard the change from p-type to n-type as discontinuous. In ger-manium doped with Group III and Group V impurities the ionization potential is so small that at room temperature we can regard the donors and acceptors as fully ionized (that is, we are in the exhaustion region).

The energy level situation shown in Fig. 17.16 is not stable and could only exist if the n-type and p-type material were in separate crystals. There is an excess hole concentration in the p-type, and excess electron concentration in the n-type, so that when they are in the same crystal holes will diffuse to the right and electrons to the left, each producing a positive current to the right. This gives a positive potential to the n-type material, so that the energy levels of its electrons are lowered because of their negative charge. This process continues until the Fermi levels of the two halves are equalized, as in Fig. 17.17. The difference of potential between the two halves means that strong electric fields exist near the junction, and these sweep out the mobile carriers in the vicinity of the junction, giving a barrier layer of high resistance. The whole of the potential drop occurs across this barrier layer; removal of the holes in the barrier layer on the p-type side leaves a negative space charge of $-eN_a$ per unit volume, and removal of electrons on the n-type side leaves a positive space charge of $+eN_d$ per unit volume, as shown in Fig. 17.18. We now apply Poisson's equation to the barrier layer, making the same

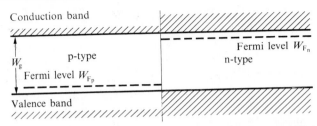

FIG. 17.16. Energy bands and Fermi level at p–n junction, before equilibrium is established.

kind of simplifying assumptions as in the treatment of the metal–semiconductor junction. Let the change from p- to n-type occur discontinuously at the plane $x = 0$, and let the barrier thickness be $\delta_p$ and $\delta_n$ on each side respectively. Then, taking $V = 0$ at $x = 0$ we have, using the boundary conditions $dV/dx = 0$ at $x = -\delta_p$ and at $x = +\delta_n$:

| $x < 0$ | $x > 0$ |
|---|---|

$$\frac{d^2V}{dx^2} = +\frac{|e|\,N_a}{\epsilon} \qquad\qquad \frac{d^2V}{dx^2} = -\frac{|e|\,N_d}{\epsilon}$$

$$V = \frac{|e|\,N_a}{2\epsilon}\{(x+\delta_p)^2 - \delta_p^2\} \qquad V = -\frac{|e|\,N_d}{2\epsilon}\{(x-\delta_n)^2 - \delta_n^2\}$$

Thus the over-all potential difference $V_0$ between the n-type material and the p-type material is the difference between $V_n$, the value of $V$ at $x = \delta_n$, and $V_p$ at $x = -\delta_p$. This is

$$V_0 = V_n - V_p = \frac{|e|}{2\epsilon}(N_d\delta_n^2 + N_a\delta_p^2). \qquad (17.36)$$

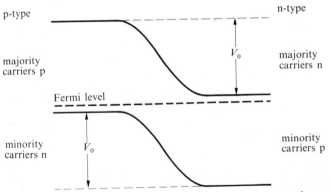

FIG. 17.17. Electron energy bands and the Fermi level at a p–n junction, after equilibrium is established. The n-type becomes positively charged, so that its electrons have lower potential energy.

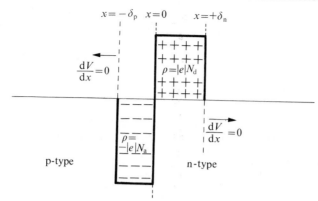

FIG. 17.18. Space-charge density at a p–n junction, on the simplified model used in the text.

From the fact that $dV/dx$ must be continuous at $x = 0$ we obtain the additional condition

$$N_a\delta_p = N_d\delta_n. \qquad (17.37)$$

This result makes the total charge on either side equal but opposite in sign, as must be the case if no lines of electric field emerge from the barrier layer. Finally, comparison of Figs 17.16 and 17.17 shows that

$$|e|(V_n - V_p) = W_{Fn} - W_{Fp} \approx W_g, \qquad (17.38)$$

where $W_{Fn}$, $W_{Fp}$ are the Fermi levels in n- and p-type as shown in Fig. 17.16; in the exhaustion range these lie close to the bottom of the conduction band and top of the valence band respectively, so that the energy difference is nearly equal to the energy gap. If we take

$$N_a = N_d = 10^{24} \text{ m}^{-3},$$

then for germanium the width of the barrier $(\delta_n + \delta_p)$ is found to be about 50 nm.

In the p-type material, holes form the 'majority carriers' but there are a small number of electrons, known as the 'minority carriers'. Similarly in the n-type material, electrons are the majority carriers and holes the minority carriers. If $p$ is the density of the majority carriers (holes) in the p-type material, then the density of holes (minority carriers) in the n-type material is $p_0 = p \exp(-|e| V_0/kT)$. There is a similar ratio for the electrons as minority and majority carriers, and the product of the hole and electron densities on each side is the same (see § 17.3).

The fact that the barrier layer has a very much higher resistivity than the bulk material means that any external voltage applied appears almost wholly across the barrier. In the bulk material the potential variation is practically zero, and currents near the barrier are the result of diffusion, being determined by the gradients in the numbers of holes and electrons. If an external voltage $V$ is applied which makes the n-type material less positive, the voltage across the barrier is reduced to $(V_0-V)$ and there is a larger number of holes with sufficient energy to cross from left to right. Since the minority carrier density is normally only about $10^{-5}$ of the majority carrier density, the latter does not change appreciably as a result of any transfer, and can be assumed to be constant. On the other hand, the minority carrier density is increased, so that the hole density at the right-hand edge of the barrier becomes

$$p_V = p \exp\{-|e|(V_0-V)/kT\} = p_0 \exp(+|e|V/kT).$$

A similar equation holds for electrons at the left-hand edge of the barrier, as shown in Table 17.4.

TABLE 17.4

*Densities of holes and electrons at barrier edges (Fig. 17.18), without and with an applied voltage*

|  | p-type edge | n-type edge | Applied voltage |
|---|---|---|---|
| Density of holes | $p$ | $p_0 = p \exp(-|e|V_0/kT)$ | 0 |
| Density of electrons | $n_0 = n \exp\{-(|e|V_0/kT)\}$ | $n$ | 0 |
| Density of holes | $p$ | $p_V = p \exp\{-|e|(V_0-V)/kT\}$ | $V$ |
| Density of electrons | $n_V = n \exp\{-|e|(V_0-V)/kT\}$ | $n$ | $V$ |

The densities $p_V$ and $n_V$ at the edges of the barrier are greater than the equilibrium densities $p_0$ and $n_0$ in the bulk material, and it follows from eqn (17.30) that there is a density gradient

$$\frac{dp}{dx} = \frac{-(p_V-p_0)}{L_h} = -\left(\frac{p_0}{L_h}\right)\left\{\exp\left(\frac{|e|V}{kT}\right)-1\right\}. \tag{17.39}$$

The current density resulting from diffusion of the holes is

$$J_h = -|e|D_h\left(\frac{dp}{dx}\right) = \frac{|e|D_h}{L_h}p_0\left\{\exp\left(\frac{|e|V}{kT}\right)-1\right\},$$

and there is a similar equation for electrons diffusing across the barrier to the left. Hence the total current density across the junction is

$$J = J_e + J_h = |e| \left( \frac{n_0 D_e}{L_e} + \frac{p_0 D_h}{L_h} \right) \left\{ \exp\left( \frac{|e| V}{kT} \right) - 1 \right\}, \qquad (17.40)$$

where $D_h$, $L_h$ are the diffusion coefficient and diffusion length for holes in the n-type material and $D_e$, $L_e$ the corresponding quantities for electrons in the p-type material. The entire current density across the junction arises from the flow of minority carriers, thus emphasizing the importance of their role.

This equation, due originally to Shockley, is similar to that for the metal–semiconductor junction, and the current–voltage characteristic (Fig. 17.15(b)) differs from Fig. 17.15(a) only in that the reverse current is much lower and the forward current much higher. There is a limit to the back voltage which can be applied because of breakdown (see § 18.3). The materials normally used for p–n junctions are silicon and germanium. Germanium has a smaller energy gap, giving a larger carrier density; the mobility is also rather higher, so that the forward resistance is lower than for silicon, but the reverse current is greater. Silicon has the advantage that it can be operated at higher temperatures before thermal ionization across the energy gap increases the back current and drops the rectification efficiency.

*Capacitance of a p–n junction*

The depletion layer with its high resistance acts as a capacitance (see Problem 17.4). The charge distribution shown in Fig. 17.18 assumes an abrupt change in the impurity concentration, and eqn (17.36) gives a voltage drop that varies with the square of the barrier thickness. In fact the process used in making p–n junctions produces an impurity density which changes almost linearly with distance, and the voltage across the barrier varies more nearly with the cube of the barrier thickness. Conversely the barrier thickness depends on the voltage, so that the barrier behaves as a capacitance which is proportional to $V^{-\frac{1}{3}}$. A p–n junction, with reverse bias to keep the flow of current small, acts as a non-linear capacitance, and is used in applications such as automatic frequency-control of radio receivers.

At high frequencies the barrier capacitance reduces the rectification efficiency of a p–n junction, since it takes a finite time to alter the charge density within the depletion layer when the applied voltage is reversed.

## References

HOUGHTON, J. T. and SMITH, S. D. (1966). *Infra-red physics*. Clarendon Press, Oxford.

ZIMAN, J. M. (1960). *Electrons and phonons*. Clarendon Press, Oxford.

## Problems

17.1. Show that the energy levels of an 'atom' consisting of a positive charge equal to that of the proton, with an effective mass $m_h^*$, and a negative charge equal to that of the electron, with an effective mass $m_e^*$, moving in a medium of relative permittivity, $\epsilon_r$, are those given by eqns (17.18) and (17.19).

Show that the binding energy of a donor impurity level in silicon ($m_h^* = \infty$, $m_e^*/m = 0.4$, $\epsilon_r = 11.5$) is about 0.041 eV, and that the Bohr radius is about 6.5 times the interatomic distance (0.235 nm).

17.2. In many semiconductors the energy gap varies linearly with $T$ over a fair temperature range, though it tends to a constant value at low temperatures. Show that if $W_g = W_g^0 - aT$, then a plot of $\ln(R_H T^{\frac{3}{2}})$ against $T^{-1}$ gives a straight line whose slope gives the value of $W_g^0$. Show also that the effect of the temperature variation of $W_g$ on eqn (17.22) is to replace $W_g$ by $W_g^0$ and to increase the apparent value of the products $(m_e^* m_h^*)$ by $\exp(2a/3k)$.

17.3. Using eqn (3.16) show that any abnormal charge distribution in germanium where $\epsilon_r = 16$, $\sigma = 10\ \Omega^{-1}\,m^{-1}$ vanishes in a time of order $10^{-11}$ s.

A sample of germanium contains $10^{18}$ holes per $m^3$. Show that the presence of a net space charge equivalent to 1 per cent of the hole concentration would give rise to an electric field gradient of about $10^7\ V\,m^{-2}$.

17.4. When an external bias voltage $V$ is applied to a p–n junction with a high resistance barrier, the voltage across the barrier in eqn (17.36) becomes $V_n - V_p = V + V_g$, where $|e|\,V_g = W_g$. Show that the total charge $\pm Q$ in each of the space-charge regions of Fig. 17.18, when $N_a = N_d = N_t$, is given by

$$Q = (\epsilon e N_t)^{\frac{1}{2}}(V + V_g)^{\frac{1}{2}} \text{ per unit area.}$$

This is a function of the applied voltage, and the barrier acts as a capacitance for a.c. voltages of magnitude

$$C = (dQ/dV) = \tfrac{1}{2}(\epsilon e N_t)^{\frac{1}{2}}(V + V_g)^{-\frac{1}{2}} \text{ per unit area.}$$

Verify that the capacitance is the same as that of a parallel-plate capacitor with plate separation equal to the barrier thicknesss and filled with dielectric of permittivity $\epsilon$.

17.5. In a crystal the velocity of a particle is given by the relation $\hbar \mathbf{v} = \text{grad}_k\, W$, where $\text{grad}_k = \mathbf{i}_x(\partial/\partial k_x) + \mathbf{i}_y(\partial/\partial k_y) + \mathbf{i}_z(\partial/\partial k_z)$, in which $\mathbf{i}_x$, etc., are unit vectors along the $x$, $y$, $z$ directions. Show that in general $\mathbf{v}$ is not parallel to $\mathbf{k}$ unless it is along one of the axes, for a particle whose energy $W$ is given by eqn (12.25).

Show that with respect to axes $(x', y', z')$ which are derived from the $(x, y, z)$ axes by a rotation through an angle $\theta$ about the $y$-axis, eqn (12.25) becomes

$$W = W_0 \pm \tfrac{1}{2}\hbar^2 \left\{ k_x'^2 \left( \frac{\cos^2\theta}{m_x^*} + \frac{\sin^2\theta}{m_z^*} \right) + \frac{k_y'^2}{m_y^*} + k_z'^2 \left( \frac{\sin^2\theta}{m_x^*} + \frac{\cos^2\theta}{m_z^*} \right) + \right.$$
$$\left. + 2k_x' k_z' \sin\theta \cos\theta \left( \frac{1}{m_x^*} - \frac{1}{m_z^*} \right) \right\}.$$

Hence show that the effective mass for a particle for which $\mathbf{k}$ is along the $z'$-axis is

$$\frac{1}{\hbar^2}\frac{\partial^2 W}{\partial k_z'^2} = \frac{\sin^2\theta}{m_x^*} + \frac{\cos^2\theta}{m_z^*}$$

($m^{*-1}$ is a tensor quantity, and cross-product terms such as $k'_x k'_z$, etc., are absent only when a suitable choice of axes (usually dictated by the crystal symmetry) is made).

17.6. For a semiconductor in the infrared, where $\omega\tau \gg 1$, the effective conductivity (see eqn (10.51)) becomes $\sigma = \sigma_0/(1 + \omega^2\tau^2)$. Show that an electromagnetic wave will fall in intensity inside the semiconductor as $W = W_o \exp(-\alpha x)$, where

$$\alpha \sim \frac{\sigma}{n\epsilon_0 c(1+\omega^2\tau^2)}$$

provided that $k \ll n$ in the complex refractive index $n - jk$. Calculate the value of $\alpha$ for a sample of Ge in which $n = 4$, $\sigma_0 = 10\ \Omega^{-1}\,\mathrm{m}^{-1}$, at a frequency where $\omega\tau = 100$.

(*Answer*: $\alpha = 10^{-1}\,\mathrm{m}^{-1}$, approximately.)

# **18.** Solid-state diodes

## 18.1. Introduction

ALMOST the first reasonably reliable detectors used in the reception of radio waves were 'crystal rectifiers', where a rectifying contact was made by a flexible wire 'cat's whisker' held lightly in contact with a semiconducting crystal. Good performance was obtained only from 'sensitive' spots on the crystal surface, requiring frequent readjustment of the contact. The invention of the thermionic diode by Fleming in 1904, and the development of a vast family of thermionic vacuum tubes, led to the crystal rectifier being entirely superseded, apart from laboratory work with very short wavelengths where thermionic devices failed, mainly because of the transit time of the electrons. Shortly before 1940, the invention of the cavity magnetron and klystron led to the widespread use of centimetre waves, where the crystal rectifier (§ 17.6) once more came into its own. The solid-state diode, with a junction between two different semiconducting substances, was developed just before 1950, and in turn has superseded the thermionic diode, being more versatile, smaller, cheaper, and more rugged.

Solid-state electronic devices are constructed from semiconducting materials, and their operation is dependent on the unique properties of semiconductors, discussed in Chapter 17. In this and the following chapters the operation of such devices is outlined in terms which do not require a detailed knowledge of the theory of semiconductors, apart from the introductory sections §§ 17.1–17.2. On this basis we start with a summary of the nomenclature commonly used and an elementary explanation of the action of the p–n junction. This is essential to the understanding of transistor action, and is a simplified version of the treatment in § 17.7.

A 'hole' is created when one electron is absent from a covalent bond of the pure crystal. A hole behaves as a positively charged carrier of current.

The 'intrinsic' energy gap is the energy required to remove an electron from such a chemical bond in a 'pure' semiconductor, creating one conduction electron and one hole. At room temperature, for silicon this energy is about 1·1 eV; for germanium it is about 0·7 eV. In a 'doped' semiconductor, a fraction of the atoms are replaced by atoms of a different valency. In silicon and germanium (valence 4), the addition of 'donor' atoms of valence 5 (such as phosphorus) produces an n-type semiconductor, while 'acceptor' atoms of valence 3 (such as aluminium) produce a p-type semiconductor. The fraction of impurity atoms varies from 1 in $10^9$ to 1 in $10^6$. The 'extrinsic' energy gap is the energy required to remove the excess electron from a donor atom (creating a conduction electron) or to move an electron to an acceptor atom (creating a hole) in a 'doped' semiconductor. For silicon this energy is of order of 0·04 eV, depending on the doping impurity.

In all semiconductors some carriers of opposite sign ('minority carriers') are present as the result of thermal excitation. In n-type semiconductors, electrons are the majority current carriers and holes the minority carriers. In p-type semiconductors, holes are the majority carriers and electrons the minority carriers. Both n- and p-type semiconductors ·are uncharged, each atom being on average electrically neutral.

## 18.2. The p–n junction

The p–n junction has been discussed in detail in § 17.7 and is shown diagrammatically in Fig. 18.1, where the region of the change in chemical composition from p-type to n-type is assumed to be infinitely narrow. The electron concentration on one side of the boundary and the hole concentration on the other side are each assumed to be constant. The electrons on the right and the holes on the left will each diffuse across the boundary until a potential difference is built up between the two sides, such that its effect on the charge carriers exactly balances the effect of the concentration gradient across the boundary. Then the net flow of charge (that is, the current) across the boundary is zero.

The electric potential V as a function of distance from the boundary is shown in Fig. 18.2. The region in which V is changing is only about 50 nm in width, so that in this region a strong electric field exists which sweeps away all mobile charges; this region is known as the 'depletion layer'. On either side of the boundary, but within the depletion layer, the lines of electric field terminate on layers of charge due to the fixed ions, positive in the n-type and negative in the p-type material. (The charge density $\rho$ is related to V by Poisson's equation $d^2V/dx^2 = -\rho/\epsilon$, so that the curvature of V has opposite sign in the two charge layers.) Outside the depletion layer the conductivity is sufficiently high that the charge density

| Majority carriers: | Holes | Electrons |
|---|---|---|
| | p-type | n-type |
| Minority carriers: | Electrons | Holes |

FIG. 18.1. Schematic diagram of a junction between a p-type semiconductor and an n-type semiconductor. The boundary is here assumed to be infinitely sharp; in practice it is about 10 nm thick.

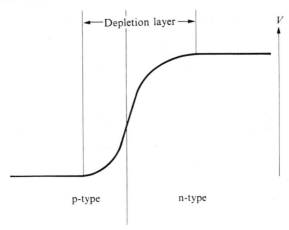

F$_{IG}$. 18.2. Variation of the electrostatic potential near the boundary between p-type and n-type material in a p–n junction. Outside the depletion layer the potential is constant when no current is flowing. The potential energy is $+|e|\,V$ for holes and $-|e|\,V$ for electrons. A positive voltage applied to the p-type material reduces the barrier height (forward bias).

is zero and $V$ is constant except for a small ohmic drop when current is flowing.

In equilibrium there is no net current, but both electrons and holes are crossing the boundary in each direction, giving a dynamic equilibrium. We consider first the holes. In spite of the potential barrier, some holes in the p-type material (coming predominantly from acceptor impurities) have sufficient energy to cross into the n-type material, where they disappear eventually by recombination with electrons. This constitutes the hole 'annihilation current'. In the n-type material, electron–hole pairs are being created, at a constant rate on average, through thermal excitation. In this region, the holes are minority carriers, and on arriving at the barrier they fall over the energy drop into the p-type material, giving a 'generation current'. Under equilibrium conditions the annihilation current and the generation current of holes are equal and opposite. Similar generation and annihilation currents exist for the electrons, which also balance in equilibrium, so that there is no *net* flow across the barrier either of electrons or of holes.

If the energy jump at the barrier is $W_B$, the chance that an electron has sufficient energy to cross it is proportional to $\exp(-W_B/kT)$, and the annihilation current $I_A$ is therefore proportional to this quantity. If in thermal equilibrium the annihilation current is $I_A^0$ and the energy jump is $W_B^0$, then we have

$$\frac{I_A}{I_A^0} = \exp\left(-\frac{W_B}{kT}\right)\Big/\exp\left(-\frac{W_B^0}{kT}\right) = \exp\left\{-\frac{W_B - W_B^0}{kT}\right\}. \qquad (18.1)$$

On connecting an external battery, the extra potential appears almost wholly across the high-resistance depletion layer, so that $W_B = W_B^0 - |e| V$ (see Fig. 18.3) for electrons. The barrier is smaller when $V$ is positive, and the annihilation current becomes

$$I_A = I_A^0 \exp(|e| V/kT). \qquad (18.2)$$

The presence of the exponential means that the annihilation current can change enormously under the influence of an external voltage. On the other hand, the generation current is determined solely by the rate at which electron–hole pairs are generated (which is determined by the temperature) and not by the energy change near the boundary, since the carriers are falling over into a region of lower energy. Hence $I_G$ is independent of $V$, and since $I_G = I_A^0$ when $V = 0$, the net electron current for finite $V$ is

$$I_e = I_A - I_G = I_G\{\exp(|e| V/kT) - 1\}.$$

A similar argument is valid for holes, giving a net current

$$I_h = I_G'\{\exp(|e| V/kT) - 1\}$$

and a total current

$$I = I_e + I_h = I_0\{\exp(|e| V/kT) - 1\}. \qquad (18.3)$$

The way in which $I$ varies with $V$ is shown in Fig. 17.15(b) for both positive and negative values of $V$. The exponential relation gives a very steeply rising curve for positive $V$, but the reverse current for negative $V$ is very small and substantially independent of $V$. The reason for this is that when $V$ is negative the exponential becomes vanishingly small and the current is simply the generation current arising simply from the random production of electron–hole pairs, giving charge carriers which fall over the energy drop at the barrier. On the other hand, the

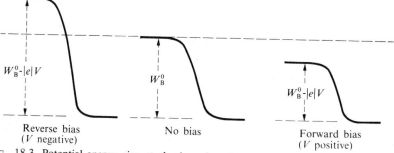

$W_B^0 - |e| V$

$W_B^0$

$W_B^0 - |e| V$

Reverse bias                    No bias                    Forward bias
(V negative)                                               (V positive)

FIG. 18.3. Potential-energy rise at the boundary in a p–n junction, with and without biasing voltages. For an electron, with negative charge, the potential-energy rise occurs when moving to the left, as shown. For a hole, with positive charge, the potential-energy rises when moving to the right, as in Fig. 18.2.

forward current for positive $V$ depends on the fraction of current carriers which have sufficient energy to surmount the barrier; this depends exponentially on $W_B$ and is very sensitive to the value of $V$. Thus the junction acts as an efficient rectifier. Unlike the thermionic diode, the reverse current is finite, but normally is only a fraction of a microampere.

When the forward voltage is sufficient to make the exponential the dominant term in eqn (18.3), we can write to a good approximation that

$$r_0 = \frac{dV}{dI} = \left(\frac{kT}{e}\right)\frac{1}{I} = \frac{25 \cdot 6}{I_{mA}} \,\Omega \text{ at } T = 300 \text{ K.} \tag{18.4}$$

This expression was derived by Shockley for the input resistance of a junction diode in the forward direction, and it is found to hold for any p–n junction provided the current does not become so large that it is limited by ohmic resistance. Clearly, the forward resistance is very low, being just a few ohms for a current of about 10 mA. This forward resistance is only realized provided that the external connections to the semiconductor needed to complete the circuit are good 'ohmic' contacts, that is, they must have no rectifying action and negligible resistance. This is achieved by heavy doping of the area around the contact, using impurity concentrations of $\sim 10^{26}$ atoms per cubic metre (over one part in $10^3$), so that electrons can readily tunnel between semiconductor and metal.

The symbol used to denote a solid-state diode is shown in Fig. 18.4, together with the batteries needed for forward and reverse bias. Fitted with large metal cooling fins, with forced air or water cooling to keep the junction temperature within safe limits, solid-state diodes can be used for currents up to 1000 A, and they have largely replaced thermionic diodes. An external resistance is sometimes used in series with the diode to limit the forward current to a safe value, since at 1000 A the value of $r_0$ in eqn (18.4) is $25 \times 10^{-6} \,\Omega$, and the power dissipation is about 25 W.

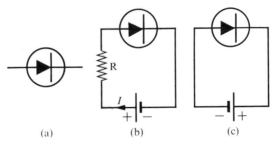

(a)                    (b)                    (c)

FIG. 18.4. The solid-state diode rectifier: (a) the symbol; (b) forward-bias connection, with resistance $R$ to limit the forward current; (c) reverse-bias connection. The direction of 'easy' current flow is that indicated by the direction of the apex of the triangle regarded as an arrow head.

## 18.3. The Zener diode

At low reverse voltages the current in a p–n junction is substantially constant; for silicon it is of order $0.001 \ \mu A$ and for germanium $0.1 \ \mu A$. However, if the reverse voltage exceeds a certain critical value, the current suddenly increases enormously and the resistance of the diode falls to a very low value. The current–voltage characteristic in this region is shown in Fig. 18.5. Two separate processes are involved in this rapid increase in the current and it is not always clear which is predominant.

The first effect which causes a rapid increase in current is an 'avalanche process'. In the depletion layer at a p–n junction, if the current carriers are accelerated by a high electric field, they gain sufficient energy to cause 'ionization' and produce further electron–hole pairs. The critical energy for this process is somewhat greater than the minimum energy required to create an electron–hole pair, since momentum must be conserved as well as energy in the collision. When a few electrons have produced new pairs of electrons and holes, these are accelerated to produce further ionization, giving an exponential increase of current by an 'avalanche' effect similar to that in the Townsend discharge in a gas (§ 3.8). The energy acquired by an electron between collisions is determined by the product of the electric field and the mean free path, which at room temperature is substantially independent of the degree of doping, and the avalanche occurs at a certain critical field which is of the order of $10^7 \ V \ m^{-1}$. By adjusting the doping to control the barrier thickness, the breakdown voltage can be as much as $100 \ V$, but it is temperature-dependent because the mean free path varies with temperature. To allow sufficient ionizing collisions to occur, the barrier thickness must be large, and it is increased by using only light doping (see eqns (17.36)–(17.37)).

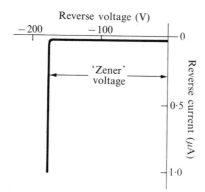

FIG. 18.5. Current–voltage characteristic of a Zener diode.

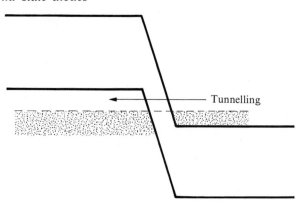

FIG. 18.6. With sufficient reverse bias the valence energy band in the p-type material is raised above the bottom of the conduction band in the n-type material, and electrons can tunnel from the latter into the former.

The second process is a 'tunnel effect' which was suggested by Zener in 1934. Electrons can tunnel through a thin barrier provided that an empty state of the same energy exists on the other side. This occurs if the applied potential is large enough to raise the top of the valence band on one side of a junction above the bottom of the conduction band on the other side, as in Fig. 18.6. The probability of tunnelling is greatly increased if the barrier is very thin, and it then predominates over the avalanche effect because electrons do not gain sufficient energy to cause ionizing collisions within the depletion layer. The breakdown voltage is rather low (a few volts), and a thin barrier is produced by using heavy doping of either side.

Above the critical voltage the current increases very rapidly but the voltage drop across the diode remains essentially constant. 'Zener diodes' (a generic title which includes those utilizing the avalanche effect) are used as control elements, either as switches to control a circuit where the voltage must not exceed a certain value, or to produce a constant reference voltage. The symbol used to distinguish a Zener diode from an ordinary diode is shown in Fig. 18.7, and this diagram illustrates how a

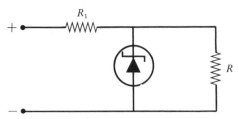

FIG. 18.7. A Zener diode, as used to keep a constant voltage across the resistance R. The resistance $R_1$ protects the diode from excessive current flow.

Zener diode is used to maintain a constant voltage across the load resistance R.

## 18.4. Rectification

If an alternating voltage is applied to a diode, a high current flows in the forward direction, but there is very little current in the reverse direction provided that the voltage amplitude does not exceed the break-down value. A diode thus produces a current flow almost wholly in one direction, and acts as a rectifier. A single diode gives only 'half-wave rectification' (as in Fig. 18.8(a)), and the output contains large a.c. components at the applied frequency, so that considerable smoothing is needed to produce a steady output. A much better circuit, shown in Fig. 18.9, consists of four diodes arranged in the form of a bridge. Both halves of the alternating waveform are rectified (see Fig. 18.8(b)) in this 'full-wave bridge rectifier'. An alternating voltage $V_0 \cos \omega t$, whose amplitude is a little larger than the required steady output voltage, is applied across the terminals PQ in Fig. 18.9 from a suitable transformer. When P is positive and Q negative, diodes $D_1$ and $D_3$ conduct and current charges the large capacitance C. When P is negative and Q positive, diodes $D_2$ and $D_4$ conduct, so that current again charges the capacitance in the same direction. Charge continues to accumulate on the capacitance until the voltage across it becomes equal to $V_0$. In alternate half-cycles the peak reverse voltage across each diode, when it is not conducting, reaches $V_0$, and the diodes must be able to withstand this reverse voltage without breakdown.

If the capacitance C has no leakage and the reverse current in the diodes is zero, C would remain charged to the greatest peak voltage applied to the system. Measurement of this steady voltage determines the amplitude of the alternating voltage, but such a 'peak voltmeter' requires

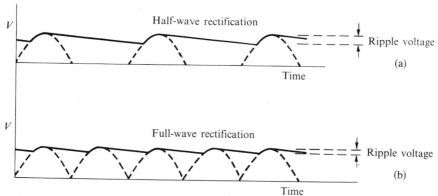

FIG. 18.8. Half-wave and full-wave rectification. $V$ = voltage across load resistance $R$.

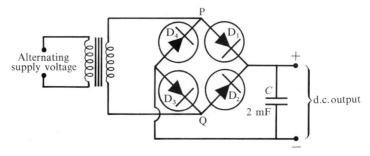

FIG. 18.9. Full-wave bridge rectifier circuit. Normally the d.c. output terminals are connected to a smoothing circuit to eliminate the ripple voltage.

a thermionic diode with zero reverse current. If the voltmeter used has a finite resistance $R$, which is in parallel with the capacitance $C$, the circuit is able to follow changes in the peak voltage so long as they do not occur within a time of order $RC$, the time constant of the $R-C$ combination.

In general the circuit of Fig. 18.9 is used to deliver direct current into a load $R$. Under these conditions the capacitor will discharge slightly through the resistance during that part of the cycle when the diode is not conducting, being recharged to the peak voltage when the diode conducts. The voltage across the load is therefore not constant, but contains a component fluctuating at the frequency of the applied alternating voltage, as in Fig. 18.8. This component is not sinusoidal, owing to the asymmetrical nature of the capacitor charge–discharge system. The 'ripple voltage', as the fluctuating component is generally called, becomes larger if $R$ is reduced, since the time constant of the $R-C$ combination is then smaller, and the capacitor discharges to a lower voltage before being recharged. The ripple voltage is therefore more serious when the diodes are on load. The ripple voltage must be eliminated or very considerably reduced if the system is used to supply d.c. power for an amplifier or other electronic device, since any alternating voltage applied to the early stages will be greatly magnified at the output.

The advantage of full-wave rectification over half-wave rectification can be seen from a comparison of the two when the capacitance $C$ is in parallel with a load resistance $R$. With the full-wave rectification the capacitor is charged to the peak voltage by the passage of current through one pair of diodes during the first half of the cycle, and is then again charged during the second half-cycle by current flowing through the second pair. If the time constant of the $R-C$ combination is long compared with the period of the supply voltage, so that the voltage drop during the discharge period is only a small fraction of the initial peak voltage, then the voltage on $C$ will fall nearly linearly with time during the discharge interval. With full-wave rectification, this interval is only half a cycle, and the size of the ripple is thus only half as great as with half-wave rectification (see Fig. 18.8). In addition its fundamental frequency is now twice the supply frequency, making the filtering problem easier.

A simple type of filter consists of a low-pass section as shown in Fig. 9.1. For supply frequencies of 50–60 Hz this requires $L \sim 25$ H and $C \sim 10$ $\mu$F. Such an inductance is heavy, bulky, and expensive, and the filter circuit is often replaced by a voltage regulator such as a Zener diode whose breakdown voltage is 1 or 2 V lower than $V_0$, using the circuit of Fig. 18.7.

## 18.5. Amplitude modulation and detection

The transmission of intelligence by a radio wave requires some form of modulation. The radio wave is normally at a high frequency known as the carrier frequency; the modulation is at a much lower frequency, such as an audio-frequency, when the information to be conveyed is speech or music. One system used is called amplitude modulation, since the amplitude of the carrier signal is made to vary with the period of the audio-frequency, and by an amount which is proportional to the strength of the audio-frequency information. For simplicity we shall consider only a single audio-frequency of constant strength. The amplitude-modulated radio signal may then be written in the form

$$V = A(1+m \cos pt)\cos \omega t. \tag{18.5}$$

Here $A$ is the amplitude of the carrier signal in the absence of modulation, and $\omega/2\pi$ its frequency. The constant $m$ is known as the depth of modulation, and cannot be greater than unity, and $p/2\pi$ is the audio-frequency.

The nature of an amplitude-modulated signal can be seen from Fig. 18.10(a), which shows the variation of the voltage $V$ with time. Its amplitude fluctuates slowly between a maximum value of $A(1+m)$ and a minimum of $A(1-m)$, the period of a complete cycle of this fluctuation being $2\pi/p$. Manipulation of eqn (18.5) shows that it may be rewritten as

$$V = A \cos \omega t + \tfrac{1}{2}mA \cos(\omega+p)t + \tfrac{1}{2}mA \cos(\omega-p)t. \tag{18.6}$$

This indicates that the modulated signal may also be regarded as composed of the carrier signal $A \cos \omega t$, together with two other frequencies, higher and lower by $p/2\pi$, which are known as side-bands and whose amplitude is proportional to the product of the carrier strength and the depth of modulation. The presence of these side-bands shows that any receiver with r.f. circuits must be designed to have a pass band which will accept the frequencies $(\omega \pm p)/2\pi$ as well as the carrier frequency $\omega/2\pi$; otherwise the audio-frequency modulation will be cut out. The presence

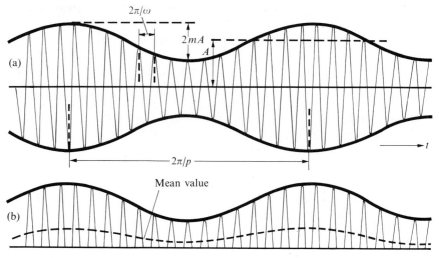

F<small>IG</small>. 18.10. (a) Amplitude-modulated signal, before detection; (b) amplitude-modulated
signal, after detection. Normally $\omega$ is much greater than $p$.

of the side-bands may be demonstrated by applying the modulated signal
to a sharply tuned frequency-meter, which will show responses at the
three frequencies $(\omega-p)/2\pi$, $\omega/2\pi$, and $(\omega+p)/2\pi$.

In general the modulation will not consist of a single audio-frequency,
but of a whole range of frequencies. For speech or music these cover the
range from about 50 Hz to several kilohertz, while for television over a
million pieces of information are transmitted per second, and a band of
several megahertz is required. By Fourier analysis any modulation can
always be resolved into a set of sinusoidal oscillations, and our analysis
can therefore proceed in terms of one such frequency, bearing in mind
that the various parts of a receiver must have the bandwidth required to
accommodate all modulation frequencies up to the highest.

The mean value of the signal voltage $V$ is zero over any period long
compared with that of the carrier frequency, and it will therefore produce
no effect in a receiver designed to accept only audio-frequencies. If the
signal is passed through a rectifier stage so that the portions where $V$ is
negative are wiped out, as in Fig. 18.10(b), the mean value of the
resultant is not zero and fluctuates at the audio-frequency rate corresponding to the modulation. This process is known as detection, since the
information which is conveyed by the modulation can now be detected by
the ear if the signal from the rectifier, after suitable amplification, is
applied to headphones or a loudspeaker. An obvious requirement of a
detector is that its output signal shall be nearly as possible a true
reproduction of the original modulation, that is, the output voltage should

be linearly proportional to the depth of modulation *m*, and the constant of proportionality should be the same for all modulation frequencies.

A circuit using a diode for the detection of amplitude-modulated waves is shown in Fig. 18.11. It is a half-wave rectifier circuit, and certain limitations must be placed on the values of *R* and *C* to obtain efficient and distortionless detection. These may be summarized as follows:

1. The load resistance *R* should be large compared with the effective output resistance *r* of the diode. The latter is approximately equal to the reciprocal of the slope of the diode characteristic, and forms a voltage divider with *R*. Since *r* varies with the size of the applied signal, the condition $R \gg r$ not only makes the fraction of the possible output voltage appearing across *R* nearly unity (high efficiency) but also makes this fraction nearly independent of *r* and hence of the magnitude of the applied signal (low distortion).

2. The time constant of the *RC* circuit should be long compared with the period of the carrier voltage, to avoid voltages of this frequency appearing in the output (that is, $1/\omega C \ll R$).

3. An upper limit to *RC* is set by the requirement that the voltage across the capacitance *C* shall change sufficiently rapidly to follow the modulation. This requires $1/pC < R$, but this is only approximate; a more accurate treatment shows that the inequality must depend on the size of *m*, because the rate of change of the carrier amplitude depends on the depth of modulation.

4. *C* should be several times as large as the electrode capacitance $C_{ca}$ of the diode, since *C* and $C_{ca}$ form a voltage divider for the r.f. voltage applied to the diode.

The condition $R \gg r$ (see point 1, above) makes it necessary for the size of the input voltage to be of the order of a volt or so, in order to work on

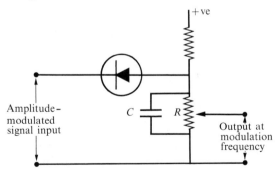

FIG. 18.11. Diode circuit for detection of amplitude-modulated signal. For audio-frequency modulation, typical values of *R*, *C* are $R = 10\,\text{k}\Omega$, $C = 10\,\text{nF}$, giving a time constant of $10^{-4}\,\text{s}$. Thus *C* behaves as a low-impedance by-pass to *R* for the radio-frequencies but not for the audio-frequencies. The single diode behaves as a linear 'half-wave' rectifier provided that the amplitude of the input signal is sufficient (see § 18.5).

a portion of the diode characteristic where the slope is fairly high. In the reception of broadcast signals ranging from millivolts down to microvolts, it is therefore necessary to amplify the signal before detection.

*Linear and square-law detection*

The use of the diode described above, where the applied signal is large enough to operate the diode on the straight part of its characteristic, is known as 'linear detection', since the output voltage is linearly proportional to the amplitude of the input voltage. If the input voltage is very small, as would be the case if a broadcast signal were applied to the diode directly without previous amplification, the diode is operated only over a very tiny portion of its characteristic, and detection or rectification results from the curvature of the characteristic of this region. The output current or voltage is proportional to the square of the input voltage, and the process is known as 'square-law detection'. An approximate analysis may be made by assuming the load resistance is small compared with the mean output resistance of the diode; the latter is very high when the applied signal is small. Then the current through the diode when a small signal voltage $v$ is applied may be written as

$$I = I_0 + \left(\frac{\partial I}{\partial V}\right)v + \frac{1}{2}\left(\frac{\partial^2 I}{\partial V^2}\right)v^2 + \ldots = I_0 + av + bv^2 + \ldots, \qquad (18.7)$$

where $I_0$ is the current flow (if any) when $v = 0$, and $a$ and $b$ are determined by the slope and curvature of the characteristic near the point $I = I_0$. If $v = v_1 \cos \omega t$, then

$$i = I - I_0 = av_1 \cos \omega t + \tfrac{1}{2}bv_1^2(1 + \cos 2\omega t) + \ldots, \qquad (18.8)$$

showing that there is a change $\tfrac{1}{2}bv_1^2$ in the mean current, which is proportional to the square of the applied signal. If the latter is modulated, so that $v_1 = B(1 + m \cos pt)$, then the low-frequency current change is $\tfrac{1}{2}bB^2(1 + 2m \cos pt + \tfrac{1}{2}m^2 + \tfrac{1}{2}m^2 \cos 2pt)$, showing that the detected signal will have harmonic distortion owing to the presence of the term in $\cos 2pt$. For this reason, and because of the very low efficiency, square-law detection is not used in radio reception.

## 18.6. Frequency changing

Since square-law detection is very inefficient compared with linear detection (see Problem 18.1), it is always desirable that a signal be amplified sufficiently, before being applied to the detector, to work the latter in its linear region. Often it is undesirable, and sometimes impossible, to provide sufficient amplification for this purpose at the carrier frequency. A device known as frequency-changing is then used, in which, as the name suggests, the carrier frequency is altered to another more convenient frequency, the modulation being preserved intact. In eqns (18.7) and (18.8) for a modulated signal, this means that $\omega$ is changed to another value, but that the terms in $m$ remain the same.

This change of frequency is accomplished by adding to the original signal an alternating voltage of another frequency $(\omega_1/2\pi)$ generated locally, and passing the two into a rectifying stage known as the mixer. The output from the mixer then contains voltage components which fluctuate at, apart from the modulation frequencies, $(\omega_1 \pm \omega)/2\pi$ (see below). If the difference frequency $(\omega_1 - \omega)/2\pi$ lies in the audible range it may be amplified and made to work headphones or a loud-speaker. This system is known as heterodyne reception. In a superheterodyne system, the sum and difference frequencies are outside the audible range, and one of them is selected and amplified. This frequency is known as the intermediate frequency (i.f.) and the i.f. amplifier magnifies the signal, with its original modulation, to a level at which it can be detected by a diode operating in the linear region. Since the mixing stage must incorporate a non-linear device, it is often called the 'first detector', while that following the i.f. amplifier is called the 'second detector'.

The operation of frequency-changing (or 'frequency-conversion') can be readily understood as follows. Suppose an alternating voltage $v_1 \cos \omega_1 t$ is supplied by a 'local oscillator', and to this is added a small signal voltage $v \cos \omega t$, where $v \ll v_1$ and $\omega_1$ is close to $\omega$. Then the total amplitude of the alternating voltage will fluctuate between $(v_1 + v)$, when the two components are in phase, and $(v_1 - v)$ when they are out of phase. The time interval between instants at which the two are in phase is $2\pi/(\omega_1 - \omega)$; the amplitude therefore fluctuates sinusoidally at a frequency equal to the difference of the two original components, and the size of the fluctuation is the same as that of the signal voltage $v$. This constitutes an amplitude-modulated voltage which can be detected as described in the last section, the difference being that the 'modulation' frequency is determined by the difference between the signal and local oscillator frequencies. The amplitude of the local oscillator voltage $v_1$ may be adjusted so that the detector is worked on the linear portion of its characteristic, and the 'modulation' of the local oscillator voltage pro-duced by the signal appears in the output as a component at the i.f. frequency whose amplitude is proportional to that of the original signal. Any slow fluctuation of the latter, such as that due to an audio-frequency amplitude modulation, is preserved, and the signal at the i.f. amplifier differs from the original only in the frequency of the carrier voltage.

For mathematical analysis, the action of the local oscillator voltage on the detector may be assumed to produce a periodic variation of its slope conductance $dI/dV$ which may be analysed as a Fourier series of the form

$$g = g_0 + g_1 \cos \omega_1 t + g_2 \cos 2\omega_1 t + \ldots . \qquad (18.9)$$

The effect of adding a small voltage $v \cos \omega t$ is to change the detector current by an amount

$$gv \cos \omega t = g_0 v \cos \omega t + g_1 v \cos \omega_1 t \cos \omega t + \ldots$$

$$= g_0 \cos \omega t + \tfrac{1}{2} g_1 v \{ \cos(\omega + \omega_1) t + \cos(\omega - \omega_1) t \} + \ldots, \quad (18.10)$$

showing that there are Fourier components at both the sum and difference of the signal and local oscillator frequencies. The i.f. amplifier may be tuned to accept either of these; components at $(n\omega_1 \pm \omega)$ also exist, but they are usually small because the coefficients $g_n$ decrease in magnitude as $n$ increases. This analysis shows that either the sum or the difference frequency may be used, although in the previous discussion only the difference term was considered. In work at very high frequencies the difference is generally used, since this is more convenient in building an i.f. amplifier. In addition, where selectivity is required, it is easier to get a narrow pass band from circuits at the lower frequency. For example, a $Q_F$ of 100 would give a pass band of about 10 kHz in the circuits of an i.f. amplifier at 1 MHz, whereas to obtain the same limited band at a radio-frequency of say 100 MHz, would require a $Q_F$ of $10^4$. In a superheterodyne receiver using such frequencies, the local oscillator might be at 99 MHz or 101 MHz, and the r.f. circuits would have to be sufficiently sharply tuned to reject any signal at 98 MHz or 102 MHz respectively, which would produce the same beat frequency. This requirement is known as 'second channel suppression'. In general it means that the tuned r.f. circuits for the signal frequency $\omega/2\pi$ must be sufficiently selective to reject any unwanted signal at the 'image frequency' $(2\omega_1 - \omega)/2\pi$, which is separated from the local oscillator frequency $\omega_1/2\pi$ by the same amount, and hence would also be accepted by the i.f. amplifier. A circuit using a diode rectifier for frequency-changing is shown in Fig. 18.12.

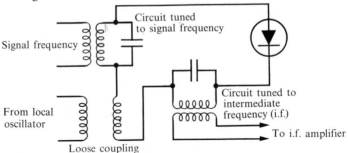

FIG. 18.12. Diode frequency-changer circuit.

## 18.7. Frequency modulation

'Frequency modulation' is an alternative method of conveying intelligence by means of a radio wave, in which the signal wave has a constant amplitude, but its frequency is varied periodically in accordance with the modulating signal. The amount of frequency variation is proportional to the amplitude of the modulating voltage, and the rate of variation is proportional to the modulating frequency. The unmodulated carrier wave, for which $\omega$ is constant, may be written as

$$V = A \cos \phi(t) = A \cos \omega t,$$

where the function $\phi(t) = \int \omega \, dt$. If the frequency of this carrier wave is modulated by a single audio-frequency $(p/2\pi)$ of constant amplitude, then the instantaneous angular frequency becomes

$$\omega_i = \omega + \Delta\omega \cos pt,$$

where $\Delta\omega$ is the maximum deviation of $\omega_i$ from $\omega$, the frequency of the unmodulated carrier. Then for the frequency-modulated wave

$$V = A \cos\left(\int \omega_i \, dt\right) = A \cos\left(\omega t + \frac{\Delta\omega}{p}\sin pt\right) = A \cos(\omega t + m_f \sin pt).$$

$$(18.11)$$

The quantity $m_f = \Delta\omega/p$ is called the modulation index. For example, if the unmodulated carrier wave has a frequency $\omega/2\pi = 10^8$ Hz, and the modulation is at a frequency $p/2\pi = 500$ Hz, and the modulation index is $m_f = 0.04$, then the modulated carrier wave will vary in frequency from $(10^8 + 20)$ Hz to $(10^8 - 20)$ Hz and back again 500 times a second. On the other hand, if the modulation index is 20, the frequency of the carrier varies from $(10^8 + 10^4)$ Hz to $(10^8 - 10^4)$ Hz and back again 500 times a second.

Since a frequency-modulated wave is not a simple sine wave, it contains side-bands, which are more complicated than those for an amplitude-modulated wave. By Fourier analysis it may be shown that the voltage waveform of eqn (18.11) can be written as

$$V = AJ_0(m_f)\cos \omega t + AJ_1(m_f)\{\cos(\omega + p)t - \cos(\omega - p)t\} +$$

$$+ AJ_2(m_f)\{\cos(\omega + 2p)t + \cos(\omega - 2p)t\} + \ldots$$

$$= A\left[J_0(m_f)\cos \omega t + \sum_{n=1}^{\infty} J_n(m_f)\{\cos(\omega + np)t + (-1)^n \cos(\omega - np)t\}\right].$$

$$(18.12)$$

Here the numerical coefficients $J_n(m_f)$ can be found from tables of Bessel functions, for $J_n$ is a Bessel function of order $n$. Although the side-band frequencies stretch to infinity, the more distant side-bands have small intensity. If the modulation index $m_f = 0.5$, the first-order side-bands $(\omega \pm p)$ have amplitude $0.24$, and the second-order side-bands $(\omega \pm 2p)$ have amplitude $0.03$ relative to the unmodulated carrier; higher-order side-bands are negligible. If $m_f = 5$, the amplitudes of the side-bands are larger, as shown in Fig. 18.13; the amplitude of the carrier is markedly reduced, and most of the energy is in the side-bands (the total energy is independent of $m_f$). This represents an economy in transmitter power over amplitude modulation, where the carrier wave is fixed in amplitude and carries half the energy even with 100 per cent depth of modulation. As a rough rule the width of the frequency band over which the side-bands have appreciable amplitude is approximately

$$2\left\{\left(\frac{\Delta\omega}{2\pi}\right) + \left(\frac{p}{2\pi}\right)\right\}.$$

In a typical system for transmitting speech and music, the maximum frequency deviation $(\Delta\omega/2\pi)$ is $\pm 75 \text{ kHz}$, and the maximum audio-modulation frequency 15 kHz, so the bandwidth required is

$$2(75+15) = 180 \text{ kHz.}$$

Though the bandwidth required is thus considerably greater than for transmission of an amplitude-modulated wave with the same maximum audio-frequency, a frequency-modulation system has the great advantage in that it cuts out all amplitude-modulated disturbances caused by interference and noise, and so gives much improved reception. The carrier frequencies used for frequency-modulation transmission are high $(\approx 100 \text{ MHz})$, partly because the fractional frequency deviation $(\Delta\omega/\omega)$ is then small and easier to realize in transmission, and partly because only the direct ray from the transmitter is then received. Any ray received

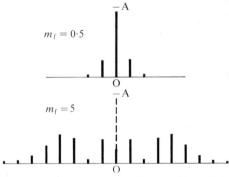

FIG. 18.13. Side-bands in a frequency-modulated wave with modulation index 0·5 and 5 respectively. OA = amplitude of unmodulated carrier.

indirectly (for example, by reflection from the ionosphere) would be more seriously distorted by selective fading (unequal transmission of different frequencies) than in an amplitude-modulated system, because of the greater bandwidth required.

In the reception of frequency-modulated transmission it is necessary to convert the frequency modulation into an amplitude modulation, and this is accomplished by a 'discriminator'. Several types of discriminator are in use, the essential ingredient being a circuit whose impedance depends on frequency. The system shown in Fig. 18.14 depends on the variation in phase of the voltage across a circuit, tuned to a centre frequency $\omega$, as the frequency departs from this value. The secondary winding (inductance $L$) of a transformer is tuned to frequency $\omega$ by the capacitance $C$, across which the voltage is $V_S$. If $I_P$ is the primary current, and $M$ the mutual inductance between primary and secondary, then in the secondary circuit

$$0 = j\omega M I_P + \mathbf{Z} I_S = j\omega M I_P + j\omega C \mathbf{Z} V_S,$$

so that

$$V_S = -(M/C\mathbf{Z})I_P. \tag{18.13}$$

If $M$ is kept below the value for critical coupling (§ 7.4), the secondary resonance curve has just a single peak (see Fig. 7.10). For simplicity we neglect the impedance reflected from the primary into the secondary circuit; then near resonance, where $(\delta\omega/\omega)Q_F$ is small, we have, from eqn (7.20),

$$\mathbf{Z}^{-1} = R^{-1}\{1 - 2j(\delta\omega/\omega)Q_F\},$$

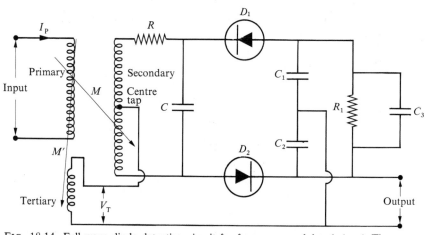

FIG. 18.14. Full-wave diode detection circuit for frequency-modulated signal. The secondary winding of the transformer is tuned by capacitance $C$ to the centre frequency $\omega$, and at this frequency the voltage across the secondary $V_S$ is in quadrature with the voltage $V_T$ from the tertiary coil (see text and Fig. 18.15). Unwanted amplitude-modulation signals are suppressed by $R_1$, $C_3$ for which typical values are $R_1 = 20\ \text{k}\Omega$, $C_3 = 10\ \mu\text{F}$, giving a time constant $R_1 C_3 = 0.2\ \text{s}$.

giving

$$V_S = -(M/CR)\{1-2j(\delta\omega/\omega)Q_F\}I_P. \qquad (18.14)$$

This result shows that the secondary voltage has an imaginary component which varies linearly with $\delta\omega$ and vanishes when $\delta\omega = 0$.

In Fig. 18.14 the secondary winding is centre-tapped, so that equal and opposite voltages $\pm\frac{1}{2}V_s$ are applied to the diodes $D_1$, $D_2$. A third circuit, coupled to the primary by the mutual inductance $M'$, develops a voltage $V_T = j\omega M'I_P$ which is applied equally to the two diodes, being connected to the centre-tap. At frequency $\omega$, the net voltages $V_1$, $V_2$ across the diodes are equal, as in Fig. 18.15(a). When $\delta\omega \neq 0$, the voltages $V_1$, $V_2$ vary in opposite directions as in Fig. 18.15(b), oscillating with $\delta\omega$. Since $\delta\omega = \Delta\omega \cos pt$, the variation is at the modulation frequency $p$. From the relation

$$V_1^2 = (V_T + \tfrac{1}{2}\delta V_S)^2 + (\tfrac{1}{2}V_S)^2$$

we have for small changes

$$2\,\delta V_1 = (V_T/V_1)\,\delta V_S = \frac{V_T}{V_1}\left(\frac{M}{CR}\right)2j\left(\frac{\delta\omega}{\omega}\right)Q_F I_P, \qquad (18.15)$$

showing that the voltage variation is linear with $\delta\omega$.

The voltages $V_1$, $V_2$ are rectified by the diodes, and the modulation-frequency signal is taken from across $C_1$ or $C_2$, as in Fig. 18.14. If an unwanted amplitude modulated signal is present, its effect is periodically to increase and decrease both $V_T$ and $V_S$ in phase; the circuit $R_1$, $C_3$ is arranged to have a long time constant to suppress such modulation in the output signal.

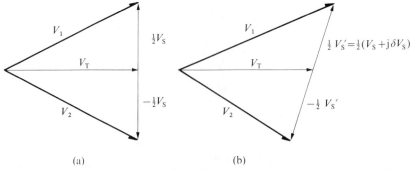

(a)                                    (b)

FIG. 18.15. Vector diagram showing the voltage $V_T$ from the tertiary coil and the voltages $\pm\frac{1}{2}V_s$ from the centre-tapped secondary coil of Fig. 18.14. In (a) $\frac{1}{2}V_s$ is exactly in quadrature with $V_T$ when the secondary circuit is tuned exactly to resonance with the centre frequency $\omega$. In (b) the secondary voltage has a component $\frac{1}{2}\delta V_S$, proportional to the frequency deviation $\delta\omega$; $\frac{1}{2}\delta V_S$ is in quadrature with $V_S$ and hence in phase with $V_T$. $V_1$, $V_2$ are the voltages applied to the two diodes.

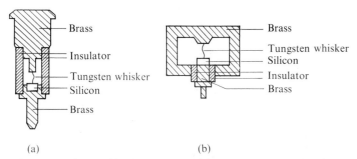

(a)                                        (b)

FIG. 18.16. Crystal diodes. (a) Capsule type: (b) waveguide mounting for millimetre wavelengths.

## 18.8. The point-contact diode

As mentioned in § 17.7, the barrier capacitance of a p–n junction reduces its efficiency at high frequencies. For a barrier thickness of 50 nm and an area of $1 \, \text{mm}^2$ the capacitance is $\sim 1000 \, \text{pF}$, which has an impedance of about $1 \, \Omega$ at a frequency of 100 MHz, so that little voltage appears across the barrier. By using a fine tungsten wire in light contact with the semiconductor surface, the capacitance can be reduced to $0 \cdot 1 \, \text{pF}$ or less, and efficient rectification can be obtained up to about $10^5 \, \text{MHz}$. With such a small contact area there is an appreciable 'spreading resistance' $(\sim 20 \, \Omega)$ to current from the contact flowing into the bulk semiconductor. This limits the slope conductance in the forward direction to values much smaller than in the p–n diode (see Fig. 17.15); the back current is also considerably higher. For good rectification efficiency, about a milliwatt of power is needed, and for small signals the diode is normally used in a frequency-changing circuit. Fig. 18.16 shows two typical mountings; a capsule type for wavelengths of 100 mm and above (an alternative coaxial construction is preferred for wavelengths $\sim 10 \, \text{mm}$), and a waveguide mounting for millimeter waves. The theory of the metal–semiconductor junction is discussed in § 17.6.

## Problem

18.1. An amplitude-modulated voltage signal $v = B(1 + m \cos pt)\cos \omega t$ is applied to two different receivers: (1) a diode detector with the characteristic given by eqn (18.7), working in the square-law region, followed by an audio-frequency amplifier with overall amplification $A$; (2) a signal frequency amplifier with overall amplification $A$, followed by the same diode working in the linear region. Show that the output voltages (assuming that the diode works into a resistance $R$ in each case) from the two systems are in the ratio $bB:a$, and show that if $B = 10^{-5} \, \text{V}$, $a = 10^{-3} \, \text{A V}^{-1}$, $b = 10^{-5} \, \text{A V}^{-2}$ the ratio is $10^{-7}$. This illustrates the inefficiency of square-law detection.

# 19. Solid-state triodes

## 19.1. The junction transistor

THE junction transistor is a single crystal of semiconductor material in which different regions have been doped by a diffusion process in such a way that a thin layer of p-type material is produced between two regions of n-type material, or vice versa. For convenience, we shall restrict our discussion to the former, but it may readily be applied to the latter by interchanging the roles of holes and electrons. It is clear from Fig. 19.1 that a transistor may be regarded as two p–n junctions back-to-back, the material (p-type in our example) in the thin central section being common to both junctions. It is important that this central section (known as the 'base') be so thin that nearly all the carriers which cross one junction also cross the other; in practice this involves a thickness which may be as small as $0 \cdot 1$–$0 \cdot 2 \, \mu$m. The transistor is not normally symmetrical, that is, the two n-regions (known as the 'emitter' and the 'collector') have different impurity concentrations; also, the collector is usually more highly doped than the base.

The symbols used to denote npn junction transistors are shown in Fig. 19.2, where the letters are used to label the electrodes and the arrows

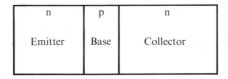

| n | p | n |
|---|---|---|
| Emitter | Base | Collector |

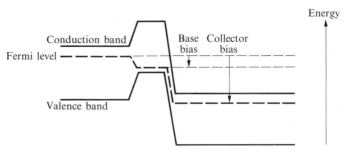

FIG. 19.1. The npn junction transistor, showing the electron energy levels under normal bias. The emitter–base diode is an n–p junction with forward bias, while the base–collector diode is a p–n junction with reverse bias.

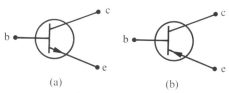

FIG. 19.2. Symbols used to denote (a) the npn transistor, (b) the pnp transistor. b base; e emitter; c collector. The arrow denotes the direction of positive current flow at the emitter electrode.

show the direction of current flow at the emitter. In operation the device is connected to batteries, as shown in Fig. 19.3, the larger voltage being applied to the collector. The emitter–base junction is a forward-biased diode, and a reduction in the voltage applied to the base produces a large increase in the emitter current, electrons flowing into the base from the emitter. The current is given by eqn (18.3). Because the base is very thin, practically all the electrons reach the collector provided that it is at a positive voltage with respect to the emitter. The current is primarily determined by the base voltage, and depends very little on the collector voltage. If a load $z_L$ is placed in the collector–emitter circuit, as in Fig. 19.3, a voltage $I_c z_L$ is developed across it whose value is primarily determined by the voltage $V_b$ applied to the base.

In the base region, which is p-type material, the electrons are minority carriers. They are therefore in 'enemy territory', and in danger of being lost by recombination with the majority carriers which are holes. The electrons move across the base by diffusion to the collector junction, where they fall down the energy drop which is in their favour, and eventually arrive at the collector electrode. The currents at each electrode are $I_e$, $I_b$, and $I_c$, as shown in Fig. 19.3, and from conservation of charge

$$I_e = I_b + I_c. \tag{19.1a}$$

To keep $I_b$ small, the thickness of the base must be small compared with the diffusion length $L_e$ for electrons (cf. eqn (17.30)). Normally $I_b$ is less than 2 per cent of the emitter current; this is the fraction lost by recombination in the base, into which, to maintain charge neutrality, a current of holes flows from the base electrode to balance the charge lost by recombination.

If, as in § 7.6, small changes in the three currents are denoted by $i_e$, $i_b$, and $i_c$, then clearly

$$i_e = i_b + i_c. \tag{19.1b}$$

For the circuit shown in Fig. 19.3, the current gain is defined as the ratio of the current change through the collector to that through the base, that is,

$$\beta = i_c / i_b. \tag{19.2}$$

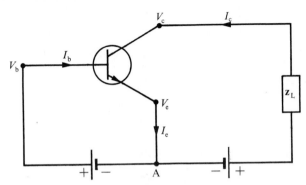

FIG. 19.3. The common emitter (CE) connection for a transistor with a load $z_L$ in the collector–emitter circuit. The arrows indicate the direction of positive current flow for an npn transistor, with the biasing batteries, whose values are about $V_b - V_e = 0.6$ V (for silicon), $V_c - V_e = 10$ V. A is the common terminal.

A typical value of $\beta$ is 50 or more, so that the current gain is high. The arrangement shown in Fig. 19.3 is primarily a 'current amplifier'. Since the emitter electrode is common to the input and output circuits, it is called the 'common emitter' connection, often denoted by the letters CE.

An alternative circuit is shown in Fig. 19.4; the common electrode is the base, and this 'common base' connection is denoted by CB. In this case the current gain is defined as the ratio of the current change through the collector to that through the emitter. That is,

$$\alpha = i_c/i_e, \tag{19.3}$$

and since $i_b$ is very small compared with the other currents it follows that the current gain $\alpha$ in this case is close to unity. Clearly there is a third alternative, the 'common collector' connection (CC), often called the emitter follower circuit because of its similarity to the cathode follower circuit used with triode vacuum tubes; this connection is discussed in § 19.3. The batteries which maintain the required steady voltages on emitter, collector, and base are shown in Figs 19.3 and 19.4; for a pnp transistor the batteries must be reversed.

By combining the equations above we obtain the relations

$$\beta + 1 = 1/(1-\alpha); \qquad \alpha = \beta/(\beta+1); \qquad \beta = \alpha/(1-\alpha). \tag{19.4}$$

The quantity $\beta + 1$ occurs frequently in transistor calculations and is often denoted by $\beta'$; clearly $\beta'$ is very nearly equal to $\beta$. $\alpha$ is substantially constant and can be taken as equal to unity, but $\beta$ can vary considerably between transistors bearing the same type number because of small differences in $(1-\alpha)$. An essential feature of transistor circuit design is to achieve an over-all performance which is not appreciably affected by this variation. The quantities $\alpha$, $\beta$ listed in transistor handbooks are the ratios $i_c/i_e$, $i_c/i_b$ measured with zero impedance in the collector lead (sometimes known as 'short-circuiting load'). Since $(1-\alpha)$ is the fraction of mobile carriers lost in the base through recombination, it is practically independent of the size of the current flowing, and so are $\alpha$ and $\beta$.

The battery connections in Fig. 19.4 emphasize that the junction transistor consists of two diodes back to back; the emitter–base diode, with forward bias; and the base–collector diode, with reverse bias. It is the combination of these two diodes with forward and reverse bias that is the essence of transistor action: electrons swept into the collector by the forward-biased emitter–base diode have a marked effect on the collector current, but the latter is almost independent of the collector voltage because the base–collector diode has reverse bias. This means that the output resistance of the transistor is extremely high (several megohms), whereas the input resistance is that of a forward-biased diode.

From eqn (18.3) we have

$$I_c = \alpha I_e = I_0\left\{\exp\left(\frac{|e| V_{eb}}{kT}\right)-1\right\},$$

where the second term in the bracket is normally very small compared to the exponential term. Neglecting it, we have for the input conductance

$$\partial I_e/\partial V_{eb} = I_e(|e|/kT) \sim 40I_e,$$

so that the input resistance is $(40\,I_e)^{-1}$, or about 25 $\Omega$ when $I_e = 1$ mA. In the same approximation we have

$$g_m = \partial I_c/\partial V_{eb} = I_c(|e|/kT) \sim 40I_c.$$

The quantity $g_m$ will be used in the circuit analysis of § 19.3, but qualitatively the action of the transistor can be described in a simple way.

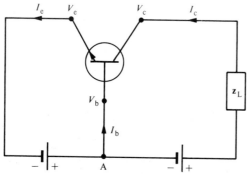

FIG. 19.4. The common base (CB) connection for a transistor with a load $z_L$ in the collector–base circuit. the arrows indicate the direction of positive current flow for an npn transistor, with the biasing batteries, whose values are about $V_b - V_e = 0.6$ V; $V_c - V_e = 10$ V.

It translates a small change in the emitter current flowing through the small input resistance into a nearly equal change in the collector current; when this flows through a high resistance load, both voltage and power amplification are obtained. The ratio $i_c/i_e$ or $i_c/i_b$ gives the current gain when the input is applied respectively to the emitter electrode and the base electrode, and the load is zero. The load in practical circuits is of the order of 1–10 k$\Omega$, but because of the high output impedance, the actual current gain can still be taken as simply $\alpha$ or $\beta$ to a very good approximation.

## 19.2. Characteristics of the junction transistor

The behaviour of a transistor can be assessed from its static characteristics. Since we have three electrodes, there are a number of these, but the most useful is the relation between the output current and the output voltage for various values of the base current in the CE connection. A circuit for obtaining such a characteristic is shown in Fig. 19.5. The resistances $R_1$, $R_2$ form a potentiometer by means of which $V_{eb} = V_e - V_b$ is adjusted, thus altering the value of the base current $I_b$; by varying the value of $V_0$, a series of curves such as those given in Fig. 19.6 is obtained. Here $V_c = V_0 - I_c R_L$; the right-hand battery must have sufficient voltage to allow for the drop in $R_L$, and $V_c$ can be calculated from the other quantities.

In operation the values of $V_0$, $V_{eb}$, and $R_L$ are chosen so that in the absence of any alternating voltage the operating point (or 'quiescent point') Q is near the centre of a characteristic where $I_c$ varies almost linearly with both $V_c$ and $V_b$. This permits relatively large swings in $V_c$ without distortion, that is to say, the change in output voltage or output current is linearly related to the change in the input. All changes must be

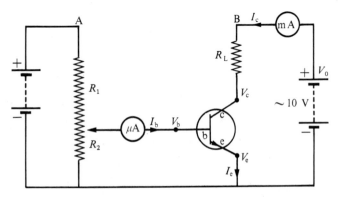

FIG. 19.5. Circuit for measuring the static characteristics of an npn transistor in the CE connection.

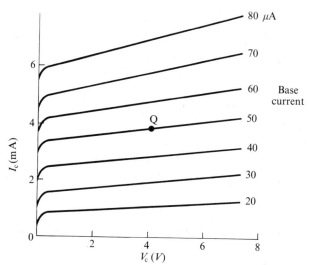

FIG. 19.6. Characteristics of a junction transistor; collector current versus collector voltage for various values of the base current. The straight portions are well represented by $I_c = h_{21}I_b + h_{22}V_o$ (cf. eqn (7.50b)), with $h_{21} \sim 100$ and $h_{22} \sim 5 \times 10^5 \ \Omega^{-1}$.

confined to the 'straight' portion of the characteristic, since distortion will occur if they extend into the curved region.

In practice, transistor handbooks give suitable values of the currents which enable an approximate Q point to be selected. The steady bias voltages are then adjusted until a satisfactory alternating output waveform is obtained. For example, in the circuit of Fig. 19.5 using the transistor whose characteristic is shown in Fig. 19.6, to obtain the operating point Q, $I_c$ must be 3·8 mA when $I_b$ is 50 $\mu$A. If $R_L$ is 500 $\Omega$, the voltage drop across it is 1·9 V, so that the battery voltage $V_0$ must be 5·9 V to make $V_c = 4$ V.

In a silicon npn transistor the voltage drop across the emitter–base diode is usually about 0·6–0·7 V; in germanium the corresponding figure is about 0·2 V (see Fig. 17.15(b)). In a practical circuit the left-hand battery is eliminated by joining the points A and B in Fig. 19.5, and adjusting the resistances $R_1$, $R_2$ to give the correct voltage at $V_b$, allowing for the fact that a stabilizing resistance is also needed in the emitter lead (see § 20.2).

*Frequency-dependence of $\alpha$ and $\beta$*

In a transistor, a sudden increase in emitter current is followed by a corresponding increase in the collector current $I_c$, but $I_c$ does not attain its final value until (for an npn transistor) electrons from the emitter have crossed the base and holes have entered the base to replace those lost by recombination within it. This time-lag is very short, but becomes important at high frequencies, where it results in $\alpha$ and $\beta$ becoming functions of the frequency. A simple treatment indicates how they are affected.

We assume that the electron density varies linearly over the base as in Fig. 19.7, from a value $n_f$ at the emitter edge to zero at the collector edge. The density gradient is $n_f/d$, where $d$ is the width of the base. The rate of flow across the base because of diffusion is $D(n_f/d)$ where $D$ is the diffusion coefficient for electrons. If $A$ is the cross-sectional area, the collector current is

$$I_c = A \, |e| D (n_f/d). \tag{19.5}$$

From Fig. 19.7 it is obvious that the total electric charge in the base is

$$Q_b = A \, |e| (\tfrac{1}{2} n_f) d, \tag{19.6}$$

since the average charge density in the base is $\tfrac{1}{2} n_f$. Combining these two equations we have

$$Q_b = I_c (d^2/2D) = I_c \tau_0, \tag{19.7}$$

where $\tau_0 = (d^2/2D)$ has the dimensions of time. It is called the mean carrier transit time and is determined by the values of $d$ and $D$.

From the conservation of charge and eqn (19.1a) we have

$$dQ_b/dt = I_e - I_b - I_c, \tag{19.8}$$

and hence from eqn (19.7)

$$\frac{dQ_b}{dt} + \frac{Q_b}{\tau_0} = I_e - I_b = \alpha_0 I_e, \tag{19.9}$$

where $\alpha_0$ is the ratio of $I_c$ to $I_e$ in the steady state. The solution of this equation is

$$Q_b/\tau_0 = \alpha_0 I_e \{1 - \exp(-t/\tau_0)\}$$

and

$$I_c = \alpha_0 I_e - (dQ_b/dt) = \alpha_0 I_e \{1 - \exp(-t/\tau_0)\}, \tag{19.10}$$

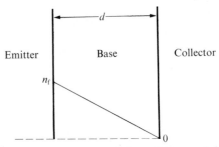

FIG. 19.7. Variation of electron density over the base, thickness $d$, from $n_f$ at the emitter edge to 0 at the collector edge.

showing that there is a time-delay in the build up of the current at the collector to its steady-state value $I_c = \alpha_0 I_e$.

If there is an alternating variation $i_e \exp(j\omega t)$ in the value of $I_e$, the corresponding change $q_b$ in $Q_b$ is given by the differential equation

$$\frac{dq_b}{dt} + \frac{q_b}{\tau_0} = \alpha_0 i_e \exp(j\omega t), \qquad (19.11)$$

whose steady-state solution is

$$\frac{q_b}{\tau_0} = \frac{\alpha_0 i_e \exp(j\omega t)}{1 + j\omega\tau_0}. \qquad (19.12)$$

This is similar to the steady-state solution $Q_b/\tau_0 = \alpha_0 I_e$ of eqn (19.9), except that we have

$$\alpha(\omega) = \frac{\alpha_0}{1 + j\omega\tau_0} = \frac{\alpha_0}{1 + j(\omega/\omega_\alpha)} \qquad (19.13)$$

instead of $\alpha_0$. Here $\omega_\alpha$ is the angular frequency at which $\alpha$ falls to $\alpha_0/\sqrt{2}$; in practice it is not exactly equal to $1/\tau_0$ because we have neglected the change in the distribution of electron density across the base when the currents are fluctuating. It is easy to show that

$$\beta(\omega) = \frac{\alpha(\omega)}{1 - \alpha(\omega)} = \frac{\beta_0}{1 + j(\omega/\omega_\beta)}, \qquad (19.14)$$

where $\beta_0 = \alpha_0/(1 - \alpha_0)$ and

$$\omega_\beta = (1 - \alpha_0)\omega_\alpha = \left(\frac{\alpha_0}{\beta_0}\right)\omega_\alpha. \qquad (19.15)$$

Here $\alpha_0$, $\beta_0$ are the values of $\alpha$, $\beta$ in the steady state, that is, at angular frequencies low compared with $\omega_\alpha$ and $\omega_\beta$. The corresponding frequencies are $f_\alpha = \omega_\alpha/2\pi$, $f_\beta = \omega_\beta/2\pi$, and $f_\beta$ is lower than $f_\alpha$ by the factor $\alpha_0/\beta_0$. A typical value for $f_\beta$ is 1 MHz, above which $\beta$ decreases and is frequency-dependent, but CB transistors can be used in the microwave region, although the gain is lower and frequency-dependent. The transit time across the base is $10^{-9}$–$10^{-10}$ s, and must be small compared to $1/f_\beta$ for constant gain. This low value is achieved by making the base as thin as possible and by increasing the mobility of the minority carriers in the base by varying the doping concentration, in order to produce an electric field in the base. This electric field accelerates the carriers, thus increasing their average speed and reducing the transit time across the base.

Since $f_\alpha$ is considerably higher than $f_\beta$, the current gain of the CB transistor is constant up to higher frequencies than that of the CE transistor, but in the microwave region effects due to small capacities and other frequency-dependent circuit elements can cause instability and loss of gain.

*Temperature-dependence*

In a CE transistor the current to the collector is actually

$$I_c = \beta I_b + I_{ce0}.$$

The last term, which has previously been omitted because it is very small compared with $\beta I_b$, represents the current flowing from base to collector when the emitter is on open-circuit. It arises from minority carriers produced in the base by thermal ionization, and increases as the temperature rises. So also does $I_b$; this is again controlled by the minority carrier density, which varies as $T^{\frac{3}{2}} \exp(-W_g/2kT)$ (see § 17.5). For germanium $W_g$ is about 0·66 eV at 300 K, and an increase in temperature can result in instability and 'thermal runaway': a temperature rise caused by ohmic heating produces an increase in current, giving a further temperature rise; this may continue to a point where the transistor is destroyed. For silicon $W_g = 1·12$ eV; the rate of current rise is much smaller and silicon transistors can operate up to 420 K. If stabilization is required, a suitable resistance is included in the emitter lead to produce negative feedback (see § 20.2).

## 19.3. Equivalent circuits for the junction transistor

A general analysis of a four-terminal active network was given in § 7.6. The transistor is a three-terminal device, with a common terminal between input and output, marked A in Figs 19.3–19.4 and in Fig. 19.8, which shows the equivalent circuit for a transistor. Here the box is the active network, and the common input and output terminals are each marked A. This circuit can be drawn in terms of eqns (7.50) which use the **h**-parameters; both these and the **y**-parameters of eqns (7.51) can be related to the physics of the transistors. The quantities

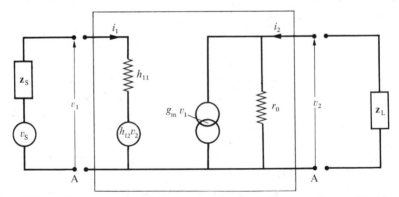

FIG. 19.8. Equivalent circuit for the junction transistor in terms of eqns (19.16) and (19.17).

which are most useful in circuit design are those which can most conveniently be measured. One such quantity is the mutual conductance $g_m$ defined in eqn (7.72), which is the same as the parameter $y_{21}$. In Fig. 19.8 the equivalent circuit is drawn to represent the eqns (7.50a) and (7.51b), which we repeat as

and

$$v_1 = h_{11}i_1 + h_{12}v_2,$$                                  (19.16)

$$i_2 = y_{21}v_1 + y_{22}v_2 = g_m v_1 + v_2/r_0.$$              (19.17)

Here the parameters $h$, $y$ are real and need not be written as vectors.

The resistance $h_{11}$ includes two contributions. For the CE and CB connections the first is the resistance of the forward-biased input diode, given by eqn (18.4), while the second is the 'spreading' resistance within the base (normally 10–100 $\Omega$), encountered by current flowing to or from the base electrode. The second term $h_{12}v_2$ represents the influence exerted by the output circuit on the input circuit. $g_m v_1$ is the current generator in the output circuit, proportional to the input voltage $v_1$, in parallel with an internal conductance $y_{22}$ which is the reciprocal of the resistance $r_0$ of the back-biased output diode. This is normally of the order of megohms, and large compared with the load, for which a typical value is a few kilohms.

### The CE connection

If $r_0$ is very large compared with $z_L$, it follows from eqn (19.2) and Fig. 19.3 that the current gain $A_i$ is simply

$$A_i = i_2/i_1 = \beta = +i_c/i_b.$$

On combining this with eqns (19.16) and (19.17) we obtain

$$v_1(y_{22} + h_{12}g_m) = i_1(h_{11}y_{22} + \beta h_{12}).$$

Since $r_0$ is very large it is a good approximation to put $y_{22} = 0$, in which case the input resistance is just

$$z_i = v_1/i_1 = \beta/g_m.$$                                   (19.18)

The output voltage is $v_2 = -i_2 z_L$, measured from the common terminal as zero, and the voltage amplification $A_v$ is

$$A_v = v_2/v_1 = -i_2 z_L/i_1 z_i = -\beta(z_L/z_i) = -g_m z_L.$$   (19.19)

This agrees with the result expressed in eqn (7.66b); the presence of the negative sign shows that the voltages at the input and output terminals vary in opposite directions with respect to the common terminal. If the output resistance is required, it can be found from eqn (7.56); the assumptions made above are equivalent to treating it as infinite.

*The CB connection*

With the same approximation $r_0 = \infty$, and using eqn (19.3) applied to Fig. 19.4 the current gain is

$$A_i = i_2/i_1 = -\alpha = -i_c/i_e,$$

where the negative sign appears because the input current $I_e$ in Fig. 19.4 flows in the opposite direction to $I_b$ in Fig. 19.3. Following similar algebra as for the CE connection, it is found that the input impedance is

$$\mathbf{z}_i = v_1/i_1 = \alpha/g_m, \tag{19.20}$$

and the voltage amplification is

$$A_v = +\mathbf{z}_L i_2/\mathbf{z}_i i_1 = +\alpha \mathbf{z}_L/\alpha g_m^{-1} = +g_m \mathbf{z}_L. \tag{19.21}$$

Since $\alpha$ is nearly unity while $\beta$ lies between 50 and 300 we see that the CB connection has a very low input resistance compared with the CE connection, and this is one reason why it is used much less frequently. The change in sign of $A_v$ compared with the CE connection is a simple consequence of the fact that the input currents $I_b$ and $I_e$ flow in opposite directions in Figs 19.3 and 19.4.

*The CC connection*

The third possible arrangement is with the collector as the common electrode. This is known as the common collector connection; it is shown in Fig. 19.9. An important difference from the other two connections is that the large differential resistance of the back biased diode between base and collector is now in series with the input, so that we anticipate that the input impedance will be high.

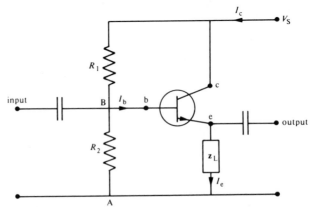

Fig. 19.9. Circuit diagram for the CC (emitter follower) connection. The correct voltage on the base is normally attained by making $R_1$, $R_2$ resistances of the correct size to give the normal base voltage after allowing for the small base current.

The voltage gain can be calculated in a simple way as follows. The voltage across the load is

$$V_L = z_L I_e = z_L I_c (1 + \beta^{-1}), \qquad (19.22)$$

while the voltage applied between B and A is $V_{bc} = V_L + V_{eb}$. For small changes the voltage amplification is $\partial V_L / \partial V_{bc}$, and hence

$$A_v^{-1} = \frac{\partial V_{bc}}{\partial V_L} = 1 + \frac{\partial V_{eb}}{\partial V_L}$$

$$= 1 + \frac{\partial V_{eb}}{\partial I_c} \frac{1}{z_L (1 + \beta^{-1})}$$

$$= 1 + \frac{1}{g_m z_L (1 + \beta^{-1})}.$$

Since $\beta^{-1}$ is very small, we have to a good approximation,

$$A_v = g_m z_L / (1 + g_m z_L). \qquad (19.23)$$

This is always less than unity, but can be made to approach it by using a load $z_L$ of a few kilohms. The current gain is large, since

$$A_i = i_e / i_b = 1 + \beta. \qquad (19.24)$$

The input resistance is given by

$$\partial V_{bc} / \partial I_b = (1 + \beta)(\partial V_{bc} / \partial I_e) = (1 + \beta) z_L, \qquad (19.25)$$

and is therefore very high. We can find the output impedance by noting that eqn (7.66a) can be written as

$$|A_v| = \frac{g_m z_L}{1 + z_L / r_0}$$

and from comparison with eqn (19.23) it is clear that $r_0 = g_m^{-1}$ and is comparatively low. The CC connection can therefore be used for matching a high-impedance source to a low-impedance load; it is much cheaper for this purpose than a transformer, and has the additional advantage of giving a power gain $\sim (1 + \beta)$.

### Comparison of CB, CE, and CC connections

The three methods of connection give considerable flexibility in the design of transistor circuits, as can be seen from the summary of their properties in Table 19.1. Provided that the impedance of the source is small, both CB and CE connections have high voltage gain, but the current gains are $\alpha$ and $\beta$ respectively, so that the *power gain* of the CE connection is much higher than that of the CB connection. Since it also has a much higher input resistance, the CE circuit is much more widely used than the CB circuit. The performance depends on the transistor parameters, particularly $\beta$; this is a drawback because $\beta$ may vary widely from one transistor to another. This is overcome by the addition of negative feedback, as described in § 20.2.

At high frequencies the performance of the CE circuit becomes less good than that of the CB circuit because of feedback through the interelectrode capacitance between input and output. This capacitance is $C_{cb}$ for the CE circuit, and is larger than the corresponding capacitance $C_{ce}$ for the CB circuit, so that the Miller effect (§ 20.2) becomes relatively more important at any given frequency.

The CC circuit is used to match a high-impedance source to a lower-impedance load. Its voltage gain is less than unity but it can have a considerable power gain.

TABLE 19.1

*Comparison of current gain $A_i$, voltage gain $A_v$, input resistance $z_i$, and output resistance $r_0$ for the CE, CB, and CC connections (the approximations assume $|z_L|$ is a few kilohm).*

|  | CE | CB | CC |
| --- | --- | --- | --- |
| $|A_i|$ | $\beta$ | $\alpha(\sim 1)$ | $1+\beta$ |
| $A_v = (z_L/z_i)A_i$ | High (negative) | High | $\sim 1$ |
| $z_i$ | $\beta/g_m$ | $(\alpha/g_m) \sim (1/g_m)$ | $(1+\beta)z_L$ |
| $r_0$ | High | High | Low |
| Phase change with resistive load | $\pi$ | 0 | 0 |

## 19.4. The field-effect transistor, or FET

Field-effect transistors were discovered at about the same time as junction transistors, but manufacturing problems for the junction transistor were solved much more rapidly than for the former. It was only in 1970 that the field-effect transistor (FET) and the insulated gate device (IG), also known as the metal oxide–semiconductor device (MOS), became sufficiently consistent in their characteristics from mass production for them to be widely used commercially. They differ from the junction transistor described previously in that the current is carried only by one type of carrier—the majority carrier; they are therefore known as unipolar devices, whereas the junction transistor is a bipolar device in which both types of carriers are involved.

The FET consists of two regions of p-type semiconductor (normally silicon) on either side of a slice of n-type semiconductor, the two regions being connected together as in Fig. 19.10, and forming the gate G.

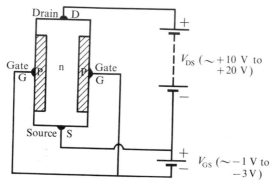

FIG. 19.10. Schematic diagram of a field effect transistor (FET), showing the biasing batteries for a device with an n-type channel (the batteries must be reversed when the channel is p-type and the gate n-type).

Connections known as the source S and the drain D are made to each end of the n-type slice. The voltage applied to the drain is positive with respect to the source, as in the figure, while that applied to the gate is negative and normally considerably smaller in magnitude than that on the drain. The n-type slice provides a channel in which electrons can flow from the source to the drain, but the current flow is restricted to a central section whose width depends on the voltage applied to the gate. Thus the resistance of the conducting channel, and the current flowing through it, can be controlled by the gate voltage. In this respect the device resembles a thermionic triode, the gate being analogous to the grid, the source to the cathode, and the drain to the anode.

Field-effect transistors can equally well be made with p-type channel and n-type gate, giving additional flexibility in circuit design; the bias voltages must then be reversed. The symbol for a junction FET is shown in Fig. 19.11, where the arrow indicates whether the channel is n-type or p-type. For simplicity we shall describe only the action of an FET with an n-type channel.

The surface between gate and channel forms a p–n junction, with a 'depletion layer' at the boundary. There are no free carriers in this layer apart from a few electron–hole pairs formed by dissociation under the influence of the thermal energy, so that there is a 'space charge' arising from the donor ions in the n-type and the acceptor ions in the p-type material. The n-type channel is lightly doped, while the p-type regions are heavily doped. It therefore follows from eqn (17.37) for the relative widths of the depletion layers that $\delta_n \gg \delta_p$, that is, the depletion region lies almost entirely in the channel and very little in the p-type region. In the depletion region in the n-type material the space charge density is $\rho = |e| N_d$, where $N_d$ is the number of positively charged donor

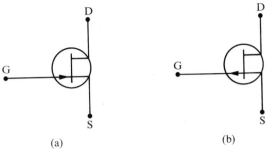

FIG. 19.11. Symbols for a junction FET. The direction of the arrow distinguishes between (a) n-type channel, (b) p-type channel; it shows the direction of conventional current flow when the gate-channel diode is forward-biased.

atoms per unit volume, and the variation of the electric potential $V$ is given by Poisson's equation

$$\nabla^2 V = -\rho/\epsilon = -|e|N_d/\epsilon. \tag{19.26}$$

We choose a set of Cartesian coordinates in which the $z$-axis is parallel to the line joining the source to the drain, while the $x$-axis is normal to the two gate electrodes, assumed to be parallel planes $x = \pm\frac{1}{2}H$. The channel is thus a rectangular parallelepiped, extending indefinitely in the $y$ direction. Hence $\partial^2 V/\partial y^2 = 0$ and, when there is no voltage difference between source and drain, $\partial^2 V/\partial z^2 = 0$. The value of $V$ is then given by the solution of

$$\partial^2 V/\partial x^2 = -|e|N_d/\epsilon, \tag{19.27}$$

and if $N_d$ is uniform throughout the channel, the solution for positive values of $x$ is

$$V - V_G = \left(\frac{|e|N_d}{8\epsilon}\right)(H - 2x)(2x + H - 2h). \tag{19.28}$$

Here the constants of integration have been chosen to make $V = V_G$ at $x = \frac{1}{2}H$ (that is, we have ignored the depletion layer inside the p-type region), and $\partial V/\partial x = 0$ at $x = \frac{1}{2}h$. This equation for $V$ holds for $\frac{1}{2}H > x > \frac{1}{2}h$; the region $x = 0$ to $\frac{1}{2}h$ is the conduction channel, in which $V$ is constant. The variation of $V$ is shown in Fig. 19.12, which includes the mirror image region for negative values of $x = 0$ to $-\frac{1}{2}H$. In the conducting channel $V$ has the same value as at $x = \frac{1}{2}h$, that is,

$$V - V_G = (|e|N_d/8\epsilon)(H - h)^2 \tag{19.29}$$

from which we see that the value of $h$ depends on the potential difference $V - V_G$. Thus the width of the conducting channel depends on this potential difference, as can be seen from Fig. 19.12(a) and (b).

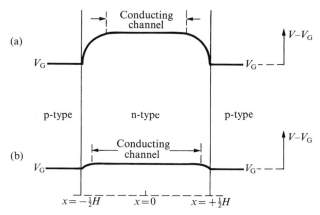

FIG. 19.12. Variation of the electric potential $V$ in the n-type channel of an FET. $V$ rises roughly quadratically within the channel near the surface of the gate, becoming constant within the conducting channel. This channel is narrower in (a), where $V-V_G$ is higher than in (b).

When a small voltage difference $V_{DS}$ is set up between D and S, initially the current flow rises linearly with $V_{DS}$. However, as $V_D$ rises, the width of the conducting channel decreases markedly nearer the drain, because the potential in the conducting channel is large there, and, from eqn (19.29), an increase in $V$ causes a reduction in $h$. Thus in Fig. 19.12, (b) corresponds to a point near the source, while (a) corresponds to a point near the drain. In the conducting channel the current is independent of $z$, but the current density and the field strength $E_z = -\partial V/\partial z$ driving it must be larger where the channel is narrower. The potential $V$ must therefore vary non-linearly with $z$, and no simple solution for $V$ as a function of $x$ and $z$ can be obtained. The way in which the widths of depletion layer and conducting channel vary with $z$ is shown in Fig. 19.13. As the conducting channel becomes narrower, the resistance between source and drain increases, and the current rises less rapidly than the voltage $V_{DS}$, as shown in the characteristic curves of Fig. 19.14. Essentially the size of the current is determined by the electric field strength in the depletion layer, giving the name 'field-effect transistor'.

*Pinch-off*

In eqn (19.29), if $(V-V_G)$ becomes equal to $(|e|N_d/8\epsilon)H^2$, then $h$ must fall to zero; the conducting channel vanishes, and so must the current flow. In practice this does not happen; as the conducting channel narrows, the current density increases and so does the electric field driving it. The value of $I_D$ becomes almost constant, nearly independent of $V_D$ but not of $V_G$, as shown in the characteristic curves of Fig. 19.14. In this region the value of $I_D$, for a given value of $V_{DS}$, is given approximately by the

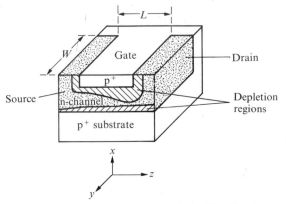

FIG. 19.13. Cross-section of an n-type channel junction FET. The upper gate and lower substrate are connected internally, both in front of the paper (not shown) and behind. The depletion regions extend further into the n-channel nearer the drain.

expression

$$I_D = (I_D)_0 \left( 1 - \frac{V_{GS}}{V_p} \right)^n,$$  (19.30)

where the exponent $n$ is close to 2. Thus a plot of $I_D^{\frac{1}{2}}$ against $V_{GS}$ (both $V_{GS}$ and $V_p$ are negative for an n-type channel) gives a straight line cutting the axis $I_D = 0$ at $V_{GS} = V_p$, which is known as the 'pinch-off' voltage.

The mutual conductance (or transconductance) $g_m = (\partial I_D / \partial V_{GS})$, $V_{DS}$ being kept constant. In the pinch-off region, from eqn (19.30) with $n = 2$,

$$g_m = -\frac{2(I_D)_0}{V_p} \left( 1 - \frac{V_{GS}}{V_p} \right)$$

$$= (g_m)_0 \left( 1 - \frac{V_{GS}}{V_p} \right),$$  (19.31)

where $(g_m)_0$ is the value of $g_m$ when $V_{GS} = 0$ (since $V_p$ is negative for an n-channel device, $(g_m)_0 = -2(I_D)_0 / V_p$ is positive). To make $g_m$ large, we need to make $(I_D)_0$ as large as possible; since it is proportional to the width $W$ of the channel in the $y$ direction and varies inversely with the distance $L$ between source and drain, this means increasing the ratio $W/L$. However, the capacitance between the gate and the conducting channel increases with the area $WL$, so that in practice $L$ is made as small as possible and $W$ adjusted accordingly.

A typical value of $(g_m)_0$ is $0 \cdot 4$–$1 \cdot 2$ mA V$^{-1}$. This is high, but it depends on the characteristics of the device, which in manufacture vary from sample to sample. As with junction transistors, the practice is to introduce sufficient feedback to make the amplifier gain independent of the actual

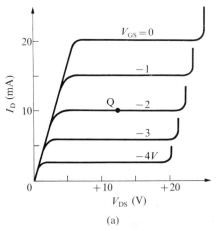
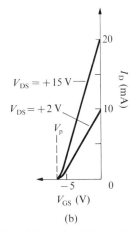

(a)                                      (b)

FIG. 19.14. Characteristics of an n-type channel junction FET. (a) $I_D$ against $V_{DS}$ for five values of $V_{GS}$; (b) $I_D$ against $V_{GS}$ for $V_{DS} = +2$ and $+15$ V. The value of the pinch-off voltage $V_p$ is $-6$ V. the curves of $I_D$ turn up sharply at the higher values of $V_{DS}$ because of avalanche breakdown of the p–n junction between gate and channel; in operation this region must be avoided, to prevent damage. Q is a typical quiescent point.

value of $g_m$. Clearly, in the pinch-off region where $I_D$ is practically independent of $V_{DS}$, the output resistance is very high (typically, the output conductance is about $10^{-2}g_m$). The input resistance is also high, because the gate is reverse-biased with respect to the source. This contrasts strongly with the junction transistor, which has a low input impedance because the input diode is forward-biased.

The structure of a planar junction FET is shown in Fig. 19.15. Starting with a single crystal, which is highly doped p-type material to form the substrate, the n-type channel region is grown on epitaxially (that is, as a single crystal whose axes have the same orientation as in the substrate). The other zones are formed by diffusing in from the surface the required impurity. These include the upper $p^+$ (highly doped) gate (as in Fig. 19.13), and the highly conducting $n^+$ regions which form source and drain. Metal is deposited by evaporation onto these to give good ohmic contacts, surface insulation being provided by the oxide layer of silica ($SiO_2$). The channel length $L$ in Fig. 19.13 is only a few $\mu$m in length, while the width $W$ can be made as large as required by extending the structure in the plane out of the paper, and by the use of repeated structures such as SGDGSGD..., where the source fingers are all connected together at one end and the drain fingers at the other. The channels between them are then in parallel with one another.

The advantage of the planar construction is that it gives a high component density; many transistors can be constructed on a single slice of semiconductor crystal. The slice may be sawn up to produce individual

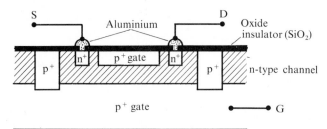

FIG. 19.15. Structure of a planar junction FET with n-type channel. The layer of silica (SiO₂) is used only to insulate the electrodes.

transistors, or made into an 'integrated circuit' where both active and passive elements are manufactured by successive diffusions. The latter are widely used in digital circuitry where a large number of elements are required, since the packing density can be high and the circuit connections are very reliable. The individual components can be made with fewer processing steps and smaller dimensions for unipolar transistors than for bipolar transistors (see Hittinger 1973).

### 19.5. The MOS field-effect transistor, or MOSFET

The insulated-gate FET is an important variant of the junction FET described in the last section. The difference is that the metal gate electrode is insulated from the semiconductor crystal by a thin film (~0·1 μm thick) of Silica (SiO₂), as illustrated in Fig. 19.16.

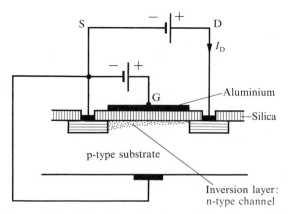

FIG. 19.16. Structure of a metal–oxide–semiconductor field effect transistor (MOSFET). The silica layer, about 0·1 μm thick, insulates the metal gate electrode G from the semiconductor. When G is made positive, electrons are attracted to the interface between oxide and semiconductor, providing a conducting channel in which electrons can flow from source to drain when a voltage $V_{DS}$ is applied. When $V_{GS} = 0$, there is no conducting channel in the 'normally-off' mode.

The name of MOSFET is derived from the sequence metal–oxide–semiconductor. In construction, two strongly doped (n') zones are diffused into the p-type substrate to form the source and the drain, connected to metal electrodes. The metal gate electrode is on top of the oxide film; it forms, with the p-type substrate, a parallel-plate capacitor, in which the silica is the dielectric. The presence of this oxide layer makes the leakage current to the gate extremely low (of the order of $10^{-9}\mu A$), but the gate voltage still controls the current flowing underneath between source and drain. When the voltage on the gate is positive with respect to the substrate, electrons in the latter are attracted towards, and holes repelled from, the interface between the substrate and the oxide layer. This produces an n-type layer (the 'inversion layer' or 'n-channel') through which the current $I_D$ flows between source and drain when a voltage $V_{DS}$ is applied. Variation of the voltage on the gate changes the electron density in the n-channel, and controls the size of the current $I_D$. Basically the action is the same as in the junction FET, but the gate voltage can be of either sign, whereas in the junction FET a gate voltage of the wrong sign will cause a large gate current to flow when the gate–channel p–n junction becomes forward-biased.

The additional flexibility in the gate voltage shows itself in the two normal methods of operation for the MOSFET. These are as follows.

### 'Normally-off' or 'enhancement' mode

At zero gate voltage there is no conducting channel. If a positive voltage is applied to the drain, the p–n junction between drain and substrate becomes reverse-biased so that no current (apart from leakage) flows between source and drain. It is necessary to apply a positive gate voltage in order to create a conducting channel with mobile charge carriers, as described above. The positive voltage between gate and source at which appreciable current starts to flow is called the pinch-off voltage $V_p$ (or the threshold voltage). A typical set of characteristics is shown in Fig. 19.17 (a), (b), and the symbol is shown in (c). The electrode B ('bulk') is the substrate terminal, which is generally connected to the source (often internally). The arrow shows the direction of current flow in the p–n junction.

### 'Normally-on' or 'depletion' mode

When positively charged impurity ions are present in the silica insulating layer, an electric field is set up, even when $V_{GS} = 0$, in the same direction as if there were a positive voltage on the gate. An n-type inversion layer is therefore present, giving a conduction channel except when the gate voltage is made sufficiently negative. The pinch-off voltage at which the current is cut off is thus negative, and the characteristics are

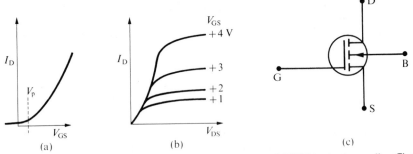

FIG. 19.17. (a), (b) Characteristics, and (c) symbol for a MOSFET in the 'normally-off' or enhancement mode. The value of $V_p$ is positive, and no current flows when $V_{GS} = 0$. The arrow indicates the direction of current flow in the p–n junction, the direction shown being for an n-channel device.

as shown in Fig. 19.18(a), (b), and the symbol as in (c), where the vertical line is now continuous instead of being broken.

These diagrams apply only to the MOSFET with n-type channel. With a p-type channel the arrow in the symbol is reversed, and so must be the biasing voltages applied to the electrodes. An n-channel device is faster in action than a p-type channel device, corresponding to the greater mobility of electrons compared with holes. In practice, the MOSFET is normally used in the enhancement mode, which allows greater flexibility in circuit design.

Because of the high input impedance, the equivalent circuit of an FET has the simple form shown in Fig. 19.19. At high frequencies the various interelectrode capacities must be included; also, a resistance $R_G$ is normally connected between gate and source to stop charge accumulating on the gate. This allows the gate current to drain away and maintains $V_{GS}$ close to zero. The drain current varies with temperature for two reasons:

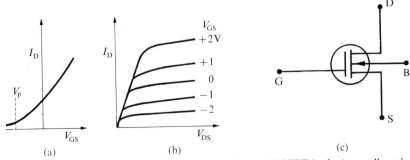

FIG. 19.18. (a), (b) Characteristics, and (c) symbol for a MOSFET in the 'normally-on' or depletion mode. The value of $V_p$ is negative, and the device can be used with $V_{GS}$ either positive or negative. The arrow indicates the direction of current flow in the p–n junction, and the direction shown corresponds to an n-channel device.

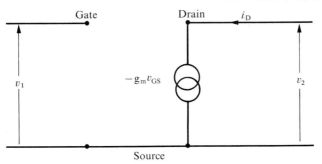

FIG. 19.19. Equivalent circuit of an FET.

(1) the mobility of the charge carriers in the channel decreases as the temperature rises; (2) the pinch-off voltage $V_p$ decreases with increasing temperature. The first effect causes $I_D$ to fall with increasing temperature, while the second (see eqn (19.30)) causes it to rise. At a certain (negative) value of $V_{GS}$ the two effects balance, but in practical circuits it is difficult to stabilize the operating point so as to make use of this property. At higher values of $I_D$, effect (1) predominates, so that there is no danger of thermal run-away with an FET.

### 19.6. Summary of transistor properties

1. The junction transistor behaves as a current amplifier. Its input impedance is low, being that of a forward-biased p–n junction.

2. The FET has a high input impedance, being that of a reverse-biased diode for the junction FET, and that of the oxide layer for a MOSFET. The high input impedance simplifies the design of circuits with a large number of stages.

3. Distortion is smaller for the FET and MOSFET, because of the square-law relation of $I_D$ to $V_{GS}$, than for the junction transistor with its exponential relation.

4. Variations in manufacture and with temperature are smaller for the FET than for the junction transistor; in either case they are overcome in amplifier circuits by the use of negative feedback (see § 20.2).

5. Power consumption for the FET is smaller than for a junction transistor, because for the latter a biasing potentiometer (see Fig. 20.4) is not required.

### Reference

HITTINGER, W. C. (1973). *Scient. Am.* **229**, 48.

# 20. Amplifiers and oscillators

### 20.1. Negative feedback

NEGATIVE feedback is an essential constituent in the stabilization of any system employing automatic control. The system may be biological (for example, the concentration of a particular cation in the blood must be maintained within certain limits), mechanical, or electrical. Here we shall be concerned with electronic circuits. The essential requirement is that an adverse change in the output must trigger a correction at the input whose effect is to offset the original deviation. The more sensitive the trigger mechanism to small changes, and the more rapid its response, the closer will be the limits within which the output is maintained. For such stability it is of course essential that the correction be in the direction which reduces any unwanted change; hence the name 'negative feedback'. A correction in the wrong direction will produce a further departure from equilibrium; this is 'positive feedback', which leads to instability and oscillation.

A circuit using feedback must contain an amplifier, together with components that transfer a fraction of the output power back to the input. Such components are linear circuit elements such as resistors, or inductances and capacitances (frequently an interelectrode capacitance), whose impedances are frequency-dependent.

There are two types of feedback; (1) series-voltage feedback, in which a feedback *voltage* proportional to the output is injected in *series* with the input voltage; and (2) parallel-current feedback, in which a feedback *current* proportional to the output is injected in *parallel* with the original current. The two classes are illustrated in Figs 20.1 and 20.2, and are discussed in more detail below. Positive feedback occurs if the sign (or

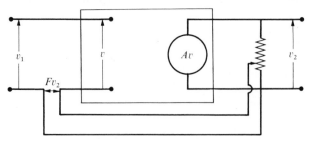

FIG. 20.1. Series voltage feedback. A fraction $Fv_2$ of the output voltage $v_2$ is fed back in series with the input voltage.

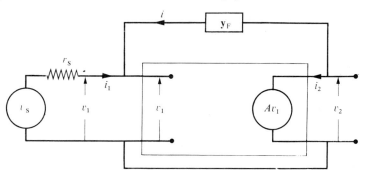

FIG. 20.2. Parallel current feedback. A current $i$ is fed back to the input through an admittance $\mathbf{y}_F$ connected between output and input terminals.

phase) of the feedback is such as to augment the signal; the over-all amplification is thereby increased, and with sufficient feedback the circuit will oscillate spontaneously without any external input signal being applied. Positive feedback is essential in a self-oscillating system. Systems in which a stable level of amplification is required make use of negative feedback, where the feedback voltage or current is of such phase as to reduce the injected signal.

### Series-voltage feedback

An example, in schematic form, of series-voltage feedback, is shown in Fig. 20.1. The four terminal network denotes an amplifier, whose (voltage) gain in the absence of feedback is $A$; a fraction $F$ of the output voltage $v_2$ is fed back in series with the input circuit. The output voltage is now $v_2 = A(v_1 + Fv_2)$, giving for the gain

$$G = v_2/v_1 = A/(1-AF). \tag{20.1}$$

Differentiation gives

$$\frac{\delta G}{G} = \frac{\delta A}{A} \cdot \frac{1}{(1-AF)}, \tag{20.2}$$

a relation which shows that a fractional change in the amplification factor $\delta A/A$ produces a fractional change in the gain $\delta G/G$ which is smaller by the factor $1/(1-AF)$. $F$ is called the feedback parameter, and $AF$ the loop gain; both $A$ and $F$ may be functions of frequency. When $AF$ is negative, the gain is reduced, but so also is the distortion. A distortion voltage appearing at the output is fed back to the input, and is reduced in the same ratio as the amplification. If now the input voltage is increased by means of a preceding amplifier until the output is restored to its former level, the distortion voltage will remain at the reduced level provided that the preceding amplifier is free from distortion. This condition is usually fulfilled, because the signal in the preceding amplifier is at a lower level, and distortion is most likely to occur where the signal level is high, in the final stages of amplification.

If the quantity $(-AF)$ is made large compared with unity, the gain becomes simply $1/F$; it depends only on the feedback ratio and not on $A$, so that the gain is independent of the characteristics of the active components. This is an important consideration in use of the transistor, which is inherently a non-linear device.

The series feedback voltage clearly alters the effective input impedance of the network. If the feedback voltage has the same sign as $v_1$ (positive feedback) then a smaller value of $v_1$ is needed to drive the same current $i_1$ and the input impedance is smaller. If the feedback voltage is of opposite sign to $v_1$, a larger value of $v_1$ is required to maintain the same $i_1$, so that negative feedback increases the effective input impedance. The current gain $A_i = i_2/i_1$ is unchanged, but the voltage gain $A_v = v_2/v_1$ is decreased by negative feedback since a larger value of $v_1$ is required to maintain the same value of $v_2$.

If the quantity $AF$ in eqn (20.1) becomes equal to $+1$, the gain becomes infinite. This means that a finite output exists when no external input voltage is present, since the series voltage feedback in Fig. 20.1 is just sufficient and in the correct phase to produce the output voltage. The system becomes unstable and, if $AF > +1$, the currents increase to a point at which they are limited by non-linearities in the circuit; that is, the quantity $AF$ is a function of current and, assuming it falls as the current increases, the latter attains a steady value such that the mean value of $AF$, averaged over a cycle, is just equal to $+1$. In general both $A$ and $F$ may be complex, so that the quantity $AF$ is complex, and also a function of frequency. The criterion for stability, which has been derived by Nyquist, is as follows. If the real and imaginary parts of $AF$ are plotted against one another, a curve is obtained from the points for different frequencies: the condition for stability is that this curve must not enclose the point $(+1, 0)$ on the real axis. In practice this may be ensured by making the gain, at any frequency where the feedback is real and positive, too small to make the real part of $AF$ greater than $+1$.

*Parallel-current feedback and the Miller effect*

An example of parallel-current feedback is presented by the circuit shown in Fig. 20.2, where an admittance $\mathbf{y}_F$ is connected between the output and input terminals. The current flowing through this admittance is

$$i = \mathbf{y}_F(v_1 - v_2) = \mathbf{y}_F(1 - A)v_1,$$

where $A$ is the voltage amplification factor of the four terminal network. The total input current is now $i_1 + i$, and the additional current $i$ is equivalent to an admittance in parallel with the input admittance of the network, of magnitude

$$i/v_1 = (1-A)\mathbf{y}_F. \tag{20.3}$$

If this is positive, the extra admittance results in a decrease in the over-all input impedance. If $A$ is large and negative, the extra admittance can greatly exceed $\mathbf{y}_F$; the reason for this is clearly that the voltage across $\mathbf{y}_F$ is $(v_1-v_2) = (1-A)v_1$ instead of just $v_1$.

In general, the input admittance in the absence of feedback will contain a capacitive component, so that it will be of the form $\mathbf{y}_i^0 = g_i + j\omega C_i^0$. If the feedback admittance is capacitive, $\mathbf{y}_F = j\omega C_F$, then the total input admittance becomes

$$\mathbf{y}_i = g_i + j\omega\{C_i^0 + (1-A)C_F\} = g_i + j\omega C_i,$$

where we have assumed that $A$ is real. The effective capacitance at the input has been increased to

$$C_i = C_i^0 + (1-A)C_F,$$

which can be much greater than $C_i^0$ if $A$ is negative. An effect of this kind occurs both in vacuum tubes and in solid-state devices through the interelectrode capacitance, which appears as a larger capacitance across the input terminals. At high frequencies this represents such a small impedance that only a fraction of the voltage from a source with a finite source resistance may actually be applied to the following input of an amplifier. The resultant drop in over-all gain is known as the 'Miller effect'.

If the input is fed from a voltage source $v_S$ with an internal resistance $r_S$, as in Fig. 20.2, the voltage $v_1$ applied to the input terminals can be written as

$$\frac{v_1}{v_S} = \frac{r_i}{r_i+r_S}\left\{\frac{1}{1+j(\omega/\omega_0)}\right\}. \tag{20.4}$$

Here $\omega_0 = (C_i r_i')^{-1}$, where $r_i' = r_i r_S/(r_i+r_S)$ and $r_i = g_i^{-1}$, so that $\omega_0$ is determined by $C_i$ and the resistance formed by $r_i$ and $r_S$ in parallel. $\omega_0$ is the angular frequency at which $|v_i/v_S|$ falls by a factor $1/\sqrt{2}$ below its value at $\omega = 0$, and is sometimes known as the 'mid-band' frequency.

### Feedback and bandwidth

If the input is to an amplifier whose gain at $\omega = 0$ is $A_0$, then the effective gain at a finite frequency is

$$A = \frac{A_0}{1+j(\omega/\omega_0)}. \tag{20.5}$$

If feedback is introduced, with $F$ independent of frequency, the gain becomes

$$A_F = A/(1-AF)$$

$$= \frac{A_0}{(1-A_0F)}\left\{\frac{1}{1+\mathrm{j}(\omega/\omega_1)}\right\}, \tag{20.6}$$

where $\omega_1 = \omega_0(1-A_0F)$. Thus the bandwidth is increased by a factor $(1-A_0F)$, while the gain is reduced by the same factor; the product (gain × bandwidth) is unchanged by the introduction of feedback.

Circuits using resistive components with feedback through a resistance (§§ 20.2–20.3) are suitable for wide-band amplifiers ( see also § 20.8).

### 20.2. Feedback and stabilization of a transistor amplifier

Fig. 20.3 shows a circuit in which a resistance $R_e$ is included in the emitter lead of a CE circuit. The voltage across $R_e$ is common to the input and output circuits, and provides an example of feedback. If the value of $\beta$ for the transistor is high, the base current is a very small fraction ($\leqslant 1$ per cent) of the collector or emitter currents, and we can take $i_c = i_e$. The voltage fed back to the input is $-i_eR_e = -(v_2/R_L)R_e = -(R_e/R_L)v_2$, and the feedback ratio $F = -(R_e/R_L)$ is negative.

As indicated in Fig. 20.3, the resistance $R_e$ may be shunted by a capacitance $C$, so that the combination is a low impedance for alternating current. This has no effect at zero frequency, so that the d.c. current and voltage are still stabilized by the presence of $R_e$, and the Q point remains

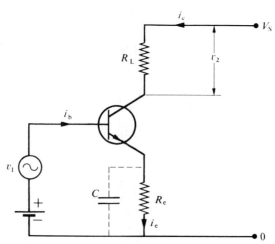

FIG. 20.3. A transistor circuit with a feedback resistance $R_e$ in the emitter lead.

the same, but the a.c. gain is not reduced if $C$ is suitably chosen. If $\omega_1$ is the lowest frequency which it is desired to amplify, the condition which $C$ must satisfy is not simply $\omega_1 C R_e = 1$. Analysis of the circuit shows that other parameters are involved, and for thermionic vacuum tubes or an FET we have

$$\omega_1 C = (g_m + r_s^{-1}),$$

while for junction transistors

$$\omega_1 C = h_{21}/(h_{11} + r_s),$$

where $r_s$ is the source resistance.

The effect of $R_e$ on the input impedance can readily be found by comparison with the equivalent circuit of Fig. 19.8. It provides a negative-feedback voltage $R_e i_e$ in the input circuit, so that

$$h_{12} v_2 = R_e i_e = R_e(i_b + i_c) = R_e(1 + \beta) i_b = R_e(1 + \beta) i_1.$$

This shows that the presence of $R_e$ results in an extra resistance $(1 + \beta) R_e$ in the input circuit, and the input impedance is increased by this amount. Taking as typical values $\beta = 100$, $R_e = 3\ \text{k}\Omega$, we find an input resistance of order $300\ \text{k}\Omega$, a value which is much larger than for the simple CE connection in which $R_e = 0$.

The effect of $R_e$ in stabilizing the performance can be seen from an analysis of the circuit of Fig. 20.4, which includes the provision of bias for the base by means of the potentiometer formed by $R_1$, $R_2$. For a silicon transistor, the voltage $V_{be}$ across the base–emitter diode is $0\cdot6$–$0\cdot7$ V; to avoid the effect of this spread, it is desirable that the voltage across $R_e$ should be large compared with $V_{be}$, a value of about $3$ V being used in

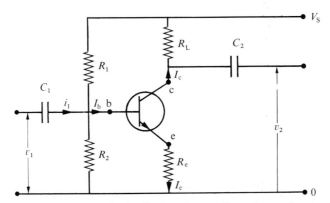

FIG. 20.4. A transistor circuit with feedback resistance $R_e$, and potentiometer formed by $R_1$, $R_2$ to provide the correct voltage on the base. With the correct choice of resistances, the current gain, voltage gain, and input impedance can be made virtually independent of the transistor characteristics.

practice. If the emitter current at the quiescent point is $1\,\text{mA}$, then $R_e$ must be $3\,\text{k}\Omega$. The voltage required on the base is then about $3\cdot6\,\text{V}$, and if the supply voltage $V_S = 7\,\text{V}$, roughly equal values of $R_1$, $R_2$ will provide the required base voltage.

We now examine how the value of the emitter current $I_e = (1+\beta)I_b$ depends on the circuit parameters. It is readily shown that

$$I_e\left\{R_e + \frac{1}{(1+\beta)}\left(\frac{R_1R_2}{R_1+R_2}\right)\right\} = \left(\frac{R_2}{R_1+R_2}\right)V_S - V_{be}. \tag{20.7}$$

If the quantities on the right-hand side are constant, the variation of $I_e$ with $\beta$ is given by

$$\frac{dI_e}{I_e} = -\frac{d(1+\beta)^{-1}}{(R_e/R_B)+(1+\beta)^{-1}} \tag{20.8}$$

showing that, if $(1+\beta)^{-1} = 0\cdot01$, we need $R_e/R_B$ to be large compared with $0\cdot01$ to lessen the dependence of $i_e$ on $\beta$. Here $R_B = R_1R_2/(R_1+R_2)$, and if $R_1 = R_2 = 15\,\text{k}\Omega$, so that $R_B = 7\cdot5\,\text{k}\Omega$, and $R_e = 3\,\text{k}\Omega$, the fractional variation in $I_e$ is reduced by a factor 41 compared with its variation when $R_e = 0$. A similar factor applies to its dependence on $V_{be}$. Thus the circuit fulfils two of the requirements: it supplies stabilization against fluctuations in $\beta$ and in $V_{be}$, and it helps to reduce the dependence of $I_e$ on $V_S$, though $I_e$ still drops from $1\,\text{mA}$ to $0\cdot37\,\text{mA}$ if $V_s$ drops to half its normal value.

For an alternating current the battery maintaining the supply voltage $V_S$ is essentially a short-circuit, and to a signal source $v_1$ the resistances $R_1$, $R_2$ are in parallel. The input impedance is therefore $R_B$ in parallel with $r_i'$, where $R_B = 7\cdot5\,\text{k}\Omega$ while $r_i' = (1+\beta)R_e = 300\,\text{k}\Omega$. Since $r_i'$ is proportional to $\beta$, it shows the same variation as $\beta$, but for the circuit of Fig. 20.4 the input impedance is virtually just $R_B$, which is independent of $\beta$.

Of the current $i_1$ drawn from the source, only a fraction $R_B/(R_B+r_i')$ flows through the base, and the current gain is therefore reduced from $i_c/i_b = \beta$ to

$$A_i = \frac{i_c}{i_1} = \beta\frac{R_B}{R_B+R_e} \sim \frac{R_B}{R_e}, \tag{20.9}$$

where we have used the approximations of neglecting $\beta^{-1}(R_B+R_e)$ in comparison with $R_e$, and taking $r_i' = (1+\beta)R_e$. In the same approximation the voltage gain $A_v$ is

$$A_v = \frac{i_cR_L}{i_br_i'} = \frac{\beta R_L}{(1+\beta)R_e} = \frac{R_L}{R_e}. \tag{20.10}$$

Thus we see that to a good approximation the current gain, the voltage gain, and the input impedance are all independent of the characteristics of the transistor.

For alternating voltages, the emitter voltage is $v_e = i_e R_e = i_b(1+\beta)R_e$ $= i_b r_i'$, which is the voltage applied to the base. Hence this circuit behaves in the same way as the 'emitter follower' circuit, and the approximation that $v_b = v_e$ is often useful in circuit analysis.

## 20.3. The small-signal low-frequency amplifier

A small-signal amplifier is one where the input voltage is sufficiently small that the resulting change in output current is only a small fraction of the mean output current; usually, a parameter such as $g_m$ can be assumed to be constant within the operating range. To provide more amplification than is available from a single transistor, several stages are coupled together 'in cascade'. The coupling must allow the alternating voltages to be transferred with minimum loss, while preserving the correct steady potential on the transistor electrodes. For audio-frequencies, the most common method employs resistance–capacitance ($RC$) coupling. In Fig. 20.4, the output voltage is fed through the capacitance $C_2$ to the input of the next stage (that is $C_2$ of one stage is identical with $C_1$ of the next stage). To ensure that uniform amplification is maintained over a range of frequencies, the susceptance $1/\omega C$ of the coupling capacitor must be small compared with the input impedance of the next stage down to the lowest frequency at which the amplifier is required to operate. The high-frequency limit is usually set by the Miller effect (§ 20.1), which reduces the input impedance of each stage. As a result of the coupling, the load for each stage is now $R_L$ in parallel with the input impedance of the following stage, which for Fig. 20.4 is $R_1$, $R_2$ in parallel ($R_B$ in the previous section). By making $R_B$ large compared with $R_L$, nearly all the output voltage is transferred to the input of the next stage.

*Direct coupling*

Solid-state devices do not require large differences between the steady potentials on the electrodes, and it is possible to couple two stages together directly, provided that each transistor can be maintained at a suitable Q point. No capacitor is required, reducing the number of components with no loss of gain, and making d.c. amplification possible. Two direct-coupled amplifiers are shown in Figs 20.5 and 20.6, the first using two npn transistors in the CE connection, the second an npn followed by a pnp transistor. It is advantageous to arrange that the major part of the current output of the first transistor flows into the second transistor; this requires the input impedance of the second stage to be small compared with the load resistance $R_L$ in the collector lead of the first transistor. Then, for the pair, the over-all current gain $A_i$ is equal to the product of the individual current gains. The voltage gain $A_v$ $= A_i \times$ (load resistance of second circuit/input resistance of first circuit).

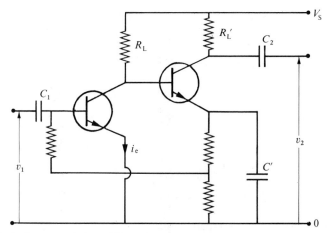

FIG. 20.5. A direct-coupled amplifier using two npn transitors in the CE connection. The capacitance $C'$ reduces the negative feedback due to the resistors in the emitter lead of the second transistor.

The base voltage for the first stage is provided by the potentiometer in the emitter circuit of the second stage. The resistance of the potentiometer also gives d.c. stabilization, but the capacitor $C'$ shunted across it means that there is no negative feedback for the alternating current.

The ability to use either npn or pnp transistors gives flexibility in the design. With two identical stages, as in Fig. 20.5, whether the phase change in each be 0 or $\pi$, the input and output voltages must be in phase. It is readily verified that the phase change is also zero for Fig. 20.6. In

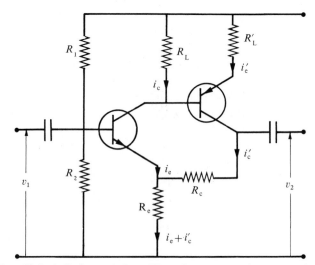

FIG. 20.6. A direct-coupled amplifier using an npn and a pnp transistor.

this circuit, the resistance connected between the collector of stage 2 and the emitter of stage 1 provides negative feedback. As we have seen, the 'emitter follower' nature of stage 1 means that its base voltage follows its emitter voltage, so that $R_c$ is effectively connected between the collector of stage 2 and the base of stage 1, and the feedback is carried across two stages.

### The Darlington compound

A simple and much used combination of transistors is the 'Darlington compound' circuit of Fig. 20.7, consisting of two transistors in which the second forms the load for the first. The over-all current gain is approximately $\beta^2$, and can be as high as several thousand. The base current of the second transistor is the emitter current of the first, and it follows that the first must be operated at comparatively low emitter current, while the second has a much higher emitter current. The circuit is therefore suitable for applications requiring a fairly high output current.

The Darlington compound is used in a circuit very much like that for a single transistor, but with a higher current gain. In the circuit of Fig. 20.7, negative feedback is provided by the load $R$, which plays a role similar to $R_e$ in Fig. 20.4. Essentially the transistors are in two 'emitter follower' circuits, so that the voltage gain is near unity, provided that $R$ is large compared with the input resistance of the emitter–base diode given by eqn (18.4). The current gain $\sim R_B/R$, where $R_B = R_1 R_2/(R_1 + R_2)$.

### Direct coupling with the MOSFET

The very low gate current drawn by a MOSFET makes it suitable for direct coupling. A simple amplifier is shown in Fig. 20.8, in which negative feedback is carried back over three stages by means of the resistance $R_F$. This resistance carries no direct current, so that the gate

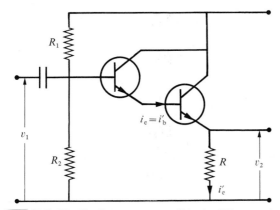

FIG. 20.7. The 'Darlington compound' in the CE connection.

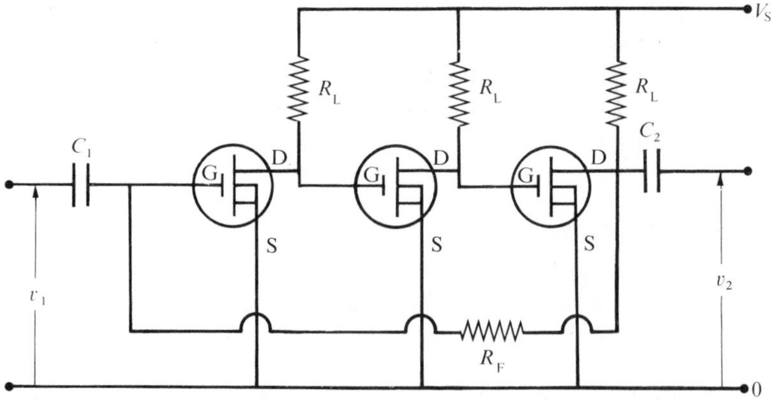

FIG. 20.8. A direct-coupled amplifier using three MOSFETs with negative feedback.

voltage of the first FET is equal to the drain voltage, which is the same for all three FETs provided that $R_L$ is identical in each stage. The MOSFETs are used in the normally-off (enhancement) mode.

Essentially the circuit is the same as that of Fig. 20.2; the feedback admittance is a pure conductance, and amplification is provided by the three FETs. From eqn (20.3) the input admittance is $(1-A)/R_F$ in parallel with the input impedance of the first stage. The latter is so high that the input resistance is effectively $R_F/(1-A)$, or approximately $R_F/(-A)$, even though the value of $R_F$ may be as high as 50 MΩ. There is a phase change of $\pi$ at each section, since an increase in the input voltage causes the current flowing through $R_L$ to rise, so that the drain voltage falls. To give a positive input resistance, an odd number of stages is required; in practice three must be used to give sufficient amplification, which may be several thousand.

## 20.4. Amplification at radio-frequencies

At high frequencies the Miller effect, giving a low input impedance, becomes very important. Because of the capacitance between the electrodes, the input impedance is usually capacitative; inductance is added in parallel to produce a parallel tuned circuit, which has a fairly high impedance (see § 7.3). Other electrode capacitances, and stray capacity of lead wires are treated similarly; a typical circuit is the tuned-gate, tuned-drain amplifier shown in Fig. 20.9. The use of tuned circuits means that only a narrow range of frequencies near the resonance frequency is amplified, since the admittance of each tuned circuit is approximately (see Problem 7.1)

$$\mathbf{Y} = G + 2\mathrm{j}\,\Delta\omega C,\tag{20.11}$$

where $G$ is the conductance at resonance, $\Delta\omega$ is the departure of $\omega$ from the resonance value $\omega_0 = (LC)^{-\frac{1}{2}}$. If the FET has mutual conductance $g_m$ and output resistance $r_0$, the amplified voltage across the output tuned circuit is

$$\frac{-g_m v_G}{r_0^{-1} + G + 2j\,\Delta\omega C_2} = A v_G. \tag{20.12}$$

The amplification therefore falls by a factor $1/\sqrt{2}$ at frequencies deviating from the resonant frequency such that

$$\pm 2\,\Delta\omega C_2 = r_0^{-1} + G.$$

If the selectivity is defined as $f/(2\,\Delta f) = \omega/(2\,\Delta\omega)$, so that it is analogous to the $Q_F$ of a resonant circuit, we see that the selectivity is that of the tuned circuit shunted by the output resistance of the FET. The presence of more than one tuned circuit increases the over-all selectivity, since each reduces the response at frequencies away from resonance, and the over-all selectivity must be adequate to accommodate the range of frequencies comprised in a modulated signal (see §§ 18.5 and 18.7).

If $C_{GD}$ is the electrode capacitance between gate and drain, it follows from eqn (20.3) that the input admittance of the FET is

$$(1-A)j\omega C_{GD}.$$

From eqn (20.12) above, the amplification is of the form $(-A_0 + jA'\,\Delta\omega)$, and the input admittance therefore contains a real part $+\omega C_{GD}(A'\,\Delta\omega)$ which vanishes at resonance ($\Delta\omega = 0$) but is negative when $\Delta\omega$ is negative. At frequencies below the resonant frequency there is a negative conductance in parallel with the input-tuned circuit, and if this is sufficiently negative to outweigh the positive conductance corresponding to the losses, the circuit will break into oscillation. This is avoided by balancing the feedback through the interelectrode capacitance by providing another

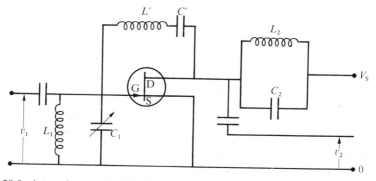

FIG. 20.9. A tuned-gate, tuned-drain amplifier for 100 MHz, with neutralizing feedback circuit.

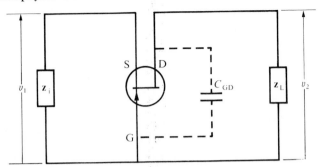

Fig. 20.10. A high-frequency amplifier using an FET in the 'common gate connection'. $z_L$ and $z_i$ are parallel tuned circuits, and may be quarter-wave tuned lines.

feedback path, a process known as 'neutralization'. In the circuit of Fig. 20.9, the inductance $L'$ provides this extra path, $C'$ being simply the capacitance of a 'blocking capacitance' of low impedance to maintain the required steady bias voltages on drain and gate. The inductance $L'$ forms a parallel tuned circuit with the electrode capacitance $C_{GD}(\sim 1\ \text{pF})$, making the feedback admittance small and primarily inductive rather than capacitative. Any negative component in the input admittance is thus very much reduced, and oscillation is avoided.

Amplification at still higher frequencies can be obtained by using an FET in the 'common gate' connection illustrated in Fig. 20.10. Here the gate–drain capacitance $C_{GD}$ becomes part of the tuned circuit forming the load $z_L$. As in the CC connection for the junction transistor (§ 19.3), negative feedback is provided because the output current flows through $z_i$ in the source–drain circuit, and the voltage amplification is simply $z_L/z_i$, since the gate current is negligible. Both $z_L$ and $z_i$ are parallel tuned circuits, and may be quarter-wave tuned lines (see § 9.5).

## 20.5. Power amplifiers

Because of the non-linear characteristics of active devices, the output will always contain some distortion. If in eqn (18.8) we take $i$ to be the current output and $v$ the signal input, it is clear that distortion produces harmonics of the signal frequency in the output current. For an r.f. amplifier whose load is a tuned circuit, the harmonic content of the *voltage* output is small because the load has a low impedance for the harmonics and a high impedance for the fundamental frequency. In amplifiers for a range of frequencies the even harmonics can be eliminated by the use of a 'push–pull' circuit, consisting of two matched devices to which the same input voltage is applied but with opposite

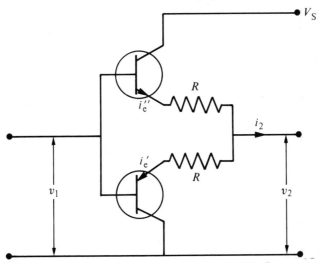

Fɪɢ. 20.11. Push–pull amplifier circuit using a pair of matched transistors, one npn and the other pnp.

phase. The output currents then have the form

$$i' = g_0 + g_1 v + g_2 v^2 + g_3 v^3 + g_4 v^4 + ...,$$
$$i'' = g_0 - g_1 v + g_2 v^2 - g_3 v^3 + g_4 v^4 + ...,$$

(20.13)

and by arranging the circuit so that the output current is the difference between these two currents, we obtain an output

$$i' - i'' = 2(g_1 v + g_3 v^3 + ...),$$

(20.14)

from which the even harmonics are missing. A simple circuit, using a matched pair of transistors, one npn and the other pnp, is shown in Fig. 20.11. The same signal voltage $v_1$ is applied to the base of each transistor, but because of the reversed polarity, the difference $(i''_e - i'_e)$ between the two emitter currents gives the current $i_2$ through the load, and all even harmonics are missing. The most important remaining harmonic is the third; this is much smaller for FETs, with their square-law characteristics, than for junction transistors with exponential characteristics.

*Efficiency of power amplifiers*

In an amplifier the ultimate source of the a.c. output power delivered into the load is the supply; this is not obvious, because the current drawn from the supply does not change when a signal is being amplified if there is no distortion. If the load is a resistance $R_L$ in the collector lead carrying a current $I_c$, the power drawn from the supply (voltage $V_s$) is $I_c V_s$, of which an amount $i_c^2 R_L = I_c(V_s - V_c)$ is dissipated in the load. The remainder $I_c V_c$, where $V_c$ is the collector voltage, is dissipated in the transistor and appears as heat. When a signal voltage is applied to the base, the collector current becomes $I_c + i_c \sin \omega t$, and the collector voltage becomes $V_s - R_L(I_c + i_c \sin \omega t)$. The mean power dissipated in the transistor is given by the product of these two, and contains terms in $\sin \omega t$ which average to zero over a cycle, the remainder being

$$V_s I_c - R_L I_c^2 - R_L i_c^2 \sin^2 \omega t.$$

The first part of this expression, independent of $i_c$, is the power dissipated in the transistor in the absence of a signal, while the last, which averages to $\frac{1}{2} R_L i_c^2$, represents the reduction in transistor heating when a signal is present. It is just equal to the signal power in the load. In physical terms, the reduction in heating of the transistor arises from the fact that the current through it increases when the collector voltage decreases, and vice versa, so that on average $V_c I_c$ is diminished.

The efficiency of an amplifier is equal to the ratio of the a.c. power output to the power drawn from the supply. In a typical example with 5 W output, $I_c = 1$ A, $i_c = 0\cdot9$ A, $V_s = 12$ V, and $R_L = 15$ Ω. The efficiency $\frac{1}{2} R_L i_c^2 / I_c V_s$ is close to 50 per cent.

*Class A, B, and C operation*

An amplifier in which the transistor is operated on the linear portion of the characteristic is known as a Class A amplifier. Higher efficiency can be obtained by biasing the transistor to a point such that the main current flow occurs when the output electrode is at the lowest voltage point during the alternating voltage swing. In a Class B amplifier, the transistor is biased close to the point where the steady current from the output electrode is zero. Current then flows only during one half of the r.f. cycle, giving higher efficiency but considerable distortion. This is reduced by using a pair of transistors in push–pull, one transistor providing current during one half of the cycle and the other in the other half.

In Class C operation the transistor is biased even further back, so that current flows only for a small fraction of a cycle. The amplitude of the applied alternating voltage must be slightly greater than the amount by which the input electrode is biased beyond cut-off. The efficiency is high, but the output is very distorted, consisting only of short pulses of current repeated at the signal frequency. Class C amplifiers are used only for power output stages; a large drive voltage is needed, and distortion in the output voltage is reduced by use of a tuned circuit for the output load. If harmonics of the input signal are required, the load can be tuned to a multiple of the input frequency; the distorted output current has a high harmonic content, each harmonic being an exact multiple of the fundamental. Care must be taken to avoid unwanted oscillations at such higher frequencies because of positive feedback through interelectrode and stray capacity.

## 20.6. Oscillators

If the feedback in an amplifier is such that the factor $AF$ in eqn (20.1) is positive and greater than unity, the circuit becomes unstable. The power fed back to the input is enough to compensate for all losses in the input circuit, so that the net resistance becomes negative. Reference to § 5.3 shows that any transient signal in the circuit then increases exponentially instead of decaying, power from the d.c. supply being converted into a.c. power. No external input signal is needed; the impulse from switching on the supply will start oscillations if sufficient feedback is present.

The design of the oscillator depends on the desired waveform, frequency, and output; some typical examples are described below. Most solid-state oscillators are adapted from circuits devised for thermionic vacuum tubes, and some still bear the names of their originators. At radio-frequencies a tuned circuit is used, either in the input or the output. At high frequencies sufficient feedback may be present through the Miller effect, but more generally it is provided by mutual inductance coupling. An equivalent circuit for such an oscillator, with the tuned circuit at the output, is shown in Fig. 20.12. For the input circuit, if $r_i$ is large, the flow of current in the circuit can be neglected, with considerable simplification in the analysis. We have then

$$v_1 = M(di/dt)$$

and

$$i_2 = g_m v_1 + v_2/r_0,$$

so that                                                                     (20.15)

$$g_m M(di/dt) = i_2 - v_2/r_0.$$                                            (20.16)

For the output circuit

$$i_2 - i = -C(dv_2/dt),$$

which can be combined with (20.16) to eliminate $i_2$, giving

$$g_m M(di/dt) = i - v_2/r_0 - C(dv_2/dt).$$                                 (20.17)

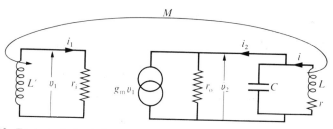

FIG. 20.12. Equivalent circuit for an oscillator with mutual inductance coupling between input and output.

Finally we have also

$$v_2 + L(di/dt) + ri = 0, \tag{20.18}$$

so that $v_2$ can be eliminated to give

$$LC\left(\frac{d^2i}{dt^2}\right) + \frac{di}{dt}\left(rC + \frac{L}{r_o} - g_mM\right) + i\left(1 + \frac{r}{r_o}\right) = 0. \tag{20.19}$$

This differential equation represents oscillations which will be sustained provided that

$$g_mM = rC + \frac{L}{r_o} = L\left(\frac{1}{R} + \frac{1}{r_o}\right), \tag{20.20}$$

where $R = L/Cr$ is the parallel impedance of the tuned circuit, with which the output impedance $r_o$ is in parallel. The frequency of oscillation is given by

$$\omega^2 LC = 1 + r/r_o, \tag{20.21}$$

and hence depends slightly on the value of $r_o$. In practice $g_mM$ must exceed the value given by (20.20), and the level of oscillation builds up until curvature in the characteristic of the active device reduces the average value of $g_m$ to the level required by (20.20). The resonance frequency is also influenced by the coupling (not included here) needed to extract power from the output circuit, and it may drift if the electrical constants of the components change with temperature. For high-frequency stability, a quartz-crystal oscillator (see § 22.5) is used. This is a low-power oscillator (for example, a few watts), which is followed by r.f. amplifiers (Class C for efficiency) to increase the level to the required output.

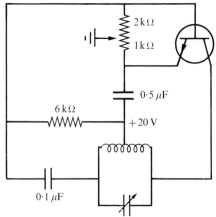

FIG. 20.13. Basic circuit for a Hartley oscillator.

*The Hartley oscillator*

When a large power output is required, but good frequency stability is not essential, an oscillator run under Class C conditions is used. Like the Class C amplifier it gives high efficiency, the input electrode being biased back well beyond cut-off. The excitation voltage must have sufficient amplitude to carry the device into the conducting region at the positive peaks, a convenient circuit being the Hartley oscillator shown in Fig. 20.13. A tapped inductance, used as an autotransformer, supplies the feedback, the emitter of a junction transistor being connected to a point near the centre of the inductance through a blocking capacitor which provides a low-impedance path for the radio-frequency. The other two electrodes are connected to opposite ends of the inductance, where the alternating potentials are in antiphase and provide feedback of the right sign. The Q point is fixed by the emitter and base resistors, which may have to be adjustable so that oscillations can start under Class A conditions, since sufficient input swing to make current flow may not be available for starting under Class C conditions.

*Resistance–capacitance oscillators*

At low frequencies the inductance needed for a tuned circuit becomes tiresomely large, and it can be avoided by the use of a resistance–capacitance oscillator, using a circuit for which feedback in the correct phase is obtained only at one frequency. In the network shown in Fig. 20.14, the voltage ratio $v_2/v_1$ is

$$\frac{v_2}{v_1} = 1 + \frac{R}{R'} + \frac{C'}{C} + j\left(\omega C'R - \frac{1}{\omega CR'}\right) \qquad (20.22)$$

and the phase difference is zero at the frequency which satisfies the relation $\omega^2 CC'RR' = 1$. If $v_2$ is the output and $v_1$ the input of an amplifier with zero phase shift between input and output, oscillations will take place at the frequency just given provided the gain is greater than $(1 + R/R' + C'/C)$, a condition which is readily satisfied for, say, $R = R'$, $C = C'$. By switching in a number of capacitors, increasing in size by a factor 10, and varying $R$, $R'$, oscillations can be obtained over a range of

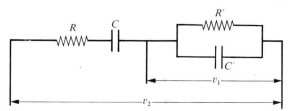

FIG. 20.14. A resistance–capacitance network for providing feedback in a low-frequency sine-wave oscillator.

frequencies from about 10 Hz to over 100 kHz, while the feedback ratio can be kept constant by using ganged resistors to keep the ratio $R/R'$ constant. This circuit, known as a 'Wien bridge' oscillator from its resemblance to the Wien bridge (§ 22.5), is often used in audio-frequency signal generators.

### 20.7. The multivibrator

The oscillators considered so far produce sinusoidal oscillations whose frequency is controlled almost entirely by the circuit parameters; even in a Class C oscillator the voltage across the tuned circuit is nearly sinusoidal in form, though the current is not.

The question arises, what happens if feedback is introduced into an untuned amplifier, of sufficient size and the correct sign to produce instability. The device shown in Fig. 20.15 consists of a two-stage aperiodic amplifier with feedback from the output of the second stage to the input of the first stage. If the voltage amplification of each stage is $A$, the over-all amplification is $A^2$. The feedback factor $F$ is practically unity, so that if $A^2 > 1$, the device should be unstable.

In analysing this system we must take account of the electrode capacitance and other stray capacitances in the amplifier; they are represented by the small capacitance $C$ shunting the load resistance $R$ in each stage. The output current of the first stage is

$$i_1 = g_m v_1 + v_2/r_0,$$

where lower-case symbols indicate we are dealing only with fluctuating components. Here $v_1$ is the voltage change applied at the input of stage 1, and its output voltage change $v_2$ gives the voltage change at the second

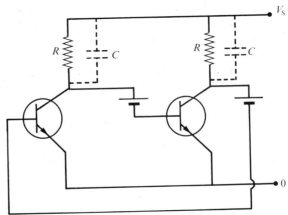

Fig. 20.15. Two-stage aperiodic amplifier with feedback.

input. Since the current $i_1$ flows through $R$ and $C$ in parallel, we have

$$-i_1 = C(dv_2/dt) + v_2/R.$$

Elimination of $i_1$ gives a relation between $v_1$ and $v_2$, and on applying an identical analysis to the second stage a similar relation is obtained with $v_1$ and $v_2$ interchanged. Hence, writing $R'$ for the resistance formed by $R$ and $r_0$ in parallel,

$$-g_m v_1 = C(dv_2/dt) + v_2/R'$$
$$-g_m v_2 = C(dv_1/dt) + v_1/R' \tag{20.23}$$

The voltage amplification in each stage is $A = -g_m R'$, and the equations can be written as

$$Av_1 = \tau(dv_2/dt) + v_2,$$
$$Av_2 = \tau(dv_1/dt) + v_1. \tag{20.24}$$

Here $\tau = CR'$ is the time constant of the capacitance $C$ in parallel with $R$ and $r_0$. The solution of these equations is

$$v_1 = -v_2 = v_0 \exp\{-(A+1)t/\tau\}. \tag{20.25}$$

If the value of $-A$ for each stage is greater than unity, this solution shows that any disturbance of the input potential will increase exponentially. The input to one stage rises while the other decreases at the same rate; current increases in the former until it 'saturates', and decreases in the other until it cuts off. The situation remains until a short pulse applied in the input of the stage which is cut off brings it into the conducting region; then the exponential rise in its input voltage carries it to saturation while the other stage is changing back from saturation to cut-off. The time required for the reversal is very short. If $-A \gg 1$, the time constant in the exponential of eqn (20.25) is approximately $\tau/(-A) = C/g_m$, which is of order $10^{-8}$ s for $C = 15$ pF, $g_m = 1 \cdot 5$ mA $V^{-1}$.

Essentially the circuit acts as a very fast current switch; the two stages are respectively 'on' and 'off', but a rapid change to 'off' and 'on' can be triggered by either a short positive pulse to the input of the 'off' stage or a negative pulse to the 'on' stage. In the multivibrator, direct coupling between the stages is replaced by $RC$ coupling, as in Fig. 20.16, and a periodic reversal now occurs automatically. Suppose that at a given instant stage 1 is 'on,' and 2 is 'off'. The voltage at input 2 is negative, but it is falling in size as the current through $R_1$ required to charge the capacitor $C_1$ declines towards zero. When conduction in stage 2 becomes possible, the exponential voltage 'landslide' described above takes place. The voltage at point Y in Fig. 20.16 drops suddenly, and this drop is transferred through $C_2R_2$ to the input of stage 1, which therefore switches from 'on' to 'off'. The voltage across $R_2$ now decays as $C_2$ recharges to its

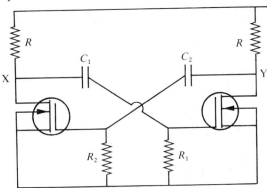

FIG. 20.16. The multivibrator. With a MOSFET in each stage, and $R_1 = R_2 = 68\ M\Omega$, $C_1 = C_2 = 1\ \mu F$, square waves with a period of 2 min can be obtained.

equilibrium value, and the reverse landslide occurs when it has fallen sufficiently to permit conduction in stage 1. The cycle is now complete. The point Y is alternately at the supply voltage when the current through the second FET is zero, and at a low voltage determined by the resistance R and the current through it; point X undergoes a similar cycle in antiphase. The output waveform is very nearly rectangular in shape, and the period of a complete cycle is mainly determined by the sum of the times required for $C_1$ to charge through $R_1$, and $C_2$ through $R_2$.

The multivibrator is readily synchronized with a sinusoidal signal applied to the input of one stage, provided its natural period is close to that of the sine wave. The effect of the latter is to delay the return of the input voltage to the conducting point if it is arriving early, and to speed it up if its natural period would cause it to return too late. This property of synchronization is of use in frequency measurement; the square waves have a high harmonic content, and the harmonics occur at exact multiplexes of a standard frequency. The multivibrator may also be used for frequency division, for it will synchronize with a signal whose period is close to an exact fraction of its own; that is, a frequency of 5–10 times its own.

By changing the methods of coupling between the two stages a useful variety of switching units can be produced. The 'monostable' circuit has a stable state in which stage 1 is normally 'on' and stage 2 is 'off', but a negative pulse to the input of stage 1 will trigger a short-lived state of pre-determined duration in which 1 is 'off' and 2 is 'on'. In the 'bistable' circuit or 'flip–flop' both the 'on–off' and the 'off–on' states are stable; a transition from one to the other is triggered by a negative pulse to the 'on' stage. A complete cycle requires two pulses, one steered to each stage, and a succession of such units can form a 'scaler' in which the number of pulses is divided by 2 in each unit.

## 20.8. The operational amplifier

The operational amplifier consists of a very high gain, d.c.-coupled differential amplifier with single-ended output. Typically, it contains about 20 transistors and a dozen resistors, incorporated as an integrated circuit in a single chip. Ideally, all parameters should be independent of frequency, temperature, supply voltage, time, etc. Practical and ideal voltage amplifiers are compared in Table 20.1.

TABLE 20.1

*Comparison of the ideal parameters for an operational amplifier with those achieved in practice*

|                              | Ideal    | Practical           |
| ---------------------------- | -------- | ------------------- |
| (a) gain A                   | Infinite | $10^5$ to $10^7$    |
| (b) input impedance          | Infinite | 2–50 megohm         |
| (c) output impedance         | Zero     | 60–300 ohm          |
| (d) input bias current       | Zero     | 0.03–80 nA          |
| (e) input offset current     | Zero     | 0.005–20 nA         |
| (f) input offset voltage     | Zero     | 0.02–5 mV           |
| (g) slew rate                | Infinite | 0.3–10 V per $\mu$s |
| (h) bandwidth at unit gain   | Infinite | 0.6–3 MHz           |

As in all transistor amplifiers, the input stages need to be biased. The bias currents into the two terminals are not the same, and the difference between them $I_{b+} - I_{b-}$ is called the input offset current. It is usually an order of magnitude smaller than the input bias current (see Table 20.1). Ideally the d.c. output should be zero for zero d.c. input, but the flow of the input bias current through the external circuits results in a small voltage at the input, causing a voltage offset at the output. To ensure zero voltage at the output requires the application of a small input voltage, known as 'the input offset voltage'. Note that in (b), (c) a voltage amplifier is assumed: for a current amplifier the input impedance should be zero and the output impedance infinite (cf. Fig. 3.6). In some amplifiers, FETs are used for the first stages: in a voltage amplifier the input impedance is then nearly $10^5$ megohm, and the input bias current is effectively zero.

In (g), ideally any sudden large change in input should cause an instantaneous change in output: this would be an infinite 'slew rate'. The actual rate is determined mainly by the time required to change the charges on the internal capacitors, and in practice amplifiers are not limited by the slew rates shown in Table 20.1.

In an amplifier, one input terminal is usually earthed and the signal is applied to the live terminal. In an operational amplifier, neither input terminal is earthed, and signals can be applied to each input terminal. Only the difference is amplified—hence the name 'differential amplifier'. If input voltages $V_+$, $V_-$ are applied to the non-inverting $(+)$ and inverting $(-)$ inputs (see Fig. 20.17), the differential output voltage $V_{out}$ is given by

$$V_{out} = A[(V_+ - V_-) + \tfrac{1}{2} A_{CMRR}(V_+ + V_-)].$$

Here A is the open loop differential gain, and is commonly a function of frequency as well as of $(V_+ - V_-)$ and of $(V_+ + V_-)$. The quantity $A_{CMRR}$ is the 'common mode rejection ratio': it is generally less than $10^{-6}$ and sometimes less than $10^{-12}$, and is often omitted.

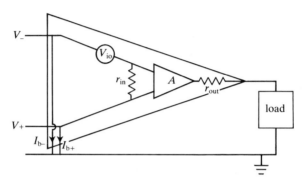

FIG. 20.17. Schematic diagram for an operational amplifier. The input bias currents are $I_{b+}$, $I_{b-}$, and their difference is the input offset current: $V_{io}$ is the input offset voltage: the input and output resistances are represented by $r_{in}$, $r_{out}$.

In general, A is inversely proportional to frequency above about 10 Hz, and the 'unity gain frequency' is that at which A falls to 1. To counteract the drop in A, a resistive network is added to the amplifier to give feedback, denoted by a factor F. This is negative in sign, but modulus signs ($|F|$) are included below to avoid confusion. A plot of the logarithm of the power gain (i.e. the gain expressed in decibels) against the logarithm of the frequency is known as a 'Bode Plot'. An example is shown in Fig. 20.18, together with the corresponding phase shift; the performance can be predicted as follows. On open circuit, the 'closed loop' gain with negative feedback is denoted by $A_{CL}$, and its value is given by

$$A_{CL} = A/(1 + |F|A).$$

Hence, if $|F|A \gg 1$, the value of $A_{CL}$ tends to $1/|F|$, and the phase angle between output and input to 0. For the opposite extreme where $|F|A \ll 1$, the value of $A_{CL}$ tends to A, and the phase angle to $-90°$. In Fig. 20.18(a), the closed loop gain is almost entirely given by the two lines $BC = 20 \log_{10} |F|^{-1}$ at low frequencies ($|F|A \gg 1$), and $CD = 20 \log_{10} A$ at high frequencies ($|F|A \ll 1$). These two lines intersect at a frequency $\omega_b$, known as the 'breakpoint

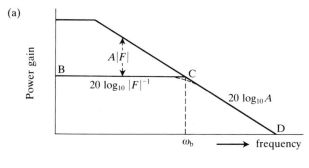

(a)

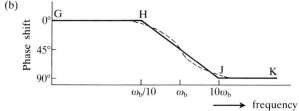

(b)

FIG. 20.18. Bode Plots. (a) Power gain v. frequency, both on logarithmic scales. The 'closed loop' gain is close to the lines BC, CD. (b) Phase angle v. frequency, both on logarithmic scales. The phase shift is close to the lines GH, HJ, JK. $|F|$ is the (negative) feedback factor, and $\omega_b$ the breakpoint frequency.

frequency': the actual gain is illustrated by the broken line. The curve of phase shift against frequency, shown by the broken line in Fig. 20.18(b), is close to the three straight lines GH, HJ, JK.

A range of applications may be illustrated with a simple circuit involving an operational amplifier with different impedances to provide feedback, as in Fig. 20.19. Some examples are:

(a) $Z_1 = 10$ kilohm resistance, $Z_2 = 100$ kilohm resistance: a simple inverting amplifier with voltage gain $= Z_2/Z_1 = 10$. A resistance equal to $R_1$, $R_2$ in

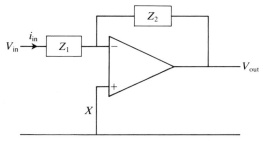

FIG. 20.19. Simple multi-purpose circuit using an operational amplifier with feedback impedances $Z_1$, $Z_2$, that may be resistances or capacitances. When $Z_1 = R_1$, $Z_2 = R_2$, a resistance equal to $R_1$ and $R_2$ in parallel may be inserted at X to balance the leakage currents through $R_1$, $R_2$.

parallel (e.g. 9.1 kilohm) is sometimes included in the positive lead at X to balance the leakage currents through $R_1$ and $R_2$.

(b) $Z_1 = R = 1$ megohm, $Z_1 = C = 1$ microfarad capacitance: integrating circuit, with $CR = $ 'time of integration'.

$$V_{out} = - (1/CR) \int^{\cdot} V_{in} \, dt.$$

(c) $Z_1 = 10$ nanofarad capacitance, $Z_2 = 1$ megohm resistance: differentiating circuit.

(d) $Z_1 = R_1$, with $Z_2 = C$, $R_2$ in parallel: low pass filter.

(e) $Z_1 = C$, $R_1$ in series, with $Z_2 = R_2 = 1$ megohm resistance: high pass filter.

Another useful circuit is the voltage follower shown in Fig. 20.20. It has 100 per cent feedback, and the gain is unity over a wide range of frequency. When a source has an output impedance comparable with that of the load, the voltage follower can be used as a buffer stage before the load: because of its very low output impedance, it gives a much better transfer of voltage to the load.

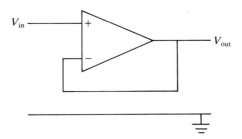

FIG. 20.20. Voltage follower circuit. This circuit has unity gain over a wide range of frequency and is used as a buffer stage after a circuit with output impedance comparable to that of the following load.

# 21. Thermionic vacuum tubes

THE initial development of modern electronics followed the invention of the thermionic vacuum tube. In the diode electrons are emitted from a heated cathode; they are attracted to an anode maintained at a positive voltage with respect to the cathode, under high-vacuum conditions inside a glass, ceramic, or metal envelope. In the triode the flow of electrons is controlled not only by the potential on the anode but, more importantly, by that on a metal grid inserted between cathode and anode. A heater supply is required for the cathode, consuming appreciable power. Thermionic vacuum tubes are more expensive to make, larger in size, and less rugged than solid-state devices, which have therefore replaced them except in certain special applications.

## 21.1. The thermionic diode

The emission of electrons from hot solids was discussed in § 11.5. The emission current per unit area of a cathode surface at temperature $T$ (in degrees Kelvin) is given by eqn (11.34),

$$J = AT^2 \exp(-\phi/kT), \tag{21.1}$$

where the constants $A$ and $\phi$ depend on the material, some representative values being listed in Table 11.1 (p. 335). The temperature at which adequate emission is obtained is determined primarily by the value of the work function $\phi$. Practical considerations, such as long life, limit the materials used in general to tungsten (2500 K), thoriated tungsten (1900 K), and a barium–strontium oxide mixture (1100 K), where the temperatures are those required for current densities of order $10^4 \, \mathrm{A \, m^{-2}}$.

The oxide-coated cathode, with its low working temperature, has the additional advantage that it can be heated indirectly, giving an equipotential cathode surface. However, it is easily 'poisoned' by the presence of gas, especially oxygen, and it is essential to maintain a good vacuum ($<10^{-4}$ Pa) for the whole life of the tube. Most surfaces contain occluded gas, which is gradually evolved in a vacuum at room temperature, but is rapidly evolved at high temperatures. Nickel is commonly used for the anode and larger electrodes, and tungsten wire for grids; they are 'out-gassed' by heating to ~1300 K *in vacuo* before the tube is assembled.

At high values of the anode voltage, all electrons emitted from the cathode reach the anode; the current density is determined solely by the

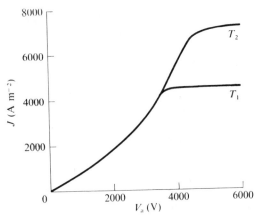

FIG. 21.1. Current–voltage curve for a diode with plane-parallel electrodes, 10 mm apart, for two different cathode temperatures $T_1$ and $T_2$ ($T_2 > T_1$).

temperature of the cathode (see Fig. 21.1) and is given by eqn (21.1). At lower anode voltages the current is smaller because of the presence of 'space charge' near the cathode (see § 3.7), and instead of being 'temperature limited' it is said to be 'space-charge limited'. Under these conditions, which correspond to the part of Fig. 21.1 nearer the origin, the current density is given by Child's law,

$$J = bV_a^{\frac{3}{2}}/d^2. \tag{21.2}$$

For a diode with plane parallel electrodes the value of $b$ is given in eqn (3.45) and $d$ is the separation between cathode and anode. From dimensional considerations a similar relation is expected for other geometries, with $d$ an appropriate distance.

### 21.2. The thermionic triode

In the thermionic triode a third electrode, known as the grid, is interposed between the cathode and the anode. If the electrodes are planar, the grid takes the form of a coarse wire mesh; if the electrodes are cylindrical the grid is wound in the form of a helix. The total current drawn from the cathode will now depend on the potentials of both grid and anode, since both control the field at the cathode. In the absence of space charge, the charge on the cathode, assumed to be at zero potential, is equal to

$$-(C_{cg}V_g + C_{ca}V_a) = -C_{cg}\left(V_g + \frac{V_a}{C_{cg}/C_{ca}}\right) = -C_{cg}\left(V_g + \frac{V_a}{\mu}\right),$$

where $C_{cg}$, $C_{ca}$ are the coefficients of capacitance between cathode and grid (at potential $V_g$) and anode (at potential $V_a$) respectively. These capacitances are in the same ratio as the division of lines of force from the cathode between the grid and the anode when $V_a = V_g$. The ratio $\mu = C_{cg}/C_{ca}$ is known as the 'amplification factor' of the triode; (the values of $C_{cg}$, $C_{ca}$ given in tube manuals do not normally fit this relation because they include the capacitances of the leads). If the current from the cathode is limited by space charge, so that the field at the cathode (as in the assumption made in deriving Child's law) is zero, then no lines of electric field reach the cathode, but all terminate on the space charge. Since the latter is mainly located very close to the cathode, it is affected equally by lines of force from the grid and from the anode, and the current leaving the cathode depends on the equivalent voltage $V_a' = (V_g + V_a/\mu)$. Experimentally it is found that the current increases very nearly with the three-halves power of this equivalent voltage in the region of complete space charge limitation, corresponding to the fact that near the cathode the potential distribution is the same as in an 'equivalent diode'. We may therefore write

$$J = b(V_g + V_a/\mu)^{\frac{3}{2}}/d^2, \tag{21.3}$$

where $d$ is known as the 'equivalent diode spacing'; for triodes with fairly high values of $\mu$ (that is, when nearly all the lines of force terminating on the space charge come from the grid) $d$ is nearly equal to the cathode–grid spacing in a plane triode. The constant $b$ has the same numerical value as for the diode. In general the triode is used with the grid voltage negative, so that no current flows to it, and the total current leaving the cathode (to which eqn (21.3) applies) is also the anode current. Current will leave the cathode so long as the equivalent voltage $(V_g + V_a/\mu)$ is positive, and a small negative grid voltage must be combined with a large positive anode voltage. This has the advantage that a source of voltage applied to the grid will influence the anode current without any current being drawn from the source by the grid.

For half a century the thermionic triode was the work-horse of electronics, but for small-power applications it has been superseded by solid-state triodes, its main use now being for high-power applications, such as radio transmitters and industrial purposes such as eddy-current heating. One set of characteristic curves, anode current against grid voltage, is shown in Fig. 21.2 for a power triode; since the anode current depends on the quantity $V_g + V_a/\mu$, we expect all the curves to be similar, but displaced to more positive values of $V_g$ as $V_a$ is increased, the shift in $V_g$ being $-(1/\mu)$ of that in $V_a$. A second set of characteristics is formed by plotting $I_a$ as a function of $V_a$ for various values of $V_g$, and a third set by plotting $V_g$ against $V_a$ for various values of the anode current. These characteristic curves are given in vacuum-tube manuals, and must be used in a numerical analysis when working at full power. In regions where they

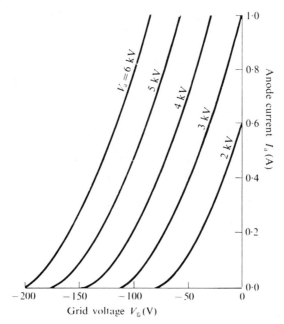

Fig. 21.2. Anode current against voltage characteristics for a power triode with $\mu = 32$. $g_m = 17$ mA V$^{-1}$ at $V_a = 4$ kV, $I_a = 1$ A. The cathode, a thoriated tungsten filament, requires 400 W heating power. The output power is 3–5 kW, depending on conditions of operation.

are approximately straight lines, the quantities

$$g_m = (\partial I_a/\partial V_g)_{V_a}; \qquad r_0 = (\partial V_a/\partial I_a)_{V_g}; \qquad -\mu = (\partial V_a/\partial V_g)_{I_a},$$

introduced in § 7.6, are meaningful parameters so long as the voltage swings are not excessive; they are also quoted in the manuals. In general the value of $\mu$ ranges from 20 to 35. Since $\mu = C_{cg}/C_{ca}$, it is substantially independent of the working conditions, being determined by the closeness of the winding of the grid and the relative distances of grid and anode from the cathode. If the grid is closely wound, the number of lines of force reaching the cathode from the anode is small compared with the number from the grid, and the magnification factor $\mu$ is high. If the grid is rather openly wound, $\mu$ is low.

Values of $g_m$ range roughly from 3 mA V$^{-1}$ to 40 mA V$^{-1}$, and depend on the working conditions. For example, at $V_a = 4$ kV and $I_a = 0.12$ A the value of $g_m$ may be 3·3 mA V$^{-1}$, while at $V_a = 1$ kV, $I_a = 2.3$ A, $g_m = 10$ mA V$^{-1}$ for the same triode. This variation with anode current is in line with the three-halves power law; from eqn (21.3) we expect that $(\partial I_a/\partial V_g)_{V_a}$, and hence also $g_m$, should vary as $I_a^{\frac{1}{3}}$.

The most important use of the triode is as an amplifier or oscillator, with the load in the anode lead as in Fig. 21.3. Small changes in the anode current obey an equation such as (7.71).

$$i_a = g_m v_g + v_a/r_0 \qquad (21.4)$$

and the voltage amplification (see eqn (7.66a)) is given by

$$A_v = \frac{v_a}{v_g} = \frac{-i_a \mathbf{z}_L}{v_g} = -g_m \frac{r_0 \mathbf{z}_L}{r_0 + \mathbf{z}_L}. \qquad (21.5)$$

Maximum power output is obtained by making $\mathbf{z}_L = r_0$. With a parallel tuned circuit as load, the maximum voltage swing can be equal to the supply voltage $V_s$ without driving the anode negative at any point. The power into the load, with $\mathbf{z}_L = r_0$, is then $\frac{1}{2}V_s^2/r_0$, showing that a triode with a low value of $r_0$ is required to obtain the highest power.

For low values of the grid swing, no electron current flows to the grid, but this may not be the case in high-power applications. For the triode whose characteristics are shown in Fig. 21.2, a typical working condition is with $V_g(\text{mean}) = -300$ V, but with a grid swing of amplitude 600 V. The average grid current is about $0.3$ A, with a grid dissipation of about 200 W, and an output power of $3.5$ kW. The Hartley circuit (Fig. 20.13), with minor modifications, can be used for a power oscillator.

At frequencies above about 100 kHz the effects of the interelectrode capacitances denoted by broken lines in Fig. 21.3 become important. The cathode–anode capacitance $C_{ca}$ (which is usually less than 1 pF) may be regarded as part of the load $\mathbf{z}_L$, but the effect of the grid–anode capacitance is magnified because it has the amplified signal voltage across it (see

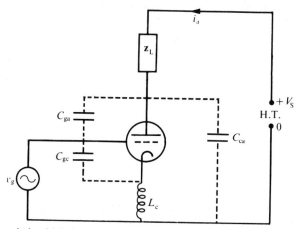

FIG. 21.3. The triode with its interelectrode capacitances. Since the high-tension (H.T.) supply must have zero impedance at the radio-frequency, the cathode–anode capacitance $C_{ca}$ is effectively in parallel with $\mathbf{z}_L$ and can be considered part of it. The small cathode lead inductance $L_c$ is important only at very high frequencies (see Problem 21.1).

Fig. 21.4). This is an example of the Miller effect, discussed in §§ 20.1 and 20.4 but originally derived for the thermionic triode; it produces an input admittance

$$y_{gc} = j\omega C_{gc} + j\omega C_{ga}(1-A_v).\qquad(21.6)$$

For a triode with $\mu = 30$ and $z_L = r_0$, $A_v = -15$, typical values are $C_{gc} = 5$ pF, $C_{ga} = 10$ pF, so that

$$C_g = 5+10(1+15) = 165 \text{ pF}.$$

At a frequency of 10 MHz, this corresponds to a capacitative input impedance of only 100 Ω.

The undesirable results of the Miller effect can be avoided by neutralization (cf. § 20.4), or by using the triode in the 'grounded grid connection' which is equivalent to the 'common gate' connection of Fig. 20.10. An alternative method is to reduce the value of $C_{ga}$ by introducing another electrode between grid and anode, maintained at a fixed potential, which intercepts the lines of force between grid and anode. By this means $C_{ga}$ can be reduced to less than 1 pF, and in many cases to less than 0·1 pF. In the 'beam power tetrode' the additional electrode is a 'screen grid' whose wires are exactly in the 'shadow' of the wires of the control grid, so that electrons flow in beams between the wires, and a minimum of current is intercepted. A special geometry is used to ensure that any secondary electrons emitted from the anode return to the anode and do not reach the screen grid, which is typically at half the mean anode potential, and may be at a higher potential than the anode for part of the cycle. Such tetrodes can be used in amplifiers up to frequencies of about $10^3$ MHz in specially designed circuits. Dielectric losses in the tube holder are reduced by the use of insulating materials with a low loss tangent; radiation losses are avoided by the use of coaxial lines, all leads being kept as short as possible. It is also necessary to keep the cathode lead as short as possible, since otherwise its inductance, combined with the cathode–grid capacitance, reduces the input admittance (see Problem 21.1). Another problem is the finite transit time of the electrons, which is discussed in the next section.

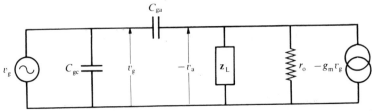

Fig. 21.4. Equivalent circuit of the triode with its interelectrode capacitances. The cathode–anode capacitance is not shown, but regarded as part of $z_L$.

## 21.3. Effect of transit time on input conductance

While an electron is leaving one electrode of a tube and approaching another it induces a charge on each of these electrodes. As it moves, the induced charge on the electrode which it has left diminishes, while that on the electrode which it is approaching increases. This can be seen quite simply by considering two plane parallel electrodes which are maintained at voltages 0 and $V_a$ respectively by means of a battery. Suppose a charge $-q$ is emitted from the plane of zero voltage. It will be accelerated towards the other plane, and when it has moved through a potential $V$ the work done on the electron will be $qV$. This work must be supplied by the battery, whence it follows that a charge $q(V/V_a)$ must have flowed through the battery. The direction of flow is such that the plane at potential $V_a$ will have acquired a charge $+q(V/V_a)$, while the charge on the other plane, which was $+q$ at the moment after the electron was emitted, is reduced to $+q(1-V/V_a)$. Thus the movement of the charge is accompanied by changes in the induced charges on the two planes, corresponding to the change in the number of lines of field from the electron which terminate on either plane (see Fig. 21.5). The duration of these changes is equal to the transit time of the charge between the two planes, and a current pulse flows for this length of time.

Similar arguments hold if one plane is replaced by a grid, and the passage of a charge through a grid therefore causes a momentary flow of charge to the grid which reverses in sign as the charge passes through. It is not necessary for the charge to hit the grid to create an induced charge, and the current flow accompanying the passage of the charge is shown in Fig. 21.6. The area under the curve up to any point represents the charge induced at that moment. The total area (that is, the net accumulated charge) is zero, since the positive and negative sections annul one another provided that all the charge flows through the grid and none is intercepted. The low-frequency components of the induced current are negligible provided the period is much longer than the transit time. This is about $10^{-9}$ s for electrons in a normal tube, and at frequencies up to 10 MHz the induced current is virtually zero. At higher frequencies,

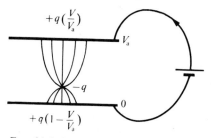

FIG. 21.5. Induced charges on electrodes.

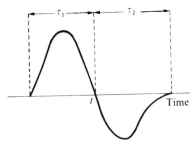

F<small>IG</small>. 21.6. Current flow to grid during transit of an electron. $t$ is the instant at which the electron passes through grid. Transit time is $\tau_1+\tau_2$.

where the transit time is an appreciable fraction of an r.f. cycle, the effect of the passage of the electrons in inducing an r.f. current to flow to the grid is appreciable. Full analysis of the effect is complicated, but an estimate of its order of magnitude can be obtained by the following method.

At a time $t$, let the voltage applied to the grid be $V = V_0 \sin \omega t$. Let the transit time from cathode to grid be $\tau_1$, and that from grid to anode be $\tau_2$. Then the current induced in the grid by the electrons approaching it will be approximately

$$I_1 = g_m V_0 \sin \omega(t-\tau_1)$$

since the size of the current is determined by the value of the grid voltage at the time $(t-\tau_1)$ when the electrons left the cathode (or, more strictly, the space-charge region). Similarly the current induced in the grid by the electrons leaving for the anode may be written

$$I_2 = -g_m V_0 \sin \omega(t-\tau_1-\tau_2),$$

the minus sign arising from the reversal of the current for departing electrons. The net current is therefore

$$I_1+I_2 = 2 g_m V_0 \cos \omega\{t-(\tau_1+\tfrac{1}{2}\tau_2)\}\sin \tfrac{1}{2}\omega\tau_2$$
$$= 2 g_m V_0 \sin \tfrac{1}{2}\omega\tau_2\{\cos \omega t \cos \omega(\tau_1+\tfrac{1}{2}\tau_2)+\sin \omega t \sin \omega(\tau_1+\tfrac{1}{2}\tau_2)\}$$
$$= 2 g_m \sin \tfrac{1}{2}\omega\tau_2\left\{\frac{\mathrm{d}V \cos \omega(\tau_1+\tfrac{1}{2}\tau_2)}{\mathrm{d}t} \; \frac{}{\omega}+ V \sin \omega(\tau_1+\tfrac{1}{2}\tau_2)\right\}.$$

This contains both a capacitative and a resistive component. The latter is more important since it causes a loading of the input circuit. If both $\omega\tau_1$ and $\omega\tau_2 \ll 1$ the input conductance may be written

$$G = g_m\omega^2(\tau_1\tau_2+\tfrac{1}{2}\tau_2^2). \tag{21.7}$$

A full analysis by North shows that for tubes of common size the input conductance $G$ is approximately equal to $g_m\omega^2\tau_1^2/10$.

If $g_m = 5$ mA V$^{-1}$ and $\tau_1 = 0.5 \times 10^{-9}$ s, the value of $G$ is found to be roughly $5 \times 10^{-21} f^2$ $\Omega^{-1}$ at a frequency $f$, and it is therefore comparable with the input conductance due to cathode-lead inductance (see Problem 21.1).

## 21.4. The klystron

We have seen already that the finite time which an electron takes to pass from cathode to anode causes difficulty in the operation of conventional tubes below a metre wavelength. At centimetre wavelengths the problem of reducing the cathode-grid clearance so as to keep the transit time down to a small fraction of a cycle becomes practically insuperable. One obvious method of reducing the transit time is to shoot the electrons at a large velocity through a grid which then acts as an effective cathode. For example, if electrons are accelerated by a potential of 2500 V, their velocity is about $3 \times 10^7$ m s$^{-1}$, and they will traverse a distance of 1 mm in $3 \times 10^{-11}$ s, which is only one-tenth of a period at a wavelength of 0.1 m. Thus, if the cathode–grid system is replaced by two grids, between which the high-frequency voltage is imposed, and the electrons are shot through these grids at a high velocity, the transit time can be kept short.

Such an arrangement must depend on some different principle for its working from that of a conventional vacuum tube, where the grid voltage influences the space charge in the potential minimum just in front of the cathode, and thus causes a change in the number of electrons flowing to the anode. When an alternating voltage is applied between grid and cathode a periodic 'density modulation' is set up in the electron stream flowing to the anode. If, now, the cathode, which emits electrons with an average energy corresponding to about 0.1 eV, is replaced by a grid through which electrons are injected at high voltage, there will be practically no space charge between this pseudo-cathode and second grid. Application of a small r.f. voltage between the two grids will not cause any change in the density of electrons leaving this space. Instead, it produces a 'velocity modulation'; some of the electrons will be accelerated by the r.f. field, while others which go through this field half a period later, when it is reversed in sign, will be retarded.

The principle of 'velocity modulation' rather than 'space-charge modulation' is fundamental in the working of the klystron and other centimetre-wave tubes. Velocity modulation is not of itself sufficient to produce amplification or oscillation, since for this we require a density modulation of the electron beam. However, if a velocity-modulated beam is allowed to 'drift' along in a field-free space, a density modulation will be set up in the following way. The electrons which were accelerated by the r.f. field will gradually overtake the slower electrons in front of them, which were retarded by the field. In this way 'bunching' of the electrons

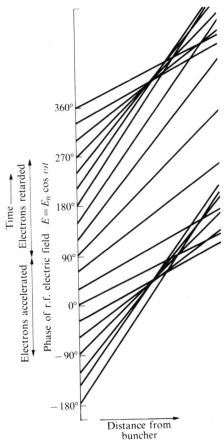

FIG. 21.7. Bunching of electrons after velocity modulation.

will occur, as illustrated in Fig. 21.7. Here the distances covered by a number of electrons, initially uniformly spaced in the beam, are plotted against time. Lines corresponding to fast electrons overtaking slow electrons converge, while at points appropriate to half a period earlier or later in the r.f. field, the lines diverge. The former gives a 'bunch' since the convergence of the lines means that more electrons occupy a given volume, while the latter gives a 'rarefaction'. If now the beam traverses a second pair of grids, between which an r.f. field of the same frequency is applied in such a phase that a bunch is retarded by the field, while a rarefaction is accelerated, energy will be transferred from the beam to the field because more electrons are slowed down than are speeded up. This constitutes a conversion of energy from the high-tension supply used for the initial acceleration of the beam into energy in the alternating electromagnetic field, in a similar manner to that in a conventional radio tube.

There, in an ordinary amplifier or oscillator, the denser current flow to the anode coincides with the moments at which the anode potential is low, so that these electrons are slowed up by the alternating component of the electric field in front of the anode, thus doing work against this field (cf. § 20.5).

A schematic diagram of a klystron is shown in Fig. 21.8. Electrons accelerated and formed into a beam by a suitable gun pass through a resonator B where a velocity modulation is imposed on them by the r.f. field. They then travel through the field-free 'drift-space' in which bunching occurs, and enter a second resonator C called the 'catcher', tuned to the same frequency as B. Finally, the electrons are collected on an electrode at high-tension potential, which plays no essential role in the action. If some of the r.f. signal in the catcher is fed back through a coupling loop to the buncher, oscillations will be set up if the phase is correctly adjusted, and if more energy is extracted from the beam by the catcher than is dissipated in the combined resistances of the catcher and buncher. In this connection it should be noted that the velocity modulation of the beam by the r.f. field in the buncher requires no net power if the transit time is short, for as many electrons are slowed down as are speeded up. For high efficiency, it is necessary to develop the greatest r.f. electric field in the resonator at the point where the beam traverses it, with the smallest dissipation of power in the resistive walls. The cavity resonator gives the best type of circuit in this respect, and its shape is determined primarily by the requirement of a short transit time for the electrons. If the latter travel with one-tenth of the velocity of light, and their transit time is to be not more than one-tenth of the period of oscillation, the gap which they traverse must be about one-hundredth of a wavelength. This requires an indented cavity of the shape shown in Fig. 21.8. It may be regarded either as a short section of coaxial line, whose inductance resonates with the capacitance across the gap, or as an indented waveguide resonant cavity.

As with most oscillators, the full mathematical theory is rather complex, but it is possible to derive the starting condition for oscillations from elementary considerations as follows. We assume that the electrons leave the gun with potential $V_0$, and the path they traverse in the resonant cavity B has a potential difference $V \cos \omega t_0$ across it at the instant $t_0$. Then if $V \ll V_0$, as will be the case for small amplitudes of oscillation, we may use the differential relation $\delta v/v = \frac{1}{2}(\delta V/V)$ to find the fractional change in their velocity after traversing the cavity; their final velocity may then be written as

$$v = v_0(1 + V \cos \omega t_0/2V_0), \qquad (21.8)$$

where $v_0 = (2eV_0/m)^{\frac{1}{2}}$ is the velocity with which an electron leaves the gun. An electron which leaves B at time $t_0$ reaches the second resonator

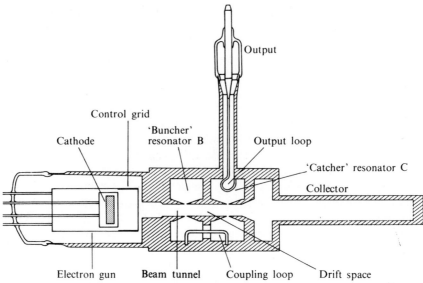

FIG. 21.8. Two-resonator klystron oscillator. Collector, buncher B, and catcher C at
high-tension positive (earth); electron gun at high-tension negative potential.

C, at a distance $x$ away, at a time

$$t = t_0+x/v = t_0+(x/v_0)(1+V \cos \omega t_0/2V_0)^{-1}$$
$$\approx t_0+(x/v_0)(1-V \cos \omega t_0/2V_0), \qquad (21.9)$$

where we have again used the approximation $V/V_0 \ll 1$.

Now the current at any point is equal to $dq/dt$, the rate at which charge
passes that point. To find the current, we shall consider a small section of
the beam containing charge $dq$, and follow it along the beam. Owing to
the velocity modulation, the front and rear portions of this section travel
at different speeds, and the time $dt$ which the section takes to pass a given
point therefore changes with the distance it has travelled. The beam
current at the time $t$ is therefore

$$I = dq/dt = (dq/dt_0)(dt_0/dt) = I_0(1+\omega xV \sin \omega t_0/2v_0 V_0)^{-1}$$
$$\approx I_0\{1-(\omega xV/2v_0 V_0)\sin \omega (t-x/v_0)\}, \qquad (21.10)$$

where we have used the relations $(dq/dt_0) = I_0$, the initial beam current,
and

$$(dt_0/dt) = (dt/dt_0)^{-1} = (1+\omega xV \sin \omega t_0/2v_0 V_0)^{-1}$$

obtained by differentiation of eqn (21.9).

Eqn (21.10) shows that we have now a density-modulated current,
whose amplitude of modulation increases linearly with $x$, the distance

travelled, so long as we restrict ourselves to small velocity modulation. If this current now passes through a second resonator, with a potential difference $V_2 \cos \omega t$, the mean power extracted from the beam will be

$$\overline{-IV_2 \cos \omega t} = -\frac{\omega}{2\pi} \int_0^{2\pi/\omega} I_0 V_2 \cos \omega t \left\{ 1 - (\omega x V/2v_0 V_0)\sin \omega\left(t - \frac{x}{v_0}\right)\right\} dt$$

$$= \frac{I_0 V V_2 \omega x}{2v_0 V_0} \frac{\omega}{2\pi} \int_0^{2\pi/\omega} \cos \omega t \sin \omega\left(t - \frac{x}{v_0}\right) dt = -\frac{I_0 V V_2 \omega x}{4v_0 V_0} \sin \frac{\omega x}{v_0},$$

since only terms in $\cos^2 \omega t$ contribute to the mean power. The power extracted from the beam will be greatest when $-\sin \omega x/v_0 = +1$, that is, $\omega x/v_0 = 2\pi(n+\frac{3}{4})$, where $n$ is an integer. In other words, if oscillations in B and C are in phase, the beam must take $(n+\frac{3}{4})$ periods to travel from B to C. The time of travel can be adjusted by altering the initial accelerating voltage $V_0$.

So far we have considered the power extracted from the beam when r.f. signals from an external source are fed into resonators B and C. It is clear, however, that if a little of the power that is fed from the beam into C is returned to B to act there as the source of signal, oscillations can be sustained so long as the power extracted from the beam is greater than that dissipated in resistive heating of resonators B and C. We may represent this dissipation by a resistance $R$ for each resonator. The oscillations will be sustained if

$$\frac{I_0 V V_2 \omega x}{4v_0 V_0} > \frac{V^2}{2R} + \frac{V_2^2}{2R}.$$

If $V = \alpha V_2$, where $\alpha$ is small, the power dissipated in the buncher may be neglected, and this relation may be expressed in the form

$$\text{minimum starting current } I_0 = \frac{2V_0(v_0/\omega x)}{\alpha R}$$

$$= \frac{V_0}{\alpha R \pi (n+\frac{3}{4})}. \qquad (21.11)$$

This shows that the smallest beam current required for sustained oscillations is of the order of the beam voltage divided by the parallel impedance of the resonator. The current required is diminished if $\alpha$ is increased, or if the time of drift $x/v_0$ between the resonators is increased, since the density of the bunches reaching the second resonator is enhanced in either case.

With large electrode voltages and currents, the klystron is an efficient and powerful oscillator, and can be used as a transmitter, but its principal

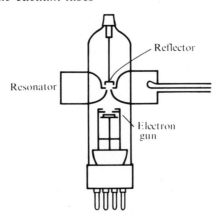

FIG. 21.9. Reflex klystron. Electrons from the gun pass through the gap in the resonator, and are returned back through the gap by the reflector which is at a potential negative with respect to the cathode.

use is as a low-power local oscillator, for which an output of a few milliwatts is sufficient, in a superheterodyne receiver. For this purpose the klystron must be tunable, and this is not feasible when two separate resonators of high $Q_F$ (several thousand) must be adjusted simultaneously. A single resonator or reflex klystron is therefore employed, as in Fig. 21.9, where the electron stream after passage through the resonator is confronted by an electrode whose potential is negative with respect to the cathode potential. The electrons are thereby halted and reflected back through the same resonator. Bunching occurs because the electrons which are speeded up in the first passage through the resonator travel further towards the reflector electrode and so return later (as in the case of a ball thrown into the air), together with the electrons which passed through the resonator half a period later and were retarded by the r.f. field. Thus the bunches occur at points in the returning beam which the faster electrons reach later rather than earlier, but as they are retarded on their return passage through the resonator when the r.f. field is directed away from the cathode instead of towards it (as in the two-resonator klystron), the required transit time for oscillation is still $(n+\frac{3}{4})$ periods. The minimum starting current is now given by eqn (21.11) with $\alpha = 1$. In an early type of klystron for $0 \cdot 1$ m wavelength, $V_0 = 1200$ V, $R = 70$ kΩ, $(n+\frac{3}{4}) = 1\frac{3}{4}$, giving a starting current of about 3 mA. The running current is about 8 mA, and the power output of an average tube $\approx 300$ mW, corresponding to an efficiency of 3 per cent. Later types run at a beam voltage of 300 V, and a current of 20 mA, with a rather lower power output.

Reflex klystrons of this type have been made to oscillate at wavelengths down to about 5 mm, which seems to be about the limit. The main difficulties in making such oscillators for shorter wavelengths arise from the smaller size of the resonant cavity, with its correspondingly lower parallel impedance $R$, which means that a higher beam current is required to start oscillations. This higher current needs to be sent through a smaller hole in the cavity, but the current density which can be obtained in the beam is limited by the mutual repulsion of the electrons.

## 21.5. The magnetron

Oscillations of high power at centimetre wavelengths are produced by the magnetron; an outline diagram of a typical tube is shown in Fig. 21.10. Electrons are emitted from a central cylindrical cathode, and are accelerated towards a coaxial cylindrical anode consisting of a solid copper block with a number of resonant cavities. These may have the shape shown in the figure, but other shapes are also possible. Essentially they form a set of quarter-wave resonant lines or cavities, the open end of the line being at the inner surface of the anode block. Thus, when oscillations take place, a strong r.f. electric field is set up at the inner

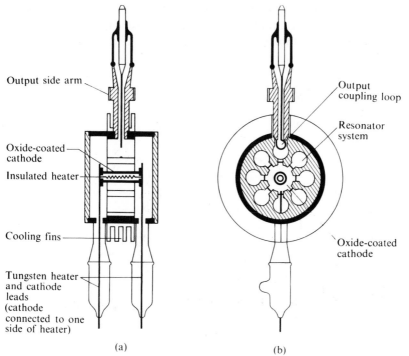

FIG. 21.10. A typical magnetron: (a) from side, (b) along axis (parallel to external magnetic field). (After Willshaw *et al.* 1946.)

surface, the field lines running mainly in the circumferential direction across the open end of the cavity, as shown in Fig. 21.11.

The magnetron operates with a strong axial magnetic flux density of a few tenths of a tesla, which is normally provided by a permanent magnet, and a potential of 10–50 kV on the anode. An electron on its way from the cathode to the anode experiences a magnetic force perpendicular to its direction of motion and a radial electric force. Its trajectory under the action of these forces can best be pictured by reference to a case with simple plane geometry (see Problem 4.10). The trajectory is then a cycloid, the path traced out by a point on the circumference of a cylinder rolling along a plane with angular velocity $\omega = eB/m$; the radius of the cylinder is $E/\omega B = mE/eB^2$, and the linear velocity of its axis is $v = E/B$.

In the case of cylindrical geometry, the electron orbit is approximately an epicycloid generated by rolling a cylinder on the cylindrical cathode, and so is rather similar to the case of the parallel planes if we imagine the latter to be given a curvature. The approximation arises because we are neglecting the radial decrease in the electric field which occurs with cylindrical geometry as we go from cathode to anode. In addition we are neglecting the mutual repulsions of the various electrons ('space charge') in either case. If the difference between the radius $a$ of the cathode and the radius $b$ of the anode is small compared with either, at a point midway between cathode and anode we may write the angular velocity of the electron cloud as approximately

$$\frac{v}{\frac{1}{2}(a+b)} = \frac{2E}{B(a+b)} = \frac{V}{b-a} \cdot \frac{2}{B(a+b)} = \frac{2V}{B(b^2-a^2)}, \qquad (21.12)$$

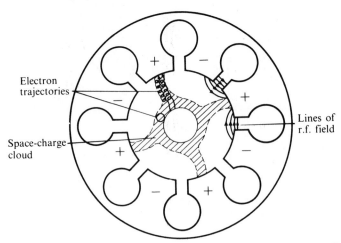

Electron trajectories

Space-charge cloud

Lines of r.f. field

FIG. 21.11. Electron trajectories and lines of r.f. electric field (arrows show direction of force on electrons) in the '$\pi$-mode' magnetron.

where V is the steady voltage applied between anode and cathode. This expression shows that to give the electron cloud a certain angular velocity of rotation, we must maintain a certain linear relation between the anode voltage and the magnetic field. The importance of this angular rotation of the electron cloud arises from the necessity of synchronizing the movements of the electrons with the alternation of the r.f. electric fields in the resonators, in order to preserve the right phase relationships. This is essential for the efficient transfer of energy from the electron cloud to the r.f. field, the basis of any oscillator.

Before considering the mechanism of this transfer, it is necessary to discuss the resonator system. Each resonant cavity behaves like a tuned circuit but the system is more complicated than that of a simple oscillator because of the presence of N such circuits, all coupled together. This coupling is partly electric, lines of r.f. electric field originating from one cavity terminating in another, but it is predominantly magnetic; the lines of r.f. magnetic field going down through one cavity are completed by returning up through another cavity. With a system of N circuits all tuned to the same frequency, and strongly coupled together, the natural frequencies of oscillation are split apart in the same way as with two circuits (see § 7.4), but analysis of the system is much more complex. The different resonant frequencies correspond to oscillations with varying change of phase between successive resonators; they can be analysed into systems of standing waves, or of travelling waves, or a mixture of the two. The simplest system is a standing wave where the phase difference between successive cavities is $\pi$ (the so-called '$\pi$-mode'), and this is also one of the most efficient modes of operation of the magnetron. At any instant the direction of the lines of force in successive cavities is exactly reversed, as would arise from a simple potential distribution where alternate segments are just plus and minus in the r.f. voltage (see Fig. 21.11).

The interaction between the electrons and the r.f. field must now be considered. Under normal conditions of operation, but in the absence of oscillation, the electrons would travel (approximately) in circles such that their farthest point from the cathode is about half-way across the cathode–anode space. At the moment when they return to the cathode, their velocity would be zero, since the magnetic field does no work on them, and that done by the electrostatic field as they move initially away from the cathode is all regained on the return path. Suppose now that an electron is just moving tangentially at the farthest point in its trajectory from the cathode. Then the force exerted on it by the magnetic field is towards the cathode, while that exerted by the electrostatic field is towards the anode. If the tangential velocity of the electron is increased at this moment through being accelerated by the fringing field of one of the cavities, the magnetic force on it (which is proportional to its velocity) will be increased, while the electrostatic force is unchanged, so that the effect is to return it towards the cathode. If, on the other hand, the electron is retarded by the r.f. field, the magnetic force is decreased and the electron will move in a path which brings it closer to the anode than it would have got in the absence of the r.f. field. If now it arrives opposite another cavity at the moment when it is again retarded, it again gives up energy to the r.f. field, and moves still closer to the anode. Note that as it does so, it moves into positions where the r.f. field is stronger and so a greater proportion of the kinetic energy of the electron is transferred to the r.f. field. On the other hand, an electron which is speeded up returns towards the cathode where its interaction with the r.f. field is smaller. Thus, if the right phase relationship can be maintained, some of the electrons will give up energy to several cavities in succession, and eventually reach the anode with kinetic energy much less than that corresponding to $eV$, while others will be returned to the cathode. On the whole, the latter take much less energy from the r.f. field than the former give to it, and the net transfer of energy will maintain oscillation.

To get the right phase relationship, the angular velocity of the electron cloud must coincide with the angular velocity of rotation of one of the Fourier components of the r.f. field system. For the $\pi$-mode, this simply means that the electron cloud must rotate through the angle $2\pi/N$, between successive cavities, in $(n+\frac{1}{2})$ cycles (where $n$ is an integer). Its angular velocity must therefore be

$$\frac{(2\pi/N)}{(n+\frac{1}{2})\tau} = \frac{2\pi f}{N(n+\frac{1}{2})},$$

where $f = 1/\tau$ is the frequency of the oscillations. Equating this to (21.12) gives

$$V = \frac{\pi f}{k} b^2 B\left(1 - \frac{a^2}{b^2}\right), \qquad (21.13)$$

where $k = N(n+\frac{1}{2})$. Assuming the ratio $a:b$ is roughly constant, it will be seen that to maintain operation at a given frequency $f$ in a given mode $k$, at a fixed flux density $B$, the anode voltage must be increased with the square of the anode diameter. If it is desired to keep the voltage fixed and to construct a magnetron of higher frequency (shorter wavelength) but with equivalent operating conditions, then the resonator system and anode diameter must be scaled in proportion to the wavelength ($b \propto 1/f$), and $B$ must be increased in proportion to $f$.

High power output from the magnetron can be achieved only if high anode voltages and high anode currents are used. By running the tube in short pulses roughly of 1 $\mu$s duration, with a repetition rate of about

1000 Hz, the power in the pulse can be made over 1000 times as great as can be obtained under continuous operation. These high powers are mainly due to three factors:

(1) the electronic conditions are such that high efficiency is attained at high level;

(2) oxide-coated cathodes can give very high currents per unit area, 100 times greater under pulsed conditions than under continuous running;

(3) the mean power dissipated on the anode is reduced, and is easily removed by conduction through the solid copper anode.

An important factor under (1) is focusing action by the r.f. field, which helps to concentrate the space charge into a number of narrow spokes (see Fig. 21.11). Each spoke then passes through the r.f. field at the moment when it is a maximum, giving the equivalent of Class C operation in ordinary triodes. The main technical difficulties have been the construction of rugged cathode surfaces, which can withstand the heavy bombardment by the returning electrons accelerated by the r.f. field, and avoiding 'mode jumping', where the frequency changes as the tube jumps from one value of $k$ to another. Power is extracted by means of a loop coupling in one of the resonators, or through a waveguide slit in one resonator.

Typical operating conditions for a medium high power magnetron operating at 0·1 m wavelength are: magnetic flux density $B = 0·28$ T, anode voltage 31 kV, anode current during pulse 35 A, output power in pulse 750 kW. In this tube the cathode diameter is 6·0 mm, and the inside diameter of the anode is 16·1 mm; the length of the anode block is 20 mm, and the over-all length of the tube is 32 mm. The dimensions of the tube are thus comparatively small, and the high power obtainable in the pulse is due to the high efficiency (70 per cent), which also reduces the dissipation on the anode block to only 30 per cent of the input power. At 100 mm wavelength, output pulse powers of a few megawatts can be achieved, but the power decreases rapidly as the wavelength is reduced, owing to a number of factors. Experimental tubes have been made to operate at wavelengths of a few millimetres, and the short wavelength limit is about the same as or a little lower than that of the klystron. Most cavity magnetrons are fixed-frequency tubes, but some magnetrons tunable over a range of 10–20 per cent in frequency have been constructed, the variation being obtained by plungers moving into the resonators from one end.

## 21.6. Travelling-wave tubes

An important class of electronic tubes for centimetre wavelengths, which we shall not discuss in detail, is that of the 'travelling-wave' tube. This is a velocity modulation device in which the electron beam interacts continuously with an electromagnetic wave, instead of only locally, as in the klystron. To make this possible the wave velocity must be reduced to coincide with the beam velocity; this is accomplished by means of a 'slow wave structure', such as a wire helix surrounding the beam, which behaves as an artificial transmission line with wave velocity $1/(LC)^{\frac{1}{2}}$ (see § 9.3). Energy is transferred to the electromagnetic wave from a slow space-charge or plasma wave on the electron beam (see Pierce 1974).

The device is untuned and amplification is possible over a wide range of frequencies. The (gain × bandwidth) product is (100 × 5000) = $5 \times 10^5$ MHz; this is 10 times better than can be achieved with a transistor amplifier, and over 100 times better than with vacuum-tube triodes. The main use of the travelling-wave tube is as a wide-band amplifier, capable of giving 100 W or more power output at a frequency of 5 GHz.

## References

WILLSHAW, E. E., RUSHFORTH, L., STAINSBY, A. G., LATHAM, R., BALLS, A. W. and KING, A. H. (1946). *J. Inst. Elect. Engrs.* **93**, 985.
PIERCE, J. R. (1974). *Almost all about waves.* M.I.T. Press, Cambridge, Massachusetts.

## Problems

21.1. Show that the effect of the inductance in the cathode lead of Fig. 21.12(a) can be represented by the equivalent circuit of Fig. 21.12(b), giving an input admittance

$$\mathbf{y} \sim j\omega C_{gc} + g_m \omega^2 L C_{gc},$$

provided that $\omega L g_m \ll 1$ and the anode current can be taken as $g_m v_g$.

If $g_m = 5$ mA V$^{-1}$, $C_{gc} = 5$ pF, $L = 5 \times 10^{-8}$ H, show that the input conductance $R_g^{-1}$ is about $5 \times 10^{-20} f^2$ Ω$^{-1}$

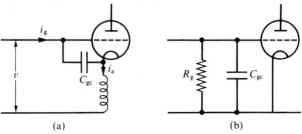

(a)                                        (b)

FIG. 21.12. Effect of cathode lead inductance at high frequencies; $R_g = (g_m \omega^2 L C_{gc})^{-1}$.

21.2. Calculate the anode voltage $V'_a$ of the equivalent diode for a triode with, plane parallel electrodes in which the grid–cathode spacing is 0·3 mm and the current density is 200 A m$^{-2}$, assuming that the equivalent diode spacing is the same as the actual grid–cathode spacing.

If $V_g = -3$ V, $V_a = +150$ V on the triode, what must be the value of $\mu$?

(*Answer*: $V'_a = 3·9$ V; $\mu = 22$.)

21.3. Prove that in a plane parallel diode the transit time of an electron is $\frac{3}{2}$ times as long under space-charge limited conditions as it would be in the absence of space charge.

What will be the transit time between cathode and grid in the triode of Problem 21.2?

(*Answer*: $7·7 \times 10^{-10}$ s.)

# 22. Alternating current measurements

## 22.1. Measurement of voltage, current, and power

IF an alternating voltage is applied to the terminals of a d.c. instrument such as a moving-coil galvanometer, the reading observed is usually zero. The movement of the galvanometer is too sluggish to follow the alternations of the applied voltage if these occur at more than a few cycles per second. The instrument therefore records only the mean value of the current over many cycles, which is zero for a symmetrical waveform. Thus the measurement of alternating currents and voltages requires the use of special instruments, which may be divided into three classes according to the principle involved in their construction. In the first class are instruments with very rapid responses so that they can follow the alternating

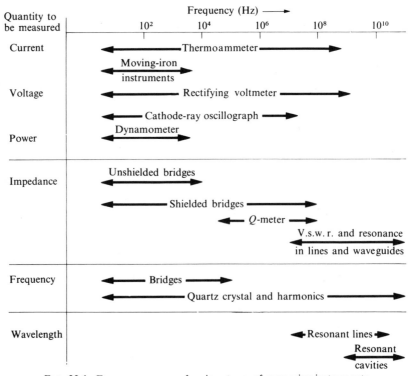

FIG. 22.1. Frequency ranges of various types of measuring instruments.

waveform; second, 'square-law' instruments, so called because they respond to the square of the current or voltage applied; and third, rectifier instruments, where the alternating voltage is converted to a steady voltage which can be measured on a d.c. instrument.

The instruments with the greatest frequency range (see Fig. 22.1) are the thermoammeter, a square-law instrument, for current; the voltmeter, incorporating a rectifier; and the cathode-ray oscillograph, a short-time-constant instrument for the display of waveform and measurement of voltage. The dynamometer is a moving-coil instrument which measures power directly, but its use is confined to supply frequencies. At radio-frequencies power is normally determined from the voltage across a known resistance, or from the current through it. In general, current and voltage are the primary quantities measured. Many commercial instruments contain electronic circuits for protection against overloading or wrong connections. Digital instruments contain solid-state electronic circuits to convert the output to digital read-out.

## The cathode-ray oscillograph

This is an instrument whereby the waveform of an alternating voltage can be displayed on a screen. It consists of a number of electrodes (Fig. 22.2) inside an evacuated glass envelope. An electron gun, with cathode, grid, and various accelerating electrodes is used to focus a beam of electrons onto a small spot on a screen. This is coated with a mixture of compounds which become fluorescent (or phosphorescent, in 'after-glow' tubes) and emit light when excited by the electron beam. The brightness depends on the beam current (of the order of $10\,\mu A$), which is controlled by the voltages on the grid and the final anode (the screen). The latter ranges from 5 kV to 10 kV on large tubes (0·1 m diameter or more) down to 500 V on small tubes. The electrodes $XX$ and $YY$ are two pairs of plates oriented at right-angles to one another, and various voltages may be applied across the plates of either pair to deflect the electron beam. These deflections in the $x$ and $y$ directions are normal to the axis of the tube and proportional to the voltages applied to the $X$ and $Y$ plates. The magnitude of the deflection, and hence the sensitivity (defined as the deflection per unit voltage applied to the $X$ or $Y$ plates), is inversely proportional to the accelerating voltage (see Problem 22.1). The sensitivity is increased by reducing the separation between the two members of a pair of deflecting plates, and they are therefore splayed in order not to intercept the beam at large deflections. Electrostatic deflection, as this system is called, causes a certain amount of distortion, and magnetic deflection, using fields generated by small coils placed outside the tube, is usual for television tubes, where very large deflection angles are employed. Electrostatic deflection is used for most laboratory work, and the

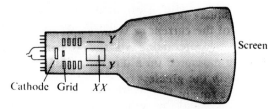

FIG. 22.2. Cathode-ray tube. A beam of electrons is deflected by the $XX$ and $YY$ plates to give a luminous spot on the fluorescent screen.

pattern observed on the screen is then determined by the voltages applied to the two sets of deflecting plates. It is usual to apply a known voltage waveform to the $X$ plates (the 'time-base'), while the unknown voltage is applied to the $Y$ plates. The most useful type of time-base is one where the spot moves to the right across the screen in the $x$ direction at constant velocity, followed by a rapid 'fly-back' to the left-hand side. This is called a linear time-base, and requires a sawtooth waveform which can be generated by special electronic circuits. Basically, a steadily increasing voltage is obtained by charging a capacitor through a constant-current device such as an FET biased to run at its saturation current.

To obtain a stationary picture, the repetition frequency of the time-base must be an exact multiple of the basic frequency of the waveform applied to the $Y$ plates. Thus the time-base must be adjustable, and 'synchronization' is usually obtained by injecting a fraction of the voltage from the $Y$ plates into the time-base circuit, so that the latter is 'locked in'. Other types of time-base include the single sweep for observing transient phenomena, the sweep being triggered by the onset of the transient; and circular or elliptical time-bases, obtained by applying sinusoidal voltages differing in phase by $\frac{1}{2}\pi$ to the $X$ and $Y$ plates. These are useful in the measurement of frequency. If an unknown frequency is applied to the anode of the cathode-ray tube, modulating the sensitivity, and if its frequency is $n$ times the time-base frequency, a stationary picture with $n$ loops is obtained (Fig. 22.3(a)). Alternatively, the unknown frequency may be applied to the grid of the tube, thus modulating the intensity. The pattern on the screen is broken up (as in Fig. 22.3(b)) into dots whose number gives the frequency ratio. Grid modulation is more sensitive than anode modulation, an amplitude of a few volts being sufficient for the unknown frequency.

The oscillograph may be used to determine the amplitude of an alternating voltage by measurement of the deflection on the screen from peak to peak. For this purpose it must be calibrated using a known d.c. or low-frequency a.c. voltage. The sensitivity of a $0\cdot1$ m diameter tube is usually of the order of a few tenths of a millimeter deflection per volt.

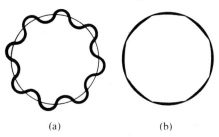

FIG. 22.3. Comparison of frequency with circular time-base. Time-base frequency = $\frac{1}{8}$(unknown frequency). (a) Anode modulation; (b) grid modulation. If the ratio of the frequencies is not exactly an integer, the pattern is not stationary but rotates.

The range may be extended by the use of an amplifier of known gain, and signals of the order of microvolts can be made to give an observable deflection. This technique may also be used for current waveform, by passing the current through a low resistance and amplifying the voltage developed across it.

The great advantage of the cathode-ray oscillograph is its ability to portray the waveform of an alternating voltage up to frequencies of a few hundred megahertz. At this point limitations arise from the difficulty of making suitable time-bases and amplifiers, as well as from inherent drawbacks in the tube itself (see Problem 22.2).

## Square-law instruments

Any d.c. instrument whose deflection depends on the square of the current or voltage can be used for a.c. measurements, and the reading obtained by calibration with d.c. will give the r.m.s. value. Thus the electrostatic voltmeter can be used for alternating voltages and currents; so can the dynamometer (§ 5.6), though it is confined to audio-frequencies. The thermoammeter has a greater frequency range; the current passes through a resistive coil which heats a copper disc; this is soldered to a thermojunction, the current from which is read on a moving-coil galvanometer. The instrument can be calibrated by d.c. or low-frequency a.c., and is used for currents of the order of milliamperes. The readings are independent of frequency up to about 1 MHz, but above that there is coupling between the coil and the thermocouple due to stray capacitance. The sensitivity can be increased by mounting the thermo-junction in an evacuated glass envelope to improve the thermal insulation; the copper disc is sometimes joined to the junction by a small glass bead, which provides thermal contact but insulates the r.f. circuit from the galvanometer. The heater is very thin, to avoid any change of resistance with frequency due to skin effect, and the deflection is very nearly proportional to the square of the current over a large range of frequencies. The frequency range may be greatly extended by using a separate thermo-junction which may be inserted in the circuit quite apart from the instrument used to measure its d.c. output voltage. Leads to the latter instrument must be carefully decoupled.

The dynamometer wattmeter can be used for measuring power, by connecting it as in Fig. 5.17. The scale reading is then proportional to the average value of $V_0 I_0 \sin \omega t \sin(\omega t + \alpha)$ over a cycle, where $V_0 \sin \omega t$ is the voltage across the load, and $I_0 \sin(\omega t + \alpha)$ is the current through it. The scale reading gives $\frac{1}{2}(V_0 I_0)\cos \alpha$, and this is the power consumed by the load, so that the instrument can be calibrated to read power directly, and no determination of the phase angle or power factor is required.

The dynamometer is suitable only for audio-frequencies up to about 1 kHz. At radio-frequencies power can readily be measured from the voltage across or the current through a known resistance. This method can be used at all frequencies where the calibration of the voltmeter or ammeter is reliable, but the direct measurement of power can be extended into the gigahertz region. For low powers ($\sim$ a watt down to microwatts) the 'thermistor' is a convenient element for this purpose. It consists of a tiny bead of various semiconducting oxides (usually of transition elements), whose resistance falls steeply with increasing temperature and hence also with power input. The resistance can be adjusted in manufacture to be $\sim 100 \ \Omega$, which can readily be matched to a coaxial line with characteristic impedance of this order. The change in resistance is found with a Wheatstone's bridge; if this is balanced with the r.f. power on, and the direct current increased to restore the balance when the radio-frequency is switched off, then the r.f. power can be calculated from the change in the d.c. power dissipated in the thermistor.

Powers of more than a few watts are measured by dissipating the power in a suitable load and observing the temperature rise; at centimetre wavelengths water, with its high absorption coefficient (see § 10.8), is a convenient load. A continuous-flow method is generally employed.

## Rectifier instruments

Rectification is the process of turning an a.c. voltage into a d.c. voltage, and clearly this may be used as the basis of methods of measuring alternating voltages. A very simple system makes use of a solid-state diode, as in Fig. 22.4, with a fairly high resistance and microammeter in the output circuit. Current flows through the meter during alternate half-cycles, and it records the mean current flow. For a sine wave this should be $I_0/\pi$, where $I_0$ is the peak current, but this is not quite true because the diode is not exactly linear when forward-biased, and it passes a small leakage current when backward-biased. If a capacitor is placed

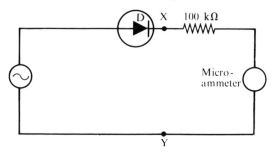

FIG. 22.4. A simple type of diode circuit for measuring alternating current.

across the terminals, X, Y, it becomes charged up to the peak voltage $V_0$ of the alternating waveform (see the discussion in § 18.4). Many commercial instruments are basically peak voltmeters, but calibrated to read $V_0/\sqrt{2}$, which is the r.m.s. value for a pure sine wave. Caution must therefore be used in interpreting such readings when the waveform is more complicated, as observed on a oscillograph.

The primary requirement of a voltmeter is an input impedance so high that it does not appreciably alter conditions in the circuit to which it is connected. If, to improve the sensitivity, the voltmeter circuit is used in combination with an amplifier, this must be chosen to have a high input impedance. A variety of commercial instruments are available which use solid-state circuitry to give a digital output instead of a pointer and scale reading.

We often require to measure the voltage of a particular frequency separate from other frequencies which may simultaneously be present. This can be done using a 'phase-sensitive detector', whose principle resembles that used in frequency changing (§ 18.6). A reference signal of frequency $\omega_1/2\pi$ is applied to a detector circuit, producing a fluctuating conductance of the form (eqn (18.9))

$$g = g_0 + g_1 \cos \omega_1 t + \ldots.$$

Suppose a small voltage $v \cos(\omega_2 t + \delta)$ is also applied to the circuit, where $\omega_2$ is close to $\omega_1$. The change in the detector current contains a low-frequency component

$$\tfrac{1}{2}g_1 v \cos\{(\omega_2 - \omega_1)t + \delta\}.$$

If $\omega_2 = \omega_1$, this represents a steady change $\tfrac{1}{2}g_1 v \cos \delta$, whose magnitude and sign depend on the phase difference $\delta$. Normally two identical detector circuits in push–pull are used; this doubles the output, and helps to eliminate spurious fluctuations. If $\omega_2$ is not exactly equal to $\omega_1$, the rectified output fluctuates at frequency $(\omega_2 - \omega_1)/2\pi$, but this is recorded only if it falls within the bandwidth of the meter or following circuits.

Thus only signals within a narrow range around $\omega_1/2\pi$ are recorded. If the over-all input due to noise or interference is large, care must be taken to ensure that any amplifier preceding the detector is not overloaded; for example, a filter circuit at the input can be used to reduce signals at unwanted frequencies. Overloading the detector can also produce spurious signals.

## 22.2. Alternating-current bridges

The measurement of resistance using direct current is usually accomplished most precisely by means of a bridge, either Wheatstone's bridge or one of its modifications. The measurement of a complex impedance is also readily achieved with high precision by means of an a.c. bridge. To determine a complex impedance fully, two quantities must be measured— its real and imaginary parts. At first sight this might seem to require two separate experiments, but in fact the balancing of an a.c. bridge requires that two separate conditions be satisfied simultaneously. These two conditions involve the real and imaginary components of the unknown impedance, and thus both are determined when both conditions are fulfilled. The reason for this extra complexity in the balancing of an a.c. bridge can readily be seen from consideration of a simple network such as the generalized Wheatstone's bridge shown in Fig. 22.5, with complex impedances in each of the arms. For a balance the voltage applied to the detector must be zero. This voltage is equal to the difference of voltage between the points A and B, which is truly zero only if the voltage at these points has not only the same amplitude but also the same phase. In other words, the voltage at each of these points must be represented by a vector with two components, and for the vectors at the two points to be identical, their components must be individually the same.

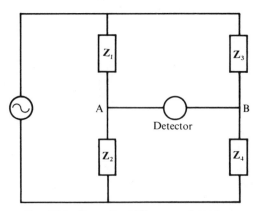

FIG. 22.5. Generalized Wheatstone's bridge.

The presence of two balance conditions which must be fulfilled simultaneously has an important bearing on the design of a.c. bridges. To avoid disturbing one balance condition when adjusting the other, it is essential that the two balance conditions shall be independent of one another. This can be achieved by choosing a bridge where each balance condition can be met by adjusting a variable impedance which does not appear in the other balance condition. The final balance can be obtained relatively quickly by first adjusting one variable until a minimum detector reading is obtained, and then the other. On returning to the first a finer balance is obtained, and so on. A second highly desirable quality is that the balance conditions shall be independent of frequency. The reason for this is that the source of power employed for the bridge never produces a pure sine wave, but contains some distortion which can be represented in a Fourier analysis of the waveform by harmonics of the fundamental frequency. The presence of quite a small harmonic content will be important if the balance conditions depend on frequency because the sensitivity of the bridge depends on the ability to detect a small fraction of the applied voltage, and this will be obscured by the harmonic content unless this is balanced out simultaneously. In practice it is often found that the harmonics do not vanish even in a bridge where the balance conditions are independent of frequency, because the impedances used may vary with frequency, usually because of stray reactances (see Problems 7.10 and 7.11). In this case it is advantageous to use either a tuned detector, such as a phase-sensitive detector with phase shifts of 0 and $\frac{1}{2}\pi$ so that signals can be observed both in phase and quadrature, or to insert a filter at the input to the detector to eliminate the harmonics.

The driving voltage for the bridge is usually provided by a small oscillator, a few volts being sufficient for most purposes. The detector consists of an amplifier followed by a voltmeter or by a cathode-ray oscilloscope. The latter has the advantage that it shows the waveform reaching the detector, and the presence of harmonics near the balance point is readily observed. To some extent it is possible to separate visually the fundamental and harmonics and to reduce the former to one-fifth or so of the harmonic. The amplifier gain must be variable (a potentiometer before the input is sufficient) in order to avoid saturation of the later stages when the bridge is far from balance. The gain is increased as the balance is approached and the amplifier has the advantage that it is not readily damaged by an overload.

When an electronic generator and amplifier are both used it may happen that one terminal of each is earthed, or has a large capacitance to the mains supply which is common to both. This would throw either a short-circuit or a large capacitance across one arm of the bridge, and an isolating transformer (preferably one with an electrostatic screen between primary and secondary) should be used between the bridge and either the generator or the detector-amplifier.

Variable impedances are required to balance the bridge, and a good general rule is that the impedances in all the arms should be of the same order of magnitude for optimum sensitivity. A great variety of bridges have been designed, of which we shall discuss only three.

### Schering's bridge

The generalized Wheatstone's bridge shown in Fig. 22.5 is the basis of most bridge circuits, and the balance condition is similar to that for the d.c. bridge:

$$\mathbf{Z}_1/\mathbf{Z}_2 = \mathbf{Z}_3/\mathbf{Z}_4. \tag{22.1}$$

The two balance conditions are contained in this complex equation, since the real and imaginary parts must be satisfied simultaneously. This will be seen in the following application to Schering's bridge, which is commonly used for the determination of capacitance.

The circuit diagram of Schering's bridge is shown in Fig. 22.6. The unknown (lossy) capacitor is represented by the series combination of $C$ and $R$. $C_1$ is a good standard capacitance, whose magnitude should be of the same order as that of the capacitor under test. $R_1$ is a fixed resistance, and $R_2$ is a variable resistance shunted by a variable capacitance $C_2$. The balance condition is

$$j\omega C_1\left(R + \frac{1}{j\omega C}\right) = R_1\left(\frac{1}{R_2} + j\omega C_2\right).$$

The real and imaginary parts of this give

$$C = C_1(R_2/R_1)$$

and                                                                          (22.2)

$$R = R_1(C_2/C_1).$$

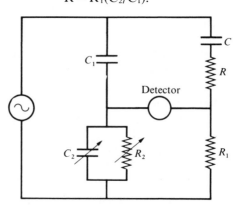

FIG. 22.6. Schering's bridge for the measurement of capacitance.

These conditions fulfil the requirements outlined earlier. They are independent of each other, provided that $R_2$ and $C_2$ only are varied, and independent of frequency. In addition the unknown capacitance is obtained in terms of a known standard capacitance and the ratio of two resistances; the variable capacitance $C_2$ enters only into the equation for the apparent series resistance of the unknown capacitor. With a good capacitor this will be small, and high accuracy in the determination of $R$ is seldom required. A small variable air capacitor usually suffices for $C_2$, and its leakage resistance under normal conditions will be so high that it does not affect $R_2$, with which it is in parallel.

## Anderson's bridge

A self-inductance has an appreciable resistance; at audio-frequencies the quality factor $Q_F$ is generally not better than about 30, while the loss tangent $(= 1/Q_F)$ of a good mica capacitor is about $10^{-4}$. It is often preferable to determine an unknown inductance in terms of standard capacitances and resistances. This can be done by Maxwell's $L/C$ bridge, which uses a network of the Wheatstone type (see Problem 22.3), but to make the two balance conditions independent of each other a standard variable capacitance is required. A modification of this bridge, due to Anderson, has the advantage that only variable resistances are required, together with a standard fixed capacitance. The circuit is shown in Fig. 22.7. The unknown self-inductance $L$, with resistance $r$, forms one arm of the bridge when placed in series with a variable resistance $S$. The fixed capacitance $C$ is in series with a variable resistance $T$, the combination

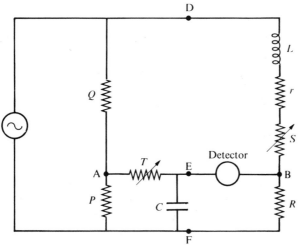

FIG. 22.7. Anderson's bridge for the measurement of self-inductance $L$ in terms of a standard capacitance $C$.

being shunted by a resistance $P$. The detector is connected from B to the junction E of $C$ and $T$, instead of to the point A. The balance condition is most readily found by calculating the voltages across FE and FB as fractions of the driving voltage $V$. The voltage across FE is a fraction $1/(1+j\omega CT)$ of that across FA, while that across FA is $V\mathbf{Z}/(\mathbf{Z}+Q) = V/(1+Q/\mathbf{Z})$, where $\mathbf{Z}$ is the total impedance between $F$ and $A$. Since

$$\frac{1}{\mathbf{Z}} = \frac{1}{P} + \frac{j\omega C}{1+j\omega CT},$$

the voltage across FE is

$$\frac{V}{\{(1+j\omega CT)(1+Q/\mathbf{Z})\}} = \frac{V}{\{(1+j\omega CT)(1+Q/P)+j\omega CQ\}},$$

while the voltage across FB is $VR/(R+j\omega L+r+S)$. Equating these voltages, we have

$$1+(j\omega L+r+S)/R = (1+j\omega CT)(1+Q/P)+j\omega CQ.$$

The real and imaginary parts of this equation give separately

$$(r+S)/R = Q/P$$

and

$$L/R = CT(1+Q/P)+CQ.$$

It is generally convenient to make $Q = P$, in which case the balance conditions reduce to

$$r = R-S,$$
$$L = CR(2T+P). \tag{22.3}$$

It is obvious from this that no balance is possible unless $CRP < L$; if this condition is being violated it will be indicated by the fact that the nearest approach to balance is obtained when $T$ is zero. Inspection of the balance conditions shows also that they are independent of frequency, and of each other if $S$ and $T$ are made the variables.

### The transformer-ratio bridge

Standard capacitors and resistors can be constructed with a reproducible accuracy of about one part in $10^4$. This is sufficient for many purposes, but not for monitoring small changes, for automation procedures, or the maintenance of standards. For such purposes toroidal transformers are used; they can be wound with such precision that a given number of turns can be tapped off to give a voltage divider accurate to one part in $10^7$. By using such transformers, with accurately calibrated ratios, in two arms of a bridge network, the ratio of the impedances of the other two arms can

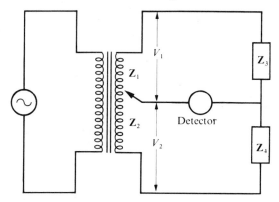

FIG. 22.8. The transformer ratio bridge.

be determined to a few parts in $10^7$. The use of a toroidal shape reduces the leakage of magnetic flux to extremely small values. The coils are wound on a special core, such as a tape of supermumetal, 25 $\mu$m thick (to reduce eddy currents), which has an initial relative permeability of about $7 \cdot 5 \times 10^4$.

A bridge of the simplest type is shown in Fig. 22.8. If $\mathbf{Z}_1$, $\mathbf{Z}_2$ are the impedances of the two sections of the secondary winding of the trans-former, the balance condition is simply given by eqn (22.1). The trans-former coils will always have some resistive loss, although this is kept as small as possible. We assume that the two sections have self-inductance $L_1$, $L_2$ and resistance $R_1$, $R_2$, and the mutual inductance between them is $M$. When the bridge is balanced, the current through each section is the same, so that

$$\frac{\mathbf{Z}_1}{\mathbf{Z}_2} = \frac{R_1 + j\omega(L_1 + M)}{R_2 + j\omega(L_2 + M)}. \qquad (22.4)$$

If we can assume that the flux leakage is zero, then $M^2 = L_1 L_2$. The turns ratio is $N_1/N_2 = n$, and the coils are wound so that $L_1 = nM$ and $M = nL_2$; if also $R_1$ is made equal to $nR_2$, then $\mathbf{Z}_1/\mathbf{Z}_2 = n$. The bridge can be balanced either by adjusting one of the impedances $\mathbf{Z}_3$ or $\mathbf{Z}_4$, or by using a transformer with a number of tappings to give an adjustable turns ratio. For example, if $\mathbf{Z}_3$ and $\mathbf{Z}_4$ are capacities, one a standard, the other can be compared with it to a few parts in $10^7$; small changes in capacity can also be measured. Since $\mathbf{Z}_1/\mathbf{Z}_2$ is real and positive, only like impe-dances can be compared. However, no capacitor is completely free from loss, and the circuit of Fig. 22.9 can then be used to obtain a balance. The T-network between points A and D provides a small extra current in the right phase. By converting the T-network to a $\pi$-network, as in Problem 7.15, it can be shown that if $R_a C_a = R_b C_b$ the resistance between A and D is $(R_a + R_b)$, and the current flowing from A to D will be in quadrature

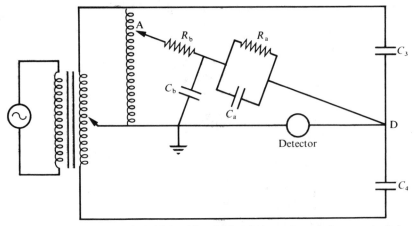

FIG. 22.9. Transformer ratio bridge with additional network to balance out-of-phase current if capacitance $C_3$ has some loss.

with that through the capacitance $C_3$. In practice the resistances $R_a$, $R_b$ are fixed, so that for a given voltage between A and D the current through the resistances is constant and independent of frequency. Balance is obtained by adjusting the turns ratio at the point A.

There are many varieties of transformer-ratio bridges, specially designed to meet different needs. At radio-frequencies great care is required to shield all components from electric and magnetic pick-up, particularly between the points of highest potential and earth. With correct design they can be used up to 100 MHz.

## 22.3. The Q-meter

The careful design and high precision needed in the manufacture of a bridge for use at radio-frequencies are unnecessary when an accuracy of the order of 1 per cent is sufficient. A cheaper and simpler alternative is the Q-meter, a general-purpose instrument whose basic circuit is shown in Fig. 22.10. A small current $I$ from an oscillator is read on a thermoammeter and then flows to a known low resistance $r$. A series tuned circuit is connected in parallel with $r$, and the voltage $V$ developed across the capacitance $C$ is read on a high-resistance voltmeter when the capacitor is adjusted for resonance. The latter is indicated by a maximum reading of the voltmeter. The $Q_F$ of the circuit under test is then equal to the ratio $V/Ir$, since $Ir$ is the voltage introduced in series with the tuned circuit. It is necessary that the resistance $r$ shall be small compared with the series resistance $R$ of the tuned circuit, in order that substantially all the current registered by the milliammeter shall flow through $r$. This requirement may be stated in another way: $r$ must be small compared with $R$ in order not

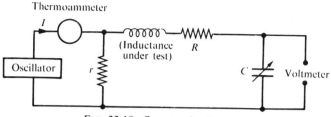

FIG. 22.10. *Q-meter circuit.*

to load the circuit under test. (It is easy to show that the measured $Q_F$ is that for a circuit whose total series resistance is the sum of $r$ and $R$.) In a commercial instrument an internal oscillator of calibrated variable frequency is provided; the current from it may be adjusted to bring the milliammeter reading always to a fixed mark, and an internal voltmeter is calibrated directly to read the $Q_F$ of an unknown coil. The variable capacitance $C$ is included in the instrument, and is calibrated so that the inductance of the unknown coil may be calculated from the known oscillator frequency and the tuning capacitance. An unknown capacitance may also be measured by the substitution method: a suitable coil $L$ is inserted and the reading of $C$ required to tune it to resonance with and without the unknown capacitor in parallel with $C$ is found. An unknown resistance can be measured by finding the effect on the $Q_F$ of a circuit when it is placed in series (if a low resistance) or in parallel (for a high resistance) with the tuned circuit.

## 22.4. Measurements on lines

At frequencies above about 100 MHz (wavelengths of 3 m and less) the leads to the impedance under test are not negligible in length compared with the wavelength and errors may be introduced because the current and voltage at the measuring instrument are not the same as those at the unknown impedance. These errors may be eliminated by making the leads part of a transmission line of known and constant impedance, the unknown impedance being placed at the end of this line and acting as its termination. The impedance $Z_0$ of the transmission line may be calculated from its dimensions (see § 9.3) and the unknown impedance is determined as a ratio to $Z_0$. This may be carried out either by determining the voltage standing-wave ratio (v.s.w.r.) on the line, or by a resonance method.

The first of these methods has the advantage that the results do not depend on the generator impedance, and the generator may therefore be connected directly to the line. From the theory of transmission lines (Chapter 9) it follows that the voltage at any point on the line may be regarded as due to an incident wave of amplitude $A$ and a wave reflected from the terminating impedance of amplitude $A_1$. The resultant voltage amplitude is a maximum $(A+A_1)$ at an anti-node where the incident and reflected waves are in phase, and a minimum $(A-A_1)$ where they are 180° out of phase, these points being a quarter of a wavelength apart. From a measurement of the v.s.w.r.

$$(A+A_1)/(A-A_1),$$

and the position of the nodes or anti-nodes, the ratio of the terminating impedance $\mathbf{Z}$ to the characteristic impedance $Z_0$ may be found using eqns (9.19):

$$\left.\begin{aligned}
\frac{A_1}{A} &= \left(\frac{Z_1^2+Z_0^2-2Z_1Z_0\cos\phi}{Z_1^2+Z_0^2+2Z_1Z_0\cos\phi}\right)^{\frac{1}{2}}, \\
\tan\delta &= \frac{2Z_1Z_0\sin\phi}{Z_1^2-Z_0^2},
\end{aligned}\right\}$$

where $\mathbf{Z} = Z_1\exp j\phi$, and $\delta$ is the difference in phase between the reflected and incident waves at the point of reflection (the termination of the line). This phase constant can be found from the position of a voltage node, this being generally more accurate than the location of an anti-node (especially if the v.s.w.r. is high) because the sensitivity of the detector can be increased as the node is approached. If the end of the line is at $x = 0$, and the voltage node at a point $x = -l$, then the phase of the incident wave at this point is $-\omega(-l/v) = 2\pi l/\lambda$, and that of the reflected wave is $\delta+\omega(-l/v) = \delta-2\pi l/\lambda$. For a node these must differ by $\pi$, whence $\delta = (4\pi l/\lambda)+\pi$. To determine $l$ accurately, it is best to replace the unknown impedance $\mathbf{Z}$ by a short-circuit and find the distance between the previous node and the new one; the latter is (electrically) exactly an integral number of half-wavelengths from the end of the line.

The v.s.w.r. can be measured by moving any loosely coupled voltage detector along the line. Since only a ratio of the maximum and minimum readings is required, the absolute calibration of the indicator is unnecessary, and a knowledge of the rectifying characteristic (d.c. current or voltage output against r.f. voltage input) is sufficient. With a coaxial line, a section of air-spaced line is made up with known dimensions, and a narrow slot is cut lengthwise along the outer conductor. Since this slot is parallel to the direction of current flow in the line, it does not disturb conditions on the line materially. In this slot (see Fig. 22.11) is inserted a small radial probe, which is parallel to the lines of electric field inside the coaxial line; it picks up a small fraction of the voltage on the line and feeds it to a detector. The intrusion of the probe is made as small as possible to minimize disturbance on the line, and for this purpose a sensitive detector is required to obtain adequate sensitivity. The probe is mounted on a movable carriage, carefully machined so that the intrusion

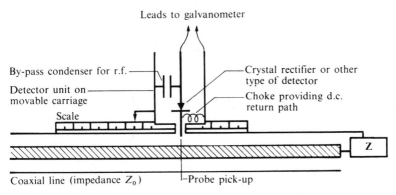

Fɪɢ. 22.11. Standing-wave detector on coaxial line.

of the probe does not change as it moves along. This may be checked by observing the constancy of the detector reading when the line is terminated by its characteristic impedance, when the v.s.w.r. should be unity. With a parallel-wire line a similar arrangement may be used with a probe near the wires, but the indicator and its leads must be shielded and kept well away from the line, since the electric and magnetic fields around the line are not now rigorously confined as they are in the coaxial line.

From the measurement of the v.s.w.r. and of δ, the ratio of the real and imaginary parts of $\mathbf{Z}$ to $Z_0$ can be found from eqns (9.19) (see above), but these are algebraically so clumsy to handle that graphical methods are normally employed. 'Impedance diagrams' can be obtained from which the real and imaginary parts of $\mathbf{Z}/Z_0$ can be read off at once.

When the impedance to be measured has only a small dissipative component the standing-wave ratio becomes very large and is difficult to measure accurately, principally because the detector law must be known over a wide range. It is often then more convenient to use a resonance method. The unknown impedance is connected across the end of a line as in Fig. 22.12, and an oscillator and detector are loosely coupled to it. A movable bridge, which should make such good contact as to be essentially a short-circuit, is adjusted until resonance is indicated by maximum deflection of the detector. If the load $\mathbf{Z}$ is represented by a resistance $R$ in parallel with a reactance $jX$, then resonance occurs when this reactance is equal and opposite to the line reactance. If the length of the line at this point is $l$, then the line reactance is

$$jX' = -jX = jZ_0 \tan 2\pi l/\lambda,$$

which may be positive or negative according to the value of $l/\lambda$. Thus the value of $X$ is determined from the resonant length. The value of $R$ may be found by measuring the sharpness of resonance; for example, by

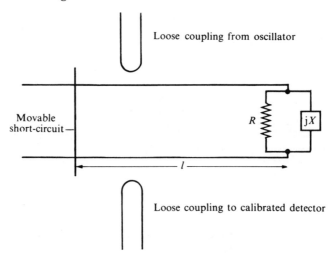

Fig. 22.12. Measurement of impedance using resonance method on transmission line.

varying the length of the line until the detector reading shows that the voltage (or current) on the line has fallen to $1/\sqrt{2}$ of the maximum. At this point the susceptance formed by $-X^{-1}$ in parallel with $Z_0^{-1} \cot 2\pi(l\pm\delta l)/\lambda$ has risen from zero to be just equal to $1/R$ (cf. the theory of the parallel tuned circuit in § 7.3). If the change in length is $\delta l$, then the value of the susceptance is

$$\left| -X^{-1} + Z_0^{-1} \cot \frac{2\pi(l\pm\delta l)}{\lambda} \right| = Z_0^{-1}\left(\frac{2\pi\delta l}{\lambda}\right)\operatorname{cosec}^2 \frac{2\pi l}{\lambda} = \frac{1}{R},$$

where $R$ can be determined. If loss in the line cannot be neglected, as has been assumed here, then it can be found by a separate measurement with the line short-circuited at both ends (or open-circuited at one end) and a correction applied. The calculation is rather complicated, but from eqn (9.28) it can be seen that the length of line can be represented by a complex admittance $\mathbf{Y}'$ which is in parallel with $1/\mathbf{Z}$. If $\mathbf{Y}'$ is separated into its real and imaginary parts $G'$ and $S'$, then the calculation proceeds as before. As the resonant lengths of line unloaded and terminated by $\mathbf{Z}$ will be different, the loss on the line must be expressed in terms of the attenuation coefficient $\alpha$ (eqn (9.24)).

The advantage of the resonance method over the v.s.w.r. method is that the detector law need only be known over a small range, the other measurements being those of lengths. Both types of measurements may also be used with waveguides at centimetre wavelengths, though here the concept of a lumped impedance loses most of its meaning.

## 22.5. Measurement of frequency and wavelength

The measurement of an audio-frequency can be made in terms of known impedances by the use of a bridge whose balance is dependent on frequency. It is obvious that the oscillation to be measured must be constant in frequency and free from harmonics, since the latter would be out of balance in the bridge (two uses of a frequency bridge are the suppression of a given frequency such as a troublesome harmonic and the analysis of harmonic content). A large number of bridges have been devised which satisfy the desiderata that the balance conditions should be mutually independent and only one of them should depend on the frequency. A simple bridge using only resistances and capacitances is due to Wien and is shown in Fig. 22.13. The balance conditions are

$$\omega^2 = \frac{1}{RSC_1C_2},$$
$$\frac{C_1}{C_2} = \frac{Q}{P} - \frac{S}{R},$$
(22.5)

which are mutually independent if the variables $R$, $S$ and also $C_1$, $C_2$ are ganged so that their ratio is kept constant. Then the second condition will remain satisfied once it has been set up and measurement of a wide range of frequencies is obtained by a single adjustment.

At wavelengths less than 1 or 2 m the measurement of wavelength directly becomes quite convenient, and high accuracy may be attained because of the high $Q_F$ of resonant transmission lines and waveguide cavities. Parallel-wire lines may be used for the longer wavelengths, but coaxial lines are better at wavelengths of 0·1 m to 1 m, and cavity resonators at centimetre wavelengths. A simple type of coaxial-line

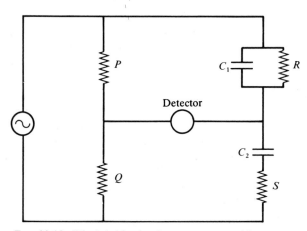

FIG. 22.13. Wien's bridge for the measurement of frequency.

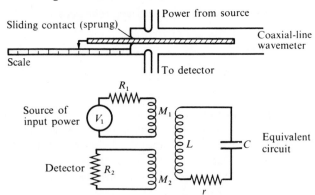

FIG. 22.14. Coaxial-line wavemeter and equivalent circuit. Note that $L$, $C$, $r$ all vary with the length of the coaxial line.

wavemeter is shown in Fig. 22.14. The centre conductor is variable in length, and moves through a spring contact which forms the closed end of the line. Power is introduced by means of a small loop which intersects some of the magnetic lines of force at the closed end, and a second loop takes power to a detector (usually a diode rectifier). These loops must be kept small to give loose coupling, and to avoid pulling the oscillator whose wavelength is to be measured through coupled-circuit effects. The equivalent circuit of the wavemeter is shown in Fig. 22.14, from which it can be seen that the detector reading is a maximum when the line is resonant. Since the loops introduce small impedances which alter the electrical length of the line, it is preferable to measure successive points of resonance, which are exactly half-wavelengths apart. The wavemeter then needs no calibration, the wavelength being found directly from a scale and vernier attached to the moving part. The accuracy is usually about one part in $10^3$, the main difficulty being in making a good contact between the moving conductor and the stationary end.

A basic type of cavity wavemeter is shown in Fig. 22.15. A section of circular waveguide is closed at one end, the other end being formed by a plunger driven by a micrometer head. Power is fed into the cavity from a waveguide through a small hole, and resonance is detected by coupling a little power out through a second hole to a detector. The holes should be kept as small as possible, subject to getting a finite detector reading, in order to avoid lowering the natural $Q_F$ of the resonator, which may be of the order of $10^4$. The wavelength in the cavity is found from the distance $\lambda_g/2$ between successive resonance points. The wavelength in free space may be calculated from the diameter of the cavity, and the mode of resonance. To avoid the difficulty of making a good contact between the moving plunger and the walls, a particular waveguide mode ($TE_{01}$ or $H_{01}$)

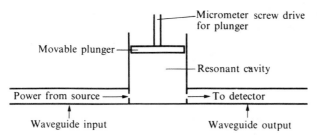

FIG. 22.15. Resonant-cavity wavemeter.

is often used, where there is no current flow across this contact (see § 9.7). Other modes of resonance may then also be present, since the TE$_{01}$ mode needs rather a large cavity diameter (the cut-off wavelength is equal to 0·82 times the diameter). These may be avoided by using special arrangements of the coupling holes (see Bleaney, Loubser, and Penrose 1947). An accuracy of one or two parts in $10^4$ may be attained, but a correction is then needed for the refractive index (1·0003) of the air in the cavity.

### The quartz-crystal oscillator

Where high accuracy of frequency control or measurement is required, use is made of the properties of a piezoelectric crystal, quartz being the most satisfactory for this purpose. A quartz crystal grows in the form of a hexagonal prism with pointed ends, the cross-section of the prism being as shown in Fig. 22.16(a). If an electric field is applied to the crystal in the X direction, the crystal contracts or elongates in the Y direction according to the sign of the electric field. Similarly, if a mechanical stress is applied in the Y direction, an electric polarization is set up in the X direction and charges appear on the faces of the crystal. These effects are reversible and are very nearly linearly related; their importance lies in the fact that they couple together an electrical and a mechanical system. If an alternating voltage is applied in the X direction, an alternating stress appears in the Y direction and the amplitude of the resulting mechanical

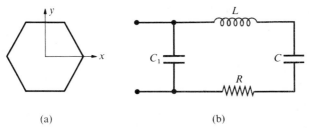

(a)                                     (b)

FIG. 22.16. (a) Quartz crystal (x, y are one of the three pairs of X, Y-axes). (b) Equivalent circuit of crystal and its electrodes.

vibrations is large if the frequency of alternation coincides with a natural mechanical vibration of the crystal. A number of different modes of oscillation exist, but those most commonly used are longitudinal and shear vibrations. The damping of the mechanical vibrations is very low, and the sharpness of the resonance makes them very suitable for use as frequency standards. The desirable properties for this purpose are:

(1) zero temperature coefficient of frequency of oscillation;

(2) high piezoelectric effect;

(3) a single mode of mechanical resonance well separated in frequency from other modes, so that there is no tendency to jump from one mode to another.

The greatest piezoelectric effect is obtained when the electrical and mechanical stresses are applied along the electrical or $X$-axis and mechanical or $Y$-axis respectively, but oscillations can be excited by any stress which has a component parallel to these axes. (Note that, owing to the high symmetry of the crystal, there are three sets of $X$- and $Y$-axes, related to one another by rotations of 120° and 240° about the $Z$-axis, the optic axis of the crystal.) An $X$-cut crystal consists of a thin slab with faces parallel to the $YZ$ plane, and the temperature coefficient of the frequency of oscillation is negative, about $-22 \times 10^{-6} \, \text{K}^{-1}$. A $Y$-cut crystal is a thin slab with its faces parallel to the $XZ$ plane, and the temperature coefficient is positive with a number of discontinuities due to couplings between different modes of oscillation. In general it is more important to obtain zero temperature coefficient than high piezoelectric activity, and intermediate cuts are used such as the AT-cut, a thin plate whose faces contain the $X$-axis and a line in the $YZ$ plane making an angle of about 35·5° with the $Z$-axis. When a voltage is applied between the large faces, a shear vibration is set up whose frequency in megahertz is 1·675/(thickness in millimetres). This is suitable for frequencies from roughly 0·5 MHz to 10 MHz. Lower frequencies may be obtained from modes where the frequency is determined by one of the long dimensions of the slab, the full range of quartz crystals being roughly from 25 kHz to 15 MHz. To apply the alternating voltage the slab is mounted between the plates of a capacitor, generally formed by sputtering a metallic film on to the large faces. This minimizes the loading on the mechanical vibrations, which is further reduced by mounting the crystal *in vacuo* between light supports touching the crystal at a mechanical node. For the highest frequency stability the crystal is kept in an oven thermostatically controlled to 0·1 K or better, because the temperature coefficient is zero only over a narrow range of temperature.

The mechanical system of a quartz crystal may be represented by the equivalent electrical circuit shown in Fig. 22.16(b). The mechanical resonance is equivalent to a series tuned circuit and this is shunted by the capacitance $C_1$ of the electrodes. Typical values are:

$X$-cut quartz (lengthwise vibration)
  *Dimensions*: rectangular bar,

$$X = 1\cdot4 \text{ mm}, \quad Y = 30\cdot7 \text{ mm}, \quad Z = 4\cdot1 \text{ mm}$$

$$R = 15 \text{ k}\Omega \qquad\qquad C_1 = 3\cdot54 \text{ pF}$$

$$L = 137 \text{ H} \qquad\qquad Q_F = 5150$$

$$C = 0\cdot0228 \text{ pF} \qquad f_0 = 89\cdot87 \text{ kHz}$$

AT-cut quartz

  *Dimensions*: disc, 25 mm diameter, thickness $1\cdot10$ mm

$$R = 24.2 \text{ }\Omega \qquad\qquad C_1 = 17.9 \text{ pF}$$

$$L = 0.119 \text{ H} \qquad\qquad Q_F = 46500$$

$$C = 0\cdot0945 \text{ pF} \qquad f_0 = 1500 \text{ kHz}$$

  (From Cady 1946)

The presence of $C_1$ causes the circuit to behave as a parallel tuned circuit at a frequency just above that of the series resonance (see Problem 22.6). The difference between these two frequencies is very small so that the phase angle of the circuit varies very rapidly. A simple circuit (due to Pierce) for maintaining the crystal in oscillation is shown in Fig. 22.17. Energy is fed back to the gate of an FET from the drain circuit through the quartz crystal; the feedback is largest when the crystal has a low impedance near the series resonance frequency. Here the equivalent

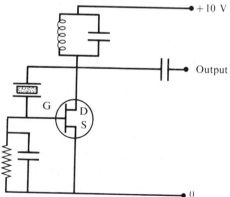

FIG. 22.17. Pierce circuit for quartz-crystal oscillator, using an FET. Feedback from drain to gate circuit takes place through the crystal at the frequency where it is a low impedance.

electrical circuit of the crystal is inductive; the condition for oscillation (gate and drain in antiphase) then requires that both the tuned circuit in the drain lead and the gain circuit should be capacitive. In such a circuit an FET with its high input impedance is better than a junction transistor because it produces less loading on the quartz crystal, improving the $Q_F$ and the frequency stability.

*Frequency standards*

The primary standard of frequency is a hyperfine transition of the caesium atom (see § 24.6), which since 1964 has been defined to be exactly

$$9\ 192\ 631\ 770\ Hz.$$

This is a more convenient and a more precise standard than previous ones based on the mean length of the solar day or year, because the motion of the earth is known to be subject to fluctuations. The excellent frequency stability of the quartz-crystal oscillator makes it extremely useful as a secondary standard in the convenient frequency region of about 100 kHz to 10 MHz, which can be extended upwards by multiplication and downwards by subdivision to control a clock. In one system of inter-comparison with the caesium standard, the output from a quartz-crystal oscillator close to 5 MHz in frequency is multiplied by electronic circuits to give a harmonic within 10 Hz of the caesium frequency, and the difference frequency is fed to a counting circuit. In this way the frequency of radio transmitters can be linked to the caesium standard, and already by 1960 standard frequency transmissions could be compared internationally and shown to be consistent to one part in $10^{11}$.

For most laboratory measurements of frequency a digital instrument is both precise and convenient. An internal crystal oscillator is used to control a 'gate' which can be opened for precise lengths of time varying from say 1 $\mu$s to 10 s. The number of cycles which occur while the gate is open, converted into pulses, is counted electronically.

## 22.6. Measurement of electric permittivity

The relative permittivity $\epsilon_r$ of a substance affords some valuable information as to the structure of its constituent molecules (see Chapter 10), and its accurate measurement is therefore of some importance. Since $\epsilon_r$ is defined by the ratio of the capacitance of a capacitor filled with the substance under test to that of the empty capacitor, it is obvious that in general two measurements of capacitance will suffice. For solids and liquids the relative permittivity varies from about 2 to 100, and any of the bridges designed to measure capacitance may be used to give accurate results. For high values of $\epsilon_r$ care must be taken to avoid stray capacitance which may seriously affect the reading obtained with the empty capacitor, if this has a rather small capacitance.

For gases the value of $\epsilon_r$ differs from unity only by about $10^{-3}$ at atmospheric pressure, and an accurate bridge such as the transformer-ratio bridge is needed to measure the difference with reasonable precision. At centimetre wavelengths Essen and Froome (1951) used the high $Q_F$ of a resonant cavity to obtain accurate measurements. The cavity was cylindrical, with a diameter of about $0.05$ m, and resonated in the $TE_{01}$ mode at about 24 GHz. The frequency of resonance was determined first with the cavity evacuated, and then filled with gas, by plotting out the resonance curve using a klystron oscillator whose frequency could be determined to one part in $10^8$ by comparison with the N.P.L. frequency standard. The frequency of resonance is given by eqn (9.34)

$$\frac{f^2}{v^2} = \frac{\mu_r \epsilon_r f^2}{c^2} = \frac{1}{\lambda_c^2} + \frac{1}{\lambda_g^2},$$

where $\mu_r$ and $\epsilon_r$ are the relative permeability and relative permittivity of the gas filling the resonator and $\lambda_c$, $\lambda_g$ are fixed by the diameter and length of the cavity respectively. Thus if $f'$ is the resonant frequency of the empty cavity, and $f''$ that of the gas-filled cavity, $(f'/f'')^2 = \mu_r \epsilon_r$. Hence the ratio of the two frequencies determines $n = (\mu_r \epsilon_r)^{\frac{1}{2}}$, the refractive index of the gas. A correction must be applied for the permeability, which differs slightly from unity for air and oxygen, since the latter is paramagnetic. A comparison of the measurements of $\epsilon_r$ at different frequencies, together with the square of the optical refractive index, is given in Table 10.4 (p. 306).

The cavity resonance method may also be used for measurement of the permittivity of liquids and solids, provided that their loss tangent is fairly small (see Faraday Society Conference on Dielectrics 1946). A partly filled cavity must be used for solids or liquids of high loss tangent, but for non-polar liquids a filled cavity was employed by Bleaney, Loubser, and Penrose (1947). A tunable cavity resonant in the $TE_{01}$ mode of the same type as described earlier (§ 22.5) was adjusted to resonance with a klystron oscillator of fixed frequency, first with the cavity empty, and then filled with liquid. By measuring a number of successive resonant points, the wavelength in the guide was found in each case, and the relative permittivity calculated from the equations

$$\frac{\epsilon_a}{\lambda^2} = \frac{1}{\lambda_a^2} + \frac{1}{\lambda_c^2}, \qquad \frac{\epsilon_r}{\lambda^2} = \frac{1}{\lambda_d^2} + \frac{1}{\lambda_c^2}, \qquad (22.6)$$

where $\epsilon_a$ is the relative permittivity of air and $\epsilon_r$ that of the liquid, $\lambda$ the wavelength in free space, and $\lambda_a$, $\lambda_d$ the wavelengths in the air- and liquid-filled cavity respectively. The loss tangent of the liquid was found from the width of the resonance curve determined by detuning the cavity. Thus only measurements of length, depending on a micrometer thread, were involved. Typical results at a temperature of 293 K are given in Table 22.1. When two measurements are given at 13·5 mm wavelength, they were made with cavities of different diameter.

TABLE 22.1

| | Relative permittivity $\epsilon_r$ | | Loss tangent ($\tan \delta$) | |
|---|---|---|---|---|
| | $\lambda = 32$ mm | $\lambda = 13\cdot5$ mm | $\lambda = 32$ mm | $\lambda = 13.5$ mm |
| Cyclohexane | 2·0244 | 2·0246, 2·0251 | 0·00005 | 0·00019 |
| n-Heptane | 1·9220 | 1·9223 | 0·00037 | 0·00076 |
| n-Hexane | 1·9016 | 1·9016 | 0·00034 | 0·00076 |
| $CS_2$ | 2·6476 | 2·6477 | 0·00024 | 0·00072 |
| $CCl_4$ | 2·2386 | 2·2390 | 0·00031 | 0·00078 |

All the samples except those of n-hexane and $CCl_4$ were specially purified. The loss tangent is considerably affected by small traces of polar impurities, but it is not certain that such impurities would account for all the dielectric loss.

## 22.7. Measurement of the velocity of radio waves

The velocity of electromagnetic radiation has long been regarded as one of the fundamental constants of physics, and much effort has been devoted to its accurate determination. Apart from one measurement of the velocity of radio waves on a transmission line by Mercier (1924), most of the early work used light waves. The results showed a good deal of scatter; since 1945 a number of new determinations have been made, of greater accuracy, which show that the true value is nearly 299 793 km s$^{-1}$. These methods have made use of radio techniques to improve the accuracy, and in some cases the wavelength of radiation used has been a few centimetres.

In § 22.5 it was pointed out that both frequency and wavelength can be measured at centimetre wavelengths. The product of these two quantities gives the phase velocity, and this has been the basis of a number of modern measurements. Frequency can be determined with an accuracy of one part in $10^{12}$ (see Bradley, Edwards, Knight, Rowley, and Woods 1972), and the main problem is the determination of wavelength. Resonant cavities have been used at centimetre wavelengths (Essen and Gordon-Smith 1948; Essen 1950), but more accurate results have been obtained by interferometer experiments, which approximate closely to a free-space method. Froome (1952) has used a microwave analogue of the

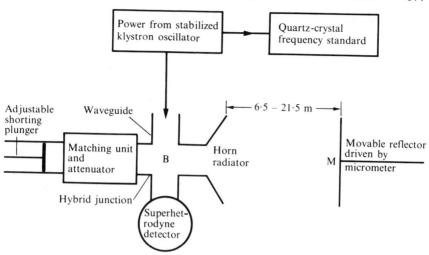

FIG. 22.18. Froome's microwave Michelson interferometer.

Michelson interferometer, as shown in Fig. 22.18. Power from a stabilized klystron oscillator flowing along a waveguide was divided into two portions at a hybrid junction B (the analogue of a half-silvered plate). One half traversed a short length of waveguide and was reflected from a shorting piston. The other was fed to a horn and launched as a wave in space. Part of this radiation was reflected back to the horn by a 0·15 m square metal plate M which could be placed at points from 6·5 m to 21·5 m away. This reflected wave on returning to the hybrid junction interferes with that reflected from the shorting piston in the second arm, and the vector sum of the two amplitudes is passed along the fourth arm to a detector (a superheterodyne receiver). The latter is used to detect when the two reflected waves are exactly in antiphase and so give a null at the detector. The metal plate M is then moved through successive null points, which occur every half-wavelength. The total distance moved was 1·62 m, and this could be measured with an accuracy of ±0·003 mm. At the same time the frequency of the klystron oscillator was measured against the quartz-crystal standard with an accuracy of one part in $10^8$. Thus the wavelength in air and the frequency were determined simultaneously. In the former case two important corrections must be applied to find the wavelength *in vacuo*:

(1) a correction for the refractive index of the air, based on the measurements of Essen and Froome (see § 22.6);

(2) a correction for the fact that the wave front reaching the mirror is not a plane, but has a small curvature, and similarly for the

reflected wave; this correction was calculated from diffraction theory, using data from different mirror distances.

The final value obtained for the velocity *in vacuo* was $(299\,792 \cdot 6 \pm 0 \cdot 7)\,\text{km s}^{-1}$. In later experiments Froome (1954, 1958) has used a four-horn interferometer of symmetrical design, first at a wavelength of 12·5 mm, then at 4 mm. The final results are

$$(299\,792 \cdot 75 \pm 0 \cdot 3)\,\text{km s}^{-1} \quad \text{(frequency 24 GHz)},$$

$$(299\,792 \cdot 5 \pm 0 \cdot 1)\,\text{km s}^{-1} \quad \text{(frequency 72 GHz)}.$$

The latter was still the accepted value in 1972, but the accuracy has since improved by a factor of about 100. Starting from a klystron oscillator, harmonics are generated by means of point-contact diodes or Josephson junctions, and the frequencies of a sequence of laser lines such as

$$HCN \quad f = \quad 890\,\text{GHz} \quad \lambda = 337 \quad \mu\text{m}$$
$$H_2O \quad f = 10\,700\,\text{GHz} \quad \lambda = \quad 28 \quad \mu\text{m}$$
$$CO_2 \quad f = 32\,176\,\text{GHz} \quad \lambda = \quad 9 \cdot 3\,\mu\text{m}$$

are determined in terms of the caesium standard. A helium–neon laser line is then used as an intermediate step in an interferometric measurement of wavelength in terms of the krypton line at $\lambda = 0 \cdot 606\,\mu\text{m}$, adopted in 1960 as the primary standard of length. Essentially both the frequency and the wavelength of one stabilized laser line are measured, and the values found for $c$ are

$$(299\,792 \cdot 4562 \pm 0 \cdot 0011)\,\text{km s}^{-1} \quad \text{(Evenson } et\,al.\text{ 1972)}$$

$$(299\,792 \cdot 4590 \pm 0 \cdot 0008)\,\text{km s}^{-1} \quad \text{(Blaney } et\,al.\text{ 1974)}.$$

## References

BLANEY, T. G., BRADLEY, C. C., EDWARDS, G. J., JOLLIFFE, B. W., KNIGHT, D. J. E., ROWLEY, W. R. C., SHOTTON, K. C., and WOODS, P. T. (1974). *Nature, Lond.* **251**, 46.

BLEANEY, B., LOUBSER, J. H. N., and PENROSE, R. P. (1947). *Proc. phys. Soc.* **59**, 185.

BRADLEY, C. C., EDWARDS, G. J., KNIGHT, D. J. E., ROWLEY, W. R. C., and WOODS, P. T. (1972). *Phys. Bull.* **23**, 15.

CADY, W. G. (1946). *Piezoelectricity*, McGraw-Hill, New York.

ESSEN, L. (1950). *Proc. R. Soc.* A**204**, 260.

—— FROOME, K. D. (1951). *Proc. phys. Soc.* B**64**, 862.

—— GORDON-SMITH, A. C. (1948). *Proc. R. Soc.* A**194**, 348.

EVENSON, K. M., WELLS, J. S., PETERSEN, F. R., DANIELSON, B. L., DAY, G. W., BARGER, R. L., and HALL, J. L. (1972). *Phys. Rev. Lett.* **29**, 1346.

FARADAY SOCIETY CONFERENCE ON DIELECTRICS (1946). *Trans. Faraday Soc.* **42A**.

FROOME, K. D. (1952). *Proc. R. Soc.* A **213**, 123.
——— (1954). *Proc. R. Soc.* A **223**, 195.
——— (1958). *Proc. R. Soc.* A **247**, 109.
MERCIER, J. (1924), *J. Phys. Radium* **5**, 168.

## Problems

22.1. A cathode-ray tube has plane parallel deflecting plates of separation $a$ and length $b$ parallel to the axis of the tube; the distance from the centre of the plates to the screen is $L$. If the electrons are initially accelerated by a voltage $V_0$, show that their deflection on the screen due to a voltage $V_1$ on the deflector plates is

$$s = \tfrac{1}{2}(LbV_1/aV_0),$$

assuming that $L \gg b$, that the field is uniform between the plates, and that edge effects can be neglected.

If $a = 5$ mm, $b = 40$ mm, $L = 0{\cdot}3$ m, and $V_0 = 1{\cdot}3$ kV, show that the deflection sensitivity is $0{\cdot}92$ mm $V^{-1}$.

22.2. If the frequency limit of the cathode-ray tube of Problem 22.1 were set by the finite transit time of the electrons through the deflector plates, show that the deflection would fall to zero at about 540 MHz.

22.3. Maxwell's bridge for comparing an inductance and a capacitor has the circuit of Fig. 22.5, with the following impedances:

$Z_1$ an inductance $L$ in series with a resistance $R_1$,

$Z_2$ a resistance $R_2$,

$Z_3$ a resistance $R_3$,

$Z_4$ a capacitance $C$ is parallel with a resistance $R_4$.

Show that the balance conditions are

$$R_1/R_3 = R_2/R_4, \qquad L = R_2R_3C.$$

To make the two balance conditions independent, $R_4$ and $C$ must be varied.

22.4. At higher audio-frequencies, resistances may possess a small inductive component; in Anderson's bridge this may be allowed for by writing the components as $P = P + jP'$, $Q = Q + jQ'$, $R = R + jR'$, $T = T + jT'$ (we neglect any inductive component in $S$ as this will be added to $L$ at all frequencies). If $P$ and $Q$ are identical impedances, show that the balance conditions are

$$r + S = R - \omega C(2RT' + 2R'T + RQ' + R'Q),$$

$$L = C(2RT + QR - 2R'T' - Q'R') + R'/\omega.$$

These equations show that it is important to make $R'$ as small as possible. If $R' = 0$, the error in the determination of $L$ is zero, while that in the resistance $r$ of the inductance is $\omega CR(2T' + Q')$.

22.5. In the equivalent circuit (Fig. 22.14) of the coaxial-line wavemeter, the source is taken to be a generator of voltage $V_1$ with internal resistance $R_1$, and the detector has a resistance $R_2$. If the series impedance of the tuned circuit by itself is $Z$, show that the ratio of the voltage $V_2$ across the detector to the input voltage is

$$\frac{V_2}{V_1} = -\frac{\omega^2 M_1 M_2}{R_1(Z + \omega^2 M_i^2/R_1 + \omega^2 M_2^2/R_2)}.$$

This equation shows that $V_2$ is a maximum when $Z$ is a minimum, that is, when the wavemeter is on tune and $Z$ is just the resistance $r$. It shows also that the coupled impedances $\omega^2 M_i^2/R_1$ and $\omega^2 M_2^2/R_2$ lower the effective $Q_F$; by writing $Z = r + 2j\,\delta\omega L$ near resonance, show that the

$$\text{'loaded } Q_F' = \frac{L^{\frac{1}{2}}}{C^{\frac{1}{2}}(r + \omega^2 M_i^2/R_1 + \omega^2 M_2^2/R_2)},$$

and that it may be measured by finding the fractional change in the frequency required to reduce $V_2$ to $1/\sqrt{2}$ of its maximum value (neglect changes in the coupled impedance when varying $\omega$).

22.6. In the equivalent circuit (Fig. 22.16(b)) of a quartz crystal, the components for a particular crystal are $L = 3 \cdot 3$ H, $C = 0 \cdot 042$ pF, $R = 4500\,\Omega$, $C_1 = 5 \cdot 8$ pF. Show that it behaves as a parallel resonant circuit at a frequency approximately 8 Hz above the series resonance frequency (the natural mechanical resonance frequency).

# 23. Fluctuations and noise

## 23.1. Brownian motion and fluctuations

THE irregular motion of small particles suspended in a fluid was first observed by Brown in 1828. This 'Brownian motion' never ceases and is a result of the random motion of the molecules both of the particles themselves and of the fluid. If the motion is observed over a long time, it is found that the average component of the velocity in any direction is zero, since positive and negative values occur with equal probability. The mean-square value of the velocity is not zero, and from classical statistical mechanics it may be shown that the average value of each of the terms $\frac{1}{2}m\dot{x}^2$, $\frac{1}{2}m\dot{y}^2$, $\frac{1}{2}m\dot{z}^2$ of the translational kinetic energy is $\frac{1}{2}kT$, where $k$ is Boltzmann's constant (approximately $1\cdot38\times10^{-23}\,\mathrm{J\,K^{-1}}$) and $T$ is the temperature in degrees Kelvin. This is a special case of the theorem of equipartition of energy: if the energy of a system can be written as the sum of a number of terms each containing only the square of a variable, then the average energy of each of these terms is $\frac{1}{2}kT$. This theorem applies just as much to macroscopic objects as to microscopic ones or molecules, but the magnitude of the fluctuations in the dynamical variable becomes smaller as the inertia of the object increases, since the average energy is independent of size. Given sufficient magnification, the motion can always be observed, and it sets a limit to the sensitivity of any measuring instrument; the fluctuations give a random signal which masks any applied signal of smaller magnitude.

If this theorem is applied to a suspension galvanometer, the following result is obtained. The suspension has one degree of freedom, a rotation measured by the angle $\theta$. The total energy may be written as the sum of two terms, the potential energy of the suspension due to work done against the restoring torque, and the kinetic energy, so that

$$W = \tfrac{1}{2}c\theta^2 + \tfrac{1}{2}\mathfrak{I}\dot{\theta}^2,\qquad(23.1)$$

where $c$ is the restoring torque per unit angle of twist and $\mathfrak{I}$ is the moment of inertia of the system. To each of these terms we must assign an average energy $\frac{1}{2}kT$, so that fluctuations in the angle $\theta$ and the angular velocity $\dot{\theta}$ will occur whose mean-square values are given by

$$\tfrac{1}{2}c\overline{\theta^2} = \tfrac{1}{2}\mathfrak{I}\overline{\dot{\theta}^2} = \tfrac{1}{2}kT.\qquad(23.2)$$

A system which is mathematically similar is the electrical tuned circuit,

consisting of an inductance, capacitance, and resistance connected together. The total electrical energy of such a system, where $I$ is the
instantaneous current and $q$ the instantaneous charge on the capacitor, is

$$W = \tfrac{1}{2}q^2/C + \tfrac{1}{2}LI^2. \tag{23.3}$$

If the theorem of equipartition of energy applies also to electrical
systems, as we would expect in view of its general nature, then the
mean-square values of the fluctuating charge and current will be given by

$$\tfrac{1}{2}\overline{q^2}/C = \tfrac{1}{2}L\overline{I^2} = \tfrac{1}{2}kT. \tag{23.4}$$

These relations give only the mean-square values of the total fluctuations, and tell us nothing about the frequency distribution of the fluctuations. If we imagine that we perform a Fourier analysis of the fluctuations,
and postulate that they are due to some random force acting on the
system, then for the electrical tuned circuit we write

$$L(d^2q/dt^2) + R(dq/dt) + q/C = V_f \exp j\omega t, \tag{23.5}$$

where $V_f$ is the amplitude of the component of the random e.m.f. causing
the fluctuations at the frequency $f = \omega/2\pi$. We now make the following
assumptions about $V_f$: its mean-square value $\overline{V_f^2}$ is independent of frequency, but voltages of different frequency are entirely uncorrelated, so
that the average value of the product $\overline{V_f V_{f'}}$ is zero. The justification for
these assumptions will not be discussed here, but it is obvious that they
are plausible in view of the random nature of the fluctuations. On solving
eqn (23.5) to find the mean-square amplitude $\overline{q_f^2}$ of the fluctuating charge
at the frequency $f$, we have

$$d(\overline{q_f^2}) = \frac{d(\overline{V_f^2})}{(L\omega^2 - 1/C)^2 + R^2\omega^2}. \tag{23.6}$$

The frequencies are continuously distributed, and the differentials are
used since this expression gives the mean-square amplitude of the fluctuations in the frequency range between $f$ and $f+df$. The total mean-square
fluctuation must be given by eqn (23.4), and hence, integrating over all
frequencies, we must have

$$\tfrac{1}{2}kT = \frac{\tfrac{1}{2}\overline{q^2}}{C} = \frac{1}{2C}\int d(\overline{q_f^2}) = \frac{1}{4\pi C}\frac{d(\overline{V_f^2})}{df}\int_0^\infty \frac{d\omega}{(L\omega^2 - 1/C)^2 + R^2\omega^2}.$$

This integral may be evaluated as follows. On making the substitution
$\omega = x(LC)^{-\frac{1}{2}}$, it becomes

$$\left(\frac{C^3}{L}\right)^{\frac{1}{2}}\int_0^\infty \frac{-d(1/x)}{(x-1/x)^2 + R^2C/L} = \left(\frac{C^3}{L}\right)^{\frac{1}{2}}\int_0^\infty \frac{dx}{(x-1/x)^2 + R^2C/L},$$

where the second form is obtained by replacing $x$ by $1/x$. Hence the integral may be written as

$$\frac{1}{2}\left(\frac{C^3}{L}\right)^{\frac{1}{2}}\int_0^\infty \frac{d(x-1/x)}{(x-1/x)^2+R^2C/L}=\frac{1}{2}\left(\frac{C^3}{L}\right)^{\frac{1}{2}}\int_{-\infty}^\infty \frac{dz}{z^2+R^2C/L}=\frac{\pi C}{2R}.$$

Hence

$$\tfrac{1}{2}kT=\frac{\tfrac{1}{2}\overline{q^2}}{C}=\frac{1}{8R}\frac{d(\overline{V_f^2})}{df},$$

or

and

$$d(\overline{V_f^2})=4kTR\,df, \tag{23.7}$$

$$d(\overline{q_f^2})=\frac{4kTR\,df}{(L\omega^2-1/C)^2+R^2\omega^2}. \tag{23.8}$$

It is easily verified (see Problem 23.1) from these results that $\tfrac{1}{2}L\overline{I^2}=\tfrac{1}{2}kT$ in accordance with eqn (23.4). The equations lead to the interesting result that, whereas the total mean-square values of the fluctuations depend only on $L$ and $C$, the expression for the distribution of the voltage fluctuations with frequency involves only $R$. The result given by eqn (23.7) may be expressed by saying that the mean-square voltage $d(\overline{V_f^2})$ of the fluctuations in the frequency range $df$ is $4kTR\,df$, and is thus proportional to the bandwidth $df$. The existence of such fluctuations was first verified by Johnson, and they are known as resistance or 'Johnson' noise. They will be considered in more detail in § 23.3.

## 23.2. Fluctuations in galvanometers

We return now to the case of the galvanometer, and consider first a moving-coil suspension galvanometer when the coil is on open-circuit. Then the equation of motion is

$$\Im(d^2\theta/dt^2)+b(d\theta/dt)+c\theta = F_f\exp j\omega t, \tag{23.9}$$

where $\Im$ is the moment of inertia, $b$ the mechanical damping constant, and $c$ the restoring torque per unit angle of twist. We assume that the fluctuations are caused by a random torque, whose Fourier component at the frequency $f=\omega/2\pi$ has the amplitude $F_f$. Our further postulates about the nature of $F$ are similar to those made about $V$ in the last section. Then the analysis is exactly similar to the previous case of the electrical tuned circuit, so that by comparison we obtain at once

and

$$d(\overline{F_f^2})=4kTb\,df \tag{23.10}$$

$$d(\overline{\theta_f^2})=\frac{4kTb\,df}{(\Im\omega^2-c)^2+b^2\omega^2}. \tag{23.11}$$

By integration it may be shown that these expressions satisfy eqn (23.2).

In general the galvanometer will be used for observing a current and will therefore be connected to a circuit whose total resistance (including the galvanometer coil) is $R$. Then we have two equations

$$\Im(d^2\theta/dt^2)+b(d\theta/dt)+c\theta = \Phi I+F_f \exp j\omega t, \\ RI = -\Phi(d\theta/dt)+V_f \exp(j\omega t+j\delta), \Bigg\} \quad (23.12)$$

where $\Phi = nAB$ and $I$ is the instantaneous current through the circuit. Two sources of fluctuations have been included; a random torque due to Brownian motion of the suspended coil, and a random voltage associated with the electrical circuit. In eqns (23.12) the Fourier components of these two sources of fluctuations at the frequency $f = \omega/2\pi$ have been used, with a phase difference $\delta$ between them. Since the two sources are independent, we do not expect any correlation in phase, and for different frequencies the phase difference $\delta$ will have random values. Elimination of the current $I$ between the two equations gives

$$\Im(d^2\theta/dt^2)+(b+\Phi^2/R)(d\theta/dt)+c\theta = (\Phi/R)V_f \exp(j\omega t+j\delta)+F_f \exp j\omega t,$$

and the square of the amplitude of the fluctuations at the frequency $f$ is found to be

$$\theta_f^2 = \frac{(\Phi/R)^2 V_f^2+F_f^2+2(\Phi/R)V_f F_f \cos \delta}{(\Im\omega^2-c)^2+(b+\Phi^2/R)^2\omega^2}.$$

On summing over a range of frequencies, $\delta$ takes all values between 0 and $2\pi$ and the mean value of $\cos \delta$ is therefore zero. Hence the mean-square angular amplitude in the frequency range $f$ to $f+df$ is

$$d(\overline{\theta_f^2}) = \frac{\{(\Phi/R)^2 \, d(\overline{V_f^2})\}+d(\overline{F_f^2})}{(\Im\omega^2-c)^2+(b+\Phi^2/R)^2\omega^2}. \quad (23.13)$$

On substituting the expressions for $d(\overline{V_f^2})$ and $d(\overline{F_f^2})$ given by eqns (23.7) and (23.10), we find

$$d(\overline{\theta_f^2}) = \frac{4kT(b+\Phi^2/R) \, df}{(\Im\omega^2-c)^2+(b+\Phi^2/R)^2\omega^2}. \quad (23.14)$$

This equation is similar to that obtained for the galvanometer on open circuit except that the total damping constant $(b+\Phi^2/R)$ appears instead of just the mechanical damping $b$. Integration of eqn (23.14) over all frequencies will obviously give the same result, $\frac{1}{2}c\overline{\theta^2} = \frac{1}{2}kT$, as for the galvanometer on open-circuit, since the result is independent of the magnitude of the damping. Thus, although there are now two independent sources of random fluctuations, and these add in the squares as shown by the numerator of eqn (23.13), each is associated with a damping term so that the total mean energy $\frac{1}{2}c\overline{\theta^2}$ stored in the suspension remains unaltered, provided that each source is at the same temperature. This argument could be extended by separating the mechanical damping $b$ into two parts, one due to imperfect elasticity of the suspension and the other to damping by the viscosity of the air. Then it follows that the total mean-square angular fluctuations have the same value whether the galvanometer is evacuated or not; the admission of air provides an extra source of fluctuations owing to the molecular bombardment, whose tendency to increase the mean-square deflection is just counterbalanced by the viscous air damping which accompanies it. The frequency distribution of the fluctuations is of course changed because of the increase in the damping, but it is important to realize that the Brownian motion is inherent in the suspended coil and is not caused by the bombardment by the gas molecules. If it were, and the suspension had an imperfect elasticity, then the molecular bombardment would result in the suspension being heated, through the dissipation of energy in it, and the gas would be cooled, even though both were originally at the same temperature. This is contrary to the second law of thermodynamics.

The processes which we regard as 'damping' in the galvanometer represent a degradation of mechanical energy into heat energy; in viscous damping, into kinetic energy of the gas molecules; in electromagnetic damping, ultimately into the vibrational energy of the lattice of the resistance in the external circuit (the coil moving in the magnetic field acts as a transducer, converting mechanical motion into electrical voltage). At the level of the molecular fluctuations, the damping processes are just the mechanisms by which thermal equilibrium is established; without them, an individual component of the system (galvanometer suspension, gas molecules, lattice of the resistor) would have no means of knowing what the temperatures are of the other components. In the electrical case, resistance arises from the conversion of electrical energy into heat energy, and at the fluctuation level is the mechanism by which the electrical fluctuations reach thermal equilibrium with the lattice fluctuations. The nature of the carriers of the electric current is no more important in this process than that of the molecules of the gas causing viscous damping.

It is convenient to define the minimum observable current for a galvanometer as that current which would produce a deflection equal to the r.m.s. value of the total Brownian angular motion. For a steady current $I$ the deflection $\theta = I(nAB)/c = I\Phi/c$, and hence the minimum observable current $I_m$ would be

$$I_m = (ckT)^{\frac{1}{2}}/\Phi. \tag{23.15}$$

In general the electromagnetic damping term $\Phi^2/R$ is much larger than the mechanical damping term $b$, and the critical damping resistance $R_c$ is given by eqn (5.50),

$$R_c = \tfrac{1}{2}\Phi^2/(\Im c)^{\frac{1}{2}},$$

while the period $\tau = 2\pi(\Im/c)^{\frac{1}{2}}$. Using these two relations the minimum observable current and voltage are conveniently expressed in the form

$$I_m = (\pi kT/R_c\tau)^{\frac{1}{2}}, \qquad V_m = (\pi kTR_c/\tau)^{\frac{1}{2}}, \qquad (23.16)$$

since $V_m = R_cI_m$ if the galvanometer is critically damped. Taking room temperature as 290 K, so that $kT = 4\times10^{-21}$ J, for a galvanometer of period 2 s and critical damping resistance 100 $\Omega$, we find that the minimum observable current and voltage are approximately $8\times10^{-12}$ A and $8\times10^{-10}$ V.

The correctness of the expressions derived above has been verified experimentally by Jones and McCombie (1952). The deflections of an ordinary galvanometer of about 2 s period (sensitivity 1 mm deflection at 1 m distance for 0·01 $\mu$A) were magnified by an optical lever. The beam of light reflected from the galvanometer mirror fell on a split photocell, so that rotation of the mirror transferred light from one cell to the other. The difference in the currents from the two photocells was observed on a second galvanometer; a deflection of 15 mm on this instrument corresponded to a voltage of about $10^{-3}$ $\mu$V (or a current of $10^{-5}$ $\mu$A) applied to the first galvanometer. To make use of this amplification, all external sources of disturbance such as vibration had to be eliminated. Typical traces obtained were similar to those shown in Fig. 23.1. With the first galvanometer on open-circuit the damping is small, and the frequency distribution of the angular deflections is large only in the region around the normal frequency of the suspension. Consequently the fluctuations resemble bursts of oscillation at the natural frequency, the number of oscillations in each being in inverse ratio to the damping (and roughly equal to the '$Q_F$' of the suspension). When the galvanometer is just critically damped, $(b + \Phi^2/R)^2 = 4\Im c$ and the denominator of eqn (23.14) can be written as $(\Im\omega^2 + c)^2$, showing that the frequency distribution of the

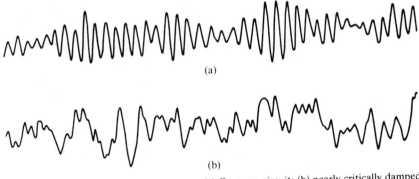

(a)

(b)

FIG. 23.1. Fluctuations of a galvanometer: (a) On open-circuit; (b) nearly critically damped. (After Jones and McCombie 1952.)

fluctuations now has its maximum value at zero frequency. The appearance of the fluctuations is now that of a random disturbance without any sinusoidal character (Fig. 23.1(b)). The voltage sensitivity of the system was found by applying a voltage of about $10^{-2}\,\mu\mathrm{V}$, obtained by attenuating a known voltage $\approx 1\,\mathrm{V}$ through a resistance chain, and a thorough statistical analysis of the results showed that the magnitude of the fluctuations agreed with the theoretical value within 1 per cent.

## 23.3. The relation between resistance noise and thermal radiation

In an evacuated enclosure containing thermal radiation at an absolute temperature $T$ the energy density in the frequency range $f$ to $f+df$ is given by Planck's law,

$$dU = \frac{8\pi h f^3\,df}{c^3\{\exp(hf/kT)-1\}},\tag{23.17}$$

where $h$ is Planck's constant and $k$ is Boltzmann's constant. For all radio-frequencies $hf \ll kT$ at room temperature, since $290k$ corresponds to a quantum of energy for a wavelength of approximately $50\,\mu\mathrm{m}$. We may therefore expand the exponential, obtaining

$$dU = 8\pi f^2 kT\,df/c^3,\tag{23.18}$$

which is simply the Rayleigh–Jeans law of classical theory. Since the polarization of the radiation is random, on the average only one-third of this energy corresponds to radiation whose electric vector is parallel to a given direction (say the y-axis), and only such radiation will induce a voltage in a short dipole aerial inserted in the enclosure parallel to the y-axis. From § 8.3 the mean-square electric field component is then given by $E_y^2 = cZ_0(\tfrac{1}{3}U)$, where $Z_0 = (\mu_0/\epsilon_0)^{\frac{1}{2}}$ is the intrinsic impedance of free space. Hence the mean-square voltage induced in an aerial of length $s$ will be

$$d(V_r^2) = s^2\,d(E_y^2) = 8\pi f^2 s^2 kTZ_0\,df/3c^2.\tag{23.19}$$

Even if the aerial consists of a perfectly conducting wire, the resulting current which flows will be finite, since energy will be re-radiated by this oscillatory current, and this energy must just be equal to that picked up by the aerial. The radiation must therefore behave as a generator of open-circuit voltage $V_r$ with an internal impedance $R_r$, as in the equivalent circuit of Fig. 23.2(a). This drives a current $I_r$ when short-circuited, and the power dissipated is $V_r^2/R_r = I_r^2 R_r$; this power is lost by re-radiation, and from § 8.9 it follows that $R_r$ is just the radiation resistance given by eqn (8.71) as

$$R_r = 2\pi Z_0 f^2 s^2/3c^2.\tag{23.20}$$

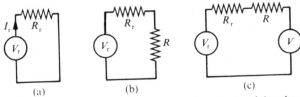

FIG. 23.2. Equivalent circuit of an aerial. $V_r$ = voltage induced by thermal radiation; $R_r$ = radiation resistance of aerial. (a) Aerial short-circuited at centre; (b) aerial with resistance $R$ at centre; (c) as (b) but showing noise voltage due to $R$.

If the aerial is not a perfectly conducting wire, and has a real ohmic resistance $R$, the equivalent circuit will be as shown in Fig. 23.2(b), and the energy dissipated in the load $R$ will be $V_r^2 R/(R_r+R)^2$. This will heat the resistance $R$, while less energy is re-radiated to the enclosure. If $R$ is initially at the same temperature $T$ as the radiation in the enclosure, the apparent result will be that $R$ is heated and the enclosure cooled, which is contrary to the second law of thermodynamics. In order that the net exchange of energy between $R$ and the enclosure be zero, we must postulate that there is a fluctuation voltage associated with $R$, .as in Fig. 23.2(c), of mean-square voltage $V^2$ and internal resistance $R$. This will send a power $V^2 R_r/(R_r+R)^2$ back into the aerial which must just equal that drawn from the enclosure and dissipated in $R$. Thus $V_r^2 R = V^2 R_r$, and in the frequency range from $f$ to $f+df$

$$\frac{d(\overline{V^2})}{R} = \frac{d(\overline{V_r^2})}{R_r} = \frac{8\pi f^2 s^2 k T Z_0\, df}{3c^2} \times \frac{3c^2}{2\pi Z_0 f^2 s^2} = 4kT\, df. \qquad (23.21)$$

This result is identical with that obtained earlier (eqn (23.7)) by considering a simple tuned circuit. The voltage fluctuations have a constant distribution with frequency so long as the energy quantum $hf \ll kT$; this limitation corresponds to our use of the classical expression (Rayleigh–Jeans law) for the energy density in the enclosure. The fluctuations associated with a resistance $R$ can be represented by inserting a voltage generator $V$, whose mean-square open-circuit voltage is given by eqn (23.21), for which $R$ acts as the internal impedance as in Fig. 23.2(c). The equivalent current generator will have a mean-square current

$$d(\overline{I^2}) = 4kT\, df/R, \qquad (23.22)$$

and it will be shunted by the resistance $R$.

Let us suppose that we are able to connect to our aerial a load of resistance $R$ which itself produces no noise (for example, a resistance kept at a temperature very close to 0 K). Then the maximum power which can be drawn from the enclosure and dissipated in $R$, obtained by making $R$ equal to $R_r$, is $d(\overline{V_r^2})/(4R_r) = kT\, df$; this is the 'available noise power'.

If $R$ is in fact a radio receiver, this power drawn from the thermal radiation incident on the aerial forms a source of 'noise', and can be heard as a hiss from a loudspeaker, or viewed on a cathode-ray oscillograph. It will obscure any signal which it is desired to receive unless the signal power in the aerial is larger than that picked up from the radiation background. This difficulty cannot be overcome by increasing the over-all amplification of the receiver, since both noise and signal will be amplified together. Thus the radiation noise sets a limit to the useful sensitivity of a receiver. If a theoretically perfect receiver is defined as one which itself introduces no noise, then the amplified noise output will be $AkT\,df$, where $A$ is the over-all amplification. The amplified signal output will be $AP$, where $P$ is the signal power incident on the aerial. Then the minimum detectable input signal may be defined as that which gives a signal output equal to the noise output, from which

$$\text{minimum detectable signal power } P_0 = kT\,df \qquad (23.23)$$

for a perfect receiver. It is clear that the only variable at our disposal here is the bandwidth $df$, and the reduction in noise obtained on narrowing the bandwidth can be seen in Fig. 23.3. This shows the noise output from an aperiodic receiver covering a wide band, before and after the insertion of a low-pass filter cutting out the higher frequencies. The change in character of the noise when the high-frequency components are absent can be seen as well as the reduction in amplitude. In general, however, the bandwidth cannot be reduced beyond a certain limit without impairing the quality of the reception, since the higher modulation frequencies will be cut out. If only audio-frequency modulation is involved, the bandwidth will be about $0\cdot01$ MHz and the minimum detectable signal power is $4\times10^{-17}$ W. In a television receiver it is necessary to have a bandwidth of say 4 MHz to include all the information necessary to form the picture, and the minimum signal power needed to equal noise in a perfect receiver is $1\cdot6\times10^{-14}$ W.

In practice, all receivers generate a certain amount of internal noise, with the result that the noise output is greater than for a perfect receiver. The signal input $P_1$ required to give a signal output equal to the noise output is therefore greater than $P_0$. The quantity $P_1-P_0$ is a measure of the internal noise generated in the receiver, and by writing

$$P_1-P_0 = kT_e\,df$$

it may be expressed in terms of the 'excess noise temperature' $T_e$ of the receiver. In an ideal receiver $T_e=0$, but in practice little is gained by making it smaller than about $T/10$, where $T$ is the temperature of the thermal radiation being received in the application for which the receiver is designed. In laboratory applications the source to which the receiver is

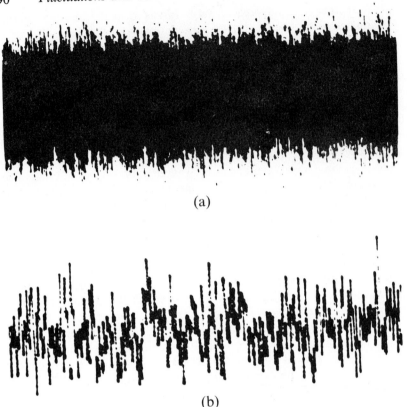

(a)

(b)

FIG. 23.3. Electrical noise output from an aperiodic amplifier. The lower oscilloscope trace shows the reduction in root-mean-square amplitude and removal of higher-frequency components produced by insertion of a low-pass filter. (Photograph by C. A. Carpenter and J. Ward.)

connected is generally at room temperature, and it is then convenient to take a value of $T = 290\,\text{K}$ to define $P_0$, making it equal to $4 \times 10^{-21}\,df$ (watts). The ratio of $P_1$ to this value of $P_0$ is then defined as the 'noise figure' of the receiver, and denoted by $F$. Since $F$ is a ratio of two powers, it is often expressed in decibels.

## 23.4. Shot noise

For most purposes it is sufficient to consider the electron current in a thermionic diode as a uniform flow of charge to the anode. Since the current consists in fact of the arrival of a finite number of electrons per second, this cannot be true. The flow of electrons is a random process, and we may expect that there will be a fluctuation in the number arriving in a given time interval, if we measure over a number of such intervals. If the arrival of the electrons consists of a succession of completely random events, then the mean-square deviation from the average number $N$ per second is proportional to $N$. These fluctuations give rise to noise in the anode circuit, known as shot noise from the obvious analogy with the random patter of shot on a target. For simplicity, we discuss first shot noise in vacuum tubes, but a similar phenomenon occurs in solid state devices (see § 23.6).

In general we are interested not in the total deviation from the mean, but in the frequency distribution of the fluctuations. To find this it is necessary to carry out a Fourier analysis of the pulse of current due to the arrival of a single electron of charge $e$. We will assume that this pulse, occurring at time $t = 0$, has some irregular shape but is entirely confined within the time interval $-\tau/2$ to $+\tau/2$. Since the total charge arriving is $e$, we have

$$e = \int_{-\tau/2}^{+\tau/2} I \, dt.$$

We do not specify anything about the duration of the pulse $\tau$ except that it is very short ($\sim$ the transit time, see § 21.3). The Fourier series representing the frequency distribution of the current due to the arrival of $e$ is written

$$I = a_0 + \sum_{n=1}^{\infty} a_n \cos \frac{2\pi n t}{T} + \sum_{n=1}^{\infty} b_n \sin \frac{2\pi n t}{T}.$$

Here $a_0$, $a_n$, and $b_n$ are coefficients to be determined, and $T$ is an undefined large interval of time. In effect we regard all the frequencies we are interested in as multiples of the fundamental frequency $1/T$. To obtain a continuous frequency distribution we should make $T$ infinite, and replace the summations in the series by integrations. For the student who is more familiar with a Fourier series than a Fourier integral we use the former, and by making $T$ large we obtain a good approximation to a continuous frequency distribution.

From the ordinary formulae of Fourier analysis

$$a_0 = \frac{1}{T} \int_{-T/2}^{+T/2} I \, dt; \qquad a_n = \frac{2}{T} \int_{-T/2}^{+T/2} I \cos \frac{2\pi n t}{T} \, dt;$$

$$b_n = \frac{2}{T} \int_{-T/2}^{+T/2} I \sin \frac{2\pi n t}{T} \, dt.$$

To evaluate the coefficients we restrict ourselves to frequencies small compared with $1/\tau$. Then, since the current is finite only in the range $-\tau/2$ to $+\tau/2$, and zero outside this range, we can write $\cos 2\pi n t / T = 1$ and

$\sin 2\pi nt/T = 0$ over the range of integration for which the current is finite. Hence $b_n$ is zero, while

$$2a_0 = a_n = \frac{2}{T} \int_{-T/2}^{+T/2} I \, dt = \frac{2e}{T}.$$

Thus we have

$$I = \frac{e}{T} + \sum_{n=1}^{\infty} \frac{2e}{T} \cos \frac{2\pi nt}{T},$$

and the mean-square value of the $n$th component is

$$\overline{I_n^2} = \tfrac{1}{2}(2e/T)^2 = 2e^2/T^2.$$

If $N$ electrons arrive in time $T$, then each will contribute an equal amount to the value of $I_n^2$. (The electrons arrive at random times, and their contributions to the Fourier series will all differ slightly in phase. Thus we must add intensities, and not amplitudes.) Then

$$\overline{I_n^2} = 2e^2 N/T^2 = 2eI_0/T,$$

where $I_0 = Ne/T$ is the mean value of the current. Now the number of Fourier components whose frequencies lie within a range between $f$ and $f+df$ is $T \, df$, since the components are equally spaced in frequency by amounts $1/T$. Adding together the mean-square values of these components gives

$$d(\overline{I^2}) = 2eI_0 \, df \qquad (23.24)$$

for the mean-square current fluctuation in the frequency range $f$ to $f+df$.

### Influence of space charge

In this derivation of the formula for shot noise the arrival of an electron is considered as a random event, completely independent of the arrival of any other electron. We expect that the emission of electrons from the cathode has this property of complete randomness, but this is not necessarily true of their arrival at another electrode. In practice it is found that the value of the shot noise is materially lower than that given by the above equation unless the current to the anode is limited only by the rate of emission from the cathode. In general the anode current is only a fraction of the emission current because of the formation of 'space charge' outside the cathode which causes a large number of electrons emitted from the cathode to be turned back to the cathode. Since this is due to the mutual interaction of the electrons, we may expect that the flow of electrons to the anode is not now a succession of completely random events. The value of the fluctuations is greatest for random events, and falls as soon as they become not completely random. Physically, the action of the space charge may be envisaged as follows. Suppose that at some instant the number of electrons emitted from the cathode rises momentarily above the average. This will cause a temporary increase in the space charge, and a number of electrons greater than average will leave the space charge region for the anode. This number is smaller than the surge from the cathode because the space charge acts as a reservoir; the effect of the increased space charge is to turn some of the excess electrons back to the cathode. Similarly, at instants when the cathode emission falls momentarily below the average, the space charge also drops and less electrons are turned back. To allow for this 'space-charge smoothing', as it is called, a factor is inserted in the equation for the shot noise. Thus

$$d(\overline{I^2}) = 2\beta e I_0 \, df. \tag{23.25}$$

$\beta$ is called the space-charge smoothing factor, and may be as low as $0.03$, showing that the smoothing effect is very considerable.

## Noise in multi-electrode tubes

The presence of grids in a tube does not affect the validity of the equations given above for shot noise so long as they do not intercept any of the current on its way to the anode. Thus eqn (23.25) is still valid for a negative-grid triode, but this is not so for a screen-grid tetrode or a pentode, for then the positive screen grid intercepts a considerable portion of the anode current. Since the chance of an electron hitting the wire of the screen grid is purely random, the screen current will have the full shot noise appropriate to its magnitude. It is obvious that similar fluctuations, though of opposite sign, must be imposed on the current that goes through the screen grid to the anode. Assuming that less than half of the total current goes to the screen, we may write approximately for the anode current

$$d(\overline{I^2}) = 2\beta e I_a \, df + 2 e I_s \, df. \tag{23.26}$$

Since $\beta$ may be less than $0.1$, while the screen current $I_s$ is $0.2$ or $0.3$ of $I_a$, the second term is often more important than the first. Hence screen-grid tetrodes and pentodes are generally more noisy than triodes. The additional noise is called 'partition noise'.

## Equivalent noise resistance

It is often convenient to define the amount of noise by referring it to an equivalent resistance $R_n$ (at 290 K) in the grid circuit. The fluctuating voltage at the grid due to $R_n$ has the mean-square value

$$d(\overline{V^2}) = 4kTR_n \, df,$$

since the grid consumes no power and the equivalent noise resistance is therefore on open-circuit. This causes a fluctuating anode current whose mean-square value is

$$d(\overline{I^2}) = g_m^2 \, d(\overline{V^2}) = g_m^2 4kTR_n \, df, \qquad (23.27)$$

where $g_m$ is the mutual conductance of the tube. If $\overline{I^2}$ is due to the shot noise, the equivalent noise resistance may be calculated by means of this formula, $T$ being taken as room temperature. The advantage of this method of specifying the noise is that the value of $R_n$, unlike that of $d(\overline{I^2})$, is independent of the bandwidth, and it facilitates comparison of the shot noise with the resistance noise in the circuits attached to the grid. If partition noise is included by replacing eqn (23.26) by (23.25) with an effective value $\beta'$ instead of $\beta$, then

$$R_n = \beta' e I_0 / 2 g_m^2 kT. \qquad (23.28)$$

An estimate of the relative importance of shot noise and resistance noise can be obtained from the formula for the equivalent noise resistance. For a typical triode, $g_m = 5$ mA V$^{-1}$, $\beta = 0{\cdot}03$, $I_0 = 10$ mA, $e = 1{\cdot}6 \times 10^{-19}$ C; this gives $R_n = 240 \, \Omega$. The value for a pentode would be somewhat higher, owing to partition noise. These figures apply at medium radio-frequencies (that is, of the order of megahertz); at higher frequencies (100 MHz upwards) $R_n$ rises owing to noise voltages induced in the grid which have period of oscillation comparable with the electron transit time (cf. § 21.3). At audio-frequencies the shot noise (particularly from tubes with oxide-coated cathodes) becomes abnormally large. This is known as the flicker effect, and is thought to be associated with changes in the state of the cathode surface which cause abnormal fluctuations in the anode current.

## 23.5. Design of receivers for optimum performance (minimum noise figure)

The correct design of a receiver is of great importance. If it is being used in an application where the signal strength is fixed, such as in r.f. astronomy or spectroscopy, then the limiting sensitivity attainable will depend entirely on the design of the receiver. In radio communications an improvement of a factor $n$ in signal–noise ratio can be achieved by increasing the transmitter power by a factor $n$, but a very much more economical method is to improve the receiver performance by the same factor instead. The following remarks illustrate only the basic principles, and do not go into any detail of receiver design.

In general all the stages of a receiver will contribute some noise, but if the amplification of each stage is high only the first stage or two is important. If stage $k$ gives noise power $N_k$, and the stage gain is $m$, then the signal–noise ratio after $n$ stages is

$$Sm^n/(N_1 m^n + N_2 m^{n-1} + ... + N_n) = S/(N_1 + N_2 m^{-1} + N_3 m^{-2} + ...).$$

$$(23.29)$$

With a stage gain of 10–100 even the second stage will contribute little to the noise output. If not, the design of the second stage should follow the same principles as that of the first stage, and only the latter need be considered.

In the circuit of Fig. 23.4(a) $S$ represents a signal source of voltage $S$ with output resistance $R_1$. $R_1$ is assumed to be noisy, at temperature $T$, and its equivalent noise voltage is represented by $V_{n1}$, in series with $S$. The source $S$ may be a signal induced in an aerial, in which case $R_1$ is the radiation resistance of the aerial and $T$ is the ambient temperature which we take to be 290 K. The source is connected to the grid of the tube, and $R_2$ is the grid-bias resistance, or the first tuned circuit, in which case $R_2$ is its parallel impedance. In general, $R_2$ will also generate resistance noise, which is represented by the insertion of a voltage source $V_{n2}$ in series with $R_2$. In the first instance we shall assume that the tube contributes no shot noise ($R_n = 0$), and that noise from subsequent stages is negligible. Then the signal–noise ratio will be the same at the grid of the first tube as at any later point in the receiver, and we need only compute the ratio of the mean-square signal voltage on the grid to the mean-square noise voltage. For simplicity $R_1$ and $R_2$ are taken to have the same temperature, which in practice will not be far from true.

Since $V_{n1}^2/R_1 = V_{n2}^2/R_2 = 4kT \, df$, and $R_2$ acts as the load for the noise generator $V_{n1}$, and $R_1$ as the load for $V_{n2}$, the mean-square noise voltage on the grid is

$$4kT \, df \left\{ R_1 \frac{R_2^2}{(R_1+R_2)^2} + R_2 \frac{R_1^2}{(R_1+R_2)^2} \right\} = 4kT \, df \times \frac{R_1 R_2}{R_1+R_2}. \quad (23.30)$$

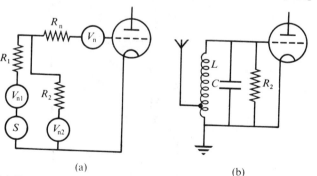

(a)                                        (b)

FIG. 23.4. (a) Equivalent input circuit of a receiver, showing noise voltages. (b) Actual input circuit, showing aerial tapped on to inductance of input tuned circuit.

Since $R_1$ and $R_2$ are random noise sources, the mean-square voltages have been added; the result is the same as that for a resistance equal to $R_1$ and $R_2$ in parallel, as we should expect.

The mean-square signal voltage on the grid is

$$S^2 R_2^2 / (R_1 + R_2)^2.$$

Hence the signal–noise ratio at the grid is

$$\frac{S^2}{4kT\,df}\frac{R_2}{R_1+R_2}\frac{1}{R_1} = \frac{P}{kT\,df}\frac{R_2}{R_1+R_2}, \tag{23.31}$$

where $P = S^2/4R_1$ is the available signal power. If eqn (23.31) is put equal to unity, we obtain the signal power $P_1$ required to give an output power equal to the noise output power of the receiver. The noise figure $F$ is defined as the ratio of this signal power to the value $kT\,df$ for a perfect receiver, and hence the noise figure is

$$F = P_1/kT\,df = 1 + R_1/R_2. \tag{23.32}$$

A simpler method of obtaining this result is as follows. In eqn (23.30) the first term is the aerial noise and the second that from the resistance $R_2$. In an ideal receiver the latter would be absent, and the only noise would be that from the aerial. Hence we can determine $F$ by dividing eqn (23.30) by its first term. This leads at once to eqn (23.32), and is the method used below. If the aerial is matched to the first circuit, $R_1 = R_2$, and $F = 2$. But if $R_2 > R_1$, $F$ is reduced and tends to its limiting value of unity as the ratio of $R_2$ to $R_1$ is increased indefinitely. Hence to obtain the highest sensitivity it pays to mismatch the aerial to the receiver, since the reduction in noise at the grid when $R_2$ is shunted by the lower resistance $R_1$ is greater than the loss of signal at the grid due to the mismatch. We see also that, in the absence of tube noise, it is possible to approach very closely to the theoretical limit of sensitivity. In a practical case $R_1$ (for a half-wave dipole aerial) would be 80 $\Omega$, while $R_2$ could be of the order of 100 k$\Omega$, giving $F = 1 \cdot 001$.

We now allow for the presence of shot noise in the vacuum tube, represented by the noise resistance $R_n$ in Fig. 23.4(a). On adding $4kTR_n\,df$ to eqn (23.30), the mean-square noise voltage at the grid becomes

$$4kT\,df\frac{R_1 R_2^2}{(R_1+R_2)^2}\left(1+\frac{R_1}{R_2}\right)+4kT\,df R_n.$$

On dividing this by the noise from the aerial we obtain

$$F = 1 + \frac{R_1}{R_2} + \frac{R_n}{R_2}\left(\frac{R_1}{R_2}+2+\frac{R_2}{R_1}\right). \tag{23.33}$$

This result shows that there is now an optimum value for $R_1/R_2$, given by

$$\frac{R_2}{R_1} = \frac{R_1}{R_2} + \frac{R_1}{R_n},\qquad(23.34)$$

at which $F$ has its minimum value. If $R_2 \gg R_1$, the optimum value of $R_1 = (R_n R_2)^{\frac{1}{2}}$; taking $R_n = 1000\ \Omega$, $R_2 = 100\ k\Omega$, this gives $R_1 = 10\ k\Omega$ and $F = 1\cdot 22$. In general, $R_1$ is fixed (for example, it is equal to the radiation resistance of the aerial), but it can be transformed to the optimum value by the use of a circuit with variable coupling such as that shown in Fig. 23.4(b).

## 23.6. Noise in solid-state triodes

In solid-state devices noise arises from a number of causes, but these are basically the same as those already discussed. At very low frequencies flicker noise, whose power varies inversely with frequency, is of practical importance; it arises from imperfections and varies widely from sample to sample. In the frequency range between about 1 kHz and many megahertz (where transit-time effects set in), the effects which are dominant in a junction transistor are as follows.

1. Johnson noise in the ohmic resistance between the base terminal and the working region; because of the thinness of the base, this resistance may be as much as $200\ \Omega$.

2. Current fluctuations (shot noise) arising from the random nature of the flow of carriers across a barrier. The flow of carriers into the base from the emitter consists of those with sufficient thermal energy to surmount the barrier at the depletion layer. The absence of space charge means that there is no 'space-charge smoothing' to reduce the size of the fluctuations, and for the emitter current we may write

$$d(\overline{I_e^2}) = 2eI_e\, df.\qquad(23.35)$$

Not all of these carriers reach the collector junction, some being lost in the base through recombination. This is again a purely random process, analogous to the interception of electrons by a grid in a vacuum tube (partition noise). It is the origin of the base current, and the current fluctuations can be written as

$$d(\overline{I_b^2}) = 2eI_b\, df.\qquad(23.36)$$

For the collector current we have similarly

$$d(\overline{I_c^2}) = 2eI_c\, df.\qquad(23.37)$$

These two sets of current fluctuations are shown as current sources in the equivalent noise circuit of Fig. 23.5. It is unnecessary to include a third

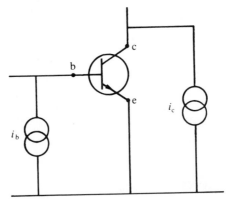

FIG. 23.5. Sources of current fluctuations for a junction transistor in the CE connection.

independent source corresponding to eqn (23.35) for the following reason. The fact that fluctuations in $I_b$ and $I_c$ are uncorrelated means that $d(\overline{I_b I_c}) = 0$. Since $I_e = I_b + I_c$, we have

$$d(\overline{I_e^2}) = d(\overline{(I_b + I_c)^2}) = d(\overline{I_b^2}) + d(\overline{I_c^2}) = 2e(I_b + I_c)\,df$$
$$= 2eI_e\,df,$$

showing that eqn (23.35) can be derived from the two following equations and does not represent an independent source of fluctuations.

For many purposes it is convenient to represent the various sources of noise by a set of noise generators in the input circuit. For a transistor in the CE connection the mutual conductance is (see § 19.1)

$$g_m = eI_c/kT.$$

The current source $i_c$ in the collector circuit can be replaced by a voltage source $v$ in the base circuit such that $\overline{i_c^2} = g_m^2 \overline{v^2}$ (cf. eqn (23.27)). Hence

$$\overline{v^2} = 2eI_c\,df/g_m^2 = (2kT/g_m)\,df. \tag{23.38}$$

The equivalent input circuit is shown in Fig. 23.6. For simplicity we neglect here the base resistance, which is small, and the resulting noise voltage has a mean-square value $\overline{v_s^2} = 4kTR_s\,df$. Across the input resistance $R_i$ this develops a mean-square voltage

$$4kTR_s\,df R_i^2/(R_i + R_s)^2.$$

Similarly the current source $i_b$, driving a current that is divided between $R_s$ and $R_i$, develops a mean-square voltage across $R_i$ equal to

$$2eI_b\,df R_s^2 R_i^2/(R_i + R_s)^2.$$

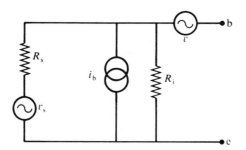

FIG. 23.6. Equivalent input noise circuit for a junction transistor in the CE connection.

In the CE connection the input resistance $R_i = \beta/g_m$, where $\beta = I_c/I_b$ is the current gain. $R_i$ is high, and we therefore assume that $R_s \ll R_i$, a simplifying assumption that will prove to be justified. Then, together with the mean-square voltage given by eqn (23.38), we have altogether a mean-square voltage across the base–emitter terminals equal to

$$4kTR_s \, df + 2eI_b R_s^2 \, df + (2kT/g_m) \, df.$$

As in the previous section, we obtain the noise figure $F$ by dividing through by the first term, which represents the noise due to the source alone, obtaining

$$F = 1 + \frac{eI_b R_s}{2kT} + \frac{1}{2g_m R_s}.$$

Since $eI_c/kT = g_m$, this can be written as

$$F = 1 + \tfrac{1}{2}\left(\frac{I_b}{I_c} g_m R_s + \frac{1}{g_m R_s}\right),$$

from which it is clear that there is an optimum value of $R_s$, such that $I_b(g_m R_s)^2 = I_c$, or $R_s = \beta^{\frac{1}{2}}/g_m$. Since $\beta$ is large, this is clearly small compared with $R_i = \beta/g_m$, as we have assumed. Taking $I_c = 1$ mA and $\beta = 100$, $g_m^{-1} = 25\ \Omega$, so that $R_s = 250\ \Omega$ and $R_i = 2500\ \Omega$, we find that the optimum value of $F$ is 1·2.

The discussion given above follows a simple treatment by Robinson (1970); its validity is discussed in Appendix B of Robinson (1974).

## References

JONES, R. V. and McCOMBIE, C. W. (1952). *Phil. Trans. R. Soc.* **244**, 205.
ROBINSON, F. N. H. (1970). *Wireless World*, 339
ROBINSON, F. N. H. (1974). *Noise and Fluctuations*. Clarendon Press, Oxford.

## Problems

23.1. Show, by differentiation of eqn (23.5) to obtain the differential equation for the current $I = dq/dt$, and following through an analysis similar to that of § 23.1, that $\frac{1}{2}L\overline{I^2} = \frac{1}{2}kT$.

23.2. A signal generator whose output impedance is 500 Ω is calibrated in terms of the power it will deliver into a matched load (that is, the available signal power). It is connected to a receiver whose bandwidth is 10 kHz, and whose first stage consists of a triode whose shot noise is negligible, with a 1000 Ω resistance connected between cathode and grid. What will the signal generator reading be when it is adjusted so that the signal output from the receiver is equal to the noise output?

$$(Answer: 6\times10^{-17} \text{ W.})$$

23.3. Referring to Problem 8.10, assume that the target is low over the sea and intercepts the power incident on an area $A_1$. This power is scattered with the same angular distribution as that of the radiation from a short horizontal dipole parallel to the transmitter dipole. Some of this scattered power falls on an aerial of effective area $A_2$ located at the transmitter, and is detected by a receiver of noise figure $F$ and bandwidth $df$. Show that the signal–noise ratio, for the signal returned from the target, is unity for a target distance

$$D = \left(\frac{36\pi^2 WA_1A_2}{FkT\,df}\right)^{\frac{1}{8}}\left(\frac{Hh}{\lambda}\right)^{\frac{1}{2}}.$$

(This formula shows how difficult it is to increase the range by increasing the transmitter power $W$, and how much better it is to reduce the wavelength.)

23.4. By following the treatment of § 23.3 using Planck's law instead of the Rayleigh–Jeans law, show that the quantum-mechanical formula for resistance noise is

$$\frac{d(\overline{V^2})}{R} = \frac{4hf\,df}{\exp(hf/kT)-1}.$$

Verify from this formula that the transition from classical region to quantum-mechanical region occurs when the number of quanta per unit bandwidth in the noise power is of the order of unity.

23.5. In the electrical tuned circuit of § 23.1, show that, over a narrow range of frequencies $df$ at the resonant frequency $f$ of the circuit, the mean-square fluctuating current is

$$d(\overline{I_f^2}) = 4kTR^{-1}\,df,$$

and the mean-square fluctuating voltage on the capacitor is

$$d(\overline{V_f^2})_C = 4kT(L/CR)\,df.$$

Verify that the total mean-square fluctuating current $\overline{I^2}$ is equivalent to taking $d(\overline{I_f^2})$ over a range $df$ such that

$$df = R/4L \quad \text{or} \quad df/f = \pi/2Q_F,$$

and that an equivalent relation holds for the fluctuating voltages on the capacitor.

# 24. Magnetic resonance

## 24.1. The magnetic-resonance phenomenon

IT was shown in § 14.1 that when an atom or nucleus with a resultant angular momentum $\mathbf{G}$ and magnetic moment $\mathbf{m}$ is placed in a steady magnetic flux density $\mathbf{B}_0$ the equation of motion (obtained from eqn (14.2) by multiplying by $\gamma$) is

$$d\mathbf{m}/dt = \gamma\,\mathbf{m}\wedge\mathbf{B}_0, \qquad (24.1)$$

where $\gamma = \mathbf{m}/\mathbf{G}$ is the magnetogyric ratio. The motion represented by this equation consists of a precession of the angular-momentum vector $\mathbf{G}$ and hence also of $\mathbf{m}$ about the direction of $\mathbf{B}_0$ with a uniform angular velocity $-\gamma\mathbf{B}_0$, which we shall denote by $\omega_L$. If the system is undisturbed it will continue indefinitely in this state of uniform precession with $\mathbf{m}$ at a fixed angle to $\mathbf{B}_0$, and it is convenient to make use of rotating coordinate systems in considering this motion. It is shown in § A.10 that the rate of change $d\mathbf{m}/dt$ of any vector quantity such as $\mathbf{m}$ in the laboratory coordinate system is related to the rate of change $D\mathbf{m}/Dt$ in a system rotating with angular velocity $\omega$ relative to the laboratory system, by the equation

$$d\mathbf{m}/dt = D\mathbf{m}/Dt + \omega\wedge\mathbf{m}.$$

Substitution of this in eqn (24.1) gives

$$D\mathbf{m}/Dt = \gamma\,\mathbf{m}\wedge\mathbf{B}_0 - \omega\wedge\mathbf{m}$$
$$= \gamma\,\mathbf{m}\wedge\mathbf{B}_0 + \mathbf{m}\wedge\omega$$
$$= \gamma\,\mathbf{m}\wedge(\mathbf{B}_0 + \omega/\gamma). \qquad (24.2)$$

This result shows that in the rotating coordinate system the apparent magnetic field is $(\mathbf{B}_0 + \omega/\gamma)$, and the apparent precession velocity is $-\gamma(\mathbf{B}_0 + \omega/\gamma) = \omega_L - \omega$. Thus the apparent angular velocity is decreased by $\omega$, as would be expected from simple considerations of relative angular velocity. If we write $\mathbf{B}' = -\omega/\gamma$, the apparent field in the rotating system is $(\mathbf{B}_0 - \mathbf{B}')$, and it is reduced if $\mathbf{B}'$ is positive (that is, $\omega$ has the same sign as $\omega_L$) as shown in Fig. 24.1. Clearly, if $\mathbf{B}' = \mathbf{B}_0$, the apparent field $(\mathbf{B}_0 - \mathbf{B}')$ and precession velocity $-\gamma(\mathbf{B}_0 - \mathbf{B}')$ are both zero, and the vector $\mathbf{m}$ is stationary in the rotating coordinate system.

We shall now consider the effect of applying a small oscillating magnetic flux density $\mathbf{B}_1 \cos \omega t$ in the plane normal to the direction of the steady flux density $\mathbf{B}_0$. This oscillating field may be plane polarized or circularly

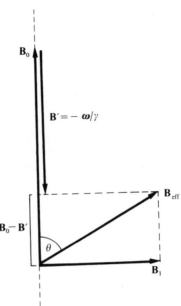

Fig. 24.1. Effective fields in a coordinate system rotating with angular velocity **ω**.

polarized; in the latter case $\mathbf{B}_1$ is simply a vector, normal to $\mathbf{B}_0$, which is constant in length but which rotates about $\mathbf{B}_0$ with angular velocity **ω**. If the oscillating field is plane-polarized, with a component say in the $x$ direction (taking $\mathbf{B}_0$ along the $z$-axis), it may be decomposed into two vectors rotating in opposite senses; thus there is no loss of generality in considering only the circularly polarized case.

If we now transfer from the laboratory system to a system rotating with the angular velocity **ω**, then the vector $\mathbf{B}_1$ is stationary in this system, and can be represented by a constant vector $\mathbf{B}_1$ as shown in Fig. 24.1. In this system, the atom or nucleus feels an apparent magnetic flux density $(\mathbf{B}_0-\mathbf{B}')$ parallel to the $z$-axis, together with the flux density $\mathbf{B}_1$ normal to this axis; thus the resultant field in this rotating system is the vector sum of these two fields, which is denoted by $\mathbf{B}_{\mathrm{eff}}$ in Fig. 24.1. To an observer in this system, the dipole moment **m** will appear to precess about $\mathbf{B}_{\mathrm{eff}}$ with angular velocity $-\gamma\mathbf{B}_{\mathrm{eff}}$, and its projection on $\mathbf{B}_0$ will change as time goes on. If **m** were initially parallel to $\mathbf{B}_0$ (as we should expect in a macroscopic system), it would precess about $\mathbf{B}_{\mathrm{eff}}$ 'and at some subsequent time would reach a maximum angle $2\theta$ with $\mathbf{B}_0$, where

$$\tan\theta = B_1/(B_0-B').$$

If $\mathbf{B}_0-\mathbf{B}' = 0$, $\mathbf{B}_{\mathrm{eff}} = \mathbf{B}_1$ and $\theta = \frac{1}{2}\pi$, so that **m** will turn right over to reach a position antiparallel to $\mathbf{B}_0$ before commencing to return. This occurs only when

$$\boldsymbol{\omega} = -\gamma\mathbf{B}' = -\gamma\mathbf{B}_0 = \boldsymbol{\omega}_L,$$

so that the frequency of the applied field is then the same as that of the Larmor precession. Thus the precession about $\mathbf{B}_{eff}$ is a forced resonance phenomenon, whose amplitude is greatest when the applied frequency $\omega$ is equal to the natural frequency $\omega_L$.

In the case of an atomic or nuclear system, the angular momentum is quantized, and so are its projections on $\mathbf{B}_0$, so that the energy $W = -\mathbf{m}.\mathbf{B}_0$ is also quantized. We shall consider first the nuclear case, assuming a nucleus of spin angular momentum $\mathbf{I}\hbar$, and magnetic moment $\mathbf{m}$, where the magnetogyric ratio $\gamma = g_N(e/2m_p)$. Then the potential energy in a state whose magnetic quantum number is $m$ is

$$W_m = -\mathbf{m}.\mathbf{B}_0 = -\gamma\hbar\mathbf{I}.\mathbf{B}_0 = -\gamma\hbar m B_0. \tag{24.3}$$

Under the influence of an oscillating magnetic field polarized in the plane normal to $\mathbf{B}_0$, transitions between states with different values of $m$ may take place according to the selection rule $\Delta m = \pm 1$. The quantum of energy required is

$$\hbar\omega = W_m - W_{m-1} = -\gamma(h/2\pi)B_0. \tag{24.4}$$

This is the same for all transitions, as shown in Fig. 24.2. From this it follows that the frequency of the radiation must be

$$\nu = -\frac{\gamma}{2\pi}B_0 = +\frac{\omega_L}{2\pi}. \tag{24.5}$$

This is the resonance condition, which is the same as that given by the classical treatment above. The minus sign is significant only if circularly polarized radiation is used. If a system of nuclei with a positive value of $\gamma$ is to absorb energy from an applied oscillatory field, the selection rule for absorption is $\Delta m = -1$, and the vector $\mathbf{B}_1$ must rotate in the left-hand sense about $\mathbf{B}_0$, while if $\gamma$ is negative, the reverse holds. This gives a method of determining the sign of $\gamma$, but for many purposes this is immaterial and linearly polarized radiation may be used. Since this can be regarded as composed of two circularly polarized components rotating in opposite senses, transitions can be induced whatever the sign of $\gamma$. In the usual spectroscopic terminology these are 'magnetic dipole' transitions, corresponding to the fact that they are caused by the interaction of an oscillatory magnetic field with the magnetic dipole moments of the system. This phenomenon is generally known as 'magnetic resonance', and it offers a method of determining $\gamma$ directly from a measurement of a frequency and a magnetic field. The order of magnitude of the frequencies required may be found from the specific charge $(e/m_p)$ of the proton,

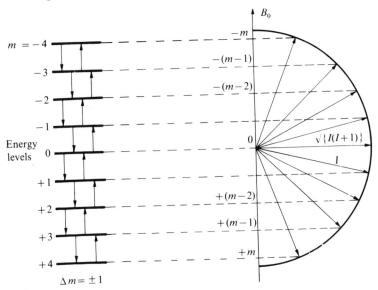

FIG. 24.2. Diagram showing the nine energy levels and the allowed transitions between them for a nuclear spin $I = 4$ in a flux density $B_0$. Transitions can be induced by an oscillating field of frequency $\nu$ if $\nu = -(\gamma/2\pi)B_0$.

if we assume that $g_N$ is around unity. The value of $e/m_p$ for the proton is nearly equal to the Faraday, that is, it is about $10^8 \, \mathrm{C \, kg^{-1}}$. Hence the frequency

$$\nu = -\frac{g_N}{2\pi}\left(\frac{e}{2m_p}\right)B_0$$

is $\approx 10^7 \, \mathrm{Hz}$ in a field of $1\,\mathrm{T}$ in the case of a nuclear magnetic moment. Electronic magnetic moments are some 2000 times larger, owing to the smaller mass of the electron, while the associated angular momentum is of the same order as in the nuclear case, so that the value of $\gamma$ and of the frequency required are higher in proportion. For an atom with a magnetic moment due only to electron spin, the frequency of the radiation required for resonance in a field of $1 \cdot 07\,\mathrm{T}$ is $30\,\mathrm{GHz}$ (a wavelength of $10\,\mathrm{mm}$).

The magnetic-resonance phenomenon has been used to investigate systems of both atomic and nuclear magnetic moments, and we shall discuss first the latter. It makes possible a direct determination of the value of $\gamma$, and hence, for a nucleus whose spin $I$ is known, of the nuclear magnetic moment. The chief experimental difficulty lies in the smallness of the effect, and we shall describe first an ingenious method due to Rabi, where the phenomenon is detected by its effect on the path of a molecule in a molecular beam.

## 24.2. Molecular beams and nuclear magnetic resonance

The use of atomic beams for the measurement of atomic magnetic moments has been mentioned in § 14.4. The method depends on deflecting the atoms by passing them through an inhomogeneous magnetic field; the deflection is proportional to the projection of the magnetic moment on the direction of the field gradient, and the initial beam is split into $(2J+1)$ beams if the total electronic angular momentum has quantum number $J$. The magnetic moment can be computed from the size of the deflections if the magnitude of the field gradient is known. The main difficulty in achieving high precision is the spread in velocity of the atoms in the beam, since the deflection is inversely proportional to the square of this velocity.

Application of this method to the determination of nuclear magnetic moments demands great refinements, since the size of the moment is some 2000 times smaller than that of an electron, and the deflection is correspondingly smaller. For a direct measurement molecules such as $H_2$ or NaCl, with no electronic magnetic moment, must be used. Owing to the great difficulty of working with purely nuclear moments, methods were devised using atoms with a hyperfine structure due to magnetic interaction between the electronic moment and the nuclear magnetic moment. These are rather complicated, and such deflection methods have now been superseded by others making use of magnetic resonance. The first of these was carried out by Rabi and his colleagues (1939).

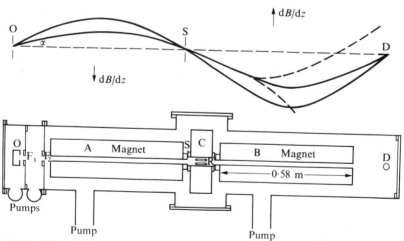

FIG. 24.3. Schematic diagram of a molecular-beam apparatus and of some molecular paths. The two full curves (above) indicate the paths of two molecules having different moments and velocities and whose moments do not change during passage through the apparatus. The dotted curves indicate possible changes in path if the component of magnetic moment is changed. The motion in the z direction is greatly exaggerated.

A schematic diagram of the apparatus is shown in Fig. 24.3. Molecules from an oven O emerge through slits $F_1$, $F_2$, moving at small angles with the axis of the apparatus. They enter a region of inhomogeneous magnetic field in the A-magnet, and are deflected by an amount proportional to the projection of their magnetic moments on the direction of the field; thus some of them will pass through the collimating slit S. Neglecting for the moment the C-magnet, we follow the molecules through the B-magnet, which produces an inhomogeneous field exactly like that of the A-magnet except that the gradient is reversed. The force on the nuclear magnets in a molecule is therefore also reversed (provided that the orientation of these magnets is the same as it was when passing through the A-magnet), and the molecules are deflected upwards and reach a detector D. The proviso about the orientation is important, because if it changes between A and B, then the deflection produced by B is different from that in A, and the molecules will not reach the detector. This gives a means of detecting a magnetic-resonance phenomenon, since it can be used to cause a change in the orientation after leaving A and before entering B. The C-magnet produces a uniform flux density $\mathbf{B}_0$, and between its pole faces is a loop R carrying an r.f. current which produces an oscillating flux density $B_1 \cos \omega t$ in a direction perpendicular to $\mathbf{B}_0$. When the resonance condition is fulfilled, transitions are induced within the Zeeman levels of the nuclear moment in the field $\mathbf{B}_0$, so that the orientation of the nuclear moment is changed. The beam current at the detector then falls, a typical example being shown in Fig. 24.4. Here the radio-frequency is kept constant, and the magnetic flux density $\mathbf{B}_0$ is varied through the resonance. The value of $\gamma$ is found from the observed values of $\mathbf{B}_0$ (at the centre of the resonance) and the frequency. From the width of the resonance curve in Fig. 24.4 it can be seen that the accuracy is very much greater than could be obtained by a simple deflection method.

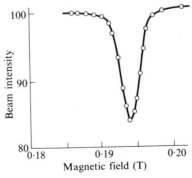

FIG. 24.4. Resonance curve for the $^{19}$F nucleus in sodium fluoride (NaF) obtained by Rabi.

An estimate of the width can be obtained as follows. If the nuclear moment is initially parallel to $\mathbf{B}_0$, then at resonance it precesses about $\mathbf{B}_1$ in the rotating coordinate frame with angular velocity $\gamma\mathbf{B}_1$, and in a time $t$ it will exactly reverse its orientation provided that $\gamma B_1 t = \pi$. This corresponds to the maximum in the resonance curve. When we are off resonance, the moment precesses about $\mathbf{B}_{\text{eff}}$, and if $\mathbf{B}_{\text{eff}}$ is at an angle $\theta = 45°$ to $\mathbf{B}_0$ the moment will only reach a maximum angle of $2\theta = \frac{1}{2}\pi$ to $\mathbf{B}_0$. If we take this to define the half intensity points on the resonance curve, they correspond to

$$B_0 - B' = \pm B_1,$$

so that

$$\Delta\nu = \pm\gamma(B_0 - B')/2\pi = \pm 1/2t, \qquad (24.6)$$

showing that the linewidth is simply related to the time $t$ for which the dipole moment is subjected to the oscillatory field.

An experiment of this kind is by no means easy, as can be seen from the fact that the deflection of a molecule is only about $0\cdot05$ mm in a magnet with a flux gradient of the order of $10^3\,\mathrm{T\,m^{-1}}$. This is using a magnet $0\cdot5$ m long with pole faces of the shape shown in Fig. 24.5; the curvature is adjusted to give a uniform value of $\mathrm{d}B/\mathrm{d}z$ over the width of the beam. Because of the small deflections the defining slits at S and the detector must be very narrow ($\approx 0\cdot01$ mm) and the beam intensity at the detector is very small, being determined by the solid angle which the detector slit makes with the oven, some $1\cdot5$ m away! Originally the difficulty of detecting a beam of uncharged molecules limited the method to hydrogen, deuterium, and the alkali metals, whose nuclear moments were measured with a precision of a few parts per thousand (see Table 24.1).

The spectra of molecules are complicated by the presence of: (1) a magnetic moment associated with rotation of the molecule as a whole; (2) interaction between the nuclear magnetic moment and the rotational magnetic moment; (3) interactions between the nuclear spins if more than

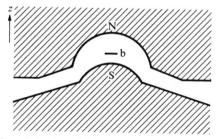

FIG. 24.5. Cross-section of the A and B magnets in Fig. 24.3, normal to the path of the beam (marked b in the figure). The curved parts of the pole pieces are cylindrical but the radius of curvature for the upper pole is larger than for the lower one.

TABLE 24.1

*Nuclear spins and magnetic moments of some common isotopes*

| Nucleus | Spin | Magnetic moment (nuclear magnetons) | |
| | | Molecular beam value | Nuclear resonance value |
|---|---|---|---|
| Neutron | $\frac{1}{2}$ | −1·913 | — |
| $^1$H | $\frac{1}{2}$ | +2·789 | +2·7927 |
| $^2$H | 1 | +0·856 | +0·8574 |
| $^3$H | $\frac{1}{2}$ | — | +2·9788 |
| Alkali metals: | | | |
| $^6$Li | 1 | +0·821 | +0·8220 |
| $^7$Li | $\frac{3}{2}$ | +3·253 | +3·2563 |
| $^{23}$Na | $\frac{3}{2}$ | +2·215 | +2·2175 |
| $^{39}$K | $\frac{3}{2}$ | +0·391 | +0·3915 |
| $^{41}$K | $\frac{3}{2}$ | +0·215 | +0·2154 |
| $^{85}$Rb | $\frac{5}{2}$ | +1·340 | +1·3527 |
| $^{87}$Rb | $\frac{3}{2}$ | +2·733 | +2·7505 |
| $^{133}$Cs | $\frac{7}{2}$ | +2·558 | +2·5789 |
| Halogens: | | | |
| $^{19}$F | $\frac{1}{2}$ | +2·62 | +2·6285 |
| $^{35}$Cl | $\frac{3}{2}$ | +0·819 | +0·8218 |
| $^{37}$Cl | $\frac{3}{2}$ | +0·681 | +0·6841 |
| $^{79}$Br | $\frac{3}{2}$ | +2·110 | +2·1056 |
| $^{81}$Br | $\frac{3}{2}$ | +2·271 | +2·2696 |
| $^{127}$I | $\frac{5}{2}$ | — | +2·8090 |

one is present; (4) electric interaction between the electric quadrupole moment of the nucleus (if $I \geqslant 1$) and the electric field gradient in the molecule.

The detection of an atom or molecule is greatly simplified if it can be converted into a charged ion. There are two principal methods: (*a*) bombardment by a beam of electrons, and (*b*) surface ionization (see § 11.5), where the molecule falls on a hot surface whose work function is greater than its ionization potential, from which it 'evaporates' as a positive ion. Ions formed in these ways are passed through a mass spectrometer to separate them from the background ion current (and sometimes from other isotopes of the same element). They are then accelerated to strike a surface from which secondary electrons are ejected on impact; these form a current which is further amplified by an electron multiplier (see § 11.4).

To obtain high precision, a narrow resonance curve is required, and in a beam experiment (see eqn (24.6)) this requires a large value of $t$, that is, a long path through the C-field. The corresponding requirement of high uniformity in the C-field is made less rigorous by the use of two separate oscillatory fields, one at each end of the C-field. With this arrangement, due to Ramsey (1949), only the mean value of the C-field over the whole path is required to be the same as that at the positions of the oscillatory fields.

## 24.3. Nuclear magnetic resonance in bulk material

The molecular-beam method of detecting nuclear magnetic resonance is experimentally very difficult; it was first used at a time when there seemed no prospect of the direct detection of the emission or absorption of quanta of the oscillatory field. At radio-frequencies the position is quite different from optical spectroscopy, since the rate of spontaneous emission of quanta is negligibly small. Except at frequencies greater than $10^4$ MHz and temperatures less than 1 K, the pair of energy levels involved have a separation $h\nu$ small compared with $kT$. This means that transitions between the two levels are continuously being induced through the effect of thermal fluctuations. An electronic oscillator is effectively a very high-temperature source, and radiation from it will be absorbed provided it is close to the resonant frequency. The problem is to observe the transitions caused by the applied oscillatory field against the general background of electrical noise associated with the thermal fluctuations.

The magnitude of the absorption in nuclear magnetic resonance can be estimated in the following way. From the theory of anomalous dispersion (§ 10.6), the imaginary part of the susceptibility $\chi''$ at the centre of a narrow line is related to the static susceptibility $\chi_0$ by the formula (see Problem 24.1)

$$\chi''/\chi_0 = \omega/2\Delta\omega = \nu/2\Delta\nu, \tag{24.7}$$

where $\Delta\nu$ is the distance from the centre of the line to a point where the intensity has fallen to half its maximum value. Now

$$\chi_0 = \frac{\mu_0 n g_N^2 \mu_N^2 I(I+1)}{3kT},$$

and in a favourable case, such as the protons in water, the value of $\chi_0$ at room temperature is approximately $10^{-8}$ (S.I. units). To estimate $\chi''$ we need an approximate value of $\Delta\nu$. The main cause of line-broadening arises from the random magnetic fields of neighbouring nuclear magnetic moments. The effect of these is to change the actual field at a given nucleus by an amount depending on the orientation and number of neighbouring magnetic dipoles: this causes a spread in the magnetic field acting on different nuclei, and so gives a finite linewidth. The spread $\Delta B$ is of the order of $\mu_0 m/4\pi d^3$, where $m$ is the dipole moment of a neighbour, and $d$ its distance. The mean value of $d^3$ is just the reciprocal of the number of protons per unit volume, and for water this gives $\Delta B$ $\sim 10^{-4}$ T. Since $\nu/\Delta\nu = B_0/\Delta B$, at a field of $0.2$ T $\chi''$ would be about $10^{-8} \times 10^3 \sim 10^{-5}$. For protons, whose resonant frequency would then be about 8·5 MHz, the power transmitted through 10 km of the substance would fall only by 1·8 per cent (see Problem 24.3). It is obviously out of the question to use a transmission method; instead, the substance is inserted in the r.f. magnetic field of a coil of a circuit tuned to 8·5 MHz. The magnetic-resonance phenomenon then causes a change in $1/Q_F$, where $Q_F$ is the quality factor of the circuit, of magnitude (see Problem 24.2)

$$\delta(1/Q_F) = \chi''/(1+\chi') \sim 10^{-5}.\tag{24.8}$$

The smallest value of $1/Q_F$, the reciprocal of the quality factor of the circuit in the absence of resonance, is $\sim 0·5 \times 10^{-2}$, so that to detect the resonance requires the measurement of a change in $Q_F$ of less than 1 per cent.

Basically, the method of detection is similar to the use of the $Q$-meter described in § 22.3. Power from an electronic oscillator is fed through a high resistance $R'$ to a tuned circuit (Fig. 24.6), whose parallel impedance at resonance is $Q_F/\omega C$. Provided that $R' \gg Q_F/\omega C$, a constant current is drawn by the circuit, irrespective of small changes in $Q_F$. A small fractional variation $\delta Q_F/Q_F$ therefore appears as a fractional change $\delta v/v$ in the voltage across the tuned circuit, which can be amplified and detected. From eqn (24.8)

$$\delta v/v = \delta Q_F/Q_F = -Q_F \, \delta(1/Q_F) \sim -Q_F \chi'',$$

and we can estimate the smallest observable value of $\chi''$ by setting the value of $\delta v$ equal to that which gives a signal–noise ratio of unity. If $R$ is the parallel resistance of the tuned circuit at resonance, and from eqn (23.7) we insert the value $(4kTR \, df)^{\frac{1}{2}}$ for $\delta v$, we have

$$\chi''_{min} = \frac{1}{Q_F v} (4kTR \, df)^{\frac{1}{2}}.\tag{24.9}$$

Taking as reasonable values $Q_F = 10^2$, $R = 10 \, k\Omega$, $df = 0·1 \, MHz$, $v = 1 \, V$,

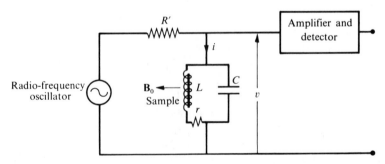

FIG. 24.6. Basic circuit for detection of nuclear magnetic resonance. $R'$ is a high resistance such that $R' \gg R = Q_F/\omega C$, the parallel impedance of the tuned circuit. This ensures that a constant current $i$ flows to the tuned circuit, and a change in $Q_F$ produces a corresponding change in $v = iR$, the voltage across the tuned circuit. The output from the amplifier and detector is taken to a display unit (oscilloscope or pen recorder).

we find that $\chi''_{min} = 4 \times 10^{-8}$, giving an ample margin of sensitivity compared with eqn (24.8).

The presence of $Q_F$ in the denominator of eqn (24.9) shows the advantage of using a tuned circuit. Only a small sample is needed to fill the coil; the oscillatory magnetic field is much larger than in an ordinary radiation field, and the magnetic absorption is correspondingly increased. (The tuned circuit is equivalent to a path length of $Q_F/2\pi$ wavelengths; see Problem 24.10.) A further advantage of a tuned circuit is that the noise voltages across it are limited to those in a band $df$ near the resonance frequency; this band has an effective width $df \sim f/Q_F$ (see Problem 23.5), and this gives the value used above. Higher sensitivity can be obtained by further limitation of the noise bandwidth; for example, a phase-sensitive detector at the modulation frequency can be combined with the system of field modulation described in the following paragraph.

In an experiment the sample is placed in a coil whose axis (the direction of the oscillating flux density $\mathbf{B}_1$) is perpendicular to the steady field $\mathbf{B}_0$. The latter is then swept through resonance, at which point the voltage $v$ across the tuned circuit momentarily diminishes. A low-frequency modulation is imposed on $\mathbf{B}_0$, so that the momentary drop in $v$ is repeated in synchronism with this modulation, giving a low-frequency modulation of $v$. After detection, the modulation can be amplified (if necessary) and applied to the $Y$ plates of an oscillograph. The $X$ sweep is derived from the low-frequency modulation, so that a given $X$ deflection corresponds to a given change in $B_0$ caused by the modulation. As the mean value of $B_0$ (the field corresponding to the midpoint of the modulation) is slowly varied, the resonance line appears at one end of the trace, moves across, and disappears at the other end.

The first experiment of this kind was carried out by Purcell, Torrey, and Pound (1946). Simultaneously, nuclear resonance was observed quite independently by Bloch, Hansen, and Packard (1946) using a slightly different principle known as nuclear induction. To understand this we shall return to the rotating coordinate system used in Fig. 24.1. If there is some damping mechanism by which energy can be transferred from the system of nuclear spins to the outside world, the equilibrium state will be one where the net magnetization $\mathbf{M}$ will be parallel to the steady flux density $\mathbf{B}_0$. When a rotating flux density $\mathbf{B}_1$ is applied, the magnetization vector precesses at an angle $\delta$ about $\mathbf{B}_0$ with the angular velocity $\omega$ of the applied rotating field. Thus it is a constant vector in the rotating coordinate system, though not coplanar with $\mathbf{B}_0$ and $\mathbf{B}_1$ (this is the steady-state solution of the forced precession of the damped system). Theory shows that, if $\Delta B$ is the linewidth, $\delta$ is given by

$$\tan \delta = \frac{B_1}{\{(B_0 - B')^2 + \Delta B^2\}^{\frac{1}{2}}}.$$

Hence $\delta$ is greatest at resonance ($\mathbf{B}_0 = \mathbf{B}'$), and the rotating component $M \sin \delta$ is then also a maximum, leading the rotating flux density $\mathbf{B}_1$ by an angle $\frac{1}{2}\pi$. This component will induce a voltage in a coil placed at right-angles to the main coil producing the driving field $B_1 \cos \omega t$. The induced voltage is an oscillatory one with the same frequency, and the detector coil must therefore be carefully oriented to reduce as far as possible any direct pick-up from the driving coil. Here again the steady flux density $\mathbf{B}_0$ is 'wobbled' at an audio-frequency, since this differentiates between such stray pick-up, which will not be modulated at the wobble frequency, and the resonance effect, which is.

## 24.4. Relaxation effects in nuclear magnetic resonance

The Purcell method is generally known as 'nuclear resonance', and the Bloch method as 'nuclear induction'; they are alternative methods of detecting the same phenomenon, and have the same ultimate sensitivity. Both have been pushed to the limits of sensitivity in applications such as measurement of the value of $\gamma$ for rare isotopes. In this connection linewidth is of great importance, since the intensity at the centre of an absorption line varies inversely as the linewidth (see eqn (24.7)), and the accuracy with which the position of the centre of the line can be determined is also higher for a narrower line. It turns out that the linewidth varies very considerably with the nature of the sample, and we shall discuss this briefly first.

In the estimate of the linewidth for $H_2O$ made in § 24.3 the field due to one neighbouring proton was found to be about $10^{-4}$ T. This was an underestimate of the width to be expected, since there is more than one neighbour, and we would expect the full width ($2\Delta B$) to be over $10^{-3}$ T (it is found to be about $1 \cdot 6 \times 10^{-3}$ T in ice at low temperatures). In liquid water, on the other hand, the resonance is extremely narrow; so narrow, that its actual width is very difficult to determine, as variation of the field produced by the external magnet over the volume of the sample is usually the limiting factor in determining the breadth. By working at low field strength, however, Brown and Purcell (1949) were able to show that the over-all width was less than $7 \times 10^{-7}$ T. The explanation for this striking difference from the width in ice is as follows. In water the molecules are not stationary, but on the average change their positions once every $10^{-11}$ s or so, this figure being given by the relaxation time of the Debye absorption discussed in § 10.8. This means that at a given nucleus the field of a neighbouring nucleus will not be constant, but will change its value every $10^{-11}$ s. This is a very much shorter time than that of the precession period in a field of, say, $0 \cdot 2$ T, which is $\approx 10^{-7}$ s. At first sight the rapid fluctuations of the random fields of the neighbours might be expected to broaden the line, but in fact the nucleus cannot respond to changing fields whose duration is less than $T_2 = (\Delta\omega)^{-1} = (\gamma\Delta B)^{-1}$, the inverse linewidth; in the words of Purcell (1948) 'the nucleus rides out the storm like a well-balanced gyroscope on perfect gymbals.' It turns out that the more rapid the fluctuations the more closely does their effect average to zero. If the rate of fluctuation is lower, as it is in a liquid of high viscosity such as glycerine, the linewidth is not so effectively reduced, and increases rapidly if the viscosity is increased by lowering the temperature. When the Debye relaxation time becomes much longer than the characteristic time $T_2$, the full linewidth is attained. In many substances the linewidth does not have the expected width due to the random magnetic fields of other nuclear dipole moments even in the solid state. This is attributed to internal motion within the solid lattice, which 'averages out' the fields of the neighbours, and nuclear resonance has been used to investigate such internal motions in many cases.

The nuclear magnetic resonance spectrum of protons in ethyl alcohol (Fig. 24.7) shows that extremely narrow lines can be obtained in liquids under conditions of high resolution. This spectrum also illustrates an important chemical application (see § 24.5); the same nuclear species (in this case protons) at different sites in a chemical compound may give lines at slightly different field strengths at the same frequency, or different frequency at the same field.

A second question of considerable importance is the rate at which energy is transferred from the nuclear spin system to the lattice containing the nuclei. In the absence of an external magnetic field $\mathbf{B}_0$, the nuclear spins will point in random directions, and $\chi_0$ will be zero. If a field $\mathbf{B}_0$ is now applied, the energy levels corresponding to different nuclear spin orientations will be split, as shown in Fig. 24.8. For simplicity, a spin $I = \frac{1}{2}$ is assumed, giving just two levels. Originally, the populations of these two levels were equal, and after $\mathbf{B}_0$ is switched on they will remain so until a

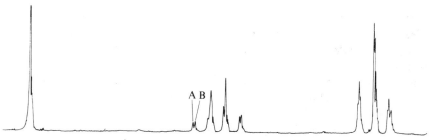

FIG. 24.7. Nuclear magnetic resonance spectrum at 30 MHz of ethyl alcohol. Left, proton signal from OH; centre, proton signal from $CH_2$ group; right, proton signal from $CH_3$ group. The separations between these main groups arise from changes in the diamagnetic shielding ('chemical shift'), while splittings within the groups are due to spin–spin interactions between the protons. The high resolution achieved is illustrated by the separation of lines A, B; this is about 0·25 Hz.

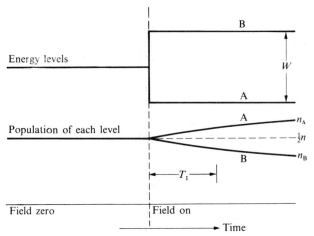

FIG. 24.8. The energy levels and the relative populations of a nucleus of spin $I = \frac{1}{2}$ before and after a magnetic field is switched on. The populations approach the new equilibrium values exponentially, with time constant $T_1$, the 'relaxation time'. In thermal equilibrium $n_B = n_A \exp(-W/kT)$ and $n_A + n_B = n$.

number are transferred from the upper to the lower state to give the equilibrium Boltzmann distribution in which

$$n_B = n_A \exp(-W/kT).$$

This involves a transfer of energy from the system of spins to the lattice, and the magnetization approaches its equilibrium value $\mathbf{M}_0$ according to the exponential law (compare the corresponding equation for electric polarization in § 10.8)

$$d\mathbf{M}/dt = (\mathbf{M}_0 - \mathbf{M})/T_1,$$

which gives

$$\mathbf{M}_0 - \mathbf{M} = \mathbf{M}_0 \exp(-t/T_1) \tag{24.10}$$

if $\mathbf{M} = 0$ at $t = 0$. The parameter $T_1$ is known as the spin–lattice relaxation time, since it determines the rate at which energy is transferred from the magnetic dipoles of the system of nuclear spins to the crystal lattice in which they are embedded. Such a transfer requires that transitions be induced between the various nuclear levels corresponding to different orientations, and these can only be caused by the presence of an oscillating magnetic field whose frequency satisfies the condition for resonance. Furthermore, this oscillating field must originate in the thermal motion of the surroundings of the nucleus. In a liquid the Brownian motion of neighbouring molecules causes the local magnetic fields of their nuclei to fluctuate rapidly, and this random fluctuation contains components of the right frequency to cause transitions. For distilled water at 293 K Bloembergen, Purcell, and Pound (1948) found the value of $T_1$ to be about 2 s.

This can be shortened by dissolving a paramagnetic salt in the water, so that the protons interact with the much bigger electronic magnetic moments. Another feature is that one would expect the desired component of the randomly fluctuating magnetic fields to be greatest (and hence $T_1$ to be shortest) when the Debye relaxation time of the liquid $\tau$ is of the order $1/\omega_0$, where $\omega_0$ is the angular frequency of the magnetic resonance. This was verified by Bloembergen, Purcell, and Pound by measuring $T_1$ in glycerine over a range of temperatures, which gives a wide range of viscosity and hence of the Debye relaxation time $\tau$. In solids at low temperatures, where there is practically no thermal motion, we should expect $T_1$ to be very long, but in practice it turns out to be nothing like as long as the values predicted by theory. This is ascribed to the presence of paramagnetic impurities, where the electron spins turn over and so give a fluctuating field. Owing to their larger magnetic moments, very few such impurity ions are required, and the thermal contact between the electron spin (or orbit) and the lattice is much more intimate than that between a nuclear spin and the lattice (see § 24.7), so that a transfer of energy to the electron spins from the nuclei is effectively a transfer to the lattice itself.

The variation of $T_1$ with the Debye relaxation time $\tau$ for three substances is shown in Fig. 24.9. In ethyl alcohol, $\tau \ll 1/\omega_0$ and $T_1$ decreases as $\tau$ increases. In glycerin $T_1$ passes through a minimum and increases again, while in ice, where $\tau \gg 1/\omega_0$, $T_1$ is rising with increasing $\tau$. This variation of $T_1$ (with a corresponding increase in the linewidth towards the value calculated for an assembly of static dipoles as $\tau$ increases) confirms the internal molecular motion in ice indicated by the Debye relaxation. Internal motions exist in many other solids, and have

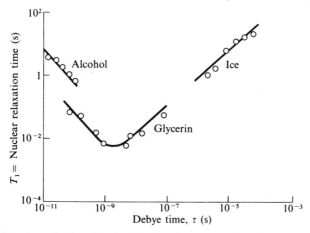

FIG. 24.9. The thermal relaxation time $T_1$ for protons in ethyl alcohol, glycerin, and ice, measured at 29 MHz, plotted against the Debye relaxation time $\tau$, obtained from dielectric dispersion data. Note the logarithmic scales.

been investigated by measurement of the relaxation time $T_1$ as well as the linewidth in nuclear magnetic resonance. Such experiments have added considerably to our knowledge of the solid state.

## 24.5. Applications of nuclear resonance

The most obvious application of nuclear resonance is the measurement of $\gamma$ for all possible isotopes. A direct measurement of $\gamma$ involves a precise measurement of the frequency, which is comparatively easy, and of the magnetic flux density $B_0$, which is difficult. For this reason it is usual to measure instead the ratio of $\gamma$ for the unknown isotope to that for a standard isotope such as $^1$H, the proton. This may be accomplished by measuring the two frequencies of magnetic resonance in the same magnetic field. To achieve high accuracy, a single sample is used, such as a solution of a substance containing the unknown isotope in water. From the discussion of linewidth in § 24.4 it will be appreciated that a solution gives the narrowest lines, and hence the greatest accuracy as well as the greatest intensity at the centre of the line. The sample is surrounded by two coils, one for each of the two resonance frequencies, which are usually arranged to be mutually perpendicular so as to minimize the mutual inductance between them. The accuracy which can be achieved in this way is illustrated by the ratio of the nuclear moments of the deuteron and the proton, found by Wimett (1953) to be

$$m_D/m_H = 0.307\ 012\ 192 \pm 0.000\ 000\ 015.$$

The measurement was made with compressed HD gas in order to avoid difficulties with 'diamagnetic shielding' (see below), and certain other small corrections.

The accuracy of absolute values of $\gamma$ is determined by results for the proton obtained in various Standards Laboratories, which lie close to $267.513 \times 10^6$ per tesla for a spherical sample of pure $H_2O$. This absolute determination for the proton is also important because nuclear resonance offers a simple and accurate laboratory method of measuring magnetic flux density (see § 6.6). Only a frequency measurement is required, which can readily be carried out to an accuracy of one part in $10^6$ or better (see § 22.5). The main requirement is that the flux density must be uniform over the proton sample in order to avoid broadening of the resonance line, but this is not often a serious limitation.

### Diamagnetic shielding

The reason why the nature of the sample is specified carefully in this measurement is because the resonance frequency is determined by the magnetic flux density at the proton, which may not be exactly the same as the external flux density. One correction arises from 'diamagnetic shield-ing', an effect closely allied to diamagnetism. It arises from the precession of the closed shells of electrons about $\mathbf{B}_0$, which sets up a small field at the nucleus with the opposite sense to $\mathbf{B}_0$, thus making the actual field acting at the nucleus slightly smaller than the external field. The apparent value of $\gamma$, if no correction is made, is therefore less than the true value by a fractional amount of about $2 \cdot 8 \times 10^{-5}$ for hydrogen, rising to $10^{-2}$ for the heaviest elements. This correction has been computed, but the probable error rises considerably in the heavy elements. This shielding effect, which makes the field in the interior of an atom different from that outside, must be distinguished from the 'demagnetizing field' (§ 4.4) and the 'local field', which can be evaluated sufficiently accurately for this purpose by the method due to Lorentz (cf. § 10.3). Each of these fields is propor-tional to the bulk diamagnetism of the sample, and in a spherical sample the demagnetizing field and the Lorentz field just cancel, so that the average local field is the same as the external field, apart from the diamagnetic correction.

### Chemical shielding

The change in the local field due to diamagnetic shielding may be written as $\mathbf{B} = \mathbf{B}_0(1 - \sigma)$. The value of $\sigma$ depends on the local density of electrons near the nucleus, and this varies slightly from compound to compound. It results in a displacement of the resonant frequency for the same isotope; this is known as a 'chemical shift' and has proved to be a powerful analytical and research tool in chemistry. 'Paramagnetic shifts', opposite in sign, arise from the induced paramagnetic moment in com-pounds with a temperature-independent paramagnetism such as $K_3Co(CN)_6$. The diamagnetic shielding may also vary between different nuclear sites in the same compound; in Fig. 24.7 the spectrum of $CH_3CH_2OH$ shows resonances from the protons in the $CH_3$, $CH_2$, and OH groups which are separated in frequency by about one part in $10^6$. Since $\mathbf{B} - \mathbf{B}_0 = -\sigma \mathbf{B}_0$, greater resolution can be obtained at higher frequen-cies, requiring higher values of $\mathbf{B}_0$, provided that the requirement of higher homogeneity in the field strength can be satisfied. For protons, frequencies up to 220 MHz, using a flux density of over 5 T from a superconductive magnet, are utilized without increase in linewidth. Spe-cial methods have been developed to give a resolution in liquids of about one part in $10^8$; these make it possible to observe further splittings in the spectrum (typically of order 10 Hz) which arise from interactions between neighbouring nuclear spins.

### Solid-state applications

In strongly magnetic compounds large shifts in the nuclear resonance frequency are observed because of the fields set up by the electronic magnetic dipoles. Such dipoles change their orientation very rapidly, either through relaxation effects due to the thermal fluctuations of the lattice or through interaction with neighbouring spins; so long as many reorientations occur in a time $T_2 = (\gamma_N \Delta B)^{-1}$ characteristic of the nuclear resonance linewidth $\Delta B$, relatively narrow nuclear magnetic resonance (n.m.r.) lines are obtained, shifted in frequency by a local field which is proportional to the time average of the electronic magnetic moment. Thus the shift is temperature-dependent and proportional to the average electronic magnetization; in a paramagnetic substance we can write for the nuclear precession frequency

$$\boldsymbol{\omega} = -\gamma_N(\mathbf{B}_0 + a\mathbf{M}) = -\gamma_N \mathbf{B}_0(1 + a\chi/\mu_0),$$

showing that the fractional change in frequency is proportional to the susceptibility. In an ordered magnetic substance, where the magnetization is finite in the absence of an external field, n.m.r. can be observed at a frequency which is very nearly proportional to the magnetization in a ferromagnetic substance, or to the sublattice magnetization in an antiferromagnetic or ferrimagnetic compound.

The nuclear magnetic moment of $^{19}$F is quite large, and signals of high intensity have been observed from nuclear magnetic resonance of this nucleus in a number of magnetic compounds. It was found by Shulman and Jaccarino (1956) that in $MnF_2$ in the paramagnetic state, the $^{19}$F

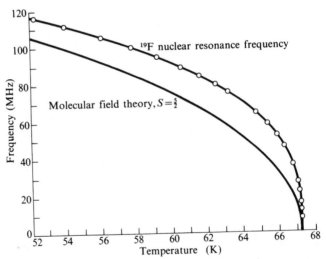

FIG. 24.10. Temperature-dependence of the $^{19}$F n.m.r. frequency in $MnF_2$ between 52 K and the Néel point 67·3 K. The lower curve shows the variation expected if the sublattice magnetization, relative to that at 0 K, followed the curve computed from molecular field theory for $S = \frac{5}{2}$. In this and other substances the resonance frequency varies near $T_N$ as $(T_N - T)^{\frac{1}{3}}$. (Heller and Benedek 1962.)

resonance was shifted by an amount proportional to the electronic paramagnetic susceptibility. However, the magnitude of the shift was greater than would be expected from the dipolar magnetic field at the $F^-$ site due to the electronic magnetic moments, assuming them to be localized on the $Mn^{2+}$ ions; the larger field is consistent with a spread of the wavefunctions of the magnetic electrons onto the $F^-$ ions, due to a small amount of covalent bonding. In the antiferromagnetic state the electronic moments are fixed in orientation, and the field which they produce at the $F^-$ nucleus is quite high even when $\mathbf{B}_0 = 0$. At 0 K in $MnF_2$ the precession frequency of $^{19}F$ in this electronic field is 159·99 MHz falling, as the temperature rises, to zero at the Néel point, 67·3 K. A comparison of the nuclear-resonance frequency, which should be proportional to the sublattice magnetization, with the magnetization calculated from the molecular field theory, is shown in Fig. 24.10.

## 24.6. Electron magnetic resonance in atomic beams

Magnetic-resonance experiments involving electronic magnetic moments can be carried out in ways analogous to experiments with nuclear magnetic moments. In general, experiments with electronic moments are easier, because the moments are so much larger and magnetic resonance is correspondingly easier to detect. In the analogue of Rabi's experiments (§ 24.2), beams of atoms with electronic magnetic dipole moments are used, for which appreciable deflections in the inhomogeneous magnetic fields A, B (see Fig. 24.3) can be achieved using relatively short magnets. The main interest of such experiments arises from the high precision which can be obtained by increasing the length of the C-field, since from eqn (24.6) the width of the resonance curve depends on the time $t$ during which the atom is subjected to the radio-frequency field. If the velocity of atoms in the beam is $10^3 \, \text{m s}^{-1}$ and the length of the C-field is 0·03 m, the value of $t$ is $3 \times 10^{-5} \, \text{s}$, and the corresponding line width $2\Delta\nu$ is about 30 kHz. To achieve such a narrow line the C-field would have to be very homogeneous, since a variation of the field by as little as $10^{-6} \, \text{T}$ would change the electronic magnetic-resonance frequency of an atom with $g = 2$ by 30 kHz. The requirement of such high homogeneity is avoided by use of the two separated oscillatory fields method of Ramsey (1949), as mentioned in § 24.2.

For an atom with an electronic magnetic moment but no nuclear moment the magnetic-resonance transitions occur at a frequency (cf. eqn (24.5))

$$\nu = (\gamma/2\pi)B_0 = g_J(e/4\pi m_e)B_0 \qquad (24.11)$$

and determination of $\nu$ in a known flux density $B_0$ gives a precise measurement of $g_J$. An important application of atomic-beam magnetic resonance is the precise measurement of hyperfine structure in atoms. As pointed out in § 14.10, in an atom with both electronic and nuclear moments the nuclear magnetic moment precesses in the electronic magnetic field of flux density $\mathbf{B}_e$, and the electronic moment in the nuclear magnetic flux density $\mathbf{B}_n$, the result being a precession of each moment about the resultant angular-momentum vector $\mathbf{F}$. This gives rise to a set of hyperfine energy levels, as illustrated in Fig. 24.11. In magnetic resonance the transitions for which $\Delta F = \pm 1$ have an intensity associated with an electronic magnetic dipole moment, and can readily be observed; from measurements of the frequencies of such transitions the hyperfine constants $A$, $B_Q$ can be determined using eqn (14.32). These require no magnetic field in the C-magnet, but in practice a small field $\mathbf{B}_0$ is generally used, whose effect (see Fig. 24.11) is to split each set of states with a given value of $F$ into $(2F+1)$ levels, provided that the Zeeman energy $g_J \mu_B (\mathbf{J} \cdot \mathbf{B}_0)$ is small compared with the hyperfine energies. This corresponds to a precession of the vector $\mathbf{F}$ about $\mathbf{B}_0$ at a rate slow compared with the precession of $\mathbf{J}$, $\mathbf{I}$ about $\mathbf{F}$. The allowed transitions are then those for which $\Delta F = 0, \pm 1$; $\Delta m_F = 0, \pm 1$, but the $\Delta m = 0$ transitions occur only if the oscillatory field has a component parallel to $\mathbf{B}_0$, while the $\Delta m = \pm 1$ transitions require a component $\mathbf{B}_1$ perpendicular to $\mathbf{B}_0$, as in Fig. 24.1.

The atomic-beam magnetic-resonance technique has been widely used to measure hyperfine constants of 'free' atoms with very high precision. A hyperfine frequency can be determined under conditions singularly free

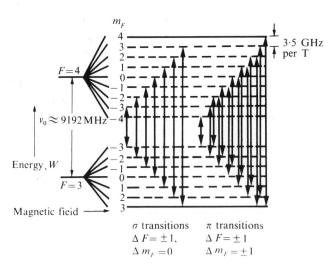

FIG. 24.11. The hyperfine structure of $^{133}$Cs, $S = \frac{1}{2}$, $I = \frac{7}{2}$, showing the Zeeman effect in a small field (note the difference in scale between the Zeeman splittings and the overall splitting) and the allowed transitions.

from external influences, and we discuss here only the example of caesium, which is used to provide an atomic standard of frequency.

The ground state of the caesium atom is $^2S_\frac{1}{2}$, and the only stable isotope $^{133}$Cs has nuclear spin $I = \frac{7}{2}$. Since the electronic ground state is $S = \frac{1}{2}$, there is no electric quadrupole interaction (see § 14.10), and in zero magnetic field there are two sets of hyperfine levels corresponding to $F = 3$ and $F = 4$, as in Fig. 24.11. Caesium is fairly volatile, so that an atomic beam of sufficient intensity can be emitted from an oven at 370 K, and this combined with the rather high atomic mass gives a low thermal velocity and increases the time required to traverse the C-field, where a distance of about 2·8 m is used between the separated oscillatory fields (see Essen and Parry 1959). A small field of $5\times10^{-6}$ T (this requires cancellation of the earth's field $\sim5 \times 10^{-5}$ T) is maintained as the C-field, and the resonance observed is the transition $(F, m_F) = (4, 0) \leftrightarrow (3, 0)$, which has only a second-order Zeeman effect

$$\nu = \nu_0+(426\times10^8)B_0^2 \text{ Hz} \quad (B_0 \text{ in tesla}). \tag{24.12}$$

As a result of international collaboration between Standards Laboratories, the measurement of $\nu_0$ can be carried out with an agreed precision of one part in $10^{11}$, a considerable improvement on previous standards. These were based on astronomical observations (for a discussion, see Essen 1969); the difficulties are illustrated by the variation in the length of the day, as measured against a caesium 'atomic clock', shown in Fig. 24.12. Following an international recommendation in 1963, the

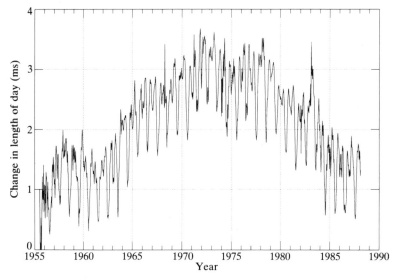

FIG. 24.12. Variation in the length of the day in terms of the atomic unit of time. (Morrison 1989.)

caesium 'clock' has become the new standard of frequency, the second being defined as the time interval containing exactly

$$9\ 192\ 631\ 770\ \text{cycles}$$

of the caesium hyperfine frequency in zero magnetic field.

## 24.7. Electron magnetic resonance in solids

Magnetic-resonance experiments on substances containing permanent magnetic dipole moments due to electrons can be carried out in a manner analogous to those on nuclear dipoles, but there are a number of significant differences. If fields of a few tenths of a tesla are used, the resonance frequency for electrons is in the vicinity of 10 GHz, corresponding to wavelengths of a few centimetres. Frequencies of this order and higher are in fact used for a number of reasons.

1. The sensitivity is high; this is partly through eqn (24.7), and partly because a better filling factor can be obtained from a small sample by using it in a tuned circuit (a cavity resonator) which has similar dimensions.

2. The electronic levels may have splittings of many GHz due to crystal field effects (see § 14.7).

3. Linewidths in the solid due to magnetic fields of neighbouring ions are of the order of $0 \cdot 01$–$0 \cdot 1$ T, and to achieve reasonable accuracy in determining the centre of a line measurements must be made using external fields as large as possible. Linewidth and line-shape are also affected by exchange interaction between the ions; this can be avoided by making measurements on 'diluted' crystals—crystals in which most of the paramagnetic ions have been replaced by diamagnetic ions. For example, a crystal of $K_2Zn(SO_4)_2.6H_2O$ containing a few tenths of a per cent of $Cu^{2+}$ $3d^9$ ions replacing $Zn^{2+}$ $3d^{10}$ ions gives a linewidth of about $10^{-3}$ T; this residual width is due mainly to the nuclear magnetic moments of the protons in the water of crystallization, and can be further reduced by growing crystals with $D_2O$ instead of $H_2O$ because of the smaller nuclear moment of the deuteron.

Another important feature of electron spin resonance in paramagnetic substances is that spin–lattice relaxation times may be extremely short. Obviously the electron, with its larger magnetic moment, will be in more intimate contact with the lattice vibrations than a nuclear dipole, but a much more important effect is that the lattice vibrations distort the local surroundings of a paramagnetic ion, and so produce a fluctuating modulation of the crystal electric field or the ligand field. The extent to which this affects the magnetic dipole depends on the degree of 'quenching' of the orbital moment: for an ion in an S state, such as $Mn^{2+}$ ($3d^5$, $^6S_{\frac{5}{2}}$) or $Gd^{3+}$ ($4f^7$, $^8S_{\frac{7}{2}}$), there is no orbital moment except through small departures from $L-S$ coupling, and the spin–lattice relaxation time $T_1$ varies from $\approx 10^{-6}$ s at room temperature to $10^{-3}$ s at liquid–helium temperatures. For other ions the values of $T_1$ are very much smaller, and in many ions of the 4f group $T_1$ is so short and the lines so broad ($\Delta\omega = T_1^{-1}$) that magnetic resonance is unobservable, except at liquid-helium or liquid-hydrogen temperatures. The value of $T_1$ always increases as the temperature falls because the lattice vibrations die out, the variation being in general of the form

$$1/T_1 = aT + bT^n + c \exp(\hbar\omega/kT). \qquad (24.13)$$

The first term $aT$ is due to 'direct transitions' in which magnetic quanta are exchanged with lattice vibrations of the same frequency as the magnetic-resonance frequency; the second term $bT^n$ (where $n = 5, 7$, or 9 according to the type of magnetic ion involved) is due to 'indirect' or 'Raman' processes in which any two lattice vibrations are involved whose frequency *difference* is equal to the magnetic-resonance frequency ($\omega_{lat} = \omega'_{lat} \pm \omega_{res}$); the exponential term is due to lattice vibrations whose quanta $\hbar\omega$ coincide with the difference in energy between the ground state and an excited state of the magnetic ion. The second two processes are weaker than the first, and at liquid-helium temperatures the first term almost always predominates.

A spin–lattice relaxation time $T_1$ will produce a linewidth of $2\Delta\nu$ between the half-intensity points of a line, where $2\pi\Delta\nu = \Delta\omega = T_1^{-1}$, and measurements of linewidth can be used to find $T_1$ when this is the dominant effect in the linewidth. For other purposes the need for high resolution in order to obtain accurate measurements makes it desirable, however, to work at temperatures where broadening due to spin–lattice relaxation is negligible, and a typical apparatus for low-temperature work is outlined in Fig. 24.13. Power from a microwave oscillator (usually a klystron) is carried by a waveguide or coaxial cable to a cavity resonator contained in a dewar vessel and placed between the poles of an electromagnet. When the latter is adjusted to resonance, power is absorbed in the paramagnetic sample, which is placed inside the cavity in a position of maximum oscillatory magnetic field. This additional power loss in the cavity produces a change in the signal reflected from the cavity, or in the signal transmitted through the cavity to another waveguide or coaxial line, which is detected by a silicon-crystal rectifier. Normally the field of the electromagnet is modulated at an audio-frequency, giving a corresponding modulation of the signal when the field for magnetic resonance is traversed; after detection, this modulation is amplified and displayed on

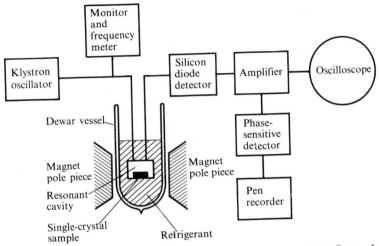

FIG. 24.13. Outline diagram of an electron spin resonance apparatus. Power from a klystron oscillator is fed through a waveguide to a loosely coupled resonant cavity containing the paramagnetic sample and immersed in a refrigerant between the poles of an electromagnet modulated at an audio-frequency. A small fraction of the cavity signal is fed through a second waveguide to a silicon diode detector. The modulation due to the absorption is amplified and displayed on an oscilloscope or fed through a phase-sensitive detector to a pen recorder.

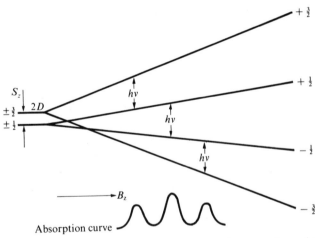

FIG. 24.14. Energy levels and absorption curve at constant frequency for an ion with $S = \tfrac{3}{2}$ and a crystal field splitting:

$$W = g\mu_B S_z B_z + D\{S_z^2 - \tfrac{1}{3}S(S+1)\}$$

when the magnetic field is parallel to the principal axis (z-axis) of the splitting term (for other directions the levels do not diverge linearly with field). The intensity of the $S_z \leftrightarrow (S_z - 1)$ transition is proportional to $\{S(S+1) - S_z(S_z - 1)\}$ giving the $3:4:3$ intensity ratio shown in the figure.

FIG. 24.15. Hyperfine structure of the $Mn(I = \frac{5}{2})$ nucleus in the electron spin resonance spectrum of an $Mn^{2+}$ ion ($S = \frac{5}{2}$, transition $S_z = \frac{1}{2} \leftrightarrow -\frac{1}{2}$). The resonance condition is $h\nu = g\mu_B(B_0 + B_N)$, where $B_N$ is the magnetic field due to the nucleus. In first approximation $B_N$ is proportional to the nuclear magnetic quantum number $I_z$; this gives a pattern of $2I + 1 = 6$ lines, equally spaced and of equal intensity, since all nuclear orientations are equally probable at the temperature of the observation. The line-shape is the derivative of the absorption curve; it is obtained by a sinusoidal modulation of $B_0$ with amplitude small compared with the linewidth. This gives a corresponding modulation of the signal (measured by a phase-sensitive detector) whose amplitude is proportional to the slope of the absorption curve.

an oscilloscope or a recorder. The sensitivity achieved is quite high, and signals from as few as $10^{13}$ electronic dipoles can be seen if the lines are narrow.

The results of electron magnetic resonance in paramagnetic salts have greatly advanced the detailed understanding of the properties of paramagnetic ions subjected to ligand field interactions in solids. In general the resonance spectrum is very anisotropic, depending strongly on the angle between the external field and the crystal axes; for this reason single crystals must be used. In dilute salts the splittings of the levels due to the external field as a function of angle can be studied, together with any crystal field splittings of the same order as the microwave frequency (see Fig. 24.14). When the nucleus of the paramagnetic ion or of a ligand ion has a nuclear moment, a hyperfine structure may be observed, as in Fig. 24:15. A number of nuclear spins and moments have been determined from hyperfine structure in electron magnetic resonance, and the degree to which the wavefunctions of the magnetic electrons overlap onto the ligand ions because of covalent bonding effects (see § 14.8) can be estimated from the hyperfine structure due to interaction with the dipole moment of the ligand nucleus. In more concentrated salts the effect of magnetic dipole and exchange interaction between neighbouring dipoles can be studied, giving one of the few direct measurements of exchange interaction.

*Ferromagnetic resonance*

In substances where the exchange forces are strong magnetic resonance may be observed in the co-operative state below the transition temperature. Since all the dipoles are coupled together by the exchange forces it is convenient to work in terms of the magnetization **M**, which is the vector sum of the individual dipole moments **m**. By performing this vector sum over both sides of eqn (24.1) we obtain the equation of motion for the magnetization

$$dM/dt = \gamma M \wedge B, \qquad (24.14)$$

where we have assumed that all dipoles have the same value of $\gamma$. Here we have written, not $B_0$ the external flux density, but $B$ the flux density within the sample, since in a ferromagnetic substance demagnetizing fields may be quite important. We shall assume that $B_0$ is along the $z$-axis, and confine ourselves to certain sample shapes such that we can write for the components of $B$:

$$B_x = -\mu_0 D_x M_x; \qquad B_y = -\mu_0 D_y M_y; \qquad B_z = B_0 - \mu_0 D_z M_z. \qquad (24.15)$$

On substituting into eqn (24.14) we obtain

$$dM_x/dt = \gamma M_y \{B_0 + \mu_0 M_z (D_y - D_z)\},$$
$$dM_y/dt = -\gamma M_x \{B_0 + \mu_0 M_z (D_x - D_z)\}, \qquad (24.16)$$
$$dM_z/dt = \gamma \{\mu_0 M_x M_y (D_x - D_y)\},$$

which are no longer linear in $M$. However, the equations can be solved in the limit of small amplitudes of precession, when the product $M_x M_y$ becomes vanishingly small and can be neglected. Then $dM_z/dt = 0$, and $M_z$ is constant, its value being equal to the static magnetization; this is large in a ferromagnetic substance, and the corrections to $B_0$ are important in determining the resonance frequency. By solving the equations for $M_x$, $M_y$ it is easily shown that the precession frequency is

$$\omega_L = -\gamma[\{B_0 + \mu_0 M_z (D_y - D_z)\}\{B_0 + \mu_0 M_z (D_x - D_z)\}]^{\frac{1}{2}}. \qquad (24.17)$$

There are three simple cases of interest:

(1) a sphere, for which $D_x = D_y = D_z$; the precession angular velocity is $\omega_L = -\gamma B_0$, the same as if there were no demagnetizing fields;

(2) a thin plane film normal to $B_0$, for which $D_x = D_y = -1$, $D_z = 0$, giving
$$\omega_L = -\gamma(B_0 - \mu_0 M_z);$$

(3) a thin plane film parallel to $B_0$, for which $D_x = D_z = -1$, $D_y = 0$ (assuming the film to be normal to the y-axis), giving
$$\omega_L = -\gamma\{B_0(B_0 + \mu_0 M_z)\}^{\frac{1}{2}}.$$

This treatment assumes that the magnetization (including the precessing components) is uniform throughout the sample; this requires that the dimensions be small compared with the wavelength in the sample, and in a conducting sample this means small compared with the skin depth.

Hence spherical samples of metal must be very small, and colloidal samples (where the particles are assumed to be spherical because of surface-tension effects in formation) have been used. Most work has been done on thin plane samples, which are attached to (but insulated from) one wall of the microwave cavity. The equations show that the magnetization must be known in order to determine $\gamma$; for simplicity it is usual to work at such high fields that the magnetization is equal to the saturation value. The phenomenon of ferromagnetic resonance was discovered experimentally by Griffiths (1946); the theory given above is due to Kittel (1948). Some values of g measured by ferromagnetic resonance are given in Table 15.2 (p. 493).

*Spin-wave resonance in ferromagnetic films*

The uniform precession mode assumed above corresponds to a spin wave with $k_s = 0$. It is possible to excite spin waves for which $k_s \neq 0$; since each spin wave corresponds to a unit change $\hbar$ in angular momentum, and hence to a change of $g\mu_B$ in magnetization, the energy required to excite a spin wave in a flux density $B$ (in the sample) is

$$\hbar\omega = g\mu_B B + Dk_s^2, \qquad (24.18)$$

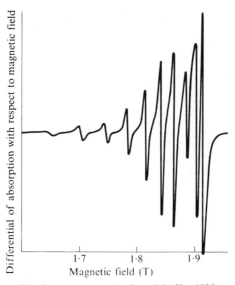

FIG. 24.16. Ferromagnetic spin-wave resonance in a thin film (600 nm thickness) of cobalt metal at room temperature and a frequency of 9370 MHz (32 mm wavelength). The line-shape is that corresponding to the differential of the absorption curve. The intensity of resonance decreases towards lower field strength (higher values of $p$) irregularly because of lack of uniformity in the film thickness. $B_0$ is normal to the surface of the film, so that rather high values of $B_0$ are needed to satisfy the resonance condition $\omega_i = -\gamma(B_0 - \mu_0 M_z)$.
(Phillips and Rosenberg 1964.)

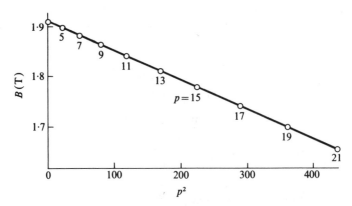

FIG. 24.17. Plot showing the linear relation between magnetic flux density and $p^2$ for ferromagnetic spin-wave resonance in a thin film of cobalt. (After Phillips and Rosenberg 1964.)

where the constant $D$ is the same as that in eqn (15.24). In a thin film of thickness $l$, the boundary conditions (assumed identical at the two faces of the film) limit the allowed values of $k_s$ to those for which the film thickness is an integral number of half-wavelengths; that is, $k_s = p\pi/l$, where $p$ is an integer. If magnetic resonance is observed at constant frequency, the value of the resonance flux density is found from eqn (24.18) above to be

$$B = B_{k_s=0} - \left(\frac{D}{g\mu_B}\frac{\pi^2}{l^2}\right)p^2, \qquad (24.19)$$

so that a series of resonances corresponding to different values of $p$ should be observed on the low field side of the ordinary ferromagnetic resonance flux density $B_{k_s=0}$. A spin wave resonance curve is shown in Fig. 24.16 for a thin film of cobalt metal, of thickness approximately equal to 600 nm. This is small compared with the skin depth, so that the oscillatory field is uniform within the sample, and the resonance intensity depends on the net magnetic moment in the direction of the oscillatory field. This is proportional to

$$\int_0^l \sin kx \, dx = \int_0^l \sin\frac{p\pi x}{l} \, dx = \left(\frac{l}{p\pi}\right)(1 - \cos p\pi).$$

This vanishes for even values of $p$, and decreases as $1/p$ for odd values, giving the intensity change shown in Fig. 24.16. The resonance field decreases accurately as $p^2$ (see Fig. 24.17), and the value of $D$ can be found if the thickness $l$ is known.

*Ferrimagnetic and antiferromagnetic resonance*

The presence of two sublattices in these substances makes the magnetic-resonance phenomena in general much more complicated. One simple case occurs in ferrimagnetic substances with strongly coupled sublattices; the two sublattices can then precess together in such a way that the relative orientation of their two magnetic moments remains unaltered. This occurs at an angular velocity $\omega = -\gamma_{\text{eff}}B$; here $B$ is the flux density in the substance and $\gamma_{\text{eff}}$ is an average value obtained from the relation

$$\mathbf{M} = \sum_i \mathbf{M}_i = \sum \gamma_i \mathbf{G}_i = \gamma_{\text{eff}} \sum \mathbf{G}_i = \gamma_{\text{eff}}\mathbf{G},$$

where the summation is over all the individual ions. In both ferri- and antiferromagnetics more complicated modes of precession occur in which the relative orientation of the sublattice magnetic moments is not preserved; the frequency of precession then depends on a number of parameters, including the exchange and anisotropy energies.

## 24.8. Cyclotron resonance with free charged particles

When a charged particle of mass $M$ and charge $q$ is moving in a uniform magnetic flux density $\mathbf{B}$, its equation of motion is

$$\mathbf{F} = q\mathbf{v} \wedge \mathbf{B}.$$

Since this force is always normal to its instantaneous velocity $\mathbf{v}$, the particle will move in a circle of radius $r$ in the plane normal to $\mathbf{B}$ with angular velocity given by the equation

$$M\omega_c^2 r = q\omega_c rB,$$

that is,

$$\omega_c = (q/M)\mathbf{B}.$$

Thus if it is possible to determine the angular velocity $\omega$ in a known flux density $\mathbf{B}$, the ratio of charge to mass of the particle may be found. The frequency $\omega_c/2\pi$ is often called the 'cyclotron frequency' since it is the frequency of the r.f. electric field required to accelerate charged particles in the cyclotron. The success of this device, which depends on resonance between the frequency of the oscillating electric field and the frequency of rotation of the particles in the field $\mathbf{B}$, suggests that a similar principle may be used to determine the ratio of $q$ to $M$.

We begin by investigating the motion of a charged particle starting from rest under the action of a uniform flux density $\mathbf{B}$ (whose direction we take to be the $z$-axis of a system of Cartesian coordinates) and an oscillating electric field of frequency $\omega/2\pi$ polarized so that the lines of electric field are parallel to the $x$-axis. Then the equations of motion are

$$M\ddot{x} = qE \cos \omega t + q\dot{y}B,$$
$$M\ddot{y} = -q\dot{x}B, \quad\quad (24.20)$$
$$M\ddot{z} = 0.$$

The last of these equations shows that the $z$-component of the motion will be independent of $E$ and $B$, and does not appear in the other equations. The second equation can be integrated once giving

$$M\dot{y} = -qxB,$$

where the constant of integration has been equated to zero, corresponding to the assumption that the particle starts at rest from the origin. $\dot{y}$ may now be eliminated from the first equation giving

$$\ddot{x} + \omega_c^2 x = (q/M)E \cos \omega t,$$

where $\omega_c = (q/M)B$. The general solution of this equation is

$$x = \frac{(qE/M)\cos \omega t}{\omega_c^2 - \omega^2} + C \cos \omega_c t + D \sin \omega_c t.$$

If the initial conditions are $x = 0$, $\dot{x} = 0$ at $t = 0$, the unknown constants are determined and we have

$$x = \frac{(qE/M)(\cos \omega t - \cos \omega_c t)}{\omega_c^2 - \omega^2}$$
$$= \frac{2(qE/M)\sin \tfrac{1}{2}(\omega_c + \omega)t \sin \tfrac{1}{2}(\omega_c - \omega)t}{(\omega_c + \omega)(\omega_c - \omega)}$$
$$= \frac{qE}{M\omega'} \sin \omega' t \left( \frac{\sin \tfrac{1}{2}\Delta\omega t}{\Delta\omega} \right), \quad\quad (24.21)$$

where $\omega' = \tfrac{1}{2}(\omega_c + \omega)$, $\Delta\omega = \omega_c - \omega$. If $\Delta\omega \ll \omega_c$, the factor $\sin \tfrac{1}{2}\Delta\omega t$ varies very slowly with respect to time compared with $\sin \omega' t$, and $\omega'$ is very close to $\omega_c$, so that using the relation $\dot{y} = -\omega_c x$ we find approximately

$$y = \frac{qE}{M\omega'} \cos \omega' t \left( \frac{\sin \tfrac{1}{2}\Delta\omega t}{\Delta\omega} \right). \quad\quad (24.22)$$

Examination of the equations for $x$ and $y$ shows that the path of the particle is a spiral with angular velocity $\omega'$ and radius

$$\frac{qE}{M\omega'} \cdot \frac{\sin \tfrac{1}{2}\Delta\omega t}{\Delta\omega}.$$

If $\Delta\omega = 0$ (that is, $\omega = \omega_c$) then the value of the factor $(\sin \tfrac{1}{2}\Delta\omega t)/\Delta\omega$ is $\tfrac{1}{2}t$, showing that the radius increases linearly with $t$. On the other hand, if

$\Delta\omega \neq 0$, the radius has a maximum value $r_0$ (when the sine is unity) equal to $qE/M\omega' |\Delta\omega|$, which is very nearly equal to $qE/M\omega_c |\Delta\omega|$ when $\Delta\omega$ is small. Hence, if a collector is placed at a distance $R_0$ from the origin, only those ions will reach it for which $r_0 \geqslant R_0$, or

$$|\Delta\omega| \leqslant qE/M\omega_c R_0 = E/BR_0.$$

This is a measure of the precision with which $\omega_c$, and hence $q/M$, can be determined. The 'resolving power' will be

$$\frac{\omega_c}{|\Delta\omega|} = \frac{qB/M}{E/BR_0} = \frac{qB^2 R_0}{ME}. \tag{24.23}$$

Hence, for a given ion and a given flux density $B$, the precision is increased by using a small amplitude of oscillating electric field strength $E$ and a large value of $R_0$. It can be shown that our expression for the resolving power is equal to $L/2R_0$ (see Problem 24.4), where $L$ is the total path traversed by the ion in its spiral journey from the origin to the collector. Thus the resolving power depends primarily on the number of revolutions which the ions make on their journey to the collector.

The apparatus used by Sommer, Thomas, and Hipple (1951) is shown in Fig. 24.18. An oscillatory voltage is applied between two parallel plates of size 3 cm×5 cm, and separation 2 cm, with a number of parallel guard rings. These rings are equally spaced, and by means of a potentiometer system a fraction of the voltage proportional to the distance from one end-plate is applied to them so that a uniform r.f. field is obtained. A small steady positive voltage of about 0·1 V is applied to the guard rings relative to the end-plates so as to retard the drift of positive ions in the direction parallel to **B**. The magnitude of **B** is determined by

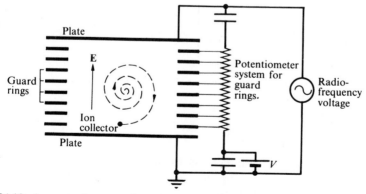

FIG. 24.18. Apparatus for measuring the cyclotron resonance frequency of the proton. **B** is normal to the plane of the paper; $V$ is a steady voltage of about 0·1 V for focusing the ion beam; **E** is an oscillatory electric field.

an n.m.r. experiment, using an r.f. coil containing a sample of oil. Ions are produced along the axis of the apparatus by firing in a narrow beam of electrons of about 70 V energy, which cause ionization by collision with the residual gas. The pressure must be kept low ($<10^{-4}$ Pa) in order to prevent scattering of the ions by collision. The whole assembly is enclosed in a glass tube of 47 mm diameter, which fits between the poles of an electromagnet.

In a typical experiment $B = 0.47$ T, and the oscillatory field strength $E$ is about $10 \text{ V m}^{-1}$ at a frequency of about 7 MHz for the $H^+$ ion. With $R_0 = 10$ mm, the ions make about 7000 revolutions and attain an energy of about 1000 eV before reaching the ion collector, which is connected to an electrometer. The ion current at the peak of a resonance is about $3 \times 10^{-14}$ A, while the background fluctuations are about $4 \times 10^{-16}$ A. Owing to the small positive voltage on the guard rings, and to space charge, a small radial electric field exists which displaces the resonant frequency slightly. In a radial field strength $E'$ the equation of motion is

$$M\omega^2 r + qE' = q\omega rB,$$

whence approximately

$$\omega = \omega_c\left(1 - \frac{E'M}{rqB^2}\right). \tag{24.24}$$

In practice it turns out that $E'$ increases linearly with $r$, and hence the shift in the resonance is independent of $r$, but proportional to $M$. Thus by making measurements both with $H^+$ and $H_2^+$ ions ($H^+$ and $D_2^+$ ions were also compared) the size of the shift can be determined.

A number of measurements of the cyclotron resonance frequency of protons have been carried out, together with analogous experiments on electrons, where the resonance frequency is about $10^4$ MHz in a field approaching 0.4 T. In all such experiments the ratio of two frequencies in the same magnetic field is determined—for example, the cyclotron resonance frequency $\nu_c$ of free protons, and the n.m.r. frequency $\nu_p$ of protons in a standard liquid sample such as water. If a correction of 28 p.p.m. is applied for diamagnetic shielding of the protons in the water sample, the ratio

$$\frac{\omega_p}{\omega_c} = \frac{2\pi\nu_p}{2\pi\nu_c} = \frac{g_N(e/2m_p)B}{(e/m_p)B} = \tfrac{1}{2}g_N$$

gives the nuclear magnetic moment of the proton in nuclear magnetons, since the spin $I = \tfrac{1}{2}$. Over the period 1965–70 (see Cohen and Taylor 1972), four independent measurements (omitting the diamagnetic correction) lie within the limits $\nu_p/\nu_c = 2.79277 \pm 0.00002$, a precision of about one part in $10^5$.

## 24.9. Cyclotron resonance in semiconductors

It was pointed out in Chapter 12 that the equations of motion of electrons (and holes) in the periodic potential of a crystal lattice are similar to those of a free particle, provided that an effective mass $m^*$ is used instead of the true mass. This holds also for motion in a magnetic field, and the cyclotron resonance frequency becomes

$$\omega_c = (q/m^*)B,$$

if the effective mass is isotropic. Determination of this frequency is thus of great importance since it gives a direct measurement of $m^*$. In principle, the experiment is similar to those described in the previous section: an oscillatory electric field is applied normal to the steady magnetic field, and either its frequency or the strength of the magnetic field is varied while the power absorbed is measured. However, the charge carriers in a solid make collisions at a rate which is usually comparable with (and often much higher than) the cyclotron resonance frequency; this gives a very important damping term, and the equation of motion may be written as (cf. eqn (3.13))

$$m^*\left(\frac{d\mathbf{v}}{dt} + \frac{1}{\tau}\mathbf{v}\right) = q(\mathbf{E} + \mathbf{v} \wedge \mathbf{B}), \tag{24.25}$$

where $q = -e$ for electrons and $+e$ for holes.

To solve this equation we assume that $\mathbf{B}$ is along the $z$-axis of a Cartesian coordinate system and $\mathbf{E}$ is an oscillatory field along the $x$-axis. We therefore write $E_x = E_0 \exp j\omega t$, and look for the steady-state solution corresponding to the driven motion at angular frequency $\omega$; we can then replace $d/dt$ by $j\omega$, and the equations become

$$\left.\begin{aligned}
\left(j\omega + \frac{1}{\tau}\right)v_x &= \frac{q}{m^*}(E_x + v_y B), \\
\left(j\omega + \frac{1}{\tau}\right)v_y &= -\frac{q}{m^*} v_x B, \\
\left(j\omega + \frac{1}{\tau}\right)v_z &= 0.
\end{aligned}\right\} \tag{24.26}$$

The last equation shows that any momentary current in the $z$ direction dies away exponentially through collisions, and we may eliminate $v_y$ between the first two equations in order to find the oscillatory velocity $v_x$ in the direction of the applied electric field. This gives

$$v_x\left(\omega_c^2 - \omega^2 + \frac{1}{\tau^2} + \frac{2j\omega}{\tau}\right) = \frac{q}{m^*} E_x\left(\frac{1}{\tau} + j\omega\right),$$

where we have written $\omega_c$ for $(q/m^*)B$, the cyclotron resonance frequency. The conductivity of the solid at angular frequency $\omega$ in the $x$

direction is $\sigma_x = nqv_x/E_x$, where $n$ is the number of charge carriers per unit volume of mass $m^*$, and is given by the relation

$$\sigma_x = \frac{nq^2}{m^*}\left\{\frac{j\omega+1/\tau}{(\omega_c^2-\omega^2)+1/\tau^2+2j\omega/\tau}\right\}$$

$$= \sigma_0\left\{\frac{1+j\omega\tau}{1+2j\omega\tau+\tau^2(\omega_c^2-\omega^2)}\right\}. \qquad (24.27)$$

where $\sigma_0 = n(q^2/m^*)\tau$ is the ordinary conductivity of the substance at zero frequency in the absence of a magnetic field. This equation shows that the high-frequency conductivity is complex; on solving for the real part $\sigma_x'$ of the conductivity, we find

$$\frac{\sigma_x'}{\sigma_0} = \frac{1+\tau^2(\omega_c^2+\omega^2)}{\{1+\tau^2(\omega_c^2-\omega^2)\}^2+4\omega^2\tau^2}. \qquad (24.28)$$

The power absorption per unit volume of the sample is $\frac{1}{2}\sigma_x'E_x^2$; since it is usual to work at fixed frequency $\omega$ and measure the power absorption as $B$ (that is, $\omega_c$) is varied, it is useful to plot the quantity $\sigma_x'/\sigma_0$ as a function of $\omega_c/\omega$ for various values of the parameter $\omega\tau$. This is shown in Fig. 24.19. When $\omega\tau$ is appreciably less than unity, the mean time between collisions is a small fraction of an r.f. period, and little change occurs until $\omega\tau$ approaches unity. However, when $\omega\tau$ is rather greater than unity, a

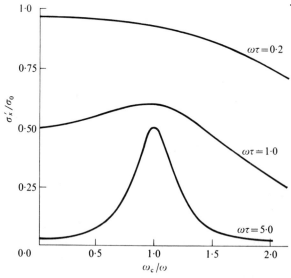

FIG. 24.19. Plot of the ratio of the r.f. conductivity to the d.c. conductivity against $\omega_c/\omega$. Cyclotron resonance measurements are usually made at constant $\omega$ and variable field; since $\omega_c$ is proportional to $B$, the curves show the conductivity against $B$ (on a reduced scale). Well-resolved resonance curves are obtained when $\omega\tau$ is rather greater than unity.

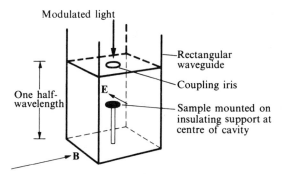

Modulated light

Rectangular
waveguide

Coupling iris

One half-
wavelength

E

Sample mounted on
insulating support at
centre of cavity

B

FIG. 24.20. Waveguide cavity resonator used in cyclotron resonance experiments, showing the sample mounted at the centre of the cavity where the oscillatory electric field is a maximum. Carriers can be excited in the sample by light passed down the waveguide and through the coupling iris linking the cavity to the guide.

distinct resonance effect is observed, with maximum power absorption at a point close to the cyclotron resonance frequency.

In a semiconductor or metal at room temperature the value of $\tau$ is about $10^{-12}$–$10^{-14}$ s, so that even at a wavelength of 10 mm, where a magnetic field at resonance of about 1 T would be needed if $m^* = m$, the value of $\omega\tau$ is about $2\times10^{-1}$–$2\times10^{-3}$. However, the electron scattering is mainly due to phonons, and is reduced at low temperatures. In a semiconductor (see § 17.5) $\tau$ should vary as $T^{-\frac{3}{2}}$, and a factor of $10^3$ is gained in going from 300 K to 3 K, making $\omega\tau \sim 2$ to 200. As can be seen from Fig. 24.19, this is sufficient for a fairly accurate determination of the resonance frequency. However, the number of charge carriers $n$, which (see § 17.5) varies as $T^{\frac{3}{2}}\exp(-W_g/2kT)$ for a pure semiconductor, becomes vanishingly small at helium temperatures. Dexter, Zeiger, and Lax (1956) overcame this difficulty by irradiating the sample with light of sufficiently short wavelength to lift electrons across the energy gap from the valence to the conduction band, thus creating both holes and conduction electrons. The main features of their apparatus are shown in Fig. 24.20. The sample, in the form of a thin disc some 3 mm in diameter and 0·5 mm thick, is mounted at a point in a waveguide cavity where the oscillatory electric field is as large as possible without producing serious carrier heating effects through acceleration of the carriers. This cavity terminates a waveguide, and the change in the signal reflected by the cavity is a measure of the increased power absorption in the sample. The cavity is immersed in liquid helium in a dewar vessel placed between the poles of an electromagnet.

A simple, satisfactory method of detection is to modulate the light beam by passing it through a rotating disc pierced by a large number of holes. The lifetime of the carriers is short and they are present only for the duration of a light pulse; the reflected microwave signal is therefore modulated at the same frequency (usually 100–1000 Hz). Instead of using irradiation by light, carriers can also be created through ionization of impurity levels by application of an electric field across the sample, or by the oscillatory microwave electric field. The latter method gives distorted line-shapes, however, since the number of secondary carriers created depends on the carrier energy and this is a maximum at resonance. It has the advantage that only electrons are created in n-type material, and holes in p-type, since the microwave energy is only sufficient to cause ionization across the small gap (~0·01 eV in germanium) of impurity levels, and not across the main gap $W_g$.

In many substances the effective mass is anisotropic (see § 12.4), and the ratio of cyclotron resonance frequency to magnetic field is a function of the orientation of the field relative to the crystal axes. For this reason a single crystal must be used, with either a special device for rotating it in the cavity, or for rotating the external magnetic field, so that a whole plane of directions relative to the external magnetic field can be explored. An absorption curve for a given orientation of germanium is given in Fig. 24.21; it is due to Dresselhaus, Kip, and Kittel (1955), who made the first observations of cyclotron resonance in semiconductors in 1953. When anisotropy is present, the equations of motion are modified and must be solved to find the relation between the cyclotron resonance frequency and the effective mass parameters; the following method for this is due to Shockley (1953).

When the energy surfaces are not spherical in **k**-space, they can be

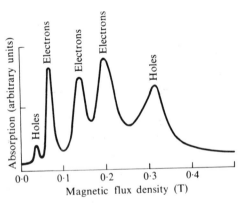

FIG. 24.21. Absorption curve for cyclotron resonance in a single crystal of germanium, at 24 GHz and 4 K. The static field is in a (110) plane at 60° from a [100] axis. Three electron lines are observed because $\mathbf{B}_0$ makes three different angles with the four [111] axes, which are the major axes of the four energy ellipses in germanium. (After Dresselhaus, Kip, and Kittel 1955.)

approximated near the band edges (see § 12.4) by equations such as

$$W = \tfrac{1}{2}\hbar^2\left(\frac{k_x^2}{m_x}+\frac{k_y^2}{m_y}+\frac{k_z^2}{m_z}\right) = \frac{1}{2}\left(\frac{p_x^2}{m_x}+\frac{p_y^2}{m_y}+\frac{p_z^2}{m_z}\right),$$

provided that the directions of the $x$-, $y$-, $z$-axes are chosen correctly. Along these axes the components of the equation of motion have their usual form, so that in a magnetic flux density with components $B_x$, $B_y$, $B_z$ we have

$$m_x(dv_x/dt) = q(v_yB_z - v_zB_y), \quad \text{etc.}$$

To find the cyclotron resonance frequency for a particular energy surface we assume a sinusoidal motion with frequency $\omega_c/2\pi$; on replacing $d/dt$ by $j\omega_c$ we obtain the set of linear equations

$$j\omega_c m_x v_x - q v_y B_z + q v_z B_y = 0,$$
$$j\omega_c m_y v_y - q v_z B_x + q v_x B_z = 0,$$
$$j\omega_c m_z v_z - q v_x B_y + q v_y B_x = 0,$$

which have an allowed solution only if the determinant

$$\begin{vmatrix} j\omega_c m_x & -qB_z & qB_y \\ qB_z & j\omega_c m_y & -qB_x \\ -qB_y & qB_x & j\omega_c m_z \end{vmatrix} = 0.$$

This gives either $\omega_c = 0$ (corresponding to $\mathbf{v}$ parallel to $\mathbf{B}$), or

$$\omega_c^2 = \frac{q^2}{m_x m_y m_z}(m_x B_x^2 + m_y B_y^2 + m_z B_z^2). \tag{24.29}$$

This equation shows that the cyclotron resonance frequency depends on the orientation of the magnetic field with respect to the crystal axes; in any given plane a plot of $\omega_c^2$ against angle gives a $(\text{cosine})^2$ variation between the maximum and minimum values. When $\mathbf{B}$ is directed along one of the principal axes, such as the $z$-axis for an ellipsoidal energy surface, $(\omega_c)_z = qB/(m_x m_y)^{\frac{1}{2}}$; thus by measurement along each axis in turn, the principal values $m_x$, $m_y$, $m_z$ of the effective mass can be determined. The results for the elemental semiconductors silicon and germanium, together with those for InSb and GaAs are shown in Table 24.2. For silicon and germanium two of the principal values of the effective mass at the bottom of the conduction band are equal; this is known as the 'transverse mass' $m_T^*$, while the third (unequal) mass is called the longitudinal mass $m_L^*$. In InSb and GaAs the minimum of the conduction band occurs at $\mathbf{k} = 0$ (see § 17.4), and the effective mass is isotropic.

The position at the top of the valence band is more complicated. Two energy surfaces coincide at $\mathbf{k} = 0$, and at points near-by in $\mathbf{k}$-space the

TABLE 24.2

*Effective masses in some semiconductors deter-
mined by cyclotron resonance, relative to the free
electron mass*

|           | Electrons | | Holes | |
|-----------|-----------|-----------|---------|---------|
| Substance | $m_L^*$ | $m_T^*$ | 'light' | 'heavy' |
| Silicon   | 0·98 | 0·19 | 0·16 | 0·5 |
| Germanium | 1·64 | 0·082 | 0·044 | 0·3 |
| InSb      | 0·014 (isotropic) | | 0·02 | 0·4 |
| GaAs      | 0·067 (isotropic) | | 0·084 | 0·45–0·6 |

energy surfaces for silicon and germanium are given by the relation

$$W = Ak^2 \pm \{B^2k^4 + C^2(k_x^2k_y^2 + k_y^2k_z^2 + k_z^2k_x^2)\}^{\frac{1}{2}}, \qquad (24.30)$$

which is also approximately correct for Group III–V semiconductors. If $C$ is small the two surfaces are nearly spherical, but with different curvature, corresponding to two different effective masses known as the 'light' and 'heavy' holes (see Table 24.2), giving resonance lines as in Fig. 24.21.

## 24.10. Azbel–Kaner resonance in metals

When a spectral line due to moving particles is observed, it is broadened through the Doppler effect, by an amount which is proportional to the random particle velocity. In a semiconductor at low temperatures, the electrons or holes have ordinary thermal velocities corresponding to energies of order $kT$, and broadening by the Doppler effect is not important. In a metal, on the other hand, the electron velocity is that at the Fermi surface; in copper, assuming $m^*/m = 1·5$ and $W_F = 4·7$ eV, this velocity is about $10^6$ m s$^{-1}$, while the phase velocity in the metal of an electromagnetic wave with a free-space wavelength of 10 mm is only about $4 \times 10^4$ m s$^{-1}$ (from eqn (8.30) it is equal to $\omega\delta$, where $\delta$ is the skin depth). Broadening through the Doppler effect thus makes it impossible to observe cyclotron resonance in metals by methods similar to those used for semiconductors. It can, however, be detected by a different method, originally due to Azbel and Kaner (1957, 1958).

As before, a reasonable degree of resolution is obtained only if $\omega\tau > 1$. This makes it essential to work at liquid-helium temperatures, using very pure samples in which the residual resistivity due to electron scattering by impurities and imperfections is as small as possible ($10^{-3}$–$10^{-5}$ of the room-temperature resistivity). In copper the radius of the electron orbit in the magnetic field required to make the cyclotron resonance frequency equal to $3 \times 10^{10}$ Hz is about $5 \times 10^{-6}$ m, and the mean path length of the electrons must be of this order in order to make $\omega\tau > 1$. This requires a conductivity greater than $10^{10}\ \Omega^{-1}\ m^{-1}$, and the 'classical' skin depth given by eqn (8.31) is about $3\times10^{-8}$ m, which is small compared with the mean path length. This is the region of the 'anomalous skin effect', where the conductivity is effectively reduced because only those electrons moving at a small angle to the surface such that their free paths lie wholly within the skin depth contribute fully to the oscillatory current (see § 12.10). However, even allowing for this, the 'anomalous' skin depth is about $2\times10^{-7}$ m, which is still small compared with the radius of the cyclotron orbit. If then a magnetic field of flux density **B** is applied parallel to the surface of the metal, a certain number of electrons moving in helical orbits about **B** will enter the skin-depth region once per cycle, and while in this region they can be accelerated by the oscillatory electric field component normal or parallel to **B**. The latter geometry is illustrated in Fig. 24.22; an important difference from the conventional cyclotron is that acceleration occurs only once per revolution instead of twice. Electrons will gain energy steadily if the frequency of the electromagnetic wave incident on the surface of the metal is synchronous with the cyclotron resonance frequency, or is an integral multiple of it. Hence the resonance condition is

$$\omega = p\omega_c = p(q/m^*)B, \tag{24.31}$$

where $p = 1$, 2, 3, etc. It is usually convenient to work at a fixed frequency, making the metal sample one end of a cavity resonator as in ferromagnetic resonance (but with **B** normal or parallel to the oscillatory electric field instead of normal to the oscillatory magnetic field), and maxima in the absorption of energy then occur at values of $B$ given by the relation

$$B_p = \frac{1}{p}\left(\frac{m^*\omega}{q}\right). \tag{24.32}$$

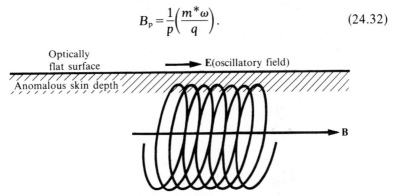

FIG. 24.22. Geometry of the steady magnetic flux density **B**, the oscillatory electric field, and the cyclotron orbits in a metal for Azbel–Kaner resonance. The electrons are accelerated by the electric field only when their orbits take them into the skin depth.

To obtain a resonance peak the electrons must complete a cyclotron orbit without experiencing a collision. This condition can be written as $\omega_c \tau \geqslant 1$, and since $\omega_c = \omega/p$ it is a more stringent condition than $\omega\tau \geqslant 1$. From eqn (24.32) it follows that this condition is more difficult to satisfy at low values of $B$, but in experimental charts similar to Fig. 24.23 peaks have been observed down to magnetic fields of about $0\cdot1$ T (values of $p$ up to 30).

Samples of high purity are needed to give good resolution; ideally they must be so flat that surface irregularities are small compared with the anomalous skin depth. If the effective mass is anisotropic, single crystals must be used, cut in special orientations so that cyclotron resonance can be observed in all the principal directions. The steady magnetic flux density **B** must be accurately parallel to the surface, or electrons will move away from the surface because of their velocity components parallel to **B**. As the electrons can drift distances $\sim\!10\ \mu$m in one cyclotron period and the skin depth $\sim\!0\cdot1\ \mu$m, the magnetic field must be aligned parallel to the surface to much better than $1°$; this is necessary to ensure that the electrons return to the same distance from the surface—otherwise they would experience a Doppler-shifted electric field. Grimes and Kip (1963) found that the effective mass is isotropic in sodium and potassium, with values of $m^*/m_e$ equal to $(1\cdot24\pm0\cdot02)$ and $(1\cdot21\pm0\cdot02)$ respectively. In copper (Koch, Stradling, and Kip 1964) the predominant absorption is due to electrons with $m^*/m_e$ about equal to $1\cdot4$, with only slight anisotropy, but other very anisotropic values ranging from $0\cdot4$ to $6$ are also observed, showing that the Fermi surface is rather complicated.

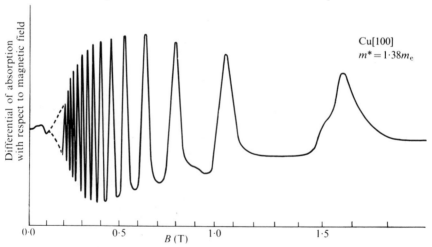

FIG. 24.23. Azbel–Kaner resonance at 4 K in a single crystal of copper. The magnetic flux is parallel to the surface and along a [100] direction; the frequency is 67 GHz (4·5 mm wavelength). (After Koch, Stradling, and Kip 1964.)

## References

AZBEL, M. YA., and KANER, E. A. (1957). *Sov. Phys. JETP* **5**, 730.
—— —— (1958). *J. Phys. Chem. Solids* **6**, 113.
BLOCH, F., HANSEN, W. W., and PACKARD, M. (1946). *Phys. Rev.* **69**, 127.
BLOEMBERGEN, N., PURCELL, E. M., and POUND, R. V. (1948). *Phys. Rev.* **73**, 679.
BROWN, R. M. and PURCELL, E. M. (1949). *Phys. Rev.* **75**, 1262.
COHEN, E. R. and TAYLOR, B. N. (1972). *Atomic masses and fundamental constants*, Vol. 4 (Eds. J. H. Sanders and A. H. Wapstra), p. 543. Plenum Press, New York.
DEXTER, R. N., ZEIGER, H. J., and LAX, B. (1956). *Phys. Rev.* **104**, 637.
DRESSELHAUS, G., KIP, A. F., and KITTEL, C. (1955). *Phys. Rev.* **98**, 368.
ESSEN, L. (1969). *Vistas Astron.* **2**, 45.
——, PARRY, J. V. L. (1959). *Nature, Lond.* **184**, 1791.
GRIFFITHS, J. H. E. (1946). *Nature, Lond.* **158**, 670.
GRIMES, C. C. and KIP, A. F. (1963). *Phys. Rev.* **132**, 1991.
HELLER, P. and BENEDEK, G. B. (1962). *Phys. Rev. Lett.* **8**, 428.
KITTEL, C. (1948). *Phys. Rev.* **73**, 155.
KOCH, J. F., STRADLING, R. A., and KIP, A. F. (1964). *Phys. Rev.* **133**, A240.
PHILLIPS, T. G. and ROSENBERG, H. M. (1964). *Phys. Lett.* **8**, 298.
PURCELL, E. M. (1948). *Science*, **107**, 433.
——, TORREY, H. C., and POUND, R. V. (1946). *Phys. Rev.* **69**, 37.
RABI, I. I., MILLMAN, S., KUSCH, P., and ZACHARIAS, J. R. (1939). *Phys. Rev.* **55**, 526.
RAMSEY, N. F. (1949). *Phys. Rev.* **76**, 996.
SHOCKLEY, W. (1953). *Phys. Rev.* **90**, 491.
SHULMAN, R. G. and JACCARINO, V. (1956). *Phys. Rev.* **103**, 1126.
SOMMER, H., THOMAS, H. A., and HIPPLE, J. A. (1951). *Phys. Rev.* **82**, 697.
WIMETT, T. F. (1953). *Phys. Rev.* **91**, 499.

## Problems

24.1. In a substance where the susceptibility is small and the Lorentz internal field can be neglected, show that eqn (10.19) can be written in the form

$$\chi' - j\chi'' = \frac{n_0 e^2}{m\epsilon_0} \frac{1}{(\omega_p^2 - \omega^2) + 2j\omega\,\Delta\omega}.$$

If $\chi_0$ is the static susceptibility, and $\chi_p''$ is the imaginary part of the susceptibility when $\omega = \omega_p$, prove that

$$\chi_p''/\chi_0 = \omega_p/(2\Delta\omega) = \nu_p/(2\Delta\nu),$$

where $\nu_p$ is the frequency at the centre of the absorption line, and $\Delta\nu = \Delta\omega/2\pi$. Although this formula was derived for electric susceptibility it is equally valid for the magnetic case.

24.2. The work done per unit volume to increase the magnetization of a substance by $dM$ in a field of flux density $B$ is $dW = B\,dM$. If $B$ is an alternating field

$$B_1 \cos \omega t = \mathrm{Re}(B_1 \exp j\omega t),$$

the magnetization may be written as

$$M = \mathrm{Re}\{(\chi' - j\chi'')(B_1/\mu_0)\exp j\omega t\},$$

where $(\chi'-j\chi'')$ is the complex susceptibility. Show that the rate of doing work per unit volume is

$$B(dM/dt) = dW/dt = -\omega\chi'(B_i^2/\mu_0)\cos\omega t \sin\omega t + \omega\chi''(B_i^2/\mu_0)\cos^2\omega t$$

and the mean power dissipated per unit volume is $\frac{1}{2}\omega\chi''B_i^2/\mu_0$.

Use the definition (6) for $Q_F$ given in § 7.3 to show that $1/Q_F = \chi''/(1+\chi')$ for a coil containing a magnetic substance in a tuned circuit with no other losses.

24.3. Adapt the results of Problem 8.6 to the case of the magnetic substance of the last problem (note that $\chi''/(1+\chi')$ is equivalent to $\epsilon''/\epsilon'$), and show that the power in an electromagnetic wave passing through such a medium would fall according to the law

$$P/P_0 = \exp(-2\pi\chi''x/\lambda)$$

if $\chi' \ll 1$.

Verify the figures given in § 24.3, that for $\lambda = 35$ m and $\chi'' = 10^{-5}$, the power will fall by about 1·8 per cent in a distance of 10 km.

24.4. Show, from eqn (24.21) and (24.22) that the instantaneous velocity of the charged particle in its spiral orbit is $(qE/M\,\Delta\omega)\sin(\frac{1}{2}\Delta\omega t)$. Hence show that the total length of path traversed by the particle in reaching its maximum radius $R_0$ when $\Delta\omega \neq 0$ is $L = 2qE/M(\Delta\omega)^2$, and verify that $L/2R_0$ is equal to the resolving power $\omega_c/\Delta\omega$.

24.5. Adapt the formula of Problem 14.5 to find the value of $g_F$ when $J$ and $I$ are coupled to form a resultant $F$, assuming that the nuclear magnetic moment can be neglected. Show that in the Zeeman splitting of Fig. 24.11,

$$g_{F=4} = -g_{F=3} = \tfrac{1}{8}g_J = \tfrac{1}{4}.$$

24.6. An atomic-beam determination of the ratio of the value of $g_J$ in the $^2P_{\frac{3}{2}}$ and $^2P_{\frac{1}{2}}$ states of the gallium atom gives it to be $2(1\cdot00172\pm0\cdot00006)$. Show, by writing $g_l = 1+\delta_l$ and $g_s = 2(1+\delta_s)$, that the ratio is equal to $2\{1+\frac{3}{2}(\delta_s-\delta_l)\}$, and hence that this result agrees within the experimental error with the accepted value $\delta_s = 0\cdot001160$ if $\delta_l$ is assumed to be zero.

24.7. A thin spherical shell of radius $r$, thickness $dr$ of electric charge density $\rho$ rotates with angular velocity $\boldsymbol{\omega}$ about a diameter. Show that the magnetic field $d\mathbf{B}$ at the centre is $+\frac{2}{3}\mu_0\rho\boldsymbol{\omega}r\,dr$.

Use this result to show that the correction at the nucleus of a hydrogen atom in a magnetic field of flux density $\mathbf{B}$ due to diamagnetic shielding (see § 24.5) is

$$\delta B/B = -\mu_0 e^2/12\pi m a_0,$$

given that the charge density at distance $r$ is

$$\rho = (-e/\pi a_0^3)\exp(-2r/a_0).$$

24.8. Show from eqn (24.28) that the conductivity at zero frequency in the direction normal to a magnetic flux density $B$ should vary as

$$\sigma_x'/\sigma_0 = 1/(1+aB^2),$$

where $a = (e\tau/m^*)^2$. This is the magnetoresistance effect, which becomes appreciable only at low temperatures where $\tau$ increases. When only one type of carrier is present, the effect vanishes because the sideways force due to the magnetic field is exactly nullified by that due to the Hall voltage; when more than one type of charge carrier is present this cancellation does not occur (see § 12.11).

24.9. Show that in a cyclotron resonance experiment where $\omega\tau \gg 1$, the value of $\sigma'_x$ at resonance ($\omega_c/\omega = 1$) approaches $\frac{1}{2}\sigma_0$, and that the loss tangent of the specimen is then (writing $\Delta\omega_e$ for $\tau^{-1}$)

$$\tan\delta_e = \frac{ne^2}{2m^*\omega\epsilon_r\epsilon_0\,\Delta\omega_e}.$$

Use the result of Problem 24.1 to show that in an electron spin resonance experiment the magnetic loss tangent at resonance for a system of $n$ electrons with $S = \frac{1}{2}$, $g = 2$ is

$$\tan\delta_m = \frac{\chi''_P}{1+\chi_0} \approx \frac{\mu_0 n\mu_B^2\omega}{2kT\,\Delta\omega_m}$$

and hence prove that (taking $m^* = m$, $\Delta\omega_e = \Delta\omega_m$)

$$\frac{\tan\delta_e}{\tan\delta_m} = \frac{4mc^2kT}{\epsilon_r(\hbar\omega)^2}.$$

Hence verify that the inherent sensitivity of a cyclotron resonance experiment is very much higher than that of a spin resonance experiment, so that fewer electrons are needed.

Discuss whether the imaginary part of the conductivity (see eqn (24.27)) can justifiably be neglected in the formula for $\tan\delta_e$.

24.10. A tuned circuit of quality factor $Q_F$ is freely oscillating. Show that in the time taken by the stored energy to fall by a factor $e^{-1}$, a wave in free space would travel $Q_F/2\pi$ wavelengths.

24.11. In cobalt metal the ferromagnetic moment per atom is $1\cdot7\,\mu_B$, the density is $8\cdot9\times10^3\,\text{kg m}^{-3}$ and the atomic mass is 59. Show that the value of $\mu_0M$ is $1\cdot80$ tesla, where $M$ is the magnetic moment per cubic metre.

A beam of polarized neutrons moves with velocity $4\times10^3\,\text{m s}^{-1}$ through a sample of cobalt metal 4 mm thick, to which is applied a field $B_0 = 1$ tesla. It is found that the angle through which the neutron spin precesses while passing through the sample is 512 radians. Show that this corresponds to the neutron 'seeing' a field $B_0 + \mu_0M$, and not $B_0$ alone. (Assume that $B_0 = \mu_0H$ inside the metal; the value of the magnetogyric ratio for the neutron is $\gamma = 1\cdot83\times10^8\,\text{rad s}^{-1}$.)

Experiments of this type confirm that the magnetic moment of the neutron arises from a current loop, not a true magnetic dipole.

# Appendix: Numerical values of the fundamental constants

| | | |
|---|---|---|
| $c$ | velocity of light *in vacuo* | $2 \cdot 998 \times 10^8 \, \text{m s}^{-1}$ |
| $N_A$ | Avogadro's number | $6 \cdot 022 \times 10^{23} \, \text{mol}^{-1}$ |
| $e$ | electronic charge | $1 \cdot 602 \times 10^{-19} \, \text{C}$ |
| $m_e$ | electron rest mass | $9 \cdot 110 \times 0^{-31} \, \text{kg}$ |
| $m_p$ | proton rest mass | $1 \cdot 673 \times 10^{-27} \, \text{kg}$ |
| $m_p/m_e$ | ratio of proton to electron mass | $1 \cdot 836 \times 10^3$ |
| $h$ | Planck's constant | $6 \cdot 626 \times 10^{-34} \, \text{J s}$ |
| $\hbar$ | Planck's constant/$2\pi$ | $1 \cdot 055 \times 10^{-34} \, \text{J s}$ |
| $h/2e$ | | $2 \cdot 068 \times 10^{-15} \, \text{J s C}^{-1}$ |
| $F$ | Faraday's constant $(N_A e)$ | $9 \cdot 648 \times 10^4 \, \text{C mol}^{-1}$ |
| $e/m$ | charge/mass for electron | $1 \cdot 759 \times 10^{11} \, \text{C kg}^{-1}$ |
| $e^2/m$ | | $2 \cdot 818 \times 10^{-8} \, \text{C}^2 \, \text{kg}^{-1}$ |
| $a_0$ | Bohr radius $= 4\pi\epsilon_0\hbar^2/m_e e^2$ | $5 \cdot 292 \times 10^{-11} \, \text{m}$ |
| $R_\infty c$ | Rydberg constant$\times c = m_e e^4/8\epsilon_0^2 h^3$ | $3 \cdot 290 \times 10^{15} \, \text{Hz}$ |
| $R_\infty$ | Rydberg constant | $1 \cdot 097 \times 10^7 \, \text{m}^{-1}$ |
| $k$ | Boltzmann's constant | $1 \cdot 381 \times 10^{-23} \, \text{J K}^{-1}$ |
| $R$ | gas constant $= N_A k$ | $8 \cdot 314 \, \text{J K}^{-1} \, \text{mol}^{-1}$ |
| $\mu_B$ | Bohr magneton $= e\hbar/2m_e$ | $9 \cdot 274 \times 10^{-24} \, \text{A m}^2$ |
| $\mu_N$ | nuclear magneton $= e\hbar/2m_p$ | $5 \cdot 051 \times 10^{-27} \, \text{A m}^2$ |
| $\alpha$ | fine structure constant $= e^2/4\pi\epsilon_0\hbar c$ | $(137 \cdot 0)^{-1}$ |
| $\epsilon_0$ | permittivity of a vacuum $= (\mu_0 c^2)^{-1}$ | $8 \cdot 854 \times 10^{-12} \, \text{F m}^{-1}$ |
| $4\pi\epsilon_0$ | | $10^7/c^2 = 10^{-9}/9$ approximately |
| $\mu_0$ | permeability of a vacuum (by definition) | $4\pi \times 10^{-7} \, \text{H m}^{-1}$ |
| $Z_0$ | intrinsic impedance of free space | $3 \cdot 767 \times 10^2 \, \Omega$ |
| eV | electronvolt | $1 \cdot 602 \times 10^{-19} \, \text{J}$ |
| $kT$ | energy for $T = 290 \, \text{K}$ | $4 \cdot 004 \times 10^{-21} \, \text{J}$ |

1 electronvolt is equivalent to:

       wavelength $\lambda = 1 \cdot 240 \times 10^{-6} \, \text{m}$

       frequency $\nu = 2 \cdot 418 \times 10^{14} \, \text{Hz}$

       wave number $\bar{\nu} = 8 \cdot 065 \times 10^5 \, \text{m}^{-1}$

       temperature $T = 1 \cdot 160 \times 10^4 \, \text{K}$

       energy $W = 1 \cdot 602 \times 10^{-19} \, \text{J}$

$1 \, \text{m}^{-1}$ is equivalent to:

       wavelength $\lambda = 1 \, \text{m}$

       temperature $T = 1 \cdot 439 \times 10^{-2} \, \text{K} = hc/k$

# Appendix: Some useful unit conversions

The system of units employed throughout this book is the International System (S.I.). However, other systems of units are still often used in the literature; these are usually m.k.s. (metre, kilogram, second) units or c.g.s. (centimetre, gram, second) units. In the theory of electricity and magnetism c.g.s. units may be either electrostatic (e.s.u.) or electromagnetic (e.m.u.).

The following is a brief list of equivalences which may be useful when reading the literature associated with topics in this book.

1 dyne $= 10^{-5}$ newtons
1 erg $= 10^{-7}$ joules
1 mm Hg $= 1$ torr $\approx 133 \cdot 322$ pascals
1 atmosphere $= 101\ 325$ pascals $\approx 10^{5}$ Pa $= 1$ bar
1 calorie $= 4 \cdot 184$ joules
1 e.m.u. of charge $= 10$ coulombs
1 e.s.u. of charge $\approx 3 \cdot 336 \times 10^{-10}$ coulombs
1 gauss $= 10^{-4}$ teslas
1 oersted $= 10^{3}/4\pi$ amperes per metre
1 maxwell $= 10^{-8}$ webers
1 electronvolt $\approx 1 \cdot 602 \times 10^{-19}$ joules
1 micron $= 1$ micrometre $= 10^{-6}$ m
1 ångström $= 10^{-1}$ nanometres $= 10^{-10}$ m
1 hertz $= 1$ cycle per second
1 Bohr magneton $= 0 \cdot 9274 \times 10^{-20}$ e.m.u.
1 Faraday $= 9 \cdot 649 \times 10^{3}$ e.m.u. mol$^{-1}$.

In S.I. units the volume susceptibility (per cubic metre) is a number larger by a factor $4\pi$ than the corresponding number (per cubic centimetre) in c.g.s. units, the latter being expressed in e.s.u. for electric susceptibility and in e.m.u. for magnetic susceptibility.

# Index